1st International Symposium on Transportation Studies for Developing Countries (ISTSDC 2017)

MATEC Web of Conferences Volume 181 (2018)

Makassar, Indonesia
4 November 2017

Editors:

Ahmad Munawar
Junyi Zhang
A.C. Sutandi
Bagus Hario Setiadji
M. A. Abdurrahman
Leksmono Suryo Putranto
Muhammad Zaly Shah
Joewono Prasetijo
Muhammad Isran Ramli

ISBN: 978-1-5108-6804-5

Printed from e-media with permission by:

Curran Associates, Inc.
57 Morehouse Lane
Red Hook, NY 12571

Some format issues inherent in the e-media version may also appear in this print version.

Printed by Curran Associates, Inc. (2019)

For additional information, please contact EDP Sciences – Web of Conferences at the address below.

EDP Sciences – Web of Conferences
17, Avenue du Hoggar
Parc d'Activité de Courtabœuf
BP 112
F-91944 Les Ulis Cedex A
France

Phone: +33 (0) 1 69 18 75 75
Fax: +33 (0) 1 69 28 84 91

contact-edps@webofconferences.org

Additional copies of this publication are available from:

Curran Associates, Inc.
57 Morehouse Lane
Red Hook, NY 12571 USA
Phone: 845-758-0400
Fax: 845-758-2633
Email: curran@proceedings.com
Web: www.proceedings.com

TABLE OF CONTENTS

Statement of Peer review

In submitting conference proceedings to MATEC Web of Conferences, I certify to the Publisher that I adhere to the **Policy on Publishing Integrity** of the journal in order to safeguard good scientific practice in publishing.

1. All articles have been subjected to peer review administered by the proceedings editors.
2. Reviews have been conducted by expert referees, who have been requested to provide unbiased and constructive comments aimed, whenever possible, at improving the work.
3. Proceedings editors have taken all reasonable steps to ensure the quality of the materials they publish and their decision to accept or reject a paper for publication has been based only on the merits of the work and the relevance to the journal.

Title, date and place of the conference

The 1st International Symposium on Transportation Studies for Developing Countries (ISTSDC)

November 4th, 2017

Makassar, Indonesia

Proceedings editor(s):

Prof. Ahmad Munawar
Prof. Leksmono Suryo Putranto
Prof. Junyi Zhang
Dr. Muhammad Zaly Shah
Dr. Anastasia Caroline Sutandi
Dr.-Ing. Joewono Prasetijo
Dr. Bagus Hario Setiadji
Dr. Eng. Muhammad Isran Ramli
Dr. Muhammad Asad Abdurrahman

Date and editor's signature

June 29th, 2018

Dr. Eng. Muhammad Isran Ramli

Bearing capacity improvement of soft soil subgrade layer with Bio Stabilized Bacillus Subtilis

Hasriana[1,*], *Lawalenna* Samang[2], *Tri* Harianto[2], *M.Natsir* Djide[3],

[1]Hasanuddin University, Doctoral Program of Civil Engineering, Makassar, Indonesia
[2]Hasanuddin University, Department of Civil Engineering, Makassar, Indonesia
[3]Hasanuddin University, Department of Farmacy, Makassar, Indonesia

Abstract. The problem that is often faced in a road construction is a soil subgrade (soft soil). If the subgrade has a low bearing capacity, causing the soil to decrease greatly, so the road construction will quickly be damaged. Method to overcome the problem is bio stabilization with bacteria utilization. The purpose of this research is to know the bearing capacity improvement of soft soil as subgrade layer with bacteria solution of bacillus subtilis. Tests include soil physical and mechanical properties, compaction parameters and CBR tests. This test is based on ASTM for each test. Variations in addition of bacterial solution were 2%, 4%, 6%, 8% and 10% of dry soil weight with 7 days curing time. The results showed that the maximum value of CBR of soft soil was found at 39% on 6% bacterial solution. Increased CBR value from 2.8% to 39% or 13 times than untreated soil. This shows the use of bacterial concentration solution on soft soil significantly increases the bearing capacity.

1 Introduction

In roadway planning, the bearing capacity of the subgrade strongly affects the thickness of the pavement, the higher bearing capacity of subgrade soil, then thickness of pavement becomes thinner to withstand the traffic load. Bearing capacity of subgrade is influenced by soil type, density, water content, and others. If a soil in the field is highly loose or very susceptible to distress and has an inconsistent index of consistency for a development project, then the soil should be improvement to meet the technical requirements. Soil stabilization using bio method with bacillus subtilis bacteria is believed to improve soil characteristics by increasing the bearing capacity of foundation soil, reducing compressibility and permeability, and reducing variation in volume change. The purpose of this research is to know the effect of bacterial addition on soft soil to increase the bearing capacity of the subgrade layer. The advantage of bacillus subtilis bacteria can reduce soil and environment pollution.

Efforts to improve soil properties has large swelling properties have been largely undertaken with soil stabilization methods, including stabilization using environmentally

* Corresponding author: hasrianahasan99@yahoo.com

unfriendly grouting methods which are usually suspense (cement, cement-clay, pozzolan, bentonite, etc.) or emulsions (Asphalt, etc.) [1]. Therefore we are looking for an alternative bio-stabilization method that is environmentally friendly, that is by utilization of microorganisms derived from bacteria because it can produce calcite/calcium carbonate crystals that can change the grain behavior. Various studies have been done before, among others, Bio-Mediated Soil Improvement: Cementation of Unsaturated Sample. This study uses Bio-Mediated Calcit Precipitation. As a soil improvement material. Further experimental Biogrouting Stabilization on Marine Sandy Clay Soil [2]. In this study presents an application that is used to increase the bearing capacity of the soil by means of stabilization. The results showed that urease catalyzing conversion of urea to ammonium and carbonate produced precipitates with calcium as calcium carbonate crystals [3]. These crystals form a bond between the sand grains increasing the strength and stiffness of the sand. The remaining ammonium chloride is extracted and removed. Soil remediation was introduced in this study to increase soil's carrying capacity as land stabilization safely and environmentally friendly.

Bacteria are microscopic organisms has 1.25 μm average diameter. Grows at a maximum temperature of 25 - 75°C, can be found in air, water, soil, animal fur, and carcasses, optimum pH grows 5.5 - 8.5 [4], bacterial size is microscopic meaning can be seen using a microscope. Active bacteria move in humid conditions. In the state of lack of water, bacteria will be inactive and can even cause death. Micro-organisms are bred by inoculating microorganisms into a new medium. The technique of inoculation used is scratching technique, with previous dilution done before the results obtained colony of pure culture. After incubation in the aerobic state for 24 hours. The culture medium used consisted of nutrient jelly and nutrient broth (NB). The role of bacteria in stabilizing the soil at the time of precipitation of calcium carbonate. The results indicate that bacteria bacillus subtilis showed that bacteria that can breed with temperature in Indonesia and produce calcite/ crystal most come from Papua region [5].

2 Material and method

The material used in this study is soft soil from Takalar District, Conventional soil sampling using crowbar and shovel, then the sample of soil is placed in the sample sack and to get the original water content then wrapped with plastic. Some of the steps that will be done are as following; first, undertake a preliminary literature review and preliminary survey to identify the problem and identification of sampling sites; second, do a preliminary test of samples that have been taken to know soft soil characteristics. Labratorium test to know physical properties including moisture content, consistency limits, and specific grafity, while mechanical properties test include compaction test and (CBR).

Table 1. Standard Methode of physical properties testing

Standard Tests	Standard Method
Water content test	ASTM D 2216-71
Specific grafity test	SNI 03-1964-2008
Liquid limit test	SNI 03-1967-1990
Plastic limit test	SNI 03-1966-1990
Sieve analysis test	SNI 03-1968-1990
Compaction test	SNI 03-1742-1989
Uncinfined compression test	SNI 03-6887-2002
Hydrometer test	SNI 03-3423-1994

2.1 Bacterial growth test

Growth tests were conducted to determine the growth properties of a bacterial species through a growth curve marked by turbidity in liquid medium with the help of shake and incubator.

(a) (b) (c)

Fig.1. a) Nutritive ingredients, b). Autoclave tool, and c) Bacterial isolates

Bacterial activation of bacillus subtilis on medium B4 (1 liter of water mixed with 20g urea, 3g Nutrient Broth, 2,12g $NaHCO_3$, 4,1g $CaCl_2.2H_2O$, 10g NH_4CI, and H_2O (Figure 1a). After that, inserted into Autoclave with temperature 121°C in 15 minute on 1 atm as in Figure 1b, after cold B4 medium was done bacterial inoculation process that mixing bacteria isolate into medium and grown room shaker for 3 days for mixing on soft soil (Figure 1c).

2.2 Test of soil physical properties

CBR measurements in this study were carried out by mixing the soft soil with water that matched the optimum moisture content of the Proctor compaction result and mixed with 2%, 4%, 6%, 8% and 10% bacteria solution, for the test after 7 days cured.

(a) (b) (c) (d)

Fig. 2. Tests procedure: a) mixing the soil with bacteria , b) making of soil samples , c) compaction process, d) Weighing the soil samples

3 Result and discussion

3.1 Physical Properties of Soft Soil

Visually observed, soft clay soil in Takalar Regency has black ash color, hard texture so it is not easy to squeeze on dry conditions, whereas if exposed to water, the soil becomes slippery and sticky in hand, easy to mold and plastic impressed. From the results of research in the laboratory as follows.

Table 2. Physical properties of soft soil

Soil Properties	**Values**
Water content (%)	78,12
Specific gravity	2,68
Liquid limit (%)	67,89
Plastic limit (%)	30,80
Plasticity index (%)	37,08
AASTHO soil classification	A-7-5
USCS Soil classification	CH
% Passing # 200	95,76
Optimum moisture conten (%)	26,5
Maximum dry density (ton/m^3)	1,48
CBR (%)	2,8

After conducting the research, has an initial moisture content of 78.12%. Proctor standard compaction test results obtained optimum water content of 26.5% with a dry volume weight (γd) of 1.48 gr/cm^3. Soft soil weight 2.68 gr/cm^3. Based on Atterberg limit test of liquid limit of 67.89%, plastic limit of 37.08%, Liquid limit greater than 50% is one of the common characteristics possessed by soft soils. The results of soil grain size distribution test, it was found that 95.76% passed sieve No.200 with plasticity index of 37.08%, so it was concluded that according to USCS classification system, classified as fine-grained soil and classified into CH type that is clay with high plasticity. According to the AASHTO classification system the soil samples are included in the group A-7-5 which is a clay soil with a moderate to bad, since the grain of soil sample passes sieve No.200 $\pm$ 35% and plastic index $\exists$ 30%

3.2 CBR value of soft soil

CBR test results were performed on soft soil stabilized with variation of bacterial solution of bacillus subtilis and 7 days treatment.

Based on CBR value results that untreated soft soil was 2.8%. The maximum of CBR value on addition of 6% bacteria. In addition of 2% bacteria, CBR value of the of 23,67% with 7 day curing, while addition bacteria of 4% produce the soil CBR value of 35,54%. Subsequently, in addition of 6% bacteria produce the soil CBR value of39%. In addition on the 8% produce CBR value of 34%, and addition of 10% bacteria 10% produce CBR value of 30%. Test results can be interpreted that untreated soft soil has low CBR value and very soft soil. These results explain that with increasing curing time and optimum condition of bacteria addition (6%), the CBR value if soil has greater (Figure 4b).

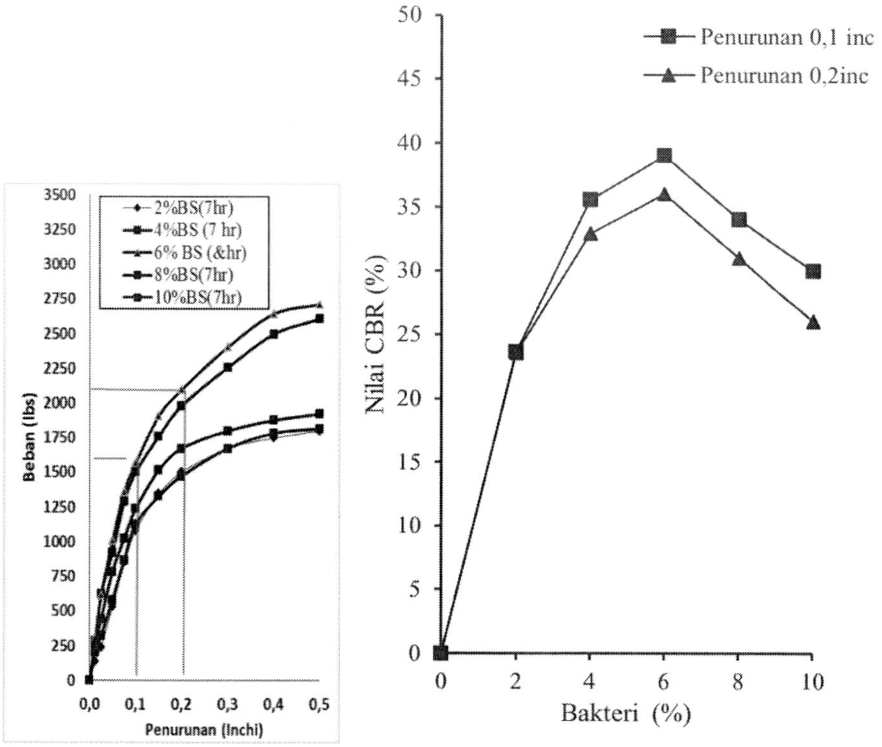

(a) (b)

Fig. 3. a. The relationship between loading and settlement on addition of bacterial solution

b. The relationship between bacterial addition and CBR value

4 Conclusion

Based on laboratory test results obtained that soft soil has an initial moisture content of 78.12%. Standard Proctor compaction test obtained optimum water content of 26.5% with a dry volume weight (□d) of 1.48 gr/cm^3. Based on Atterberg limit test, the liquid limit of 67.89% and plastic limit of 30.80%. The liquid limit greater than 50% is one of the common characteristics of soft soil, the index plasticity value of 37.08%, the soil is very sensitive to surrounding water. Testing of soil grain size distribution, it was found that 95,76% passed sieve No.200 with plastic index 37,08%, so concluded according to USCS classification system, classified as fine grained soil and classified into CH type that is clay with high plasticity. According to the AASHTO classification system the soil samples are included in the A-7-5 clusters, which are clays with moderate to poor grades, since the grain of the soil sample passes the sieve#200 □35% and the plastic index is □30%. The optimum composition of soft soil stabilization with 6% bacteria addition and 7 day curing obtained CBR value 39%. From the test results explain that maximum condition of 6% bacteria producing CBR value of 39%. In soft soil occurs biodegradation process by bacteri bacillus subtilis. Urea, $CO(NH_2)_2$ reacts with calcium through an intermediate medium produced by bacteria to form ammonium which is a cation so that it can be bonded by OH and calcium carbonate ($CaCO_3$) that precipitates and causes the soil to become hard.

References

1. Takaendengan P,P, S. Monintja, J.H.Tieoh, J.R.Sumampouw. *Pengaruh Stabilisasi Semen Terhadap Swelling Lempung Ekspansif* , Jurnal Sipil Statis.Vol1.No 8. Mei (2013).
2. Harianto T, S.Hamzah, Fadliah, WalennaPuspita ,*Biogrouting Stabilization on Marine Sandy Clay Soil International Conference on Asian ad pacific Coasts*, (2013).
3. Van Paassen, LA. *Biogrout ground improvement by micribial induced carbonate precipitation. Delft University of Technology, pp 202*. (2009).
4. Buchanan, R.E., and N.E. Gibbons, *Bergey's manual of determinative bacteriologi, 8th ed. Williams and Wilkins Co., Baltimore, Maryland Dejong,J (2009). Microbially induced Cementation to Control Sand Response to Undrained Shear. Journal Of Geotechnical and Geoenvironmental.* (1975).
5. Lisdiyanti P. *Bacterial carbonate precipitation for biogrouting*, *Prosiding Simposium Nasional Ekohidrologi*, PP 219-232. (2011).
6. Hardiyatmo, C. H.*Mekanika Tanah 1*, Gadjah Mada University Press, Jakarta.
7. Keyka A. Hamed.(2011). *Electro- Biogrouting abd Its challenges, Int.J. Electrochem. Sci., 7 2012 1196 – 1204.* (2010)
8. Hasriana, dkk. *Strength Characteristic of Soft Soil with Bacteria Activation,* 10th International Symposium on Lowland Technology September 15-17, 2016 at Mangalore, India. (2016).
9. Hasriana, dkk. *Pengaruh masa pertumbuhan bakteri bacillus subtilis terhadap peningkatan kuat tekan tanah lunak,* KNPTS 2016,. ITB Bandung. (2016).

MATEC Web of Conferences **181**, 01002 (2018) https://doi.org/10.1051/matecconf/201818101002
ISTSDC 2017

The compressive strength improvement on softsoil subgrade with Quicklime stabilization, activated Gum Rosin and Iron Oxide

Sofwan[1,*], *Lawalenna* Samang[2], *Tri* Harianto[2], and *Achmad* Bakri Muhidin[2]

[1]Hasanuddin University, Doctoral Program of Civil Engineering, Makassar, Indonesia
[2]Hasanuddin University, Department of Civil Engineering, Makassar, Indonesia

Abstract. This study is intended to increase the compressive strength and durability of subgrade with lime, on high plasticity expansive soil. The composition of the stabilizing agent used; composition-1 is soil + 5% slakelime, composition-2 is soil + 5% quicklime, and composition-3 is soil + 5% (quicklime with rosin and iron oxide activation). The testing includes swelling test, durability test, and CBR test. That the performance of stabilization with the composition-3 was able to reduce the swelling potential on 3th day by 93.89%, decrease the volume increase potential by 93.09% and decrease the weight increase potential by 94.00% in the 3 periods of wet-dry cycle, and increase the CBR value by 758.62%. That the activation of gum rosin and iron oxide on quicklime can significantly improve the performance of lime for stabilization of expansive clay as subgrade of highway.

1 Introduction

The pavement on the highway should be built on the subgrade with sufficient compressive strength and low shrinkage, therefore on the subgrade of expansive soil is required a stabilization.

Soil stabilization can be prepared by mechanical and chemical means. Mechanical stabilization is made by to mixing other soils gradation to improve the structure and technical properties (*engineering properties*). Chemical stabilization is made by adding certain chemicals to improve cohesion inter the gradation and provide resistance to water and other harmful chemicals. Lime is one of the soil stabilization material which has long been used, but the performance of lime on soft soil stabilization not provide optimal results. Previous research states that:

In the clay with lime stabilization was identified that the pure argillite underwent a slight decrease in its cohesion due to do dissolution of its clayey particles [1]. The morphological analysis suggested that's gel is formed from a result of the decomposition of the progressive clay particles in the soil reacted with the calcium ions from the lime, while XRD analysis provided no strong evidence in the formation of new phases [2]. Soil-lime

* Corresponding author: shofwanhd6@gmail.com

mixing results in cation exchange, causing flocculation and agglomeration of soil particles, effectively increases the strength, durability and workability the soil. Such treatment also improves soil compressibility, nonetheless, but lime stabilization has a number of inherent disadvantages, such as carbonation, sulfate attack, and environment impact [3].

Expansive soil stabilized with lime and dregs ash produced maximum density at 4% lime content, unconfined strength up to ash 10%, on the higher ash content the lime is not capable of binding the silicate and aluminate present in the ash. Pozzolanic reaction occurs only when there is water, and if no water then the silicate and aluminate in the ash does not mean anything so that the stabilization process is not occur [4].

Optimum lime content in the range of 5% for black cotton soil stabilization, for various lime contents and curing temperatures, it was observed that the maximum portion of strength was obtained in initial curing period of 7 days [5]. After 7 days curing period, test results show that the optimum lime content for soil stabilization is 5% [6]. Clay is stabilized with lime and rice husk ash, the highest CBR values were achieved at 6% lime content and 4% rice husk ash content, concluded that the mixture of lime and rice husk ash is not always able to increase the value of CBR, for CBR soaked or CBR un-soaked [7].

Expansive soil, first addition with 4% of lime, second addition with 8% of bagasse ash, third addition with 4% of lime + 8% of bagasse ash. After 3 curing periods, on three types of mix its CBR value decreased, although still higher than the original soil CBR. In the second cycle, soil + 4% lime achieve the highest CBR, and decreased in the third cycle [8]. The pavement layers construction of of soil-lime was flooded every rainy season, In the fourth year, stability decreased because of puddles, decrease the carrying capacity and increase soil plasticity [9].

In the previous research which has to do with lime as a stabilize, found many weaknesses. One attempt to improve lime behavior on soft soils, is $CaCO_3$ (Lime Stone) was made quicklime (CaO) as a stabilizer material, gum rosin ($C_{30}H_{50}O$) and iron oxide (Fe_2O_3) as activator on quicklime.

Iron oxide can affect the aggregate inter soil particles and cation exchange capacity, can indicate the condition of pH, redox potential, moisture, and soil temperature, can react with the calcium carbonate into iron oxide-carbonate. Flocculation of the solution can be neutralized by adding an acidic substance (ion H^+) [10-12]. Degree of acidity can decrease the permeability so that the soil is more stable [13].

Gum rosin is insoluble in water but can absorb water molecules, potentially as ion exchangers, and has the ability to penetrate the structure of the soil molecules that tend to be solid, has a high viscosity and good bonding properties inter particles, has high levels of carbon content [14].

2 Methodology

Generally, soil granules in have a high carrying capacity and excellent durability, but at a certain moisture content loses its cohesion. Therefore, the stabilization of the soil with chemicals should the imagining chemical reactions as a function of the strengthening of cohesion inter the grains in wet or dry conditions and has resistance to chemical reactions that damage. Lime stone ($CaCO_3$) was made quicklime (CaO) as stabilizator, gum rosin ($C_{30}H_{50}O$) and iron oxide (Fe_2O_3) as activator, researched to improve the performance of lime on expansive soft soil stabilization to subgrade of highway.

2.1 Testing Design

The testing includes swelling test, durability test, and CBR test. Composition-1 is soil + 5%$CaOH_2$, composition-2 is soil + 5%CaO, and composition-3 is soil + 5%(CaO with $C_{30}H_{50}O$ and Fe_2O_3 activation). For swelling test, compacted samples at the optimum moisture content, soaked in water, burdened 2,82 kPa, tested on the 3rd day, 5th day, and 7th day. For the durability test, for each soil mixture carried out wet-dry cycles 1 period, 2 period, and 3 period. 1 period is 1 time aerated for 4 days and 1 time soaked for 4 days. Swelling test and durability test do in wet-dry cycle period, CBR test do after a period of wet-dry cycles. The testing has been done in the soil mechanics laboratory of the Department of Civil Engineering Hasanuddin University Makassar. Using the testing standards ASTM D 2216-71 for the Water content test, SNI 03-1742-1989 for the Compaction test, SNI 03-1967-1990 for the Liquid limit test, SNI 03-1968-1990 for the Sieve analysis test, SNI 03-6887-2002 for the Unconfined compression test, SNI 1964-2008 for the Specific gravity test, SNI 1966:2008 for the Plastic limit & Plastic Index test, and SNI 3423:2008 for the Hydrometer test.

3 Result and Discussion

3.1 Soil properties

From the atterberg limit test, the liquid limit value (LL) = 70.86% and plasticity index (*PI*) = 39.46%. According to *Unified Soil Classification System (USCS),* soil in the category of inorganic clay with high plasticity (*CH*). From the sive analysis obtained the distribution of grain size of the soil, quantitatively dominated by clay fraction is 62.58%, followed by silt fraction is 34.75%, and sand fraction is 2.67%, that this soil can be categorized as the silty clay (Bowles,1993).

3.2. Swelling Behaviour

The Swelling behavior as in the Table 1 and Figure 1.

Table 1. The swelling behavior

Sample	***% OMC***	***% Swell* at 3rd day**	***% Swell* at 5th day**	***% Swell* at 7th day**
Original Soil	28.76	18.00	0.10	0
Soil + 5% (CaO on $C_{30}H_{50}O$ and Fe_2O_3 activation)	31.35	1.10	0	0
Soil + 5% CaO	30.02	3.94	0.45	0
Soil + 5% $CaOH_2$	29.27	5.25	0.82	0

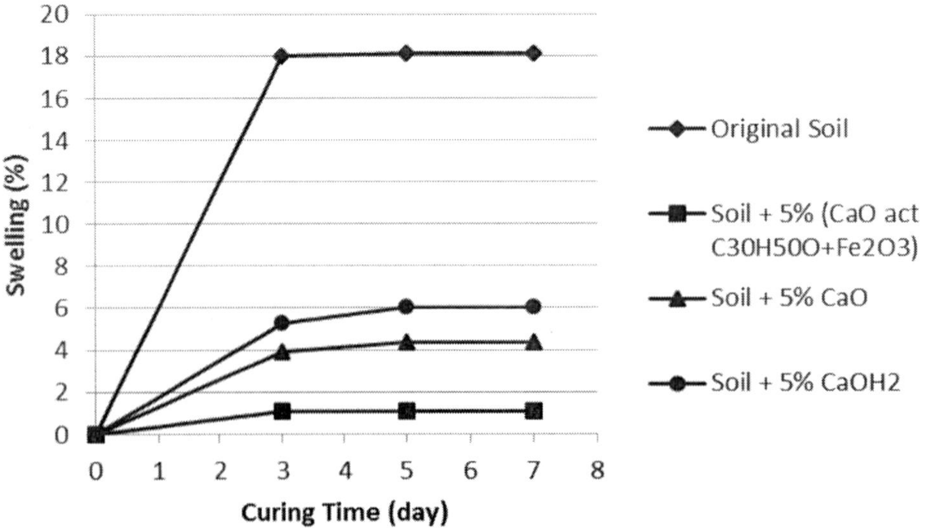

Figure 1. Correlation between % swelling and curing time

In the Table 1 dan Figure 1, swelling of the original soil is 18.00%. Mixed of soil + 5% (CaO on $C_{30}H_{50}O$ and Fe_2O_3 activation) resulting in the smallest swelling rate. On the 3rd day, soil + 5% (CaO on $C_{30}H_{50}O$ and Fe_2O_3 activation) resulting in swelling rate of 1.1% or is smaller 93.89% of the original soil, while on the 5th day and on the 7th day there is no difference in swelling rate.

3.3. Durabilities behavior

The durabilities behavior as in the Table 2 and Figure 2, the Table 3 and Figure 3

Table 2. Volume change rate (%) in the three periods of wet-dry cycles

Sample	Volume Changes rate (%)					
	Wet	Dry	Wet	Dry	Wet	Dry
Original Soil	+20.85	-2.85	+1.50	-1.40	+1.20	-1.20
Soil + 5% (CaO on $C_{30}H_{50}O$ and Fe_2O_3 activation)	+1.44	-0.34	+0.20	-0.20	0	0
Soil + 5% CaO	+4.84	-0.90	+0.80	-0.35	+0.03	0
Soil + 5% $CaOH_2$	+6.35	-1.10	+1.04	-0.23	0	0

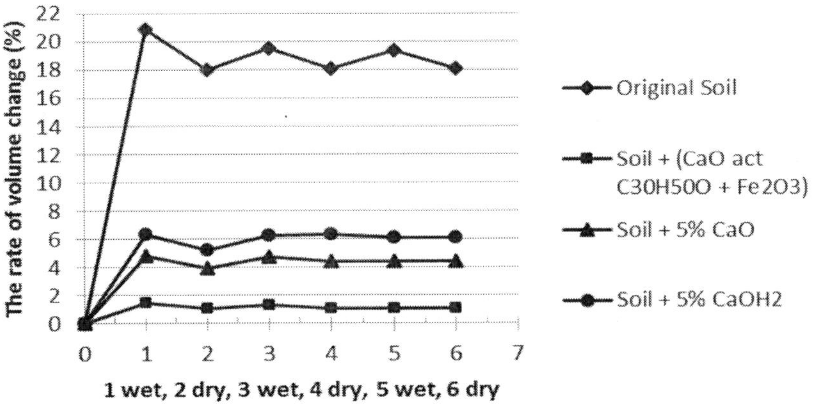

Figure 2. Correlation between % volume change rate and cycle period

Table 2 and Figure 2, after the first period of soaking the original soil increases in volume is 20.85%, and after being dried by wind the volume of original soil just reduced 2.85%. In the second period of soaking, increases in volume is 1.50%, and after drying with the wind is reduced to 1.40%. In the third period of soaking, increases in volume is 1.20%, and after drying with the wind reduced to 1.20%.

The composition of soil + 5% (CaO on $C_{30}H_{50}O$ and Fe_2O_3 activation) in three periods of wet-dry cycles, the increasing the volume only reached 1.44% from the volume before soaking, or smaller 93.09% of the largest volume increase on the original soil.

Table 3. Weight change rate (%) in the three periods of wet-dry cycles

Sample	Weight change rate (%)					
	Wet	Dry	Wet	Dry	Wet	Dry
Original Soil	+33.03	-4.51	+2.38	-2.22	+1.90	-1.90
Soil + 5% (CaO on $C_{30}H_{50}O$ and Fe_2O_3 activation)	+2.28	-0.54	+0.32	-0.32	0	0
Soil + 5% CaO	+7.67	-1.43	+1.27	-0.55	+0.03	0
Soil + 5% $CaOH_2$	+10.06	-1.74	+1.65	-0.36	0	0

Figure 3. Correlation between % weight change rate and cycle period

Table 3 and Figure 3, in the first soaking period weight of the original soil increased by 33.03%, after being dried by the wind then the weight of the original soil only decreased by 4.51%. In the second period of soaking, weight of the original soil increased by 2.38%, after being dried then decreased to 2.22%. In the third period of soaking the weight increase by 1.90%, after being dried then decreased by 1.90%.

Composition of soil + 5% (CaO on $C_{30}H_{50}O$ and Fe_2O_3 activation) in the three period of wet-dry cycles experienced weight the largest gain only 2.28% of the condition before soaking, or smaller 94.00% of the largest weight gain on the original soil.

3.4. California Bearing Ratio (CBR)

The CBR test results are shown in the Table 4 and Figure 4.

Table 4. CBR Test Result

Sample	OMC (%)	CBR (%) First Period	CBR (%) Second Period	CBR (%) Third Period
Original Soil	28.76	2.32	1.63	2.09
Soil + 5% (CaO on $C_{30}H_{50}O$ and Fe_2O_3 activation)	31.35	28.34	34.02	37.62
Soil + 5% CaO	30.02	12.25	23.92	26.21
Soil + 5% $CaOH_2$	29.27	9.93	19.03	19.92

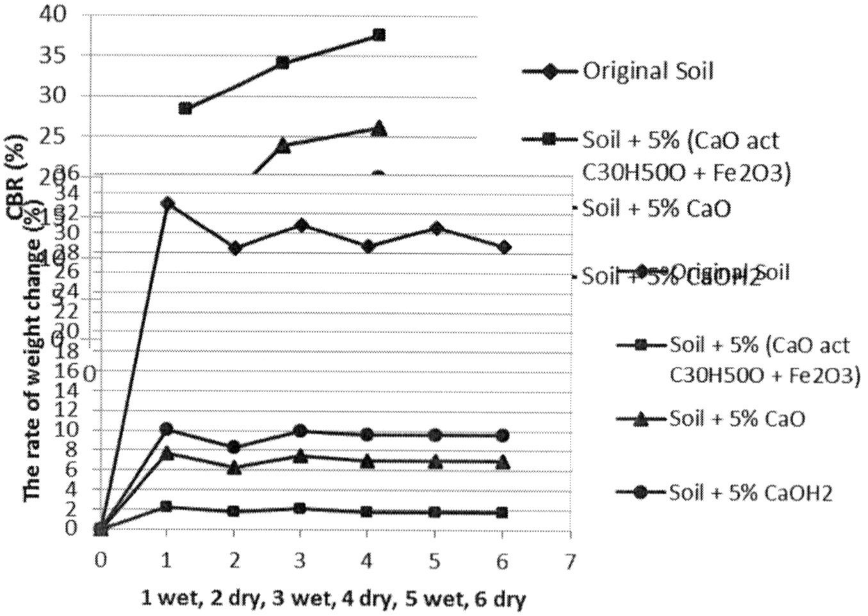

Figure 4. Correlation between CBR value (%) and cycle period

In the Table 4 and Figure 4, the CBR value of original soil is 2.32% and had a downward trend in the next period. To a mixture of soil + 5% $CaOH_2$ the CBR value in the first period was 9.93%, greater 328.02% of the original soil CBR. In the second period the CBR value was 19,03%, greater 720.26% of the original soil CBR. In the third period the CBR value was 19.92%, greater 758.62% of the original soil CBR.

To a mixture of soil + 5% CaO the CBR value in the first period was 12.25%, greater 428.02% of the original soil CBR. In the second period the CBR value was 23.92%, greater 931.03% of the original soil CBR. In the third period the CBR value was 26,21%, greater 1,029.74% of the original soil CBR.

To a mixture of soil + 5% (CaO on $C_{30}H_{50}O$ and Fe_2O_3 activation) the value of CBR in the first period was 28.34%, greater 1,121.55% of the original soil CBR. In the second period the CBR value was 34.02%, greater 1,366.38% of the original soil CBR. In the third period the CBR value was 37.62 %, greater 1,521.55 % of the original soil CBR.

4 Conclusions

That the performance of stabilization with the composition soil + 5% (quicklime with rosin and iron oxide activation) was able to reduce the swelling potential on 3th day by 93.89%, decrease the volume increase potential by 93.09% and decrease the weight increase potential by 94.00% in the 3 periods of wet-dry cycle, and increase the CBR value by 1,521.55%. That the activation of gum rosin and iron oxide on quicklime can significantly improve the performance of lime for stabilization of expansive clay as subgrade of highway.

References

1. Cuisinier, O., Deneele, D., Masrouri, F., *Shear Strength Behaviour of Compacted Clayey Soil Percolated with an Alkaline Solution*. Engineering Geology 108 (2009) 177-188, journal homepage: www.elsevier.com/locate/enggeo. (2009).
2. Wild, S., Arabi, M., and Leng-Ward, G. *Soil – Lime Reaction and Microstructural Development at Elevated Temperatures*. Clay Minerals (1986) 21, 279-292. Received 1 July 1985; revised 3 February 1986. The Mineralogical Society. (1986).
3. Jawad, I. T., Taha, M. R., Majeed, Z. H., Khan, T. A., *Soil Stabilization Using Lime: Advantages, Disadvantages and Proposing a Potential Alternative*. Research Journal of Applied Sciences, Engineering and Technology 8(4): 510-520, 2014. DOI:10.19026/rjaset.8.1000. ISSN: 2040-7459; Accepted: May 10, 2014. (1914).
4. Hatmoko, J.T., dan Lulie, Y., *UCS Tanah Lempung Ekspansif yang Distabilisasi dengan Abu Ampas Tebu dan Kapur*. Jurnal Teknik Sipil, Volume 8 No. 1, Oktober 2007 : 64 – 77. (2007).
5. Nasrizar, A.A., Muttharam, M. Ilamparuthi, K. *Role of Lime Content on Soil-Lime Reaction under Thermal Curing*. Indian Geotechnical Conference – 2010, GEOtrendz. December 16–18, 2010. IGS Mumbai Chapter & IIT Bombay. (2010).
6. Ciancio, D., Becketta, C.T.S., Carrarob, J. A. H. *Optimum lime content identification for lime-stabilised rammed earth*. Construction and Building Materials, 53, 59-65. DOI: 10.1016/j.conbuildmat.2013.11.077. September 12, 2013. (2013).

7. Ariyani, N., Nugroho, A. C., *Pengaruh Kapur Dan Abu Sekam Padi Pada Nilai CBR Laboratorium Tanah Tras Untuk Stabilitas Subgrade Timbunan*. Majalah Ilmiah UKRIM Edisi 1/th XII/2007. (2007).
8. Sauri, S., Rachmansyah, A., Zaika, Y *Pengaruh Penambahan Abu Ampas Tebu dan Kapur pada Tanah Ekspansif di Bojonegoro terhadap nilai CBR, Swelling dan Durabilitas*. Jurusan Teknik Sipil Fakultas Teknik Unversitas Brawijaya, Malang 65145. . (2016).
9. Adha, I., *Pengaruh Durabilitas Terhadap Daya Dukung Stabilisasi Tanah Menggunakan Lempung Plastisitas Rendah Dengan Kapur*. Jurnal Rekayasa Vol. 13 No. 3, Desember 2009. (2009).
10. Cornell, R.M., Schertmann, U., (1996). *The Iron Oxides: Structure, Properties, Reactions, Occurrence and Uses*. VCH Publishers: Weinheim, Germany.
11. Rossel, R.V., Cattle, S., Ortega, A., Fouad, Y., *In Situ Measurements of Soil Colour, Mineral Composition and Clay Content by Vis-NIR Spectroscopy*. Geoderma 150, 253–266. (2009).
12. Mitchel, J.K., *Fundamental of Soil Behaviour*. John Wiley and Sons, Inc., New York. (1976).
13. Nordstrom, D. K., *Negative pH and Extremely Acidic Mine Waters from Iron Mountain, California*. Environmental Science & Technology 34 (2), 254-258. (2000).
14. Doelen, V.D., Berg, V.D., Boon, J.J., a. *Comparative Chromatographic and Mass Spectrometric Studies of Triterpenoid Varnishes Fresh Material and Aged Samples from Paintings*. Studies in Conservation. 43(4). 249-264. (1998).
15. ASTM D 2216-71. *Standard Test Methods for Laboratory Determination of Water (Moisture) Content of Soil and Rock.* by Mass[1]
16. SNI 03-1742-1989. *Panduan pengujian kepadatan ringan untuk tanah Standar Nasional Indonesia. Bahan Konstruksi Bangunan Dan Rekayasa Sipil*.
17. SNI 03-1967-1990. *Metode pengujian batas cair tanah dengan alat Cassagrande*.
18. SNI 03-1968-1990. *Agregat halus dan kasar, Metode pengujian analisis saringan*.
19. SNI 03-6887-2002. *Metode pengujian kuat tekan bebas campuran tanah-semen*.
20. SNI 1964:2008. *Cara uji berat jenis tanah tanah*. Revisi dari SNI 03-1964-1990.
21. SNI 1966:2008. *Cara uji penentuan batas plastis dan indeks plastisitas tanah*. Revisi dari SNI 03-1966-1990
22. SNI 3423:2008. *Cara uji analisis ukuran butir tanah*. Revisi dari SNI 03-3423-1994.

A comparative study of Transit Oriented Development (TOD) at Yogyakarta Railway Station

Okkie Putriani[1,*], and *Ibnu* Fauzi[1]

[1]Universitas Atma Jaya, Transportation Graduate Program, Civil Engineering, Yogyakarta, Indonesia

Abstract. Optimizing the public transport and synergizing the land use can reduce the impact of urban development by attracting the development around the transit station. This situation encourages the accessibility of public transportation by creating conditions between passenger expectations realted to the concept Transit Oriented Development (TOD) between land use, mobility, and environment. This study was conducted by TOD with the area located in the center of local wisdom by cultural city, Yogyakarta Railway Station. The purpose of this study is to provide an alternative location where bus stops or Trans Jogja shelters are more easily accessible by users of rail services and facilitate the model's transfer. The method of this research is descriptive quantitative. It explains the trans it function, needs and condition of Trans Jogja as the existing public transport and the accessibility of the bus stops. The conclusion is the recommendation for the bus stop location can be relocate near the dropout East and South area of the Railway Station

1 Introduction

A future-oriented city, which has a big role in the transport sector, must be sustainable. The transportation sector should be able to provide convenience for all communities in all activities [1]. Total population of DIY in 2015 was recorded 3,679,179 with population growth in 2010 reached 1.19 percent. The total area of 3,185.80 km², the population density in DIY based on Central Bureau of Statistics (BPS) data reaches 1,115 people per km² with the highest density occurring in Yogyakarta city which is 12,699 people per km². Yogyakarta as the tourism city, culture and a center of education with the number of students scattered 116 colleges of 184,328 students and 10,833 teachers. Urban expansion (urban sprawl) and poor land use planning contribute greatly to traffic congestion, air pollution and greenhouse gas emissions [2]. Public transportation can reduce the impact of division by attracting the development around the transit station and this development also supports public transportation by creating conditions that can meet passenger expectations with the concept of transportation development in synergy with the spatial to strengthen the environment or known as Transit Oriented Development (TOD). Tugu Yogyakarta Railway Station, based on data of PT. Kereta Api Indonesia (Persero) throughout 2016 Yogyakarta

* Corresponding author: okkieandfriends@gmail.com

Station serves 5,213,767 passengers with the composition of 44% passengers from Yogyakarta Station and 56% of passengers Yogyakarta destinations served by 52 trains every day is located at the center of Yogyakarta City. These condition shows the needs of a comparative study of TOD area in Yogyakarta Station as one of the transportation system nodes and the catalyst of economic activity of Special District of Yogyakarta.

2 Literature review

2.1 Transit Oriented Development (TOD)

The concept of Transit Oriented Development (TOD) is significant way of improving the effectiveness of transit as well as supporting community goals and improving accessibility [3, 4]. It begins with the concept of human movement activity, either by vehicle or walking. Movement as one of the most widely performed activities by humans, is accommodated by the placement of activity centers integrated with transit points, so it is expected to encourage the use of public transportation. Activity centers are connected to each other in convenient and recommended walking distance in an effort to reduce intermodal shifts [5]. According to the Center for Land Transportation and Railways [6] sustainable transportation is a transportation that encourages the use of environmentally friendly technologies to meet the needs of the community. One effort that can be done is to use the concept of TOD to reduce population mobility among urban areas through an integrated system of development. TOD should be placed:

1. On the main network of mass transit.
2. In high-frequency bus / BRT bus corridors
3. In the bus feeder network that takes less than 10 minutes from the main network of mass transit

If the above requirements are not met by a region then it is necessary to take steps to connect with mass transit, besides that also need to be considered is the high frequency of public transport. The Urban Land Istitute, America, ten principles for successful development around transit [7]:

1. Make it better with a vision
2. Apply the power of partnerships
3. Think development when thinking about transit
4. Get the parking right
5. Build a place, not a project
6. Make retail development market driven, not transit driven
7. Mix uses, but not necessarily in the same place
8. Make buses a great idea
9. Encourage every price point to live around transit
10. Engage corporate attention

2.2 Local Wisdom: Cultural Aspects

The concept of culture city was formed by the King of Yogyakarta Hamengku Buwono I based on social concept, state and functional. The concept manifested in the structure, the pattern of space, and the image of the city: *Catur Sagotra* (*Catur Gotro Tunggal*) microcosm and macrocosm in one living space. The concept of the city is connected by the road space, from the functions of government, religion, economy, and culture.The concept was described by Figure1 of *Golig Gilig*, the intention of uniting all groups. Cultural concept of *Memayu Hayuning Bawono, Manunggaling Kawulo Gusti, Sangkan Paraning Dumadi*, and *Pathok Negoro*.

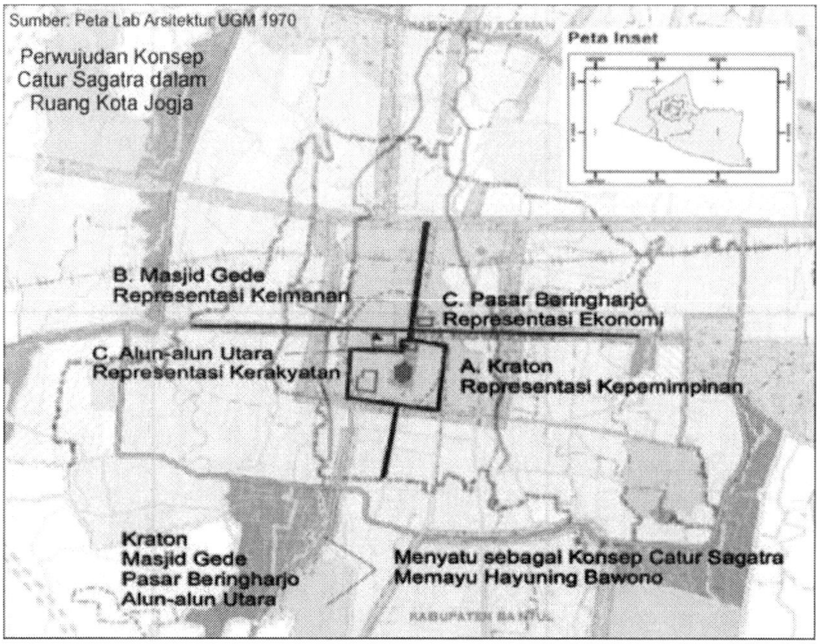

Fig. 1. Cultural Aspects of Yogyakarta City (Source: Map of Architecture Lab UGM, 1970)

2.3 Railway station

A railway station is a place at which passenger join or leave trains. It is a place of arrival and departure, for trains as well as passengers, two diverse units, causing a railway station to fall into two distinct parts: the passenger concourse and the train shed, to each of which a different measure must be applied if they are to be apt and fit for their purpose [8]. According to Indonesia Regulation Ministry of Transportation [9], types, classes and activities at railway station, there are 6 (six) criteria for classifying stations into large stations, medium stations, or small stations, namely operating facilities, number of lanes, passenger facilities, rail traffic frequency, daily passenger number, and daily amount of goods.

2.4 Bus stop

Based on the decision of the Director General of Land Transportation [10] about bus stop design standards and guidelines, bus stops are a critical part of the transit system as they serve as the first point of contact between the customer and the services. Bus stop placement throughout the community acts to promote alternative modes of transportation to the public traveling. The space, location, and design all affect the operation of the transit system, in turn, the transit patron's satisfaction. The distances between bus stops are technically different from the terms of land use, for the city center is placed at a distance of 300 to 500 m and in the suburbs between 500 and 1000 m.

2.5 Parking

The Republic of Indonesia constitution [11] about road traffic, 1st article 15th paragraph, porary vehicle because it is abandoned by the driver. In the technical guidance of the Directorate General of Land Transportation facility [12], parking is based on its layout known as on-street parking and off street parking, explained by table 1.

Table 1. General Space Requirements, Type of Vehicle and Area of Parking Space

No	Type of Vehicle	Area of Parking Space (m^2)
1.a.	Passenger Car , I classes	2.30 x 5.00
1.b.	Passenger Car , II classes	2.50 x 5.00
1.c.	Passenger Car , III classes	3.00 x 5.00
A.	Bus / truck	3.40 x 12.50
B.	Motorcycle	0.75 x 2.00

Source: General Directorate of Land Transportation

2.6 Mode choice

In urban travel demand modeling, from individual choices to general equilibrium, Oppenheim [13] stated that the main modes of transport and urban form are private vehicles, public transport (including buses and trains), including "paratransit" (taxis and others), walking and cycling. Warpani [14] stated that in some places two-wheeled vehicles are used as public vehicles (motorcycles) with free passes such as taxis. Therefore, two-wheeled vehicles can not be ignored in the loading of road network in Indonesia.

2.7 Intermodal Concept in Railway Station Design

Kandee [15] explained applying the intermodal concept in rail services heralds a number of developments, one of which is the role of railway stations. The increasing numbers of passengers has resulted in the need for modern and rational designs of station. The intermodal concepts supports the integration of related transportation modes.

3 Methodology

These descriptive research is trying to describe symptoms, events, and occurrence happenings. This research is using qualitative research method. Creswell at Noor [16] qualitative research is a descriptive analysis, the theoretical basis used as a guide to focus the research according to the fact in the field. The flow of the research stages in this study is shown in Figure 2.

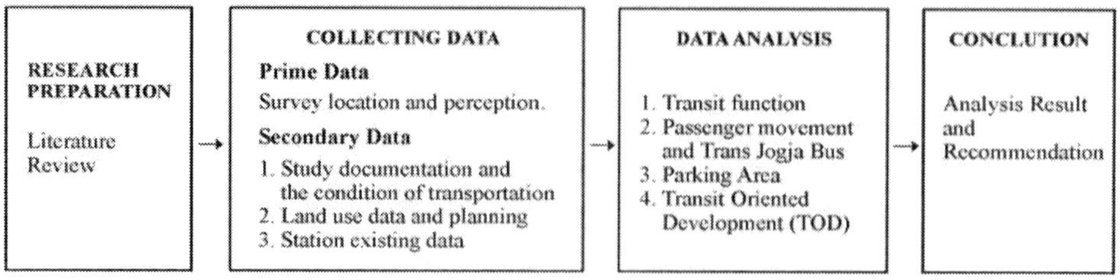

Fig. 2. Research Flowchart

4 Result

4.1 The Characteristic of Yogyakarta Railway Station (YRS)

Yogyakarta Railway Station (YRS), known as Tugu Station, is located in the center of Yogyakarta Special Region City and it is under auspices of PT Kereta Api Indonesia Region VI Yogyakarta (DAOP 6). This station showed in Figure 3 serves the departure and arrival of an executive and business class train (KA). YRS is the largest railway station in the province of Yogyakarta Special Region. This station belongs to a large class station with six main lines. In addition, in the station emplacement area there is a locomotive and railroad locomotive in the north and west of the station complex.

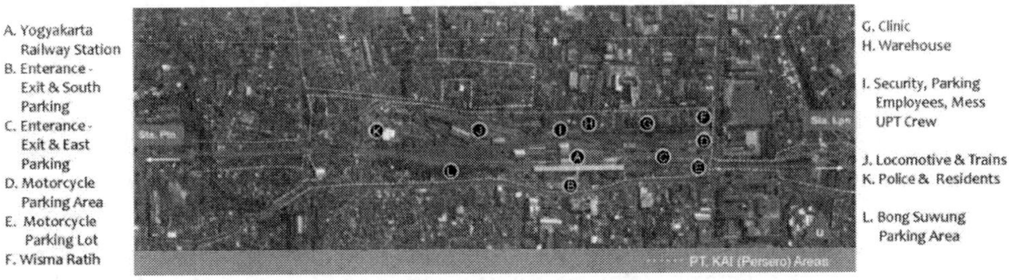

Fig. 3. Existing Area Yogyakarta Railway Station (Source : Unit Maintenance, Preservation & Architecture PT. Kereta Api Indonesia (Persero), 2016)

The station has two entrances and exits, the main entrance facing Jl. Margo Utomo (Jl.P. Mangkubumi, including Gowongan Urban Village, Jetis Sub-district) and the south entrance facing towards Jl. Pasar Kembang (Sosromenduran, Gedongtengen Sub-district).

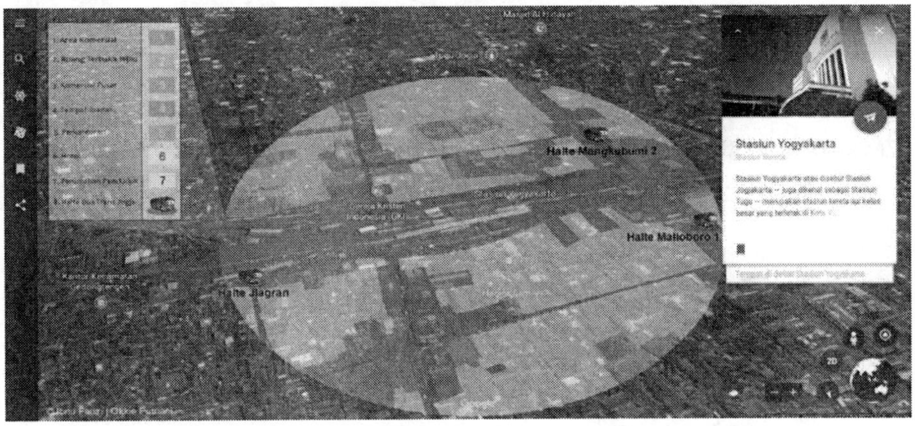

Fig. 4. Land Use of Yogyakarta Railway Station (Source: Google Earth and Survey Analysis)

4.2 Land Use of Yogyakarta Railway Station

Based on the surrounding Yogyakarta railway station Figure 4, the land use area are dominated by residential (light blue), hotel (yellow), office (orange), commercial area (blue), commercial center (pink), green open space, and Trans Jogja bus stop. A city plan should not only focus on area within its administrative boundaries, because urban lands are always expanding parallel with the growth of its population and activities.

4.3 Transit

According to Calthorpe [17], and Mirmoghtadaee [18] the transit function in the TOD region serves as a place of transit activity located to the main transit city movement network. The research area of YRS has been integrated with the main transit city movement network, Trans Jogja Bus. The accessibility to get the bus stop for the pedestrians is still very low. This can be seen in the following Figure 5 the location of Trans Jogja bus stop near Yogyakarta Railway Station.

Fig. 5. The Location of Trans Jogja Bus Stop Near Yogyakarta Railway Station

The movement pattern of the passengers at Yogyakarta Railway Station obtained from the results of a questionnaire tabulated to 100 respondents Train users in Tugu Station, presented in Table 2 below:

Table 2. Tabulation Questionnaire

Residence	40.7% Sleman; 31.1% Jogja City; 21.4% others; 6.8% Bantul
Gender	53.4% Male; 46.6% Female
Age	73.8% 17-25; 23.3% 26-35; 1.9% 36-45; 1% 46-55
Occupation	60.2% Univ. students; 21.4% Employees; 5.8% Entrepre neurs; 4.9% PNS; 1.9% Housewifes; 5.8% others
Destination	40.8% Jkt; 12.6% Solo; 11.7% Bdg; 10.7% Sby; 3.9% Purwokerto; 2.9% Cirebon; 2,9% Kediri; 10.6% Others
Accessibility to Railway Station	42.7% Easy; 19.4% Quite Easy; 14.6% Quite Difficult; 3.9% Very Difficult; 1.9% Difficult
Mode to Railway Station	46.6% Motorbike; 24.3% Rental; 14.6% Private Cars; 7.8% Taxi, 1.9% Trans Jogja Bus; 5% Others
Mode from Railway Station	40.8% Motorbike; 32% Rental; 18.4% Private Cars; 1.9% Taxi; 1.9% Trans Jogja; 5% others
Convenience of Obtaining Public Transport from Railway Station	35% Easyl 26.2% Quite Difficult; 15.5% Quite Easy, 10.7% Difficult;9.7% Very Difficult; 2.9% Very Easy
Travel Destination	33% Recreation; 24.3% Visiting Relatives; 22.3% Edu-cation; 15/5% Business; 4.9% others
Travel Time	58.3% 15-31 min; 18.4% 31-45 min; 11.7% < 15 min; 4.9% > 75 min; 3.9% 61-75 min; 2.9% 46-60 min
Railway Station Distance From Residence	31.1% 3.1-6km; 28.2% 6.1-9km; 12.6% 9.1-12km; 10.7% 12.1-15km; 10.7 1-3km; 6.8%> 15km
Punctuation of Train Schedule	95.1% on-time; 4.9% not on-time
Pedestrian Accessibility	52.4% not yet; 32% yet; 15.5 % not know
Optional Mode of Transportation	78.6% Rental; 13.6% Taxy; 7.8% TransJogja Bus

Based on the TOD strategies, the area should be reduced motor vehicles users and maximize walking of the pedestrian. From Table 1 showed 61% of respondents still use private vehicles to get to the station, and 52% of respondents used vehicles to leave the station.

4.4 Trans Jogja Bus

The data obtained from the Transportation Department of the Special Province of Yogyakarta is the number of Transjogja buses in 2017 as many as 105 buses serving 15 routes / lines are 1A, 1B, 2A, 2B, 3B, 6B, 4B, 5A, 5B, 6A, 6B , 7, 8, 9, 10 and 11. The capacity of one Transjogja bus unit is 41 passengers divided into 22 seated passengers and 19 passengers standing. Trans Jogja operational hours start at 05:30 WIB - 21:30 WIB.

4.5 People Movement and Trans Jogja Bus

4.5.1 Demand

The obtaining movement data analysis of PT. Kereta Api Indonesia (Persero) resulted:

Table 3. The Movement of Yogyakarta Station People

Activities People Mobility*		Monthly Average		Average Per 15 minutes In Rush Hours	
In	**Out**	**In**	**Out**	**In**	**Out**
1,297,856	1,595,380	216,309	265,897	165	186

* *January - June 2017*
Source: PT. Kereta Api Indonesia (Persero) proceed

The activity data of people moving in and out of Yogyakarta Station last six months, where the number of people entering either will go by train, pick up, and other necessities as much as 165 people in 15 minutes. As for the number of people coming out of the station an average of 186 people are out of the station within 15 minutes.

4.5.2 Trans Jogja Circulation

The result of data survey for 4 hours starting at 12:00 pm - 16:00 pm at two bus stops nearest Yogyakarta obtained the following data:

Table 4. Circulation of Trans Jogja Around Yogyakarta Station per 15 Minutes

TIME	MANGKUBUMI BUS STOP 2			JLAGRAN BUS STOP		
	LINE 1A	LINE 2 A	TOTAL	LINE 3A	LINE 8	TOTAL
12:00 - 12:15	1	1	2	1	0	1
12:16 - 12:30	2	0	2	0	1	1
12:31 - 12:45	1	1	2	1	0	1
12:46 - 13:00	1	1	2	1	1	2
13:01 - 13:15	2	1	3	0	0	0
13:16 - 13:30	3	1	4	1	0	1
13:31 - 13:45	1	1	2	1	1	2
14:46 - 14:00	1	1	2	1	0	1
14:01 - 14:15	1	1	2	0	0	0
14:16 - 14:30	2	1	2	1	1	2
14:31 - 14:45	1	1	3	0	0	0
14.46 - 15:00	1	1	2	1	0	1
15:01 - 15:15	1	2	2	1	1	2
15:16 - 15:30	2	1	3	1	0	1
15:31 - 15:45	2	1	3	0	1	1
15.46 - 16:00	2	1	3	1	0	1
TOTAL	**24**	**16**	**40**	**11**	**6**	**17**

Source: Surveyed, 2017

The number of Trans Jogja passing through Jl. Margoutomo (Mangkubumi 2) bus stop within 4 hours were 40 buses. Trans Jogja passes J. Jlagran Pasar Kembang bus stopo was 17 buses.

4.5.3 Public Transport Needs (Trans Jogja)

To calculate the number of additional Trans Jogja buses required can be done by calculating the number of additional buses based on the number of rit and the number of bus travel:

$$JA = \frac{JR}{JPB} \tag{1}$$

Note :

JA : number or buses (1) *JR* : number of rit (rit/day)

JPB : number of bus trips (rit/day) *WP* : travel time (minutes)

$$WP = \frac{Rate\ Distance}{Speed\ of\ Plan} \times 60\ minutes \tag{2}$$

$$JPB = \frac{Operation\ Time}{Travel\ Time} \times 60 \tag{3}$$

Table 5. Calculation of Additional Unit

Trans Jogja	Distance Rute (km)	Average Speed (km/hour)	JR	WP (2)	JPB (3)	JA (1)
Line 1A	34.7	40	96	52	18	5
Line 2A	32.1	40	64	48	20	3
Line 3A	37.6	40	44	56	17	3
Line 8	25.5	40	24	38	25	1

Based on the calculation, it was shown that the ideal journey time of Trans Jogja route 1A in one rit was 52 minutes. The number of trip bus journey 1A was 18 rit / day, then the addition of the number of Trans Jogja buses to accommodate the potential demand is 5, route 2A and 3A each 3 and route 8 is 1 Trans Jogja bus.

4.6 Condition of Existing Parking and Parking Plans at Yogyakarta Station

Parking violation is happen in the sidewalk of the station. Parking without a zone permit is because of the ignorance. East side parking area Bong Suwungwas divided into 2,844 m^2 for cars, 844 m^2 for motorcycle and 846 m^2 for motor stay. Meanwhile the parking area on the south side was divided into 1060 m^2 for the car and 687 m^2 for the motor. Existing conditions for the capacity of the car park east side Tugu Station amounted to 107 vehicles. East motorcycle parking capacity was 302 motorcycles. Motorway parking on the east side of Tugu Station can accommodate 271 motorcycles. Car parking capacity south of Tugu Station was 39 vehicles. Motor parking on the south side accommodated 264 motors.

To maximize the concept of TOD at Yogyakarta Tugu Railway Station, there are two alternatives in planning of parking capacity arrangement at Tugu Railway Station Yogyakarta.

1. Parking capacity at Tugu Station with the least capacity. The advantages of this alternative are:
   a. Parking access becomes more widespread so that the circulation in and out of the vehicle more smoothly.
   b. Users of private vehicles using parking at Tugu station switch using Transjogja mode. This arrangement can be done if it has been supported by the availability of Transjogja in fulfilling the existing demand and change of location of Transjogja

shelter closer to Tugu Station so that vehicle users are interested to switch modes to Transjogja.
   c. Private vehicle users can reduce the increasingly limited consumption of fuel oil.
   d. Reduce air pollution around Tugu Station Yogyakarta area.
2. The parking capacity at Tugu Station is maximized. The advantages of this alternative are the capacity of private vehicles parking at Tugu Station can be more optimal and parking arrangements in the Tugu Railway area are more organized. The negative effects of these alternatives are the alternative planning may lead to more people using private vehicles in the region, the number of private vehicle use, the load of roads around Tugu Station area is getting bigger, and increasing air pollution in Tugu Station area.

4.7 Transit Oriented Development (TOD)

The concept of TOD should synergize between train, pedestrian, and public transportation (Transjogja). The Transjogja connection is the need for integration in the mode transfer of trains and feeders. This require good public transport services. Transjogja should be able to accommodate the existing demands of the region, close to Tugu Station, and ease to reach the location of stops. It can be made near the dropout hall of Yogyakarja Railway Station on Jalan Mangkubumi and near entrance of Yogyakarta Station Jalan Pasar Kembang.

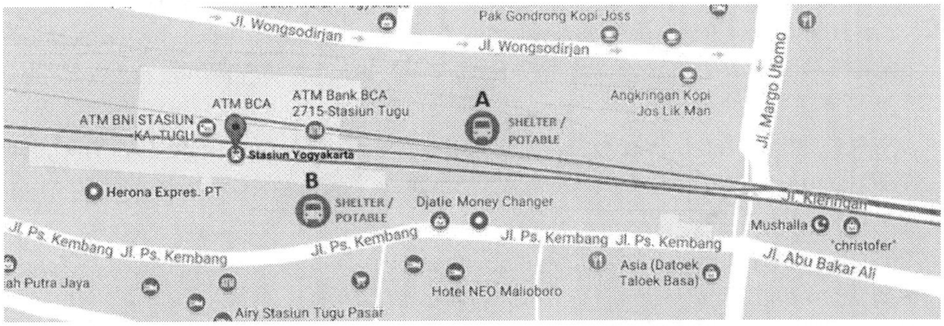

Fig. 6. The Relocation Plan of Trans Jogja Bus Stop (Source: Google Map)

The placement of the new halt plan is intended so that routes 1A and 2A can be served at bus stop A ± 100 m from the entrance of YRS hall, while the 3A and 8 routes can be served at bus stop ± 50m from the entrance of East hall. The plan will make easier for passengers to access it. They can continue walking in the pedestrian area that has been provided without having to walk too far. The adjustment of punctual time the arrival and departure schedule of trains with Trans Jogja bus needs to be arranged, so that during rush hours in Yogyakarta Station all potential demand can be covered by the existing Trans Jogja.

5 Conclusion

The number of passengers who entering the YTSwith any transportation were 165 people in 15 minutes. The number of people who coming out of the station were 186 people in 15 minutes. In order to accommodate the potential of demand, it is necessary to increase the number of Trans Jogja for 1A 5 units, 2A 3 line, 2A 3 line and make bus stop near Yogyakarta Station. The location can be made near the dropout hall of Yogyakarta Railway Station on Jalan Mangkubumi and near entrance of Yogyakarta Station Jl. Pasar Kembang. The placement of the new halt plan is intended so that routes 1A and 2A can be served at bus stop A which is ± 100 m from the entrance hall of Yogyakarta Station, while the 3A and 8 routes can be served at bus stop ± 50m from Yogyakarta Station hall entrance East. TOD in the railway station can optimize if there is continuity between the modes either train, private vehicle and public transportation (Trans Jogja). Therefore, two parking regulation alternatives are provided by increasing the parking capacity to make it easier for private vehicle users who are around Yogyakarta Station to move the modes and by minimizing the parking capacity so that the use of private vehicles in the area of Yogyakarta Station will use Trans Jogja.

6 References

1. Kusbiantoro, BS. *Memanusiakan Perencanaan Sistim Transportasi.* Kelompok Keahlian Sistem Infrastruktur Wilayah dan Kota. Bandung: ITB (2007).
2. Vincent G and Harja D. *Spatially Explicit Individual-based Forest Simulator (SexI- FS).* World Agroforestri Centre (ICRAF) and Institut de Recherche pour le Developpement (IRD). Bogor (2008).
3. Cervero, R., et al., *Transit-oriented development in the United States: Experiences, challenges, and prospects.* TCRP Report 102. Transportation Research Board.
4. Arief, A. B., Yudono, A., Akil, A., & Ramli, I. Model of Coastal Transit Oriented Development (TOD) Based on the Potential of Local Port and marine Tourism Port, Case Study: Fort Rotterdam Makassar and the Surrounding Areas. In *IOP Conference Series: Earth and Environmental Science* (Vol. 79). https://doi.org/10.1088/1755-1315/79/1/012034. (2017).
5. Wijaya, Alfred. 2009. *Penataan Ruang Yang Ramah Lingkungan Melalui Perencanaan TOD (Transit Oriented Development).* Seminar Nasional Perencanaan Wilayah dan Kota ITS. Surabaya.
6. Puslitbang Perhubungan Darat dan Perkeretaapian, *Study Transit Oriented Development (TOD) Around Tugu Station Yogyakarta*, Jakarta (2014).
7. Dunphy R., Myerson D., Pawlukiewicz M.. *Ten Principles For Successful Development Around Transit.* The Urban Land Insititute, Bank of America. Washington, D.C. ULI. http://uli.org/wp-content/uploads/2012/07/TP_DevTransit.ashx_.pdf last accessed July 30th, 2017 11.07 PM (2003).
8. Pick, Frank. The Design of Modern Railway Stations in Europe and America. *Journal of the Royal Institute of British Architects*, Vol **37**, Third Series, No 10, pages 319 to 332 (1930).
9. Republik Indonesia. *Peraturan Menteri Perhubungan Nomor 33 tahun 2011 tentang Persyaratan Jenis, Kelas, dan Kegiatan di Stasiun Kereta Api.* Jakarta.
10. Republik Indonesia. Keputusan Direktur Jenderal Perhubungan Darat Nomor : 271/HK.105/DRJD/96 Tentang *Pedoman Teknis Perekayasanaan Tempat Perhentian Kendaraan Penumpang Umum*. Jakarta.
11. Republik Indonesia. Undang-undang Republik Indonesia Nomor 22 Tahun 2009 tentang *Lalu Lintas Angkutan Jalan*. Jakarta.

12. Republik Indonesia. Keputusan Direktur Jenderal Perhubungan Darat Nomor 272 Tahun 1996 tentang *Pedoman Teknis Penyelenggaraan Fasilitas Parkir*. Jakarta.
13. Oppenheim, N. *Urban Travel Demand Modeling*, Wiley, New York (1995).
14. Warpani, S, *Pengelolaan Lalu Lintas dan Angkutan Jalan*. Bandung: ITB. (2002)
15. Kandee and Kandee, Kandee, Kandee, S. 2001. *Intermodal Concept in Railway Station Design*www.bu.ac.th/knowledgecenter/epaper/jan_june2004/somruedee.pdf last accessed July 31th, 2017 12.26 AM (2004).
16. Noor, Juliansyah. *Metodologi Penelitian : Skripsi, Tesis, Disertasi dan Karya Ilmiah*. Kharisma Putra Utama, Jakarta (2011).
17. Calthorpe, P.. The Next American Metropolis: Ecology, Community, and the American Dreams, Princeton Architectural Press, New York (1993).
18. Mirmoghtadaee, M. Challenges of Transit Oriented Development (TOD) in Iran, The Need For A Paradigm Shift. *TeMA Journal of Land Use, Mobility, and Environment, special issue*, no. **3**, pp. 35 – 46 (2016).

MATEC Web of Conferences **181**, 02002 (2018) https://doi.org/10.1051/matecconf/201818102002
ISTSDC 2017

Evaluation of transport networks as disaster evacuation routes (Case study of flood disaster in Huamual Sub-District, Maluku)

Ahmad Munawar[1*], *Joko* Sujono[1], *Adam Pamudji* Rahardjo[1], *Joko* Murwono[1], and *Ahmad* Nur Hamid[1]

[1]Gadjah Mada University, Depart. of Civil and Environmental Engineering, Yogyakarta, Indonesia

Abstract. As the global warming occurs, the effects of rainy and dry seasons in Indonesia have changed. Extreme rainfall causes some several disasters of such as flash floods and landslides which cause loss of life, damage to facilities, infrastructure and transport routes, as well as damage to the economy. The method used in this study began with a field survey conducted in Dusun Nasiri, Dusun Mange-mange, and Dusun Hatamanu to determine the condition of the road network between the centre of local government to the affected villages. In addition, laboratory tests were carried out in the Transport Laboratory, Gadjah Mada University, Yogyakarta. This study produced several maps by using ArcGIS software, and the laboratory tests produce findings on the feasibility of local materials for the restoration of the raod network. The maps showed the affected area and the routes for evacuation and the distribution of logistical, medical and paramedical aid, CFA, Makassar.

1 Introduction

Indonesia is an archipelagic country located between the Asian and Australian continents. The country had only two changes of season, rainy and dry. The extreme rainfall has caused some disasters, such as flash floods and landslides [1]. The major causes of the flash floods are heavy rainfall and sedimentation. Another cause is deforestation in the region: the land is made bare and no longer functions as a place for the absorption of rainwater [1].

The loss of life is the main damage caused by flash floods. Other impacts of flash floods are damaged facilities, infrastructure and transport routes. For example, in 2012 and 2013, flood disasters hit Maluku province. They occurred not only in the City of Ambon, but also in Central Maluku District and Western Seram District. Data from the Regional Disaster Management Agency (BPBD) of Ambon City shows that a total of 18 people died, 5,420 houses were submerged in water, and 2,170 families, or as many as 4,718 people, were evacuated. The highest level that the flood reached was seven metres, and the most severe

* Corresponding author: munawar@ugm.ac.id

flooding was in Dusun Nasiri, Dusun Mange-mange, Dusun Hatamanu, Desa Luhu, Huamual Sub-District, Western Seram District. BPBD data shows 5 people dead, 1,200 people evacuated, and 154 houses severely damaged. In addition to the loss of life, the flash floods in the region also damaged infrastructure buildings and public facilities, such as embankments, irrigation canals, road networks and bridges.

The scale of the many losses suffered calls for preventive action to minimize the adverse impacts of the flash floods [2]. Evacuation routes and evacuation activities are, therefore, the most important points to be analyzed [3], and require geospatial data and regional information for networking analysis [4]. Hazard maps and evacuation signage should be used for analysis and to provide information [5]. One preventive action which can be carried out immediately is to prepare the road network to support expeditious evacuation from flash floods. This would also be beneficial by encouraging economic growth in Maluku [6]. Therefore, it is necessary to investigate the feasibility of the road network used for evacuation and the mobilization of aid to the victims of natural disasters [7].

The evacuation process requires accessible routes to evacuate victims to safer areas. These are called the evacuation routes [8]. Evacuation routes can be reached by land, air, and sea transport [9]. The three components of evacuee or volunteer access to the evacuation places [10] are:

- The current state of the road network in the affected areas.
- The speed with which evacuees, volunteers and logistics can move to the evacuation places.
- The trip period is the time required by the evacuees, volunteers and logistics to reach the evacuation places..

The choice of route depends on travel time, travel cost, comfort and safety [11]. Transport both for the evacuation and the general economy will require a road surface able to carry existing transport loads. A road is constructed in several courses: the surface course which consists of the binder course and wearing course, the base course, the sub-base course and the subgrade. The layers of the completed construction will have some of the following functions [12]:

- To support and spread the load of the vehicles (traffic load);
- To provide some comfort with a flat surface course;
- To provide some roughness so that the vehicle does not slip;
- To provide a wearing course

To protect the underlying courses from water and weather, that can damage them.

2 Method and Analysis

There are several steps involved in the conduct of this study. The first is to carry out a field survey in the study area to determine the condition of the road network from the centre of local district government centre to the disaster- affected villages. This step includes testing in the laboratory a sample of the aggregate used in the road construction. The laboratory test will be conducted in the Department of Civil and Environmental Engineering, Faculty of Engineering, Gadjah Mada University, Yogyakarta, with the use of a guide book from the Ministry of Public Works, Directorate General of Highways.

The road network survey is conducted along the road from the Centre of Local District Government of Western Seram to the flash flood site with assistance of GPS devices and a camera.

The GPS data will be input into the ArcGIS software to evaluate the road network.

The location of the disaster-prone area in Dusun Nasiri, Dusun Mange-mange and Dusun Hatamanu in Huamual Sub-District, is a valley flanked by two hills. To determine the evacuation routes for the disaster victims, the survey first needs to know the affected areas by mapping them with the help of ArcGIS software. An example of a disaster map of Dusun Nasiri is shown in Fig. 1.

Fig. 1. Disaster map of Dusun Nasiri and possible evacuation points.

The area with the first red gradation indicates the area most severely affected by the flash flood with a water level of 2 metres and predictions of severe damage, so that people are required to immediately evacuate themselves when a high intensity flood is occurring. The area with the second red gradation shows the flash flood affected area where severe damage is predicted with a water level of 1.5 metres. The area with orange gradation shows the flash flood area categorized with a medium damage prediction, and a water level of 1 metre. The area with yellow gradation shows the affected area categorized with slight damage predicted and a water level of 0.5 metres.

The three hamlets (Dusun) affected by disaster have semi-permanent buildings, a narrow road and narrow spaces between the buildings. These are some of the factors which will cause difficulties people trying to evacuate. Flash flood disaster mitigation efforts in those hamlets have not hitherto been conducted well by the Government of Western Seram District and by the Provincial Government of Maluku located in Ambon. An evacuation route map is needed to minimize casualties from flash floods. An evacuation route is based on the easiest way to get to the safest place from the danger of flash floods and is developed with the use of ArcGIS software. Planning for flash flood evacuation locates an area in each hamlet that has a height of +15 metres and is far away from the river. The height +15 metres is considered high enough for a place safe from the flood. Figure 1 shows the evacuation points.

Table 1. The trip time from the origin point to the evacuation area of Dusun Nasiri

Point of Origin	Destination Point (Evacuation Area)	Distance (metres)	Trip Length	
			minutes	seconds
A	1	123	2	44
B	2	203	4	31
C	2	209	4	39
D	2	184	4	5
E	1	123	2	44

Fig. 1. shows the distance between the evacuation point and the evacuation area. Trip length is based on the walking speed of a group of people, which is 0.75 m / second. The distance of each point to the evacuation area can be seen on the map, and the trip length from the point of origin to the evacuation area can be calculated (Table 1). On TABLE 1 it can be seen that the trip length to reach the evacuation area from the furthest point is still short enough at 8 minutes 36 seconds. Therefore, a disaster early warning system needs to provide a flash flood warning flood 10 minutes in advance.

All the evacuation locations are nearby privately owned farms considered to be safe from the impacts of a flash flood. The routes to them are the steep uphill roads that people use to go to and from their farms. It could be said that access to the safe area does not complete the evacuation process. Facilities and infrastructure are needed. The main infrastructure that has to be constructed first is the steps as can be seen in Fig. 2. The second is the direction sign boards that can be seen in Fig. 3.

Fig. 2. Example of the steps for disaster victims to climb to get to the safest area (evacuation area)

Fig. 3. Example of a direction sign on an evacuation route

3 Mitigation and aid distribution routes

3.1 Alternative 1 (the existing route)

Access to the disaster location in Huamual Sub-District, Western Seram District of the City of Ambon, by the existing route is not effective in terms of time (see Table 2). This is because the disaster areas are located on an island and cannot be reached by land. In addition to the long distance to the location of the flash flood, both from the Centre of Western Seram Government and from the Centre of Maluku Provincial Government, there is also the condition of the roads which can no longer to be crossed by a four-wheeler, but only by using two-wheeled vehicle and then continuing on foot. The condition of existing road network and alternative routes is shown in Fig. 4.

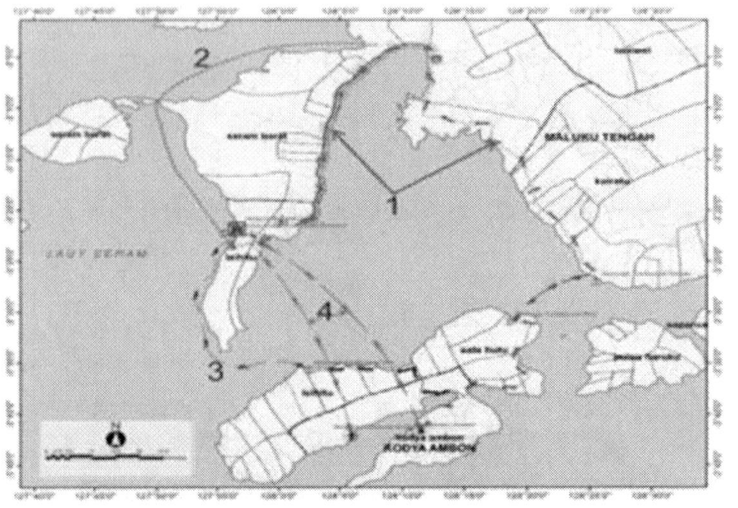

Fig. 4. Transport route alternatives to the disaster location

Table 2. Trip length from Ambon City – disaster location (alternative 1 or the existing condition)

No	Origin – Destination	Transport Mode	Distance (Km)	Trip Length Hrs	Trip Length Min
1	Ambon City – ASDP	Car	33.92	1	30
2	ASDP –ASDP	Ferry	14.24	2	0
3	ASDP – Western Seram District	Car	56.51	2	0
4	Western Seram District –Huamual Sub-District	Motor-cycle	58.68	2	40
5	Huamual Sub-District – Location	On Foot	7.75	3	15
Total			171.1	10	85
				11	25

Source: Results of the field survey in July 2014

3.2 Alternative 2 (sea + land transport)

The alternative is to change from land transport mode to sea mode. Sea mode involves using the existing docks in Masika Jaya and/or Pelita Jaya which are located ± 22.7 miles away from the Centre of Western Seram District Government in Piru City. This alternative was considered more effective and time-efficient in distributing aid to the affected area (see Table 3).

Table 3. The Required Trip Length from Ambon City – Disaster Location (Alternative 2)

No	Origin – Destination	Transport Mode	Distance	Trip Length	
			(Km)	Hrs	Min
1	Ambon City– ASDP	Car	33.92	1	30
2	ASDP–ASDP	Ferry	14.24	2	0
3	ASDP–Western Seram District	Car	56.51	2	0
4	Western Seram District – Masika Jaya or Pastika Jaya	Car	22.70	1	10
5	Masika Jaya or Pastika Jaya– Location	Water Bus (capacity: 12 passengers)	54.14	2	20
Total			181.51	8	60
				9	00

Source: Results of field survey in July 2014

3. 3 Alternative 3 (sea and land transport)

Because Maluku province is an island, transport to there is traditionally by water. One traditional dock is located in Hila Sub-District. The conditions of the road both from Ambon City and the airport to the traditional dock are adequate, since it is a national road. It is possible that to integrate sea transport with land transport modes . The planned logistical aid distribution to flood victims from Ambon to the disaster affected hamlets can be carried out by using the existing water bus and then distributing to the evacuation area by land transport already available.

The advantage of this alternative is that it does not require the provision of new facilities and infrastructure. The water bus is the daily transport for the local people who already have a traditional dock. However, there is a need to build a traditional dock in Western Seram District, especially for a disaster location. However, the existing water bus has only a small capacity, so that it cannot accommodate the large amounts of logistical aid in one trip. Thus, the government should build a traditional dock at the disaster site, and provide a water bus on standby all the time (see Table 4).

Table 4. The Required trip length from Ambon City – disaster location (alternative 3)

No	Origin – Destination	Transport Mode	Distance	Trip Length	
			(Km)	Hrs	Min
1	Ambon City – Hilla Traditional Dock	Car	32.21	1	30
2	Hilla Traditional Dock– Location	Water Bus (capacity: 12 passengers)	41.45	1	30
Total			73.66	2	60
				3	00

Source: Results of the field survey in July 2014

3. 4 Alternative 4 (air transport)

The fourth alternative is the mode of air transport. The logistical aid from the central government was distributed using airplanes from Soekarno Hatta Airport (Jakarta) to Pattimura Airport (Ambon) and then distributed by helicopter at the location of the study.

For helicopters from Pattimura Airport to land there, a heliport would need to be constructed.

The use of air transport in the distribution of logistical aid to the flash flood disaster-prone locations has some advantages in terms of trip length (see Table 5) However, applying this third alternative requires a helicopter to distribute the logistical aid from Pattimura Airport (Ambon) and a heliport in the evacuation area for the helicopter to land with the aid.

Table 5. The required trip length from Ambon City – disaster location (alternative 4)

No	Origin – Destination	Transport Mode	Distance	Trip Length	
			Km	Hr	Min
1	Ambon City–Location	Helicopter	41.00	0	25
Total			41.00	0	25

4 Comparison of alternatives

The selection of the alternatives focused more on the trip length needed to reach the disaster location than on considering the cost. It can be seen on Table 6.

Table 6. Comparison of the alternatives

Parameter	Transport Mode		
	Land	Sea	Air
Distance	Land distance: ±171.1 km	Land distance: 41.45 km Sea distance: 32.21 km	41 km
Speed of the modes used	Land Transport: 30 km/hr	Land Transport: 30 km/h Sea Transport : 30 km/h	100 km/hour
Trip length *	Land Transport: -±11 hours	Land Transport: 1.5 hours; Sea Transport : 1.5 hours Total: 3 hours	25 minutes
Infra-structure	Using a winding national road and then continuing by using the provincial and district roads as in the Table 5.1	Using a winding national road and the traditional dock in Hila Village. However, a traditional dock in Dusun Talaga is needed	No heliport in the disaster location. Heliport infra-structure is needed in Dusun Talaga as it is safe from disaster.

* Assuming no obstacles are encountered
** Assuming that the roads are navigable.

From the point of view of the time it takes to distribute logistical, medical and paramedical aid, a helicopter clearly has the fastest delivery time than the other available modes of transport. The heliport should be constructed in Dusun Talaga which lies between the affected hamlets.

The use of air transport is planned for logistical, medical and paramedical distribution only in urgent conditions. Sea transport is also used.

Dusun Talaga hamlet is safe from flash flood disaster, so the following are proposed:

- To improve the traditional dock at Dusun Talaga.
- To build a heliport near the traditional dock.

In order to accelerate the evacuation process and the distribution of aid to flash flood victims, good access the evacuation area is needed. Therefore, the transport network, especially land transport needs to be restored.

5 Conclusions and suggestions

The results of the analysis and discussion that has been conducted lead to the following conclusions:

This study has resulted in the most efficient evacuation route map for the escape from each hamlet in the event of flood disaster. In general, the evacuation route uses the village roads with a width of ± 3 metres. The evacuation route heads toward the evacuation area located on the edge of the settlements. The evacuation area is located on 15 metres above sea level. In more detail, the evacuation route for each hamlet is as follows:

Dusun Nasiri

There are two evacuation areas. The first can be reached by people with the furthest point of origin (Point B) in 8 minutes 36 seconds. The second can be reached from the furthest point of origin (Point E) in 4 minutes 21 seconds. Consequently, the early warning system for flash flood needs to be at least 10 minutes in advance.

Dusun Mange-mange

There are two evacuation areas. The first can be reached by people from the furthest point of origin (Points A and E) in 2 minutes 44 seconds. The second can be reached in 4 minutes 39 seconds. Consequently, the early warning system for flash flood needs to give warning at least 10 minutes in advance.

Dusun Hatamanu

There are three evacuation areas. The first can be reached by people starting from the furthest point of origin (Point D) in 6 minutes 49 seconds. The second can be reached by people from the furthest point of origin (Point B) in 4 minutes 36 seconds. The third can be reached by people from the furthest point of origin in 2 minutes 43 seconds. Consequently, the early warning system for the flash flood needs to give an early warning at least 10 minutes in advance.

The distribution of logistical, medical and paramedical aid during the emergency responses uses air transport which is a helicopter that can reach the area in 25 minutes from the centre of Ambon City and Pattimura Airport. This requires the heliport facility which is planned for construction in the Dusun Talaga. Non-urgent aid distribution is planned using a combination of land and sea transport modes (Alternative 3) that can reach the disaster location in 3 hours from the centre of Ambon City. If the distribution comes from West Seram District, the plan would be to use sea transport modes (Alternative 2) which can reach the location in 3 hours 30 minutes. Therefore, to facilitate aid distribution, an infrastructure is required for water buses to move closer to the beach (a dock). This infrastructure need calls for a traditional dock to be constructed in Dusun Talaga.

In general, the problems along the road from the Office of Western Seram District Government to the Huamual Sub-District Office and their solutions are as follows:

- Road damage caused by the water left by rainfall. The solution is to build drainage and then restore the damaged roads.
- Road damage caused by the landslides from the cliff. The solution is to stabilise the cliff.
- Lost bridges and those sufficiently damaged so as to be unsafe. The solution is to construct new bridges or provide a surface on the courses of the bridges and the river wall.

- Partial blocking of the road which makes the road inaccessible to cars. The solution is to clear the road of the debris.
- Roads cut by landslides and uncleared debris. The solution is to carry out reforestation in the headwaters area and also to harden the cliff after a soil test has been conducted.
- Damage to the road through the middle of the forest. The solution is to plan to plant productive trees that can encourage people's economic activity, then to rebuild the road network to accelerate economic growth, after studying and restoring the forest's water flow system.

The authors make the following suggestions for improving the quality of this study:

- The involvement of geographers is required so that the mapping will be more appropriate.
- The structure of the evacuation route steps needs to be designed.
- Further study is needed of the road structure.
- More time to be added to the survey period in order to increase familiarity with the location under study.

Acknowledgements

I would like to give my sincere thanks to Gadjah Mada University and the New Zealand Foreign Affairs & Trade Aid Programme for providing the opportunity to join the research project Community Resilience and Economic Development Programme, Capacity Development on Natural Disaster Risk Management for Communities in Small Islands.

References

1. Karnawati D. *Bencana Alam Gerakan Massa Tanah di Indonesia dan Upaya Penanggulangannya*. Yogyakarta:Penerbit Jurusan Teknik Geologi Fakultas Teknik UGM. (2005)
2. Departemen Komunikasi dan Informatika, Badan Informasi Publik. *Memahami Bencana, Informasi Tindakan Masyarakat Mengurangi Resiko Bencana*. Jakarta:Pusat Informasi Kesejahteraan Rakyat. (2008)
3. Yamashita, N., & Ohmoto T. *Local People's Responses to Flood Disasters in Flood Prone Areas of Northeast Bangladesh*. Journal of Disaster Research 109(2), 288–298. (2015)
4. Yi, C., Park, R., Murao, O. & Okamoto, E. *Emergency Management: Building an O-D Ranking Model Using GIS Network Analysis*. Journal of Disaster Research 7(6), 793–802. (2012)
5. Mizuki, C.*Flood Risk Communication from the Viewpoint of Disaster Prevention Awareness in an Urban Area of Tokyo, Kita Ward*, Journal of Disaster Research 7(5), 595– 603. (2012)
6. Susanto, S. *Disaster Management dari Negeri Rawan Bencana*. Jakarta:PT. Aksara Grafika Pratama. (2006)
7. Compos, V. *A Method for Evacuation Route Planning in Disaster Situations.* 15th Meeting of the Euro Working Group on Transport. Procedia-Social and Behavioral Science 54, pp. 503-512. (2012)
8. Variadi, V. *Kajian rencana evakuasi dalam kesiapsiagaan bencana tsunami dengan pendekatan berbasis masyarakat (Studi Kasus: Kecamatan meuraksa kota Banda Aceh).* Yogyakarta: Tesis Pasca Sarjana Fakultas Teknik Universitas Gadjah Mada. (2010)

9. Stevanov, A., & Smith, J. *Multi Objective Evacuation Routing in Transport Networks*. European Journal of Operational Research 1498, 435-446. (2009)
10. Budiarjo, A.*Evacuation Shelter Building Planning for Tsunami- Prone Area*, Case Study of Meulaboh City Indonesia'. Master's Degree Thesis, International Institute for Geo-Information Science and Earth Observations, ITC, Enschede. (2006)
11. Munawar, A. *Dasar-dasar Teknik Transportasi*. Yogyakarta:Beta Offset. (2011)
12. Sukirman, S. *Beton Aspal Campuran Panas*. Jakarta, Penerbit Granit (2003)

Housing and transportation system alignment-based on green concept

Imamah Khairunnisa[1,*], and *Shirly* Wunas[1]

[1]Hasanuddin University, Depart of Urban and Regional Planning, Gowa, Indonesia

Abstract. Land use and transportation system are the main components of the city space structure formation. Settlements as the biggest land allocation in the region play an important role in the availability of human resources and the largest generation of the city. Various problems arise along with the development of urban residential settlements, generally, occur in middle-scale cities to the larger cities i.e. development of middle-class house to elite-class that formed into gate. That fact was one of the causes of transport system problems, non-integrated in the transport system in housing area cannot be avoided. Therefore, this paper explains: 1) current condition of the house built by the developer (gated house) and the self-help house, 2) transportation systems that support the residential area, and 3) concept which supports housing and transportation system alignment in the context of green concept.

1 Introduction

The settlement is one of the largest components in the urban land use. According to the law No. 01 [1], explains that the purpose of organizing house is to supply the basic needs of the population, increased and equitable welfare population. Therefore, the settlement became very important in the quality improvement of a country's human resources.

The development of settlements land use was the most significant, primarily at medium-scale cities up to large-scale cities, even frequent the transition functions of productive farm-land becomes settlements. Problems occurred in the area of the settlement began to various, such as social problems, economic even environmental issues. This is particularly associated with the transport system. Settlement's issues that arose, one of that caused related to the development of middle-class settlement and elite class which was built following the pattern of the cul-de-sac then cause problems against the new road infrastructure alignment against previously existing access (local settlements).

Non-integrated road network infrastructure can give a negative impact on the community's economy and residential environment. Because according to the law No. 1

* Corresponding author: imamah.khairunisa@hotmail.com

[1], the purpose of housing and settlements are to ensure the realization livable and affordable house in a residential environment that is healthy, safe, harmonious, orderly, well-planned, integrated and sustainable. Settlements and transport are an absolute unity in a planning. The development of the area of the settlement should follow the road network infrastructure, so it will keep fit with the road network hierarchy [2]. Realization of settlement sustainable requires a plan that also embraced the concept of eco-friendly transport, as a base supporting the entire population mobility.

Study case focused on the Eastern Suburban area of Makassar City (Biringkanaya District and Tamalanrea District) which is one of the regions with the residential land use development was significant in the last 20 years (44%) [3].

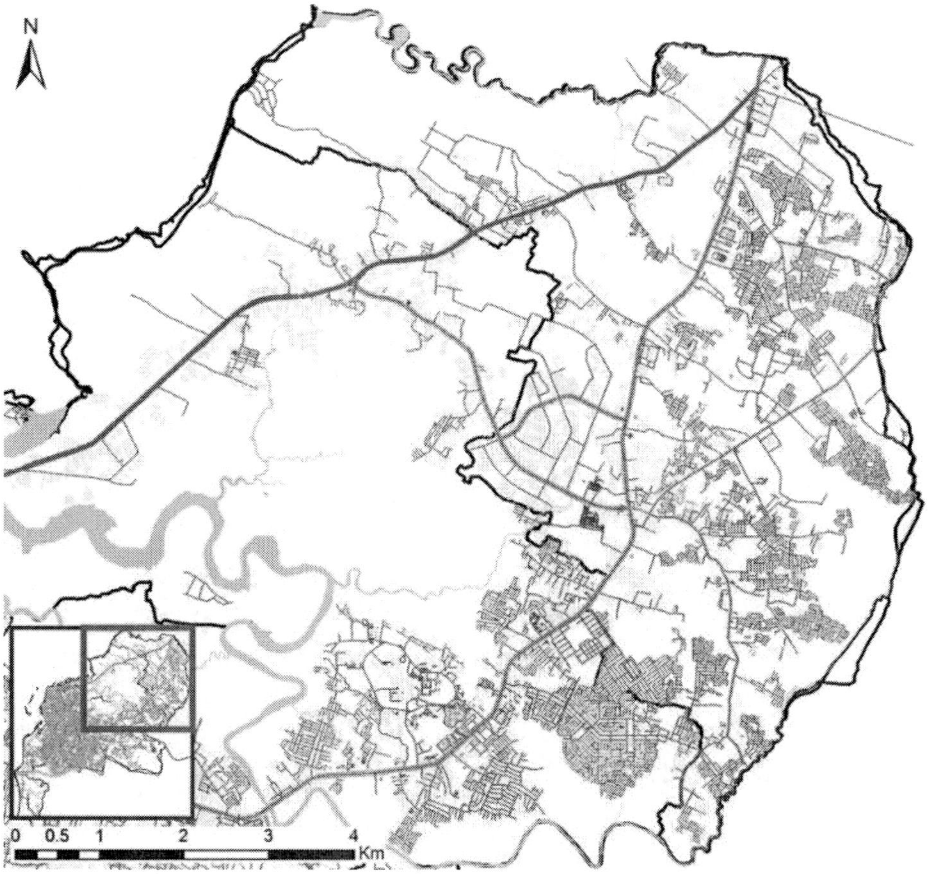

Fig. 1. Eastern suburban area of Makassar City (Tamalanrea District and Biringkanaya District)

Analysis data obtained through the field observation by using the satellite map, generate data of thematic map such as the settlement land use development map and road network development map. Methods of analysis used in this discussion to be descriptive and comparative against policies/regulations. The preparation of the concept based on alignment analyses of the transportation system and housing settlements-based on the green concept.

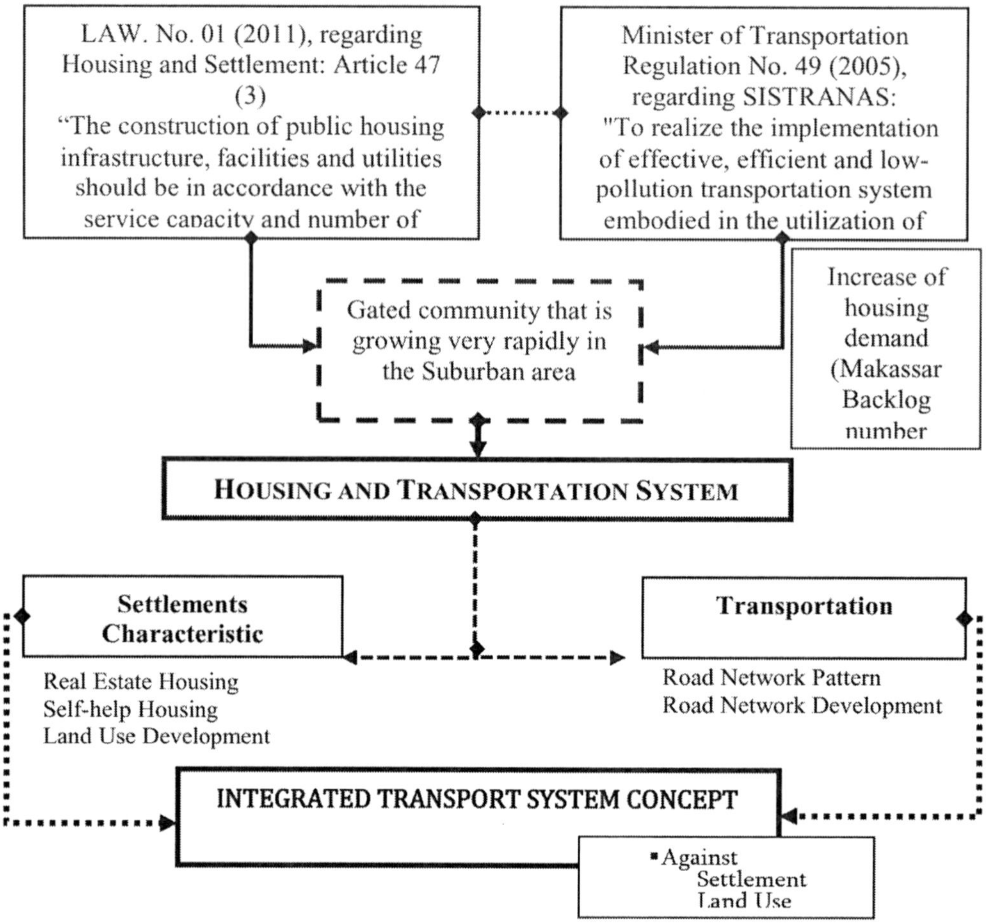

Fig. 2. Framework

This Research aims to shows current condition of the house built by the developer (gated house) and the self-help house, to identify transportation system that support the residential area and to share about concept which supports housing and transportation system alignment in the green concept context.

2 Result and discussion

This chapter discusses regarding the analysis of the housing and settlement characteristics, an analysis of the transportation network development and transportation system alignment concept.

2.1 Analysis of housing and settlements characteristics

Real estate housing is the houses in an area developed by private parties as well as Government, while self-help housing is the housing that was established through the efforts of by the community itself. The main non-governmental homes built on land owned by the legal owners singly, while real estate housing is the houses that are built above ground by private property developers. Residential streets per cluster in accordance with dimensions ownership/mastery of the land, so the types of homes and residential streets/cluster corresponding to developer's capability. This causes real estate tend to develop new settlement on the farm-small land that causes build a parapet with a cluster of houses of local residents who came to be called a community fenced or better known as gated community. The small cluster filled with new resident and often separate with local residents who earlier lived in the neighborhood (Fig. 3).

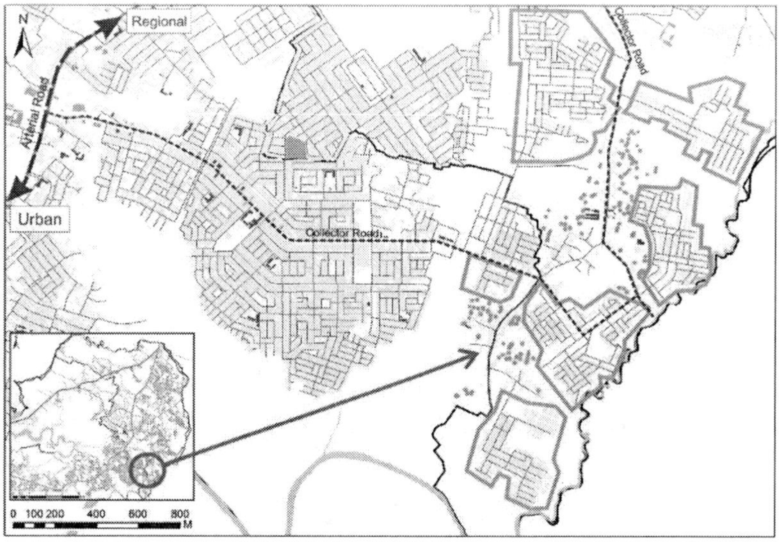

Fig. 3. Suburb settlements area map, Makassar City

Gated community housing generally followed the cul-de-sac pattern, where houses are built, not linearly and did not have connectivity with another good road network inside the gated community or outside their communities (local housing).

The development of residential land use in the suburban area, Makassar City (Tamalanrea District and Biringkanaya District) from 2003 till 2015 following the arterial road network (Perintis Kemerdekaan Street) and the toll road (Makassar Industrial District/KIMA). Settlement developments follow a linear pattern along Perintis Kemerdekaan Street (Fig. 4).

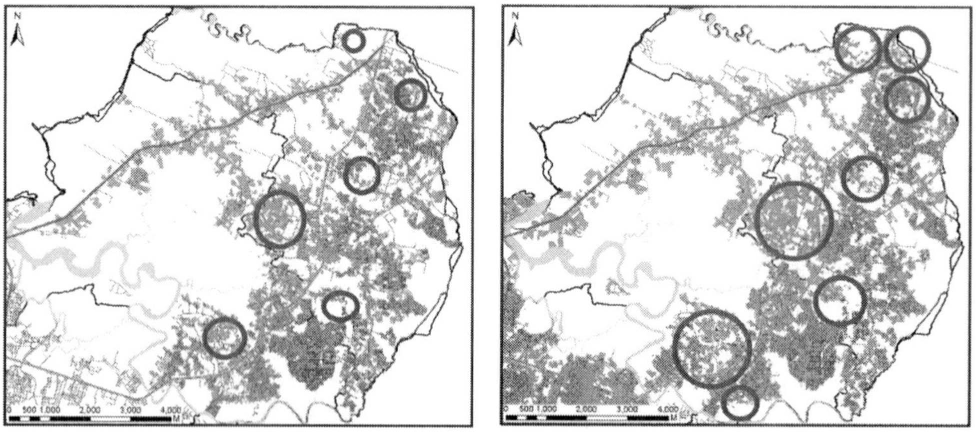

(a) (b)

Fig. 4. (a) Residential land use map (2003), (b) Residential land use map (2015) Settleme

2.2 Analysis of transport network development

Transportation systems issues within in the settlement area scope are non-integrated in the system, such as the settlement area development on the arterial road network, while the plan must be residential area should be on the secondary collector roads and the primary collector road. Small cluster housing developed using the arterial road as the main access that connects the city center with real estate. According to the Directorate General of Bina Marga [2], the arterial road network is a network of roads linking the region with each other so that the primary activity of the vehicles tend to be high with most low speed 60 Km/hours.

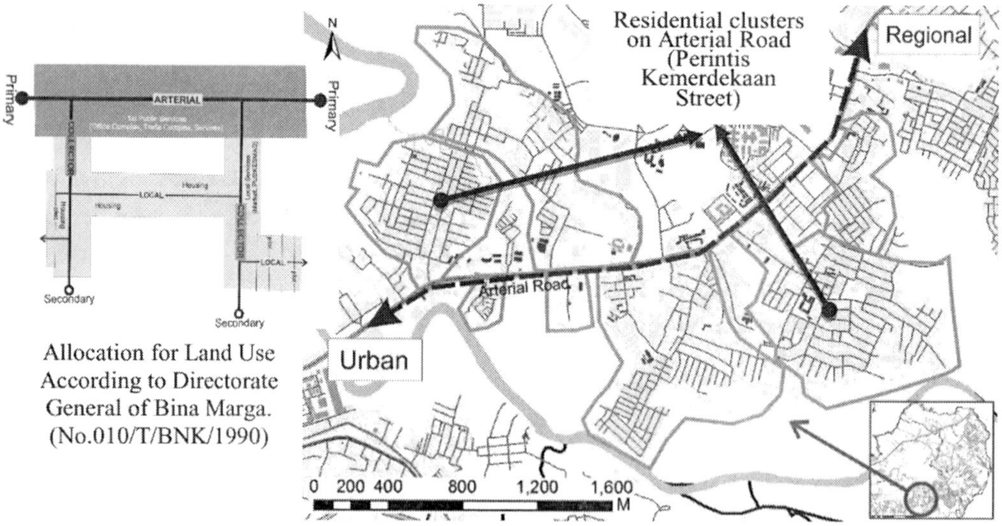

Fig. 5. Map of residental area along the arterial Road (Perintis Kemerdekaan Street)

In addition, other problems encountered was non-integrated road network in the residential neighborhood built by developers and the self-help houses around it (Fig. 3).

Road network in the residential area should be continued so as to make it easier to access especially in the self-help houses/local outside the gated.

The settlements are on the Goa Ria Street (Sudiang Sub-district) utilizing a collector road network as the main access to the public services center that have connectivity with the arterial road network (Fig. 6). The importance of road network connectivity also greatly affect the infrastructure of his supporters, such as the drainage network. Not continuously of the streets network and drainage network may cause the inundation on certain territories which certainly can lower the quality of the environment.

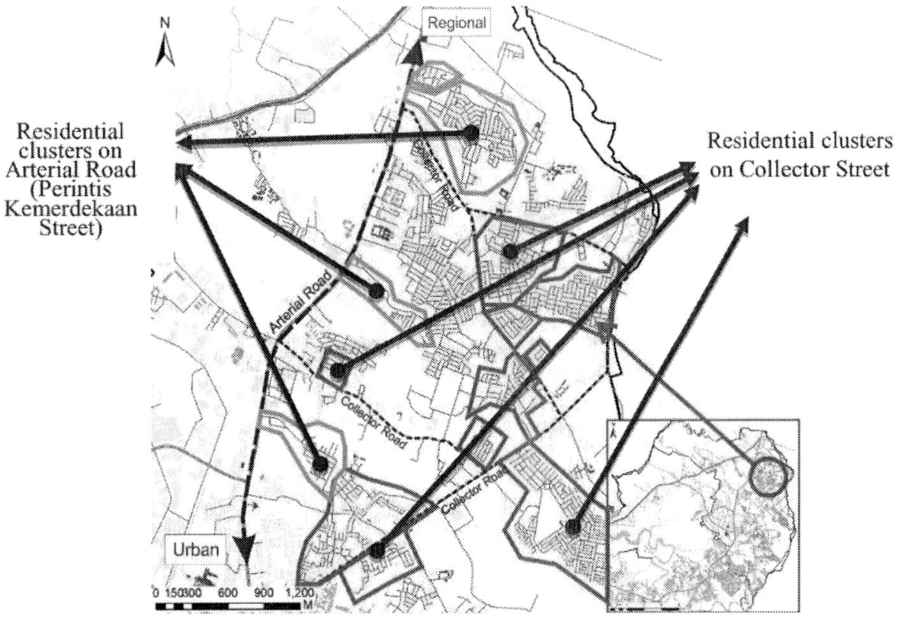

Fig. 6. Makassar suburb residental map

The development of settlement that is incompatible with the function/hierarchy of the road network may give rise to various problems of traffic on the arterial road, because of the many obstacles the side as well as the high inflows, causing congestion on arterial road network and have a direct impact on the environmental as a result of the disposal of vehicle emissions.

Fig. 7. Arterial road network (Perintis Kemerdekaan Street)

(a) (b)
Fig. 8. (a) Collector road network (Paccerakkang Raya Street) and (b) Local road network (BTP, blocks. AF)

The fulfillment of the community needs will certainly make use of private vehicles to reach the public service center is at a distance of more than 500 m (beyond the distance for pedestrians). Motor vehicles using may issue high emission vehicles in large numbers and can lower the quality of the environment especially if not supported by the green belt or green line (Fig. 7 and Fig. 8).

2.3 Transport system alignment concept

As the one continues development land use every year, housing and settlements need of referrals which are certainly integrated with the transportation system. The following illustration (Fig. 9) describes the direction of the provision of housing on the local road network and secondary collectors. The arterial road network is reserved for the primary service center that has connectivity to the city center, while the primary collector using for mixed land use like settlements and local services center.

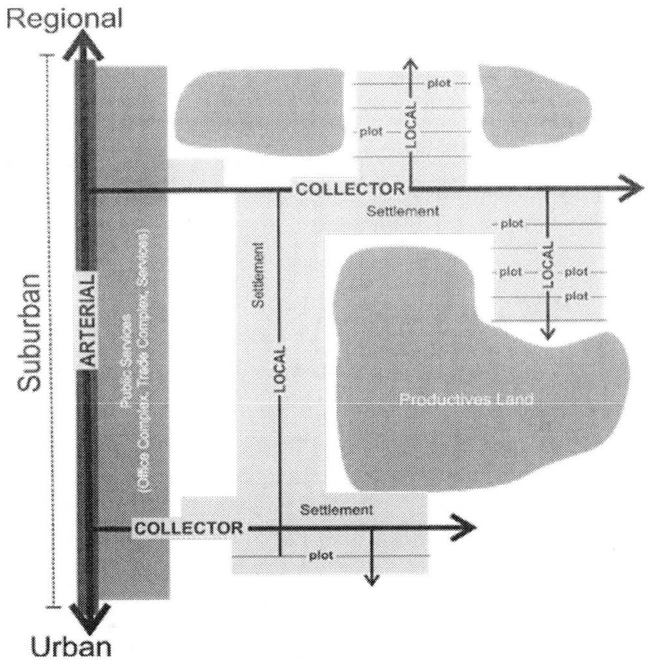

Fig. 9. Illustratiton of land procurement direction

Real estate development is directed to be on the primary collector road network which is supported by new local road network planning. Road network development is expected to remain constant so that it can keep connectivity of transport network as well as all infrastructure supporting.

Road network planning with the green concept is also expected to be realized well (Fig. 10). It is implemented as an effort to improve the quality of the environment in housing and settlements area.

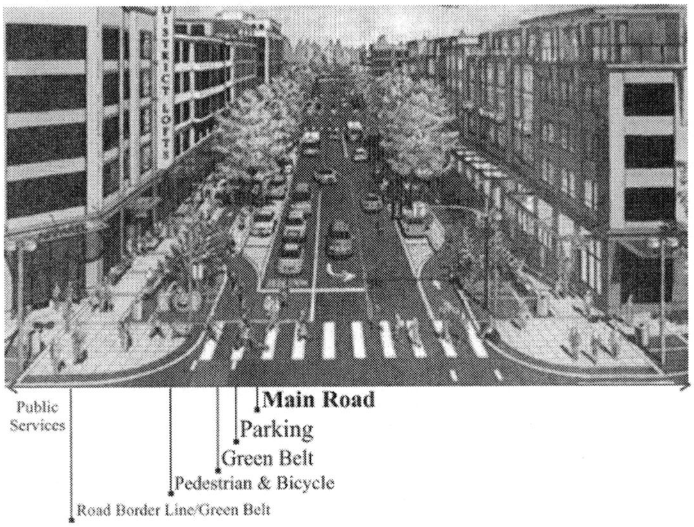

Fig. 10. Illustration of the green concept application on the road network

3 Conclusion

Settlement characteristics are divided into two types, namely residential real estate and local housing/self-help. The development of land use was spread on the suburban area and constantly developed by following arterial road network. Real estate tends to use arterial road network as the main access to the center of a city.

Road network did not continuously occur in the area of the settlements which are commonly caused by the development of real estate which forms a gated community. It also results in non-integrated of the road network that connects the real estate and self-help housing. Other issues on this discussion, road network can affect network support system such as a drainage network so it can cause problems on the environment like congestion and vehicle emissions disposal arising due to the high use of motor vehicles.

Implementation of the transport system and housing alignment concept like the direction of the residential land use development in accordance with the road network hierarchy as well as the implementation of the green concept in road network planning.

References

1. Law no. **1** on Housing and Settlement Area, (2011)
2. Directorate General of Bina Marga, Guidance on Determination of Classification of Road Functions in Urban Areas, No. 010/T/BNK/1990, (1990)
3. K. Imamah, *Land Use Development and Eco-Friendly Transportation (Pengembangan Guna Lahan Perumahan dan Transportasi Ramah Lingkungan)*, Thesis: Hasanuddin University, (2016)
4. Regulation of the Minister of Transportation no. **49** on National Transportation System (SISTRANAS), (2005)

Pedestrian crossing model in urban street (Case study on commercial area in Bandung)

Sony Sulaksono Wibowo[1*], *Rian* Wicaksana[1]

[1]Bandung Instute of Technology, Study Program of Civil Engineering, Bandung, Indonesia

Abstract Pedestrians who cross without any crossing facilities and under mixed-traffic tend to have varying responses. The responses can be analyzed by using multiple linear regression model, with pedestrian crossing delay and pedestrian crossing speed set as response variables. This research aims to develop two pedestrian crossing models based on the condition at the midblock part of urban street, in particular commercial area and without specific crossing facilities. The two models are pedestrian crossing delay model and pedestrian crossing speed model. The affecting factors are considered in linear relationship and the multiple-linear regression models are used. The principal factor in the pedestrian crossing delay model is group size of more than 3 persons, while in the model of pedestrian crossing speed, the principal factors are number of group size and pedestrian baggage. The mean of pedestrian crossing delay was about 3 seconds while pedestrian crossing speed was about 1 m/s.

1. Introduction

The complexity of urban street problems generally will occur when the land-use is a center of human activities, such as market, shops, bus stop, and so forth. When this condition takes place, then there would be crowd of people around the road who can be found. In certain conditions where the roads are already crowded, vehicular and pedestrian traffic flows can be interfering on each other. This kind of chaotic traffic condition is going to be worse if there is no clear regulation about separation of car and motorcycle. This condition is well-known as mixed-traffic. Mixed-traffic is a very common case in Indonesia, particularly in Bandung. Under this chaotic condition, the pedestrians, which is varying in gender, ability, group size, or their baggage size, will tend to response freely, especially when there is no certain crossing facility.

The middle section of the road or so-called midblock area is a very vulnerable spot for the occurrence of intervention between the traffic flow and pedestrian flow, especially if the area is the commercial activity. In this condition, pedestrians often cross arbitrarily to move from one side of the road to the other side of the road through

* Corresponding author: sonyssw@gmail.com

traffic whose patterns and compositions may vary. This phenomenon can be called "jaywalking". When pedestrians do "jaywalking", then the varying behavior and characteristics of the crossing pedestrian may appear. The responses of the crossing pedestrian generally consist of waiting time, crossing time, gap or gap received, crossing pattern, and so forth.

Given the varying characteristics of crossing pedestrian and coupled with the mixed-traffic conditions occurring at most roads in Indonesia, particularly at Bandung, then a wide range of responses from the crossing pedestrians may appear. One thing that might be concerned for this situation is to know which pedestrian characteristics that gives a significant impact toward their responses when they cross the road. In this case, the responses will be limited to pedestrian crossing delay and pedestrian crossing speed, since time and distance can be measured and observed easily. The objective of this research is to develop pedestrian crossing model in urban street, in particular of commercial area. The model took place in the midblock area with no specific crossing facilities. In this research, two models were developed, they are model for pedestrian crossing delay and model for pedestrian crossing speed. The factors affecting to the model were considered in linear relationship. Therefore, the multiple-linear regression models were used.

2. Literature review

2.1 Pedestrian characteristics

The regulation published by department of public works stated that the characteristics of pedestrians can be divided into physical, behavioural, and psychological. The physical characteristics of pedestrian include the travel motives of pedestrians, their convenience toward the weather, the availability of public transport, and land-use of the region they are crossing at. The behavioural characteristics of pedestrians will be related to how the pedestrians are doing when they cross the road. While the psychological characteristics will be more related to the tendency about what the pedestrians desire when they cross the road, so that they will cross conveniently.

Aside from the characteristics that listed on the regulation of department of public works, there are many studies have already revealed about the similar things. For example, the studies conducted by Kuttan, et al. [1] revealed that age, gender, group size of pedestrians, and crossing pattern significantly affect their behaviour in crossing the road.

Another study related to pedestrian characteristics was conducted by Gultom [2]. This research revealed that the decision of pedestrians to cross the road was affected by their gap acceptance, age, gender, pedestrian baggage, group size, and the combination of vehicles which formed the gaps.

2.2 Pedestrian crossing delay

Pedestrian crossing delay is the time required by the pedestrians to wait before they resolve to cross the road. Kuttan, et al. [1] revealed that male pedestrians have tendency to show more hazardous crossing behaviour than female pedestrians since the waiting time of male pedestrians tend to be shorter than female pedestrians. For the group size, the same research stated that the more people involved in a crossing pedestrian group, then the more waiting time they need before crossing the road. While for the age group, there are no certain tendency which age group will wait longer than the other.

2.3 Pedestrian crossing speed

Pedestrian crossing speed is a distance traveled by pedestrians divided by the time they required to cross the observed lanes of the road.

Based on the study conducted by Hummer, Allen, and Milazo (1999) in Kuttan, et al. [1], generally, the pedestrian crossing speed is about 1,2 m/s. If the proportion of elderly people in a certain region is large enough (above 20%), then the average crossing speed of pedestrians is 1,0 m/s. Based on the studies which is listed on Highway Capacity Manual (HCM) 2000 in Kuttan, et al. [1], it states that the standard pedestrian crossing speed is 1,4 m/s.

While based on the studies conducted by Montufar (2013) in Kuttan, et al. [1], the normal pedestrian crossing speed is 1,2 m/s. But, the research was not involving 40% of elderly pedestrians, 90% of disabled people, 10% of adult pedestrians (20 year-old to 64-year old pedestrians). Somehow, this research could conclude that lower designed crossing speed of pedestrian can be applied for elderly people, not only those who can walk normally, but also those who walks with walking aids.

Based on the research that was done by Montufar (2013) in Kuttan, et al. [1], then Pedestrian Crossing Control Guide, Transportation Association of Canada (TAC PCCG) modified and classified pedestrian crossing speed for facility design into these points.

- Crossing speed is 0,8 m/s, if 20% of the pedestrians are using walking aids;
- Crossing speed is 0,9 m/s, if 20% of the pedestrians are elderly;
- Crossing speed is 1,0 m/s, if all of the pedestrians are walking normally.

Knoblauch, et al. [3] conducted a study related to pedestrian crossing speed which is categorized by age and influenced by various factors and environmental conditions, such as the type of facility used. Based on the research, it was found that in the 15th percentile, young pedestrians crossing speed is 1,25 m/s, and for elderly people, the crossing sped is only 0,97 m/s. Goh, et al. [4] also conducted a research related to pedestrian crossing speed which is categorized by age and gender. In the study, it can be concluded that the younger the age of the pedestrians, then the crossing speed will also tend to be higher. Moreover, the study also concluded that male pedestrians tend to walk faster than the female pedestrians. Kuttan, et al [1] stated that male pedestrians tend to walk faster than female pedestrian, and the research also stated that the smaller the group size of pedestrians, then the crossing speed will also tend to be faster.

3. Research Methods

The method used initially in this research generally consists of finding the primary data (field survey) and secondary data (literature review). The primary data needed in this research were collected in one of the road segment in Otto Iskandardinata Street, Bandung.

The field survey was done by measuring the geometry of the road and recording the road condition. Based on this measurement, the length of the observed road segment is 100 m, and the lane width is 3,5 m. This field survey was conducted on Friday, April 28th 2017, and it took an hour long (16.15 – 17.15). The time to conduct the field survey is considered as random, or without any considerations. It means that in this field survey, the surveyor does not know whether the traffic condition is on a peak hour or not. This action was taken because there are no further field survey that aims to determine the peak hour. After the data were collected, they will be used in the making of multiple linear regression model for pedestrian crossing delay and speed. The following figure is showing the location where the data were being collected.

Figure.1 The study area

5.Discussion

After the informations and the data needed were collected, the first thing to do is making the category of the variables that are considered in this research. The pedestrian characteristics or variables of the crossing activity are shown on this following table.

Table 1. Observed variables

No	Variable	Type of Variable	Category	Code
1	Pedestrian Crossing Delay	Continuous	Time, in seconds (s)	-
2	Pedestrian Crossing Speed	Continuous	Velocity, in (m/s)	-
3	Gender	Discrete	Female	0 - Female
			Male	1 - Male
4	Vehicle Combination	Discrete	Car - Car	0 - No 1 - Yes
			Car - Motorcycle	0 - No 1 - Yes
			Motorcycle - Car	0 - No 1 - Yes
			Motorcycle - Motorcycle	0 - No 1 - Yes
5	Ability	Discrete	Vulnerable	0 - Vulnerable
			Normal	1 - Normal
6	Pedestrian Group Size	Discrete	1 Person	0 - No 1 - Yes
			2 Persons	0 - No 1 - Yes
			2 Persons	0 - No 1 - Yes
			> 3 Persons	0 - No 1 - Yes
7	Pedestrian Baggage	Discrete	Not Carrying	0 - Tidak
			Carrying	1 - Ya

Variables that have been mentioned in table 1 are divided into three categories, they are additional variables, predictor variables, and response variables. Table 2, 3, and 4 below will show the distribution of crossing pedestrian for each category of the mentioned variables.

Table 2. Distribution of crossing pedestrian based on predictor variables

No	Variable	Category	Distribution	Percentage	Total
1	Ability	Vulnerable	19	5,2%	365
		Normal	346	94,8%	
2	Pedestrian Group Size	1 Person	224	61,4%	365
		2 Persons	94	25,8%	
		3 Persons	21	5,8%	
		> 3 Persons	26	7,1%	
3	Gender	Female	96	26,3%	365
		Male	269	73,7%	
4	Vaehicle Combination	Car - Car	138	37,8%	365
		Car - Motorcycle	92	25,2%	
		Motorcycle - Car	70	19,2%	
		Motorcycle - Motorcycle	65	17,8%	
5	Pedestrian Baggage	Carrying	66	18,1%	365
		Not Carrying	299	81,9%	

Table 3. Distribution of crossing pedestrian based on response variables

No	Variable	Category	Distribution	Percentage	Total
1	Pedestrian Crossing Delay	0,1 s - 2,2 s	121	33,2%	365
		2,2 s - 4,3 s	127	34,8%	
		4,3 s - 6,3 s	55	15,1%	
		6,3 s - 8,4 s	26	7,1%	
		8,4 s - 10,4 s	9	2,5%	
		10,4 s - 12,5 s	11	3,0%	
		12,5 s - 14,6 s	5	1,4%	
		14,6 s - 16,6 s	7	1,9%	
		16,6 s - 18,7 s	0	0,0%	
		18,7 s - 20,8 s	4	1,1%	
2	Pedestrian Crossing Speed	0,4 m/s - 0,7 m/s	46	12,6%	365
		0,7 m/s -1,0 m/s	125	34,2%	
		1,0 m/s - 1,3 m/s	125	34,2%	
		1,3 m/s - 1,6 m/s	53	14,5%	
		1,6 m/s - 1,9 m/s	8	2,2%	
		1,9 m/s - 2,2 m/s	3	0,8%	
		2,2 m/s - 2,5 m/s	3	0,8%	
		2,5 m/s - 2,8 m/s	1	0,3%	
		2,8 m/s - 3,1 m/s	0	0,0%	
		3,1 m/s - 3,4 m/s	1	0,3%	

5.1 Outlier Test

The outlier test aims to investigate the extreme values of a data set. This test will be done in two out of three known methods. They are data standardization and box plot. From these two methods, the result will be taken from the box plot method. The outlier test is only available for the continuous variables. In this case, the continuous variables is only existing on the response variables category (pedestrian crossing speed and pedestrian crossing delay). This outlier test will show the amount of extereme data. Afterward, the extreme data shown by this test will be eliminated so that a new data set with less extreme values may appear. This test will be done in repetition until the extreme values could not be found in the data set. After this test is being applied on response variables, then a new data set of pedestrian crossing delay and pedestrian crossing speed with no extreme value can be exposed in the following tables.

Table 4. Distribution of crossing pedestrian based on delay time (without outlier)

No	Variable	Category	Distribution	Percentage	Total
1	Pedestrian Crossing Delay	0,1 s - 0,9 s	18	5,6%	323
		0,9 s - 1,7 s	69	21,4%	
		1,7 s - 2,4 s	43	13,3%	
		2,4 s - 3,2 s	53	16,4%	
		3,2 s - 3,9 s	55	17,0%	
		3,9 s - 4,7 s	33	10,2%	
		4,7 s - 5,4 s	17	5,3%	
		5,4 s - 6,2 s	11	3,4%	
		6,2 s - 6,9 s	12	3,7%	
		6,9 s - 7,7 s	12	3,7%	

Table 5. Distribution of Crossing Pedestrian Based on Crossing Speed (Without Outlier)

No	Variable	Category	Distribution	Percentage	Total
1	Pedestrian Crossing Speed	0,4 m/s - 0,5 m/s	10	2,8%	353
		0,5 m/s - 0,7 m/s	22	6,2%	
		0,7 m/s - 0,8 m/s	49	13,9%	
		0,8 m/s - 0,9 m/s	54	15,3%	
		0,9 m/s - 1,1 m/s	79	22,4%	
		1,1 m/s - 1,2 m/s	57	16,1%	
		1,2 m/s - 1,3 m/s	35	9,9%	
		1,3 m/s - 1,4 m/s	28	7,9%	
		1,4 m/s - 1,6 m/s	14	4,0%	
		1,6 m/s - 1,8 m/s	5	1,4%	

5.2 Formulation of Regression Model

In order to make a good multiple linear regression model, the insignificant predictor variables must be eliminated. This research will use three methods in order to eliminate those insignificant variabels. They are the elimination based on the least coefficient value (known as *stepwise* II), elimination based on the least p-value, and elimination based on the "backward" feature in SPSS.

Every methods will produce a different set of parameter. The difference of applied method will make the formed model is being different as well. Rules of a good multiple linear regression model recites that the model have to involve as many predictor variables as possible. Moreover, every coefficients in the model have to follow the plus and minus sign (+/-) that already expected before. Table 7 and table 8 below shows the mutiple linear regression model formed by different methods.

Table 6. Model of pedestrian crossing delay

No	Parameter	Expected Sign	Variables	Method 1		Method 2		Method 3	
				Coefficient	(*p-value*)	Coefficient	(*p-value*)	Coefficient	(*p-value*)
1	Intercept	(+/-)	C	3,12	(0,000)	2,97	(0,000)	2,97	(0,000)
2	Gender	(-)	X_1	0,02	(0,925)				
3	Car - Car	(+)	X_2						
4	Car - Motorcycle	(+)	X_3	0,18	(0,468)				
5	Motorcycle - Car	(+)	X_4	0,27	(0,325)				
6	Motorcycle - Motorcycle	(+)	X_5	0,05	(0,868)				
7	Ability	(-)	X_6	-0,31	(0,591)				
8	Group Size of 1 Person	(-)	X_7						
9	Group Size of 2 Persons	(+)	X_8	0,21	(0,371)				
10	Group Size of 3 Persons	(+)	X_9	0,44	(0,415)				
11	Group Size of > 3 Persons	(+)	X_{10}	1,19	(0,004)	1,23	(0,001)	1,23	(0,001)
12	Pedestrian Baggage	(-)	X_{11}	-0,21	(0,408)				
	Significance Value of F-Test			0,110		0,001		0,001	

Table 7. Model of pedestrian crossing speed

No	Parameter	Expected Sign	Variables	Method 1		Method 2		Method 3	
				Coefficient	(*p-value*)	Coefficient	(*p-value*)	Coefficient	(*p-value*)
1	Intercept	(+/-)	C	1,00	(0,000)	1,06	(0,000)	1,06	(0,000)
2	Gender	(-)	X_1	0,03	(0,359)				
3	Car - Car	(+)	X_2						
4	Car - Motorcycle	(+)	X_3	-0,01	(0,853)				
5	Motorcycle - Car	(+)	X_4	-0,06	(0,117)				
6	Motorcycle - Motorcycle	(+)	X_5	-0,04	(0,309)				
7	Ability	(-)	X_6	0,05	(0,429)				
8	Group Size of 1 Person	(-)	X_7						
9	Group Size of 2 Persons	(+)	X_8	-0,06	(0,068)	-0,07	(0,019)	-0,07	(0,019)
10	Group Size of 3 Persons	(+)	X_9	-0,13	(0,054)	-0,15	(0,012)	-0,15	(0,012)
11	Group Size of > 3 Persons	(+)	X_{10}	-0,14	(0,013)	-0,17	(0,001)	-0,17	(0,001)
12	Pedestrian Baggage	(-)	X_{11}	-0,10	(0,006)	-0,09	(0,010)	-0,09	(0,010)
	Significance Value of F-Test			0,003		0,000		0,000	

5.3Analysis of Model Selection

For the pedestrian crossing delay model, then the selected model is the one formed by method 3 or method 2. This model will be selected into the next step of analysis is because, first, despite the predictor variable involved in this model is merely one, that is pedestrian group size of more than 3 persons, the predictor variable has a strong tendency to become the involved variables in model. This case is indicated by the significant number or p-value of partial test that pedestrian group size of more than 3 persons has fulfilled the predefined requirement (< 0,1). Second, the model formed by these methods has fulfilled the requirement of simultaneous test (< 0,1) as well. Third, the sign following the coefficients of pedestrian crossing delay model just match with the expected sign. This model shows that the most influencing characteristics at determining pedestrian crossing delay in the observed road is group size of more than 3 persons. This case indicates that people who cross the road in a group of more than 3 people, tend to be waiting for the readiness of each other to cross the road. Readiness, in this case is to determine the desired combination of vehicles, or to find the appropriate gap between vehicles, which is suitable for the crossing pedestrian in a group of more than 3 persons.

The same methods (2 and 3) is selected again for the model of pedestrian crossing speed due to the same reason as it was selected in pedestrian crossing delay model. Moreover, the p-value of model formed by these methods is even better than p-value of model formed by method 1 (the more p-value getting close to 0, then the variables chance to get involved in the model are higher). Beside, the sign following the coefficients of pedestrian crossing speed model just match with the expected sign. This model shows that the most-influencing characteristics at determining pedestrian crossing speed are group size of more than 1 person, and pedestrian baggage. The reasons why pedestrians who walk in a group tend to be slower than those who do not are they have to adjust for each other physical and mental skill. It means that one member of the crossing group have to adjust his walking speed with another, usually the slowest pedestrian. The more member involved in this group, then the group walking speed tends to be slower. It is due to the increasing difficulty to adjust physical skill to the weakest pedestrian (walking speed, etc.). This statement is being backed up by the coefficients of the model which is increasing along the number of pedestrian getting involved in a crossing group. Beside the group size of pedestrian, the other significant characteristic are pedestrian baggage. In this research, baggage is limited to trolley, large goods carried on back, or baby holding. These kind of pedestrian baggage is considered as an obstruction to the sight distance and limiting the moving space of pedestrian. In general, the pedestrian who carries this kind of baggage tends to be more careful so that the baggage does not fall in the middle of the road. Therefore, their crossing speed becomes slower. There was some chances for pedestrian who walks in a group also carrying those kind of pedestrian baggage. If these two conditions meet, then the crossing speed would be much slower.

6.Conclusions

The multiple linear regression model for pedestrian crossing delay is $Y = 2{,}97 + 1{,}23\,X_{10}$, the model shows that the most influencing characteristic for determining pedestrian crossing delay is X_{10} or group size of more than 3 persons. This model can be applied by making the probability of the conditions that may appear, and determine the proportion of each conditions, so that the mean value of pedestrian crossing delay, which is going to be used as design criteria can be obtained. Based on the model, the mean value of pedestrian crossing delay at the observed road segment of Otto Iskandardinata Street, Bandung is 3,054 s.

The multiple linear regression model for pedestrian crossing speed is $Y = 1{,}06\text{-}0{,}07\,X_8\text{-}0{,}15\,X_9\text{-}0{,}17\,X_{10}\text{-}0{,}09\,X_{11}$, unfortunately, this equation is not valid for a group of crossing pedestrian because of the group size variables cannot work simultaneously. So, the equation will be divided into three conditions

- For pedestrian group size of 2 persons

$$Y = 1{,}06\text{-}0{,}07\,X_8\text{-}0{,}09\,X_{11} \tag{1}$$

- For pedestrian group size of 3 persons

$$Y = 1{,}06\text{-}0{,}15\,X_9\text{-}0{,}09\,X_{11} \tag{2}$$

- For pedestrian group size of more than 3 persons

$$Y = 1{,}06\text{-}0{,}17\,X_{10}\text{-}0{,}09\,X_{11} \tag{3}$$

where,
X_8 = Group size consists of 2 persons
X_9 = Group size consists of 3 persons
X_{10} = Group size consists of more than 3 persons
X_{11} = Pedestrian Baggage

Based on the model that mentioned above, then the mean value of pedestrian crossing speed at the observed road segment in Otto Iskandardinata Street, Bandung is 1,005 m/s.

7. Acknowledgements

Most of data and analysis was already presented in second author's final paper (unpublished) in Civil Engineering Program in Bandung Institute of Technology under first author supervision. More detail work can be obtained in university's electronic library.

References

1. Kuttan, Mannu O, et. al. *Analysis and Modelling of Pedestrian Road Crossing Pattern on Urban Undivided Roads in Mixed Traffic* in Proceedings of the 12th International Conference on Transportation Planning, and Implementation Methodologies for Developing Countries, December 2016, IIT Bombay, India. . (2016)
2. Gultom, Hansen Samuel Arberto. x *Model Perilaku Pengambilan Keputusan untuk Menyeberang Jalan di Daerah Mid-block.* Bandung: Institut Teknologi Bandung, pp. 23 – 61. Thesis for master's degree in highway engineering and development, Institut Teknologi Bandung. (2012)
3. Knoblauch, Richard, et. al. *Field Studies of Pedestrian Waking Speed and Start-up Time.* Transportation Research Board, Volume **1538**, ISSN 0361-1981. (1996)
4. Goh, Boon Hoe, et. al. *Pedestrian Crossing Speed: The Case of Malaysia.* International Journal For Transport and Traffic Engineering, Volume **2**, pp. 323-332, ISSN 2217-544X. (2012)
5. Agresti, Alanx. *An Introduction to Categorical Data Analysis.* USA: Wiley, pp. 2-3. (2009)

6. Chutani, Chirag. *L.O.S. For Pedestrian at Uncontrolled Mid-Block Crossings*. New Delhi: Transport Planning Department, School of Planning and Architecture.
7. Hidayat, Erwin. *Pemilihan Fasilitas Penyeberangan Berdasarkan Gap Kritis (Studi Kasus Jalan Dharmawangsa, Surabaya)*. Widyariset, Volume **15** No.3, pp. 585-592. (2012)
8. Munder, S., dan D.M Gavrilla. *An Experimental Study on Pedestrian Classification*. IEEE Transactions on Pattern Analysis and Machine Intelligence, Volume 28, pp 1863-1868, ISSN 0162-8828.(2006)
9. Ontario Traffic Manual. *Pedestrian Crossing Treatment*. Ontario: Ministry of Transportation, Ontario, pp. 17-18. (2014)
10. Soehartono. *Analisis Sarana Penyeberangan dan Perilaku Pejalan Kaki Menyeberang Di Ruas Jalan Prof. Sudarto, SH Kecamatan Banyumanik Kota Semarang* (Jurnal). Semarang. Repository for bachelor's degree in building engineering education, Universitas Negeri Semarang. (2013)
11. Utomo, Fajar. *Analisis Pola Pergerakan Penyeberangan Jalan di Lingkungan Sekolah Studi Kasus SDN 01/02 Cibubur – Ciracas, Jakarta Timur*. Depok: Universitas Indonesia, pp. 5. Thesis for bachelor's degree in civil engineering, Universitas Indonesia. (2011)

Pedestrian facilities evaluation using Pedestrian Level of Service (PLOS) for university area: Case of Bandung Institute of Technology

Sony Sulaksono Wibowo[1,*], and *Destri* Nurhalima R. M.[1]

[1]Bandung Institute of Technology, Study Prog. of Civil Eng., 40132 Bandung, Indonesia

Abstract. Promotion of walking activities can be done by applying the concept of walkability through infrastructure improvement of pedestrian facilities. Knowledge of how well a highway can be pedestrian friendly and secure through pedestrian assessment become important. The pedestrian level of service (PLOS) is the most common approach to assessing the quality of pedestrian facility operations. This research evaluated pedestrian facilities by using several PLOS methods. i.e. HCM (2000), Trip Quality (Jazkiewicz, 1999), Gainesville (Dixon, 1996), and Australian (Main Roads WA, 2006), for the case of university area at Bandung Institute Technology. Data collection techniques used consisted of traffic counting survey and observation. Improvement recommendations are given to poorly rated performance indicators in the PLOS evaluation. The results of PLOS evaluations is vary and improvement of pedestrian facilities are still required on some indicator such as maintenance, effective width, lighting, access and supporting facility, crossing facilities, and potential conflict with vehicle.

1 Introduction

Walking is a mode of transportation that can never be separated from a travel chain. Whatever type of transportation mode used to reach a destination must involve walking activity. Walking mode may be the key point in encouraging shifting mode from private to public in the sustainable transport. Walking assessment was already described comprehensively in [1]. This paper would like to examine more detail to the pedestrian level of service (PLOS) for the specific area such as in university area. Four method of PLOS were used i.e. HCM, Trip Quality, Gainesville, and Australian PLOS [2-6]. The case of some sidewalks and walkways in the area of Bandung Institute of Technology campus were applied accordingly. Result of the various PLOS analysis could give better pictures of the facilities issues to improvement. Similar work had been done by Sdoukopoulos for pedestrian network in Thessaloniki and Karatas in METU Campus [7].

* Corresponding author: sonyssw@gmail.com

2 Notes on Plos Methods

Generally, the PLOS method can be distinguished into two approaches. The first approach was that only concerns on pedestrian flow. HCM (2000) is a good example of the first method. In this method, the concept of traffic flow characteristic was adopted to the pedestrian flow on the specific passageway. The second approach considered into the quality of the walking environment, especially interaction between pedestrian flow and vehicle movement. On the other hand, it was about safety issue on the pedestrian facilities. Beside that, the issue of walking comfort somehow was taking into account. Example of this approach was (Walking) Trip Quality. Other methods, such as Gainesville and Australian PLOS combined those two approaches into their method. In term consideration issues, the following table described issue comparison of the PLOS methods [4].

Table 1. Issue Comparison of PLOS Dimensions

PLOS Dimension	HCM (2000)	Trip Quality (Jaskiewicz, 1999)	Gainesville (Dixon, 1996)	Australian (Main Roads WA, 2006)
User characteristics	x	x	x	x
Traffic	Volume and pedestrian speed	x	Volume and vehicle speed	Volume and mix of path user
Safety and comfort	x	Enclosure/definition, overhang, buffer, shade trees, transparency, vehicle speed, and lighting.	Potential conflict with vehicle (sidestreet and driveway), pedestrian delay, reduced turn conflict implementation, crossing width, vehicle speed, median, buffer, bench/lighting, shade trees, and VLOS.	Crossing facilities, environment (buffer), potential conflict with vehicle, and personal security (lighting and sight distance).
Land use	x	Enclosure/definition, building articulation, complexity of network, complexity of space, dan transparency.	Continuity, off-street/parallel alternative facilities, TDM and multimodal.	Connectivity
Infrastructure	Effective width and capacity	Width and surface quality	Effective width, total width, barrier, median and surface quality.	Access, width, and surface quality.

The level of service for pedestrian facilities focused more on the characteristic of the facilities. The user perception and some user effort to walk were not being considered deeply. On the contrary, in the walkability method, how the user interacted with the facilities in order to do the walking were being main consideration [1].

Further discussion the method that used in this is as follow. The pedestrian level of service (PLOS) as a result of thorough measurement of walking conditions on a route, path, or pedestrian facility [8]. The PLOS is directly related to the factors affecting mobility, comfort, and safety that reflect pedestrian perceptions of how much a facility can be pedestrian friendly.

HCM (2000) explains that the main performance to be measured in assessing PLOS is the space available for walking, in other words HCM (2000) only considering the capacity of pedestrian facilities. The parameters reviewed in the assessment are volume, average velocity, module, and pedestrian volume-capacity ratio. For the practicality of field observations, the flow or volume of passing pedestrians can be used as a service measure to assign PLOS from A to F.

The applied of Trip Quality method to define nine qualitative PLOS evaluation steps to analyze the pleasures, safety, and comfort of pedestrians when traveling on a pedestrian facility [3]. The complexity of network, building articulation, complexity of space, overhang, buffers, shade trees, transparency, and physical components were components to be considered. Score of 1 to 5 are applied to each parameter assessed, with a score of 5 reflecting excellent conditions and 1 reflecting very poor conditions. Each parameter score summed and averaged to get PLOS value from A to F.

The Gainesville PLOS evaluation method was developed to be applied in the city of Gainesville (Florida) as part of mobility plans for congestion management [4]. The basic criteria for PLOS assessment used consist of six categories, the type of pedestrian facility provided, the conflict between pedestrians and motor vehicles, the existence of supporting facilities (amenities) to accommodate the movement of pedestrians, vehicular level of service (VLOS), maintenance, and TDM and support for multimodal transport. The total score obtained from the sum of each category score determines the PLOS value with a score of 1 to 21 (A - F).

In Australian PLOS method which is an amendment of a study conducted for Main Roads Western Australia by BSD Consultants [8], PLOS assessment is done on three factors, namely physical characteristics, location factor, and user factor. The condition of the pedestrian facility is explained by PLOS from scale A (ideal pedestrian condition condition) to PLOS E (condition of pedestrian facility unsuitable), based on assessment of factors affecting PLOS. Meanwhile PLOS F provides additional information for pedestrian facilities with low access points for disabled.

2.1 Data collection

The data required in PLOS evaluation are grouped into three categories: traffic data of vehicles and pedestrians, infrastructure (road geometry and condition of pedestrian facilities), and land use (related to the condition of buildings, facilities, or spaces around the pedestrian path). All data is obtained through traffic counting survey and observation. The traffic counting survey is conducted on weekdays between Monday and Friday for two-hour from 11:00 to 13:00 (observation period per 15 minutes) which is lunch break time for each observation segment. While observation time for observation activities can be done anytime, either on weekdays or weekends. It should be noted that the PLOS rating of lighting or personal security should be done at night to represent the actual condition and evaluation of PLOS for sidewalk in this study is done on one side only that are beside BNI Bank - Salman Mosque - One Eighty Cafe. It can be said that walking facilities is reviewed within the constraint of the corridor not as a network, therefore evaluation for complexity of the network on the PLOS Trip Quality method is not done. The pedestrian facility reviewed is divided into five segments of observation.

Table 2. Segmentation of Observation Location

Segment	Location	Type of Facility	Segment Length (m)
1	Ganesha Street Corridor (beside BNI Bank)	Sidewalk	180.3
2	Ganesha Street Corridor (beside Salman Mosque)	Sidewalk	199.5
3	Ganesha Street Corridor (beside One Eighty Café)	Sidewalk	107
4	Corridor of Physics Study - Civil Engineering	Walkway	119.5
5	Corridor of Civil Parking Field	Walkway	82.5

Fig. 1 Preview of Observation Location

3 Data analysis and discussion

Basically, there are five aspects of the assessment that being considered as the important aspects to evaluate. There are:

- Dimension: related to the capacity of the pedestrian facility provided. A pedestrian facility should be able to accommodate existing pedestrian traffic;
- Safety: related to the risk of pedestrian accidents due to conflicts with motor vehicles. Pedestrian facilities should be designed using the principle of potential conflict reduction;
- Security: related to the environmental safety of pedestrian facilities. The pedestrian facilities provided should make pedestrians feel safe because pedestrians can be seen and see the surrounding environment and are designed with anti-criminal principles;
- Comfort: deals with the comfort of pedestrian facilities for use. Comfort can be felt by pedestrians from the availability of protection against weather or rest areas, and others;
- Convenience: deals with the ease and suitability of pedestrian facilities for all groups of pedestrians use.

These five aspects are considered sufficient to get an idea of pedestrian level of service so that these five aspects should be used as a basis for planning a pedestrian facility. Some assessment criteria on the Trip Quality PLOS method such as building articulation and space complexity did not include into five aspects of the PLOS assessment that are considered important by the author. The main reason is that these two assessment criteria emphasize aesthetic values rather than functional values that illustrate how attractive the environment is to enjoy. Moreover, in this case, the assessment of building articulation or the complexity of the space is not very meaningful in assessing pedestrian facilities in the campus area which is using different aspects of environmental design compared to roads in urban city.

Table 3. Mapping of PLOS Evaluation Criteria

Criteria	HCM (2000)	Trip Quality	Gainesville	Australian
Dimension	- Width	- Width	- Width	- Width
Safety	x	- Enclosure/definition - Buffer - Vehicle speed	- Driveways and sidestreets - Reduced turn conflict implementation - Crossing width - Vehicle speed - Medians - Buffer	- Crossing facilities - Potential conflict with vehicle
Security	x	- Transparency - Lighting	- Reduced turn conflict implementation	- Personal security
Comfort	- Space for walking	- Overhang - Buffer - Shade trees - Surface quality	- Buffer - Benches or lighting - Shade trees - Vehicle LOS - Maintenance	- Surface quality - Path environment - Path user volume - Mix of path user
Convenience	x	- Complexity of network	- Continuity - Offstreet/parallel alternative facilities - Pedestrian signal delay - TDM/multimodal	- Access - Support Facilities - Connectivity

3.1 Overview of evaluation results

From the PLOS evaluation results using four PLOS evaluation methods obtained varying PLOS values even for the same observation segment. This difference is indeed expected to occur due to the different assessment factors considered in the evaluation of each method. But basically the aspects of the assessment discussed in each method are almost similar.

Table 4. PLOS Evaluation Results

Segment	Location	Type of Facility	PLOS/Score			
			HCM (2000)	*Trip Quality*	*Gainesville*	*Australian*
1	1	*Sidewalk*	A / 5.51 p/min/m	C / 3.15	C / 13.5	D/F / 58
2	1	*Sidewalk*	C / 23.33 p/min/m	B / 3.6	C / 12.5	C/F / 71
3	1	*Sidewalk*	A / 4.44 p/min/m	C / 3.08	C / 13.5	C/F / 70
4	2	*Walkway*	A / 7.19 p/min/m	B / 3.6	B / 16.5	B/F / 121
5	3	*Walkway*	C / 23.67 p/min/m	A / 3.98	B / 14.5	B/F / 109

Significant differences in PLOS values were obtained from the evaluation of the HCM method (2000) with the other three methods. The most extreme example can be seen in the evaluation of PLOS in segment 1, namely sidewalk beside Bank BNI in Ganesha Street corridor. The HCM (2000) method provides excellent evaluation results of PLOS A for this segment, whereas the Trip Quality and Gainesville methods provide a much lower value of PLOS C which means fair or standard, even the Australian method provides very low evaluations of D Indicates that the condition of existing pedestrian paths is poor. The comfort of the pedestrian is minimal and there are safety issues in the pedestrian environment.

This can happen because the HCM (2000) method does not take into account the environmental conditions of pedestrian facilities and rely solely on the parameters of the capacity-based or pedestrian flow assessment. If a PLOS pedestrian facility is only assessed using the HCM (2000) method it can provide misleading results related to the actual condition of the pedestrian facility. As in the case of segment 1, actual current conditions exhibit poor surface quality or maintenance with low pedestrian volume but are rated very well by the HCM method (2000) or in segment 4, where the actual walkway conditions are convenient to use based on Trip Quality method or safe from conflicts with vehicles based on Gainesville and Australian methods but rated standard by the HCM method (2000) as the ease of pedestrians to move (related to capacity and current) is slightly disturbed because of the possibility of minor conflict between pedestrians walking side by side or low speed due to limited walkway area.

The use of the HCM (2000) method is likely more suitable for evaluating the capacity of pedestrian facilities in areas with high pedestrian volumes using pedestrian facilities provided as the main pedestrian site, not on road in roadside sidewalks. The advantages of the HCM method (2000) when compared with three other methods apart from the ease of assessment and simple data requirements are the ability to describe the use of pedestrian facilities that are subject to change over time and can be compared with the capacity provided by pedestrian facilities. The evaluation results of the Trip Quality method in the study sites tend to provide a moderate to excellent value, namely PLOS A - C. PLOS evaluation with Trip Quality method is highly qualitative considering the environmental design of pedestrian facilities and few attention to traffic factors that only involve vehicle speed criterion.

From the results of the evaluation it appears that the Gainesville and Australian PLOS evaluation methods yield the same PLOS value in almost every observation segment with values that tend to be lower than the PLOS values of other methods. This possibility occurs because both methods involve quantitative and qualitative parameters related to the pedestrian volume, infrastructure and environmental characteristics of pedestrian facilities, in contrast to the HCM method (2000) which involves only quantitative or quantitative parameters Trip Quality method is dominated by qualitative assessment parameters. The incorporation of quantitative and qualitative parameters on PLOS evaluation, as in both methods, can improve the reliability of the assessment results.

From the evaluation results can also be seen that This occurs because the walkway reviewed excels in overhang criteria with the roof usage, the physical component of which is due to its excellent surface conditions, the availability of adequate illumination and crossing facilities and the limited potential for conflict with the vehicle when compared to the sidewalk reviewed.

3.2 Performance Indicators for Improvement

Based on PLOS evaluation results, the performance indicators or evaluation criteria of PLOS that need to improve are as follows.

Table 5. Performance Indicators That Requires Improvement

Segment	Location	Type of Facility	Performance Indicator			
			HCM (2000)	*Trip Quality*	*Gainesville*	*Australian*
1	1	*Sidewalk*	x	- Surface quality - Transparency* - Building articulation* - Complexity of spaces* - Lighting	- Driveways and sidestreets - Maintenance	- Surface quality - Crossing facilities - Access - Support facilities - Potential conflict with vehicle - Personal security
2	1	*Sidewalk*	Effective width	- Transparency* - Lighting	- Minimum 1.53 meter wide and barrier free - Maintenance	- Crossing facilities - Access - Path width - Support facilities - Personal security
3	1	*Sidewalk*	x	- Transparency* - Building articulation* - Complexity of spaces* - Lighting	- Driveways and sidestreets - Maintenance	- Access - Support facilities - Potential conflict with vehicle - Personal security
4	2	*Walkway*	x	- Transparency* - Building articulation*	x	- Access
5	3	*Walkway*	Effective width	- Effective width - Building articulation*	- Minimum 1.53 meter wide and barrier free	- Access - Path width - Support facilities

*Recommendations regarding these performance indicators are not provided because they are outside the realm of the authors' studies

3.3 Recommendations for Improvement

Result of the various PLOS analysis could give better pictures of the facilities issues to improvement. One PLOS method might not capture all issue. The general recommendation for improvement that can be given are:

- Surface quality/maintenance; (1) repairing pedestrian pavement by replacing broken or broken tile or tactile indicator and tile installation on the missing part and (2) the use of high durability pavement and routine vegetation intrusion pruning on the surface of the pedestrian path.
- Effective width; (1) applying the concept of sidewalk café or parklet which is more suitable for segment condition 2, (2) provides a guardrail on pedestrian path for trading area with comparison of pedestrian width and trade area width maximum 1: 1.1 or equal to 1.4 meters of pedestrian width with an effective width of 0.6 meters if the cost component becomes a limitation in segment 2, and (3) widening effective width to segment 5 by closing the existing drainage by using a gutter inlet.
- Lighting/personal security; Providing pedestrian lights on segments 1, 2, and 3 with a light illumination level of 6 lux in the horizontal and 11 lux in the vertical direction.
- Access; (1) installation of the curb ramp at the intersection and ramp point near the entrance to the sidewalks of segments 1, 2, and 3 and (2) if the cost does not become a limitation, the installation of ramps in segments 4 and 5 and wheelchair access in Segment 5 can be done, (3) the use of sign to remind the stairs in segment 4 and signs to direct the wheelchair user to use the ramp in the entrance through the main gate south of ITB on segment 5 can be done if the cost becomes boundary.
- Support facilities; (1) installation of detectable warning at the intersection point and near the obstruction on all segments and (2) installation of guiding tactile indicators on segments 4 and 5. (3) adding signs or information boards on all segments.
- Crossing facility; Provision of zebra cross facilities in segments 1, 2, and 3.
- Potential conflicts with vehicles; The use of warning signs that many passers-by and vehicles passing in and out into the sidewalks of segments 1 and 3.

3.4 Improvement Impact to Updated PLOS Value

With assumption of all improvement had been carried out, new value of PLOS can be calculated. It was not about just to achieve the value of each PLOS became A but it can be considered as how much effort needed to achieve a better value of PLOS. Table 6 to Table 9 below show the new value after improvement. Noted that the red color indicates changes and number inside the brackets is PLOS score before improvement. From the table it appears that recommendations related to effective width improvements only affect the value of HCM method and not for the other three methods. Overall, however, the PLOS aggregate value of each method for all segments has met the desired PLOS improvement targets.

Table 6. Change of HCM (2000) PLOS Value

Segment	1	2	3	4	5
Volume (p/min/m)	5.51	16 (23.33)	4.44	7.19	16 (23.67)
PLOS	A	A (C)	A	A	A (C)

Table 7. Change of Trip Quality PLOS Value

Category	Segment				
	1	2	3	4	5
ENC: Enclosure	5	5	4	4	5
CPN : Complexity of network	-	-	-	-	-
BA : Building articulation	2	3	2	2	1
CS : Complexity of spaces	2	4	2	3	3
AWN : Overhang	3	3	3	5	5
BF : Buffer	5	5	5	5	5
ST : Shade trees	4	4	4	4	4
TR : Transparency	2	2	2	1	5
PC : Physical Condition					
(1) SC : Sidewalk condition and configuration					
a. Effective width	2	2	2	4	2
b. Surface quality	5 (1)	5 (3)	5 (4)	5	5
Average SC score	3.5	3.5	3.5	4.5	3.5
(2) VS : Vehicle speed	3	4	3	5	3
(3) L : Lighting	5 (2)	5 (2)	5 (2)	5	5
Average PC score	3.8333	4.1667	3.8333	4.8333	3.8333
Total	26.83 (25.17)	30.17 (28.83)	25.83 (24.67)	28.83	31.83
Average	3.35 (3.15)	3.77 (3.60)	3.23 (3.08)	3.60	3.98
PLOS	B (C)	B	C	B	A

Table 8. Change of Gainesville PLOS Value

Category	Criteria	Points	Segment				
			1	2	3	4	5
C1 : Pedestrian Facility Provided (Max. value = 10)	- Not continuous or non existent	0	6	6	6	4	4
	- Continuous on one side	4					
	- Continuous on both side	6					
	Minimum 1.53 m wide barrier and free	2	0	0	0	2	0
	Sidewalk eidth > 1.53 m	1	1	1	1	1	1
	Off street /parallel alternative facility	1	1	1	0	1	1
C2 : Conflicts (Max. value = 4)	Driveways and Sidestreets	1	0	1	0	1	1
	Pedestrian signal delay	0.5	0.5	0.5	0.5	0.5	0.5
	Reduced turn conflict implementation	0.5	0	0	0	0	0
	Crossing width ≤ 18.3 m	0.5	0.5	0.5	0.5	0.5	0.5
	Posted speed	0.5	0.5	0.5	0.5	0.5	0.5
	Medians present	1	0	0	0	0	0
C3 : Amenities (Max. value = 2)	*Buffer* ≥ 1 m	1	1	1	1	1	1
	Benches or pedestrian scale lighting	0.5	0.5	0.5	0.5	0.5	0.5
	Shade trees	0.5	0.5	0.5	0.5	0.5	0.5
C4 : Motor vehicle LOS (Max. value = 2)	- LOS E, F or travel lanes ≥ 6	0	2	0	2	2	2
	- LOS D and travel lanes < 6	1					
	- LOS A, B, C and travel lanes < 6	2					
C5 : Maintenance (Max. value = 2)	- Major or frequent problems	-1	2 (-1)	2 (0)	2 (0)	2	2
	- Minor or infrequent problems	0					
	- No problems	2					
C6 : TDM/Multimodal (Max. value = 1)	- No support	0	1	0	1	0	0
	- Support exists	1					
	Total		16.5 (13.5)	14.5 (12.5)	15.5 (12.5)	16.5	14.5
	PLOS		B (C)	B (C)	B (C)	B	B

Table 9. Change of Australian PLOS Value

Category	Criteria	Weight	Segment				
			1	2	3	4	5
Physical Charactheristics	Access	5	4 (1)	4 (1)	4 (1)	1	1
	Path width	4	1	1	1	3	1
	Surface quality	5	4 (0)	4 (2)	4 (3)	4	4
	Crossing facilities	4	4 (0)	4 (0)	4 (0)	3	3
	Support facilities	2	4 (1)	4 (1)	4 (1)	5 (2)	2
Location Factors	Connectivity	4	4	4	4	4	4
	Path Environment	2	4	4	4	4	4
	Potential conflict with vehicle	3	1	2	0	4	4
User Factors	Path user volume	3	0	0	0	0	0
	Mix of path users	4	4	4	4	4	4
	Personal security	4	4 (1)	4 (1)	4 (1)	4	4
	Total of weighted score		127 (58)	130 (71)	124 (70)	123 (121)	113
	PLOS		B (D/F)	B (C/F)	B (C/F)	B/F	B/F

4 Conclusions

The results of the PLOS evaluation are obtained that in the same observation segment can have more than one PLOS value when the PLOS evaluation method used is different. It should be noted that there is no single method that can be used completely to evaluate the quality of the pedestrian facility operations (in this case sidewalk and walkway) accurately. From the evaluation result, it is found that sidewalk and walkway in ITB area still need improvement on performance indicators such as maintenance, effective width, lighting, access and supporting facilities, crossing facilities, and potential conflicts with vehicles.

References

1. Wibowo, S., S. *The Development of Walking Environment Measures for Indonesia Cities*, Journal of Technology and Social Sciences, (JTSS, J. Tech. Soc. Sci.), Vol.1 No.1, ISSN 2432-5686, Japan. (2017)
2. Highway Capacity Manual (HCM), Washington D.C. *National Research Council Transportation Research Board*. (2000)
3. Jaskiewicz, F. *Pedestrian Level of Service Based on Trip Quality, Transportation Research Circular E-C019*: Urban Street Symposium, Dallas, Texas. (1999)
4. Dixon, Linda B. *Bicycle and Pedestrian Level-of-Service Performance Measures and Standards for Congestion Management Systems*. Transportation Research Record: Journal of the Transportation Research Board No 1538. (1996)
5. Australian Road Research Board. *Walkability Audit Tool*. Perth: Department of Transport, page 6-11. (2011)
6. Sdoukopoulos, E. "Methodologies for Assessing the Pedestrian Level of Service: International Experience and Adjustment to the Greek Walking Environment – The Case of Thessaloniki" in Conference of Young Researchers Seminar 2011 (Copenhagen, Denmark, 8 – 10 June 2011). (2011)
7. Karatas, Pinar. *Determination of Pedestrian Level of Service For Walkways : METU Campus Example*, Thesis Master Program, Middle East Technical University, page 5-45. (2015)Gallin, N. "*Quantifying Pedestrian Friendliness-Guidelines for Assessing Pedestrian Level of Service*", in Proceedings of Walking the 21st Century (Perth, Western Australia, 20 - 22 February 2001), page 119-128. (2001)

Queuing analysis using Viswalk for check-in counter: Case study of Lombok Praya International Airport

Sony Sulaksono Wibowo[1,*], *and Siti Raudhatul* Fadilah[1]

[1]Bandung Institute of Technology, Study Prog. of Civil Eng., Bandung, Indonesia

Abstract. Queuing analysis on check-in counter in airport is required to accommodate the passenger processing facilities, such as check-in service. The purpose of this paper is to evaluate the performance of a queuing system at check-in facilities in 2nd passenger terminal Lombok Praya International Airport in 2016 and 2036. This analysis is done by determining parameters of the queues at every check-in counter through simulations on Viswalk. Before running the simulation, the model of airport terminal must be created first. Queuing simulation starts with the traffic demand forecasting of domestic departure passenger for the next 20 years. To determine the optimal number of check-in counters, an iteration of various number is performed. Check-in services time data were adopted from previous study which became an input of the simulation using two different scenarios. The expected output are the queuing parameter and the distribution of passenger service time at each check-in counter.

1 Introduction

In 2016, the number of aircraft passengers at Lombok Praya International Airport was recoded reach to about 3.5 million people while the terminal's current capacity is only 3.2 million people [1]. An average number of passengers was 11,000-12,000 people per day that up about 34% from 2015. Recently, the Lombok Praya International Airport became the second highest passenger for Eastern Area of Indonesia, among airports under PT Angkasa Pura I (API) [1]. The increasing number of passengers at Lombok Praya International Airport is related to the role of West Nusa Tenggara as a MICE tourism location (meeting, incentive, convention, and exhibition). After Lombok is designated as the world's best halal tourist sites, the flow of tourist visits continues to increase. This is predicted to lead an increase in the number of passenger, both domestic and international.

An increase in air traffic activity will affect the level of passenger service at the airport terminal. Insufficient space capacity results in density and accumulation of passengers while processing at the terminal, one of them is at check-in service. Given

* Corresponding author: sonyssw@gmail.com

those problems, a good allocation management of check-in service is required which will affect the passenger's satisfaction level and cost effectiveness. It is judged by the number and time of passengers waiting in the queue and time of passenger service. Therefore, it is necessary to determine the queuing parameters of check-in service in passenger terminal of Lombok Praya International Airport. The objective of this research is to analysis the queue in check in counters using the simulation software of Viswalk. Study case of Lombok Praya International Airport was applied but the result could be extended to similar cases in other airport or other transportation terminals.

2 Literature review

2.1 Queuing theory

Basically, the queue is caused by traffic flow process, both human and/or vehicle, disturbed by the existence of a service activity that must be passed [2]. The purpose of making a queuing model is to minimize the total cost, i.e. the direct costs as the result of providing service facilities, and indirect costs which arise as the impact of people who waiting to be served. The queuing model is an important tool for obtaining a prosper management system by eliminating the queue.

To the less of connectivity, so that the passengers do not use the terminal/halte [3, 4]. There is a tendency on deregulation of transportation rules on intercity bus in some countries such as German, UK, USA, and Sweden not using terminal as transit facility. This bus transports or drops off passengers in predetermined curbside and can access city road [5].

This research aims to identify any factors affecting on quality of intercity bus terminal services. There is a great limitation on researches concerning intercity terminal so that in preparing these research indicators, there are literature studies having no cover on bus terminal [3, 4], it is not only caused by very limited sources, but also by transfer facilities such as interchange [3, 4]. Although in the literature study, there are broad reviews on its scope not only on terminal, but in each prepared category indicators, it has included terminal / transfer facility service quality indicator. From the literature studies, there is only Yatskiv et al. (2009) studying on intercity terminal with Regression Model analysis method [3]. From results of literature studies, there is none studying on quality of intercity bus terminal service.

The queuing system is the arrival of the customer to get service, waiting to be served if the service facility (server) is still busy, get service and then leave the system after being served. There are three components in a queue system, namely:

- Arrival or system input;

The characteristics of arrival are population size, behavior, and a statistical distribution.

- The queue discipline, or the queue itself;

Queuing characteristic include whether the number of queues, queue length, and the number of people in the queue are limited or not.

- Service facilities.

The characteristic of this component is including the design and the statistics distribution of service time.

The queue discipline shows the decision guidelines used to select individuals who enter the queue and determine who will be served first. There are three types of queuing disciplines that are commonly used. Those queue disciplines will be explained briefly below.

- First in First out (FIFO) or First Come First Served (FCFS)

This queue discipline is very often used in transportation case, where the first person or vehicle arrives at a service counter will be served first.

- First in Last Out (FILO) or First Come Last Served (FCLS)

The second queue discipline is also quite often used in transportation case, where the first person or vehicle arrives will be served last.

- First Vacant First Served (FVFS)

This queue discipline is quite different from the two previous queue disciplines, where the first person arrives will be served by the first empty counter. In this case, only a single queue will be formed, but the number of counter may be more than one.

The queuing model is created to produce tools that can be used to estimate the performance of each process of transport activities, which related to the queue including the time and length of the queue. This model is distinguished by arrival pattern, departure pattern, and queue discipline. Distribution of arrival rate and time of arrival, as well as the departure rate and time of departure between vehicles or persons will be follow these patterns below.

- Uniform time interval pattern (assuming the arrival rate is uniform)
- Exponential-negative time interval pattern (assuming the arrival rate is follow the poisson distribution)

2.2 Viswalk

PTV Vissim is the leading microscopic simulation program, developed by PTV Group (*Planung und Transport Verkehr AG*), for modeling multimodal transport operations [4, 6]. Vissim can be used to simulate pedestrians flows and vehicular traffic, and the interaction between them. With the add-on module PTV Viswalk, it is possible to simulate large numbers of pedestrians and flows, both inside and outside buildings. Viswalk can also be used to simulate more complex pedestrian movements, for example in stadiums, train stations, and traffic intersections. By using Viswalk, analysis of route selection and the congestion points which able to happened is possible.

Viswalks is based on the social force model, developed by Helbing and Molnar as in [4], which can reproduce some aspects of human behaviour [4, 6]. The social force model is part of self-driven particle models which was introduced by Vicsek et al [6]. Self-driven particle models can be used to describe the collective motion of a groups. That group is modelled by a collection of particles where each particle is autonomous. The social force model is based on the assumption that a number of different forces act on pedestrians, resulting in a single social force that describes the pedestrian's motivation to move. The social force can either be an acceleration or deceleration force depending on the pedestrian's perceived information about the environment [4].

3 Methodology

This paper work can be divided into three main parts, as follows:

- Forecasting number of passengers at peak hour until 2036;
- Determining the number of check-in counters that are opened iteratively;
- Modeling and simulating the check-in service queue using Viswalk.

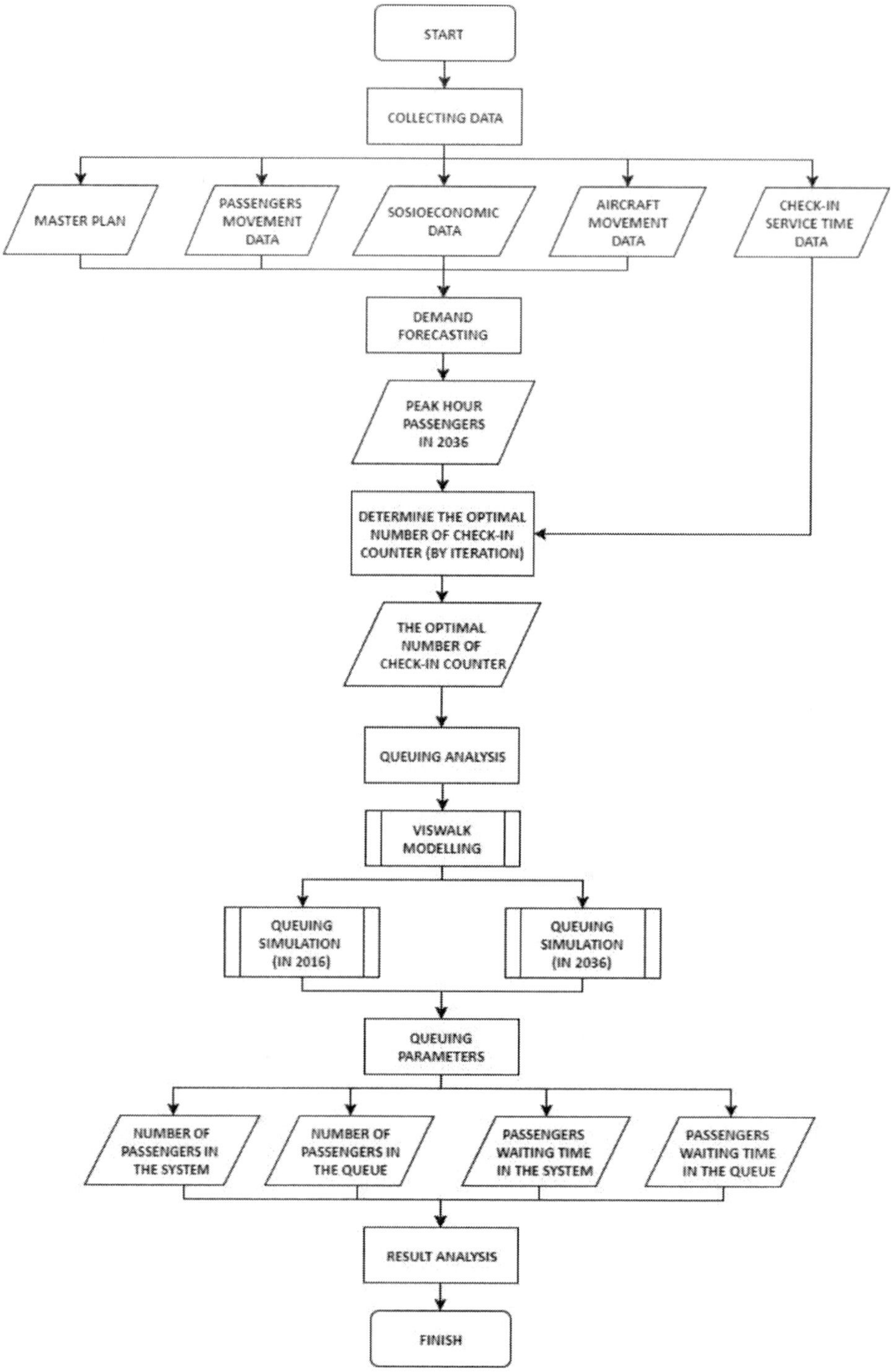

Fig.1. Flowchart of paper work

Figure 1 shows the flowchart used in this paper. In the modeling step using Viswalk, the assumptions used are:

- The behavior of pedestrians, gender, and clothing is considered the same and does not affect modeling.
- Movement aids that are modeled only ladders and escalators, while elevators are modeled as obstacles due to software limitations.
- The speed of pedestrians when up and down stairs is considered the same.
- The movement of pedestrians on the ladder is assumed to have the same speed as the movement in the horizontal plane.

As for the queuing system itself, both in the calculation and in the modeling, there are several assumptions used, namely:

- Passenger plane only consists of direct departure passengers.
- Passengers come in one by one, i.e. no passengers come in groups.
- There is no group check-in service.
- No passengers use online check-in technology.
- Unlimited queue capacity, i.e. no passengers are missing from the system, but only wait until it can be served.

4 Data analysis and discussion

4.1 Forecasting passenger movements

The assumption used in this paper is that 2nd passenger terminal of Lombok Praya International Airport is a domestic terminal. Therefore, the focus on this discussion is only on the departure area, especially the passenger and goods check-in areas, so that international departure data does not need to be processed. In simulating and calculating check-in queue parameters to determine the arrival rate of passengers at 2nd passenger terminal of Lombok Praya International Airport, forecasting passenger movement is an important thing to do. The purpose of this stage is to ensure that the check-in counters provided able to accommodate the passengers' needs for the next 20 years (2016-2036). The table below shows the data of annual number of passenger from Lombok Praya International Airport.

Table 1. Annual passengers data of Lombok Praya International Airport

Year	The Annual Number of Passengers			
	Departure		Arrival	
	International	Domestic	International	Domestic
2008	59.336	470.230	58.947	467.143
2009	78.858	625.406	65.549	520.164
2010	75.861	605.393	78.593	625.237
2011	90.275	722.234	94.563	752.786
2012	98.580	789.401	104.401	829.570
2013	107.551	862.184	116.397	924.058
2014	116.531	935.470	128.421	1.019.143
2015	125.516	1.009.559	140.415	1.115.962
2016	134.679	1.085.781	152.435	1.213.051

In this paper, the boundary of the study area or commonly called hinterland that will be used is West Nusa Tenggara. This province was chosen as hinterland because it is where Lombok Praya International Airport is located and basically this airport is not a transit airport, but as a final destination flight. Forecasting passenger movement is done using three methods, namely exponential, linear, and econometric. The best method of representing the actual condition of the check-in queuing system of Lombok Praya International Airport will be used in further analysis. Below is the result of the calculation.

Table 2. Forecasting Passenger Movement Results

Methods	Domestic Departure	
	R^2	Number of Passengers
Exponential	0,9809	12.018.893
Linear	0,9974	3.121.722
Econometric	0,9898	3.108.425

From the table above, the linear forecasting method gives the highest R^2 value compared to the other two methods. It also produces the logical and representative number of passengers for the next 20 years. So, there are 1,213,051 passengers in 2016 and increased to 3,121,722 passengers in 2036 which shown by the graph below.

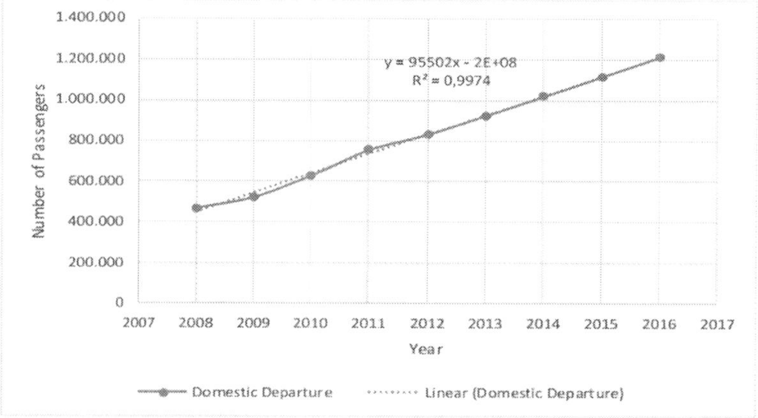

Fig. 1 Linear Forecasting Result

Next step is to calculate the daily number of passengers by dividing the annual number of passengers to the number of days in a year. It is assumed that one year consists of 365 days. Then, the data will be used in the aircraft modulation step to be distributed to all operating airlines at 2nd passenger terminal of Lombok Praya International Airport. From website www.flightradar24.com (accessed on December, 2016), a domestic flight schedule at Lombok Praya International Airport can be obtained. Based on this data, it can be seen there are five airlines who operating there, namely Garuda Indonesia, Citilink, Batik Air, Wings Air, and Lion Air.

To determine the number of passengers per hour, the data load factor, aircraft capacity, and the percentage of passengers transported, must be known first. For domestic flights, passengers can check-in 2 hours to 45 minutes before the scheduled departure, so that the peak hour occurs within three hours (06.00-09.00 a.m. on Friday). Based on the flight schedule, it was found that the peak hour of the check-in process took place at 05.00 to 06.00 a.m. with 471 passengers per hour in 2016 and 1,214 passengers per hour in 2036. These number of passengers will become an input in the queuing simulation on Viswalk.

4.2 Determination of the optimal number of check-in counters

The number of check-in counters available at 2nd passenger terminal of Lombok Praya International Airport is 24 counters. Therefore, it is necessary to divide the ownership of the check-in counter by calculating the percentage of the total number of passengers for each airline in 2016. Table bellows shown the result of the calculation.

Table 3. Distribution of Check-in Counters

Airlines	The Number of Passengers							Total	%	n
	Mon	Tue	Wed	Thu	Fri	Sat	Sun			
Garuda Indonesia	1043	1116	1043	1043	1043	1043	1116	7447	22,.53	5
Citilink	440	330	440	440	550	440	440	3080	9.32	2
Lion Air	1982	1902	1982	2824	2284	2004	2133	15111	45.73	1 1
Batik Air	648	648	777	648	648	648	648	4665	14.12	3

There are three types of check-in counters used, namely priority check-in, check-in without luggage, and general check-in. The peak hour number of passengers on domestic departures in 2016 is 471 passengers per hour. By assuming the percentage of passengers for each type of check-in counter, the first scenario criteria in the following table are obtained.

Table 4. First scenario criteria

Type of Check-in	Percentage	Number of Passengers	Service Duration (min)	Median (min)
Priority Check-in	10%	47	1.00 – 3.00	2.0
Check-in without Luggage	30%	141	1.00 – 3.00	2.0
General Check-in	60%	283	1.00 – 6.00	3.5

Table 5. The number of passengers each Airline in 2016

Type of Check-in	The Number of Passengers						Total
	Garuda Indonesia	Citilink	Lion Air	Batik Air	Wings Air	Others	
Priority Check-in	10	0	0	6	0	0	16
Check-in without Luggage	30	16	65	19	11	0	141
General Check-in	60	37	154	37	26	0	315

The determination of the optimal number of check-in counters opened by each airline is conducted iteratively through Viswalk with the results in the table below.

Table 6. The number of check-in counters used in 2016

Type of Check-in	Number of Check-in Counters						Total
	Garuda Indonesia	Citilink	Lion Air	Batik Air	Wings Air	Others	
Priority Check-in	1	0	0	1	0	0	2
Check-in without Luggage	1	1	2	1	1	0	6
Priority Check-in	3	1	6	1	1	1	13
Total	5	2	8	3	2	1	21

While in the second scenario, iteration of the number of check-in counters is done with 1,214 passengers per hour and follows the criteria below.

Table 7. Second scenario criteria

Type of Check-in	Percentage	Number of Passengers	Service Duration (min)	Median (min)
Priority Check-in	10%	122	1.00 – 3.00	2.0
Check-in without Luggage	30%	364	1.00 – 3.00	2.0
General Check-in	60%	728	1.00 – 6.00	3.5

Table 8. The number of passengers each airline in 2016

Type of Check-in	The Number of Passengers						Total
	Garuda Indonesia	Citilink	Lion Air	Batik Air	Wings Air	Others	
Priority Check-in	26	0	0	16	0	0	42
Check-in without Luggage	77	41	169	48	29	0	364
General Check-in	154	95	395	96	68	0	808

Due to the high demand of passengers, the self-service check-in facility is required in 2036 to serve passengers without baggage with the results in the table below.

Table 9. The number of check-in counters used in 2036

Type of Check-in	Number of Check-in Counters						Total
	Garuda Indonesia	Citilink	Lion Air	Batik Air	Wings Air	Others	
Priority Check-in	1	0	0	1	0	0	2
Check-in without Luggage	0	0	0	0	0	0	0
Priority Check-in	4	2	11	2	2	1	22
Total	5	2	11	3	2	1	24

Table 10 .The number of self-service check-in machines used in 2036

Type of Check-in	Number of Self Service Check-in Machines						Total
	Garuda Indonesia	Citilink	Lion Air	Batik Air	Wings Air	Others	
Priority Check-in	0	0	0	0	0	0	0
Check-in without Luggage	3	2	5	2	1	0	13
Priority Check-in	0	0	0	0	0	0	0
Total	3	2	5	2	1	0	13

3.3 Queuing simulation in 2016 and 2036

The analysis of check-in counter queue at 2nd passenger terminal of Lombok Praya International Airport is done by simulation of Viswalk for each type of airline. In queuing simulation, applied the arrival rate and number of check-in counters that have been obtained previously. This simulation is done three times within one hour real time. However, the simulation will stop immediately when the time runs out, so that not all passengers can be served. Passenger movement starts from the x-ray security area, then goes into the check-in area, and ends on the 2nd floor. There are two types of routes used, namely static and partial. Static routes are used to set start and end points. While the partial route serves to distribute passengers randomly on the check-in counter. The image below shows the check-in area at 2nd passenger terminal of Lombok Praya International Airport.

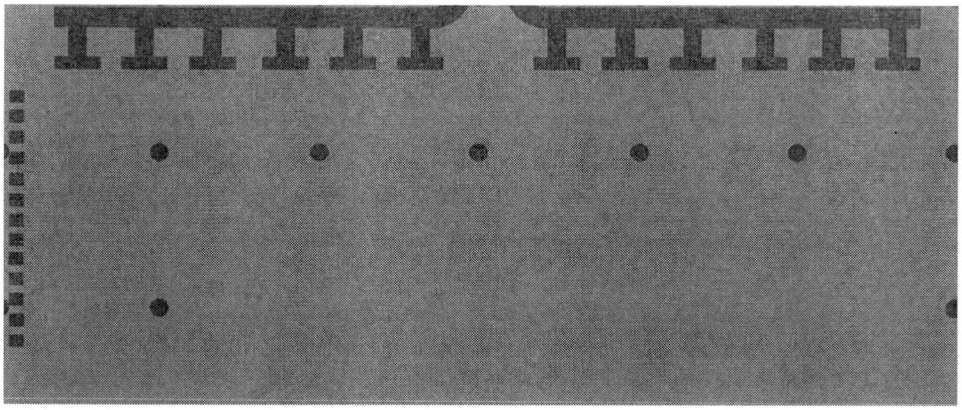

Fig. 2. Model of check-in area for simulation

Table 11 below shows the example of queuing parameters of Garuda Indonesia in 2016 as the result of simulation using Viswalk. There are four queueing parameters to be reviewed in this paper, i.e. number of passengers in the system, number of passengers in the queue, passenger's duration in the system, and passenger's duration in the queue.

Table 11. Queuing parameters of Garuda Indonesia in 2016

Type of Check-in	Number of Passengers	Number of Counters	Average Number of Passengers		Passengers' Duration		Average Service Duration
			System	Queue	System	Queue	
	(persons)		(persons)	(persons)	(minutes)	(minutes)	(minutes)
Priority	10	1	2	1	2,05	0,09	1,96
No Luggage	30	3	3	2	3,43	1,53	1,90
General	60	6	5	4	10,46	6,69	3,09
Total	100	10	10	7			

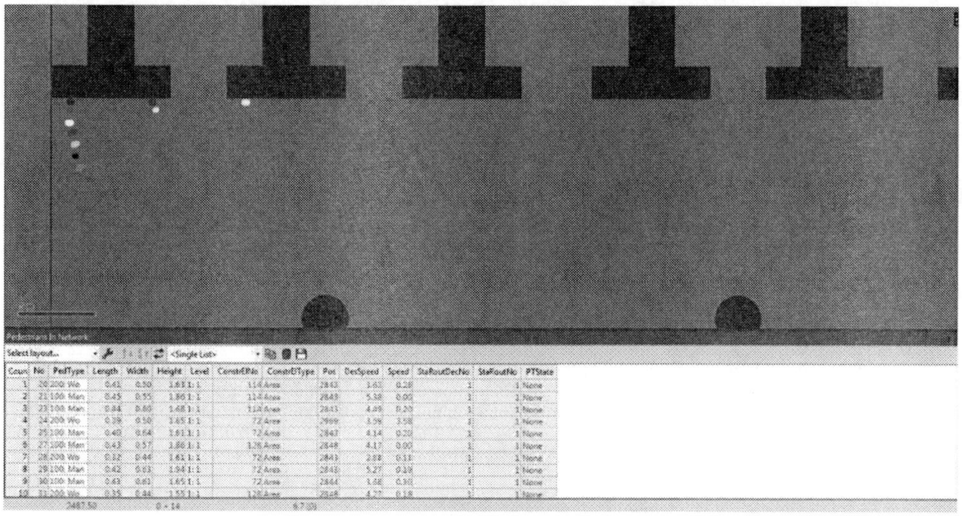

Fig.3. Example of queuing simulation using viswalk

From the simulation and analysis of check-in queue, the comparison of arrival rate and service levels of passengers in 2016 and 2036 is shown in the table below.

Table 12. Comparison of arrival and service levels of check-in counter in 2016

Airlines	Type of Check-in	Number of Passengers	Arrival Rate (Passengers/ Hour)			Level of Service (Passengers/Hour)		
			1st Sim	2nd Sim	3rd Sim	1st Sim	2nd Sim	3rd Sim
Garuda Indonesia	Priority	10	6	10	7	5	9	7
	Without Luggage	30	22	28	22	20	24	20
	General	60	49	52	56	35	35	36
	Total	100	77	90	85	60	68	63
Citilink	Without Luggage	16	9	14	10	9	14	10
	General	37	29	34	30	13	16	15
	Total	53	38	48	40	22	30	25
Wings Air	Without Luggage	11	9	11	7	6	11	7
	General	26	19	23	19	12	15	14
	Total	37	28	34	26	18	26	21
Batik Air	Priority	6	2	4	5	2	4	4
	Without Luggage	19	13	14	12	13	14	12
	General	37	29	34	30	13	16	15
	Total	62	44	52	47	28	34	31
Lion Air	Without Luggage	65	30	38	38	23	30	27
	General	154	131	130	150	76	78	80
	Total	219	161	168	188	99	108	107

Table 13. Comparison of arrival and service levels of check-in counter in 2036

Airlines	Type of Check-in	Number of Passengers	Arrival Rate (Passengers/ Hour)			Level of Service (Passengers/Hour)		
			1st Sim	2nd Sim	3rd Sim	1st Sim	2nd Sim	3rd Sim
Garuda Indonesia	Priority	26	19	21	19	18	18	18
	Without Luggage	77	54	57	14	58	63	65
	General	154	128	130	149	52	52	55
	Total	257	201	208	182	128	133	138
Citilink	Without Luggage	41	32	37	40	30	34	30
	General	95	82	86	89	27	29	27
	Total	136	114	123	129	57	63	57
Wings Air	Without Luggage	29	22	26	22	20	21	20
	General	68	59	61	65	26	28	28
	Total	97	81	87	87	46	49	48
Batik Air	Priority	16	12	14	10	12	14	10
	Without Luggage	48	37	44	48	36	38	38
	General	96	82	86	89	27	29	27
	Total	160	131	144	147	75	81	75
Lion Air	Without Luggage	169	137	147	171	118	115	125
	General	395	377	351	372	177	168	176
	Total	564	514	498	543	295	283	301

4.3 Solutions for increasing number of passengers

There are several ways to minimize the queue at check-in counter, one of them is providing the self-service check-in facility, as an alternative for passengers to check-in at the airport terminal. Passengers simply enter the reference booking number or scan the barcode printed on the itinerary, then print the boarding pass.

However, the availability of self-service check-in facility at airports in Indonesia is still lacking. There are only a few airlines that have provided such facility, such as Citilink, Air Asia, and Garuda Indonesia. In fact, the use of this facility can give some benefits for passengers, including:

- Easy to use and understand by passengers;
- A faster check-in process;
- Control held completely by passengers;
- No need to queue at the check-in counter;
- Reduce passengers' waiting time in queue;
- Passengers can choose the desired seat number.

For the airlines, the use of self-service check-in can lower the operating costs and increase revenue, while improving passenger services. This facility can provide good influences for the airline with the following reasons.

- Passenger queues on the queue line of check-in counters are fewer;
- Cheaper for long-term operations;
- Reduce the need of officers, equipment, and new check-in counter;
- Reduce operational costs;
- Serving the needs of passengers, especially during peak hours;
- Eliminated the unnecessary interactions between officers and passenger who does not need help anymore.

Other check-in service which currently developing is city check-in. This facility has been implemented in various big airports in the world, such as Hong Kong and Singapore. In Indonesia, city check-in only provided in Jakarta for several airlines, for example Garuda Indonesia. City check-in facility allow passengers to check-in at the city center, such as at certain airline marketing offices. Passengers can hand their luggage to the officer, and later the baggage will be delivered directly to the airport. Generally, city check-in opens 48 to 4 hours before the scheduled departure time.

By providing city check-in facility in Lombok, passengers do not need to come in a hurry to the airport and can enjoy the tour much longer. Therefore, this facility can reduce the check-in queue at 2nd passenger terminal of Lombok Praya International Airport.

5 Conclusions

- The annual passenger movement for domestic departure in 2016 is 1,213,051 passengers. While in 2036, passenger movement is forecast to increase to 3,121,722 passengers per year.
- Check-in counter opens from 2 hours to 45 minutes before the scheduled departure time. Peak hour occurs at 05.00-06.00 a.m. with the number of passengers in 2016 is 471 passengers per hour, and increased to 1,214 passengers per hour in 2036.
- The optimum number of check-in counters opened to serve the peak hour passengers in 2016 is twenty-one. While in 2036, it takes a total of twenty-four counters for priority check-in and general check-in, and thirteen self-service check-in machines.
- There are several attempts to minimize the check-in queue, among which are:

- Provides self-service check-in facility which would be better if equipped with an auto bag drop;
- Provides city check-in services.

Acknowledgement

Most of the data and analysis contained in this paper have basically been presented in first author's final paper (unpublished) in Civil Engineering Program in Bandung Institute of Technology under second author supervision. More detailed work can be obtained in university's electronic library.

References

1. Statistic Indonesia (BPS) of West Nusa Tenggara Province: Nusa Tenggara Barat in Figures (2008-2016)
2. Ashford, N. J., Mumayiz, S.A. & Wright, P.H. *Planning, Design, and Development of 21st* – Century Airports 4th (2011)
3. Oktorina, Yurika. *Analisis Pergerakan Penumpang di Dalam Terminal I-A Bandar Udara Soekarno Hatta*. Thesis for Master Degree with Transportation Specificity, Universitas Indonesia (2012)
4. Marten, J.B. & Henningsson, J. "*Verification and Validation of Viswalk for Building Evacuation Modelling*" (2012)
5. National Standardization Agency (BSN). *Standar Nasional Indonesia* (SNI) 03-7046-2004 tentang Terminal Penumpang Bandar Udara. (2004)
6. PTV (Planung Transport Verkehr). *PTV Vissim 8 User Manual*. (2014)

Travel mode choice preferences of urban commuters in Kuching City, Malaysia based on stated preference data

Resdiansyah [1,*]

[1]Universitas Pembangunan Jaya, Department of Civil Engineering, Bintaro, Indonesia.

Abstract. One aspect of Kuching City that has not progressed in tandem with the rest of the city is the public transport system, which is relatively old and almost non-existent. Transport and City planners seem to be at their wit's end in coming up with satisfactory solutions to Kuching's public transportation woes. In current situation, many proposals, but none have proven workable. As a result, representative buses remain a rare sight on Kuching city's roads. To achieve a sustainable public transport industry, the old buses need to be regenerated and replaced with modern buses. The objectives of the intended study are to explore the consumer's travel behaviour by employing mode choice modelling. Consequently, a study was conducted in Kuching City Area by using stated preference technique, analysed and compiled by using SPSS.17 multiple linear regressions analysis. In this context, discrete choice analysis was used to examine the relationship between independent variables (travel time, waiting time, fares and comfort) and dependent variables (choice of respondent whether to consume old bus or choose new bus services). A total of 2000 respondents were interviewed. The findings showed that for the trips purpose, fares and comfortability were the primary factors that reflected the decision or behaviour of the respondents asked. It was discovered that there is a significant relationship between the choice of the respondents and comfortability. It also appeared that longer travel time did not affect for the traveler's choice at this stage. Hence, the study suggests that the local authority and the bus operators should establish a "quality partnership" and working together in order to come out with a much better and appropriate transport policy and schemes for the existing public transportation systems, especially bus services.

1 Introduction

In Kuching City and the suburban areas, buses are the primary mode of public transportation system available. Currently, bus services in Kuching City Area are relatively old and most of the buses are in state of dilapidation. One aspect of Kuching that has not progressed in tandem with the rest of the city is the public transport system, which is almost non-existent. Transport and City planners seem to be at their wit's end in coming up with

* Corresponding author:

satisfactory solutions to Kuching's public transportation woes. In current situation, many proposals, but none have proven workable. As a result, representative buses remain a rare sight on Kuching city's roads. Ironically, statistics show that only one per cent of the people in Kuching City use buses to move around the city, 87 per cent drive to work and five per cent ride their motorcycles. This explains why traffic congestion in Kuching City is getting from bad to worse. On the other hand, the usage of the motor vehicles (especially cars) has become inevitable to the people of Kuching as the public bus transport system is too old, not frequent and timely and there are not many other public transport options available.

Thus, to achieve a sustainable public transportation industry, the old buses need to be regenerated and replaced with modern buses, so people will start thinking to shift to public transportation. This study is to explore the travel behavior by employing mode choice modelling. In this context, stated preference (SP) technique was used to evaluate the traveler's behavior. Also, the study provides a simple overview of the design and implementation of stated preference techniques to estimate the importance of the mode attributes. SP techniques are very popular since it was applied, developed and accepted for the purpose on transportation behavior study in the last three decades. In SP techniques, the study presented the methods of discrete choice analysis based on mathematical modelling to analyze the travel demand and traveler's behavior in order to come out with the best criteria for the implementation of new public transportation schemes.

1.1 Background of Public Transportation in Study Area

Briefly, public transportation in Kuching City area is only focusing on the buses, though there are other modes of public transportation such as taxis and vans. Taxis can be easily found outside the hotels or at the taxis stands. Fares are reasonable, but should always be agreed in advance. Vans can be found at the city center adjacent to the India Street area. The fares are much cheaper compared to taxis. Generally, the service is quite similar to the taxis because they are operated independently.

However, vans do not serve the consumers who wish to travel within the city area of which buses and taxis are still the best option for this particular purpose. Bus services are the best option due to cheaper fares offered. Fares depend on the travel distance. Majority of the buses are not air-conditioned. On the other hand, there is no centralized and modernly built bus interchange in Kuching city. Most of the local bus terminals are separately located within the center of the city area (adjacent to the waterfront, India Street and along the Gambier Street. Additionally, departure points also vary and depend on the destinations/zones served and individual bus operators. There are five major bus operators in Kuching which are namely Bau Transport, Biaramas Express Bus, City Public Link Transport, Petra Jaya Transport and Sarawak Transport Company.

1.2 Problem Statement and Objective

As a result of rapid changes and competitive in the modern industry, the companies survival in the public transport services is largely depends on their great extent of capability to increase the patronage and influence the motorist travelers to shift their modes of mobility from car to public transport system. Generally, the preliminary intention of this study is to provide answers to the following questions:

a) How mode choice is affected by?

i. The level of services for both modes,

ii.Travel time,

iii.Waiting time, and

iv.Fares (cost),

b) What are the travelers' behavior characteristics that influence their decision making in selecting modes of transportation?

The main objective of this study is to determine and compare the travel behavior of the existing and future bus passengers, in order to enhance its competitiveness and quality of services in the future. Specifically, the objectives of this study are arrayed as the followings:

i. To describe the characteristics of the public transportation demand by developing a model using SP technique with respect to some various scenarios for both of the buses, hence evaluate it by using discrete choice analysis.

ii. To evaluate the potential of the future exclusive and modern buses to cope with the current demand of the existing bus services in Kuching City area, where the results analysis is based on the journey purposes.

iii. To identify the response of the bus passenger according to the level of services such as comfortability, waiting time, travel time and cost offered by the bus operators.

2 Conceptual framework of study

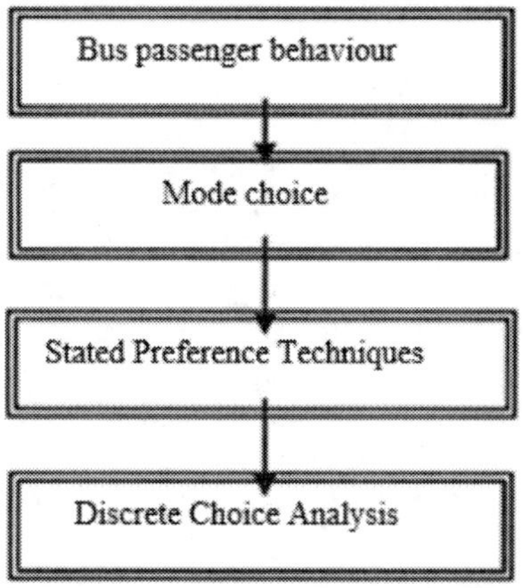

Fig 1. Conceptual Framework

The general steps of the study processes are summarised in Figure 1. The conceptual framework described here is mainly focused on the study of the bus user's behaviour. Eventually, the study will create a model, which will represent and determine the significance attributes of bus passenger behaviour that effect mode choice. A mode choice can be viewed as an outcome from the above sequential decision-making process. Thereafter, stated preference technique will be used to obtain the primary data from bus passenger. Finally, in order to describe the mode choice model, the researcher proposed to use a discrete choice analysis.

2.1 Mode choice

The choice of travel mode is probably the most important model in travel demand analysis. This is because the mode choice model is concerned with the trip-maker's behavior regarding the selection of travel mode. The reason underlying this choice varies between individuals, trip type, and the relative level of service and cost associated with the available models [1]. The objectives of mode choice models can be classified as:

2.2 Aggregate

To estimate the number of proportion of trips from each zone to each zone by purpose that take any mode.

2.3 Disaggregate

To estimate the probability that a particular trips from each zone to each zone by a specific individual will take mode. But contemporary mode choice models are almost disaggregate model. So, in this study, mode choice models are basically disaggregate mode choice model. Disaggregate as used here means that models are intended to describe the mode choice made by individuals rather than by large groups or zonal [2]. It is interesting to see how far we can get by using a series of mathematical techniques. To summarize, there is a number of mathematical models are used to predict mode choice. These generally attempt to estimate the probability of a trip-maker's movement will use a particular mode. Several mode choice models forms are used. This includes binary logit model, multinomial logit model, hierarchical logit model, multinomial logit model, choice by elimination and satisfaction, utility and disutility function, probabilistic model and many other approaches. But, in this study the researcher will only focusing in binary logit model.

2.4 Discrete choice analysis theory

A discrete choice analysis based on random utility theory, has been the backbone of transport planning since the 1970s. It predicts a decision made by an individual (choice of mode, choice of route, etc.) as a function of any number of variables. The models are used when the consumption of one good rules out the consumption of another good, such as for example in travel mode choice situations [3]. As mention earlier, the models are a mathematical function, which predicts an individual's choice based on the utility or relative attractiveness of competing alternatives. The model generally includes characteristics of the individual and relative attributes of competing choices.

2.5 Binary Logit Model

Although there are a number of mathematical utility function in discrete choice models, but the most basic mathematical utility function for the models are as follows [4]:

$$U = U(q_1, q_2, q_3), \text{ where,}$$

$$q_1 = \begin{cases} 1 & \text{If mode 1 is chosen,} \\ 0 & \text{Otherwise;} \end{cases}$$
$$q_2 = \begin{cases} 1 & \text{If mode 2 is chosen,} \\ 0 & \text{Otherwise;} \end{cases} \quad (1)$$
$$q_3 = \begin{cases} 1 & \text{If mode 3 is chosen,} \\ 0 & \text{Otherwise;} \end{cases}$$

and that only one alternative is chosen:

$$q_1 q_2 = q_1 q_3 = q_2 q_3 = 0. \quad (2)$$

For the purpose of this study, binary logit model was chosen since there were only two scenarios involved in the study. The derivation of any binary choice model is conceptually straight forward [4]. The utility function for binary logit models are as follows [4, 5, 6]:

$$P_1 = \frac{e^{v_1}}{(e^{v_1} + e^{v_2})} \tag{3}$$

Where; P1 = Probability of the passenger to choose one of the scenarios,
e = Exponential,
v = Variable.

2.6 Stated Preference Technique

SP techniques can be described as a "family" of techniques, which individual respondent's statements about their preferences in a set of transports options to estimate utility functions [7, 8]. It was designed to answer questions, which could not be answered with revealed preference (RP) techniques. This is due to the fact that it is often very difficult to undertake a "revealed preference" (RP) study and to derive meaningful results. Basically, RP techniques are based on the actual, observed behavior and applied in a real-world condition. While SP technique are based on hypothetical and stated behavior.

In spite of their known limitation and problems, SP based survey techniques have become an accepted part of the transport planning tool kit [7]. They also mention that SP techniques are implemented as stated-choice experiments, in which respondents are asked to choose between two or more alternatives described to them. The summarize of the main features of SP technique as follow [9, 10]:

i. It is based on the elicitation of respondent's statements of how they would respond to different hypothetical (travel) alternatives;
ii. Each option is represented as a 'package' of different attributes like travel time, price, headways, reliability and so on;
iii. The researcher constructs these hypothetical alternatives so that the individual effect of each attribute can be estimated; this is achieved using experimental design techniques that ensure the variations in the attributes in each package are statistically independent from one another;
iv. The researcher has to make sure that interviewees are given hypothetical alternatives they can understand, appear plausible and realistic, and relate to their current level of experience;
v. The respondent state their preferences towards each option by either ranking them in order of attractiveness, rating them on a scale indicating strength of preference or simply choosing the most preferred option from a pair or group of them;
vi. The responses given by individuals are analyzed to provide quantitative measures of the relative importance of each attribute.

2.7 Identification of attributes

In this study, attributes are generally defined as the variables, which describe and define the differences between the alternatives. The attributes may involve travel time, waiting time (frequency), cost (in terms of fares), comfortability, safety, reliability, interchange, headway and others. However, the design is simple, where there were only four attributes chosen which are the travel time, waiting time (frequency), cost (in terms of fares) and comfortability. Alternatives were the existing and future buses of which the respondents should make their option. And the levels are the values of the attributes, which are shown in Table 1.

2.8 Experimental design

At this stage, the researcher will design the way in which the options will be presented to the respondents and how they be allowed to express their preferences. The form of presentation of the alternatives will be easy to understand and within the context of the respondent's experience and constraints. For simplicity, the experimental design considered only four attributes:

i. travel time – varies (the three levels)
ii. waiting time (frequency) – varies (three levels)
iii. cost (in terms of fares) – varies (two levels)
iv. comfort –varies (two levels)

Thus, for a null factorial design, we need 2 2 × 32 = 36 option of combinations and these are described in appendix.

Table 1: Definition of levels for each attribute.

Attributes	*Levels*
	Fast 20
Travel time (min)	Medium/ Original 25
	Slow 30
	Frequent 10
Frequency (min)	Medium/ Original 15
	Infrequent 20
Fares (RM)	High 2.00
	Normal/ Original 1.60
Comfort ability	High Good
	Low/ Original Poor

2.9 Scenario design

A certain number of scenarios has been defined and simulated as follows:

Table 2: Scenario Design

Scenario	Travel Time (min)	Frequency (min)	Fares (RM)	Comfort
Current Situation	25	15	1.60	Poor
I	25	15	2.00	Good
II	25	10	2.00	Poor
III	30	15	2.00	Good
IV	30	10	1.60	Poor
V	20	20	2.00	Good
VI	20	15	1.60	Poor

3 Modelling, results and discussions

The model development in this study has covered dependent variable and independent variable. Dependent variable is the choices of the respondent whether to choose the new bus services or resume the current bus services. Independent variables were fares, travel time, frequency (waiting time) and comfortability. Additionally, the models developed in the study were also based on the different consumers' trip purposes. The model is given by:

$$U = a1 + a2\,(TC - TN) + a3\,(WC - WN) + a4\,(FN) \quad (4)$$

where;

U = Utility function
a1 = constant (independent variable)
a2 = constant for travel time (dependent variables)
a3 = constant for frequency (dependent variables)
a4 = constant for fares (dependent variables)
T = travel time
W = frequency (waiting time)
F = fares
C = current bus
N = new bus

SPSS 17.0 (multiple regression analysis method) was used to develop the models for different trips purposes, which are tabulated in table 3 below:

Table 3. Model development

Journey Purpose	Model
School	$U = -4.947 + 0.028(T_C - T_N) + 0.020(W_C - W_N) + 2.708(F_N)$
Work	$U = -4.535 + 0.023(T_C - T_N) + 0.024(W_C - W_N) + 2.519(F_N)$
Social/Recreation	$U = -4.842 + 0.026(T_C - T_N) + 0.011(W_C - W_N) + 2.763(F_N)$
Shopping	$U = -5.002 + 0.033(T_C - T_N) + 0.011(W_C - W_N) + 2.790(F_N)$
Other	$U = -6.066 + 0.029(T_C - T_N) + 0.006(W_C - W_N) + 3.450(F_N)$

Figure 2 illustrates the summary of modelling results based on five different scenarios. For the trips purpose to school, recreation, shopping and others, most of the respondents preferred to opt for scenario 5 (travel time 20 min, frequency 20 min, fare RM 2.00, and the level of comfort is good with AC system) which gave the highest demand with the percentage of 75.4%, 55.1%, 70.3% and 69.1% respectively. On the other hand, scenario 3 (travel time 30 min, frequency 15 min, fare RM 2.00, and the level of comfort is good with AC system) was chosen for the purpose of working trips with the percentage of 50.5% respectively. The study revealed that most of the respondents preferred to choose scenario 5 except for working trips purposes where scenario 3 was chosen as the highest utility function. It also appeared that comfortability influences the behaviour of the respondents in making their decision of choosing the new bus system. For scenario 3 in particular, it is impossible for the bus operators to deliver new buses with equal price and at the same time provide a greater comfortability. Consequently, it can be concluded that scenario 5 is the most appropriate condition to be adopted and implemented by the bus operators in time to come, where the bus operators should provide a better bus services which are comfortable with a reasonable fare. It appeared that longer travel time did not affect the bus passenger choice for the current condition of bus services in Kuching city area.

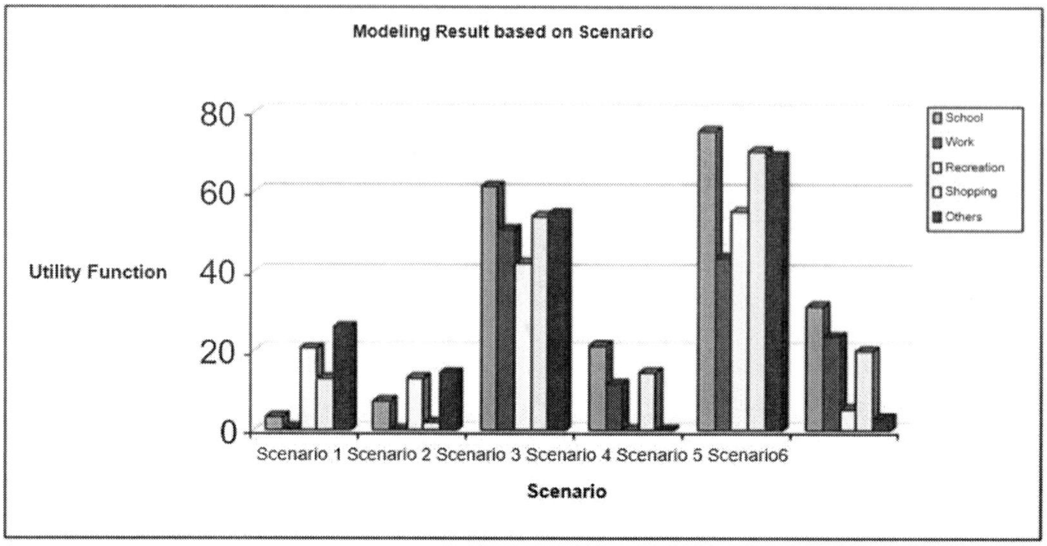

Fig. 2 Model Application Based on Scenario

4 Conclusions and recommendation

The study findings showed that for every trip purposes, fares and comfortability influence the decision making or behaviour for most of the bus consumers asked. The probability for the new bus services scheme to compete with the current bus is generally very moderate. This is due to the fact that majority of the respondents asked have chosen a good level of services especially comfortable with a reasonable price (fares). In line with the study, significant relationship between the choice of the respondents and comfortability has been discovered.

Considering wider implications of the present findings, it seems necessary to conduct a more thorough and comprehensive study regarding the travel behaviour of bus passengers, which also suggests that further studies should be conducted that involve all the public transportations users and the potential users (private car users). A bigger sample size would give further corroborative evidence about the behaviour of the public transportations users. Future researchers should consider conducting studies in other locations in order to compare the findings. Consequently, a clearer understanding regarding the public transportations users' behaviour could be achieved as a result. Importantly, the local authority and the bus operators should establish "quality partnership" and work together in order to successfully implement and achieve a better result for the public transports users which at the same time able to attract car commuters to change their mode of transportation to buses.

Acknowledgements

This research was partially funded by North Kuching City Council, East Malaysia and Universiti Malaysia Sarawak.

References

1. Papacostas, C.S. *Fundamentals of Transportation Engineering, New Jersey: Prentice Hill Englewood.* (1987).
2. Wright, P.H., Ashford, N.J., Stammer, R.J. *Transportation Engineering: Planning and Design, 4th Edition. New York: John Wiley and Sons*, Inc. (1998).
3. Nerhagen, L. *Mode Choice Behaviour, Travel Mode Choice Models and Value of Time Estimation – A Literature Review.* Dalarna University, T&S. (2000).
4. Ben-Akiva, M., Lerman, S.R. *Discrete Choice Analysis. Massachusetts*: MIT Press. (1993).
5. Nasrullah, N., Parung, H., Ali, N., Ramli, M.I. *Study of behavior transition to user moda personal transport public transport on the way in the district Takalar.* International Journal of Civil Engineering and Technology (2017).
6. Hakim, R., Ramli, M.I., Aly, S.H., Rahim, R. *A choice model on trip mode chain for inter-islands commuters in North Molucca-Indonesia: A case study of the Ternate Island – Halmahera Island trip.* International Journal of Civil Engineering and Technology (2017).
7. Danielis, R., Rotaris, L. *Analysing Freight Transport Demand Using Stated Preference Data: A survey and a Research Project for the Friuli-Venezia Giulia Region.* (2003).
8. Nur, N.K., Samang, L., Isran Ramli, M., Hamid, S. *Study of modes transformation preferences of private transport based on travel characteristics and user behavior.* International Journal of Applied Engineering Research (2016).
9. Axhausen., Koll., Stern, E., Salomon, I., Bovy, P.H.L. *Travel Behaviour: Spatial Patterns, Congestion and Modelling. Cheltenham: Edward Elgar Publishing Limited.* (2002).
10. Ortúzar, J.D., Willumsen, L.G. *Modelling Transport. Chichester: John Wiley and Sons Limited.* (2011).

Trip attraction model of central market in Poso City based on multiple linier regression model

Henny Indriyani Abulebu[1,*], *Bleiser*Tanari [1] and *Muhammad* Isran Ramli[2]

[1] Sintuwu Maroso University, Department of Civil Engineering, Poso, Indonesia

[2] Hasanuddin University, Civil Engineering Department, Gowa, Indonesia.

Abstract. Analysis of trip attraction in shopping centers is one of the most important aspects of travel demand management (TDM) in developing countries. This research aims to describe significant factors that influence people in obtaining their choice of frequency. Variables used in this research are family's socio-demographic variables, properties of trip to shopping centers, nature of selecting trip time, and ways to travel. This research adopts multinomial logit model by modeling or building causal relationship between significant variables that influence trip frequency. The findings show that on holiday, no variable has significant influence towards the trip attraction of visitor movement at the Poso Central Market, while on workday, shopping cost (X16) has become the most influential variable. Based on the regression equation, the trip attraction of visitor movement model at the Poso Central Market on holiday is Y = 2.076 + 0.997 X6 +0.276 X13 + 0.605 X15 + 0.643 X16, where R^2 = 0.004 below 5% or tend to close to 0, so it can be concluded that the ability of independent variables in explaining variation of variables is very limited. Meanwhile, the regression equation on the trip attraction of visitor movement model at the Poso Central Market on workday is Y = 3.090 + 0.250 X2 +0.158 X6 + 0.628 X15 +0.050 X16 +0.662 X17, where R^2 = 0.030 below 5% or tend to approach to close to 0, so it can be concluded that the ability of independent variables in in explaining variation of variables is also very limited.

1 Introduction

Of all variety of human activities, one of them is trip to fulfill the needs of the economy, such as shopping trip. The human needs for shopping is very important to fulfill since this plays an important role in fulfilling human being's daily needs, thus enabling high frequencies of trip to a shopping center. Recently, shopping trip has received a wide focus of researchers' attention. Shopping trip has big proportion for urban trips, especially during peak period. The journey has more individual temporal

* Corresponding author: henny_indriyani@yahoo.co.id

flexibility than travel journeys and provides more congestion and some types of environmental problems in the downtown zone [1].

Trip needs are fundamental in traffic planning. The key is how much effort is needed to meet the travel demand what method should be used. Among all factors that influence the selection of modes, the dominant factors in working is time factor. The next dominant factor is cost, and then safety, and the last factor is convenient factor [2, 3].

The function of travel-time constraint is more dominant than the function of mileage constraint and trip cost in the modeling process of transportation needs. Mode is a means used to travel. Models of mode selection aims to determine proportion of people who will use each mode. Socio-economic condition and pattern of traveler's activities is a major factor affecting traveler's decision. Before predicting transportation demand, it is necessary to determine the factors of traveler's decision, in this case how customers' behavior can be used to predict their decision in choosing transportation service. Decisions made by travel agents also greatly determine quantity, distribution of modes and routes, and time of transportation means in the transportation network as well [4-6].

Through the analysis of travel demand management (TDM), it is found that several key aspects have become people's interest to solve problems that arise; such is key aspects in four trip models (i.e. trip awakening, trip attraction, trip frequency, and others). For trip frequency, people who come to shopping places consider many factors for their trip (mall or traditional market), not only for one particular place but also for destination of shopping trip.

From various existing shopping centers in Poso, the researchers chose Poso Central Market as the research object. Formerly, this market was located at the center of Poso City, but then has been located in suburban area of Poso City. This relocation also influences the movement behaviors of travelers. This research is expected to generate model of trip frequency that can be used to determine the route for public transportation of the city.

2 Desk study

The researchers used questionnaire as technique of data collection. The data collection activities were done in Poso Central Market, where the number of respondents was obtained after a survey was conducted to find out the number of visitors entering the market zone. The sampling procedure was divided into workday and holiday, where the samples for workday were obtained on Tuesday, February 2017 and for holiday were on Saturday, February 18, 2017.

2.1 Determining Sample Size

In this research, sample size was determined based on the population size of mode transportation service users. The researchers used Slovin's formula [7], which is as follows:

$$n = \frac{N}{1 + Nd^2} \quad (1)$$

Where:

n = number of samples

d = error tolerance

N = total population

Sample size was determined based on the population number, which was the number of visitors on workday and holiday. Error tolerance for sample size was 0.05 or 95% for each population number. Table 1 shows the sample size for this research.

Table 1. Sample Size

No	Day	Number of visitor	Sample size
1	Workday	7708	381
2	Holiday	9001	383

2.2 Data Display and Analysis Method

Several statistical tests were used to analyze the data, as explained below:

2.2.1 Validity and Reliability Test

Questionnaire validity and reliability test were used to ensure that questionnaires used in this research can really measure research variables. An instrument is considered as valid if it can measure what is supposed to measure and reveal data from the examined variables.

2.2.2 Correlation Analysis

Correlation test was used to test the strength of a relationship between independent and dependent variables, and between independent variables.

2.2.3 Regression Analysis

The largest number was chosen to represent all independent variables with correlation value greater than 0.5. Later, all chosen variables were used to yield equation model.

3 Findings and Discussions

3.1 Visitors Distribution

Based on the survey of market visitors conducted on workday and holiday, the distribution of market visitor frequency in time period of 15 minutes with 10 hours of time intensity can be seen in Figure 1 and 2.

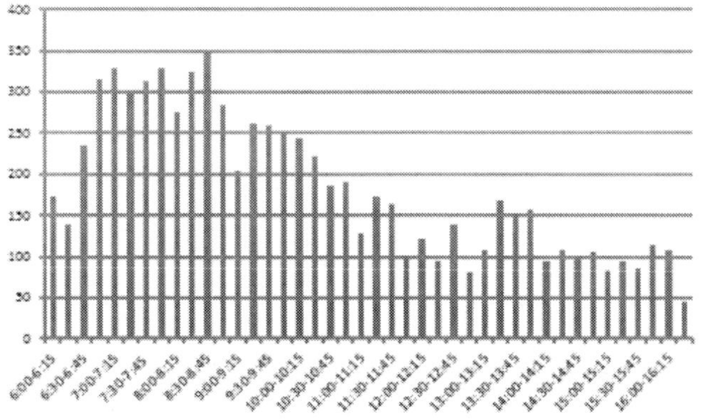

Fig. 1. Distribution Graph of Visitors on Workday

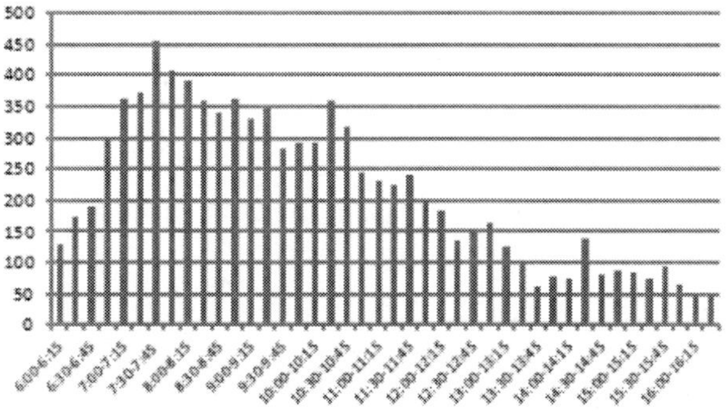

Fig. 2. Distribution Graph of Visitors on Holiday

3.2 Model of trip attraction calculation

The data obtained from questionnaire were analyzed by multiple regression through SPSS 16 software. The model of trip attraction formulation was later generated by regression analysis with the following steps:

3.3 Validity and reliability test

R table value can be seen in R table, where df = number of sample - 2, or in this case df = 383-2 = 381. From significance level 5%, R table value = 0,10085. The results obtained after analyzing samples for holiday showed that variable X2, X6, X13, X14, X15, and X16 were reliable and valid.

Table 2. Result of Validity and Reliability Test for Holiday

No	Parameter Model	Variable	r-count	Conclusion

1	X1	Age	0.084	Not valid
2	X2	Number of people at home	0.127	Valid
3	X3	Number of working people	0.035	Not valid
4	X4	Number of people at school age	0.072	Not valid
5	X5	Job	-0.019	Not valid
6	X6	Income	0.155	Valid
7	X7	Number of car	-0.108	Not valid
8	X8	Number of motorcycle	0.068	Not valid
9	X9	Car license ownership	0.072	Not valid
10	X10	Motorcycle license ownership	-0.016	Not valid
11	X11	Transportation to market	-0.196	Not valid
12	X12	Number of companion to market	-0.021	Not valid
13	X13	Distance from house to market	0.197	Valid
14	X14	Transportation cost to market	0.146	Valid
15	X15	Time needed from house to market	0.129	Valid
16	X16	Shopping expense	0.255	Valid
17	X17	Shopping duration	0.070	Not valid

R table value can be seen in R table, where df = number of sample - 2, or in this case df = 381-2 = 379. From significance level 5%, R table value = 0,10115. The results obtained after analyzing samples for workday showed that variable X2, X6, X13, X15, X16, and X17 were reliable and valid.

Table 3. Result of Validity and Reliability Test for Workday

No	Parameter Model	Variable	r-count	Conclusion
1	X1	Age	-0.042	Not valid
2	X2	Number of people at home	0.194	Valid
3	X3	Number of working people	-0.036	Not valid
4	X4	Number of people at school age	0.043	Not valid
5	X5	Job	0.003	Not valid
6	X6	Income	0.192	Valid
7	X7	Number of car	-0.002	Not valid
8	X8	Number of motorcycle	-0.019	Not valid
9	X9	Car license ownership	0.041	Not valid
10	X10	Motorcycle license ownership	0.092	Not valid
11	X11	Transportation to market	-0.026	Not valid
12	X12	Number of companion to market	-0.080	Not valid
13	X13	Distance from house to market	0.154	Valid
14	X14	Transportation cost to market	0.049	Not valid
15	X15	Time needed from house to market	0.167	Valid
16	X16	Shopping expense	0.309	Valid
17	X17	Shopping duration	0.184	Valid

3.4 Correlation analysis for variables

After performing normality test for each variable, correlation analysis was conducted to analyze the correlation between all independent variables (X) and correlation between dependent variable (Y) with independent variables (X), where the dependent variable (Y) was the intensity of shopping day.

Table 4. Matrix of Correlation between Variables on Holiday

Variable	Y	X2	X6	X13	X14	X15	X16

Y	1	-	-	-	-	-	-
X2	0.436	1	-	-	-	-	-
X6	0.866	0.900	1	-	-	-	-
X13	0.339	0.460	0.001	1	-	-	-
X14	0.241	0.783	0.109	0	1	-	-
X15	0.812	0.927	0.142	0	0	1	-
X16	0.656	0.379	0.011	0.109	0	0.017	1

Table 5. Matrix of Correlation between Variables on Workday

Variable	Y	X2	X6	X13	X15	X16	X17
Y	1	-	-	-	-	-	-
X2	0.089	1	-	-	-	-	-
X6	0.033	0.011	1	-	-	-	-
X13	0.025	0.919	0.334	1	-	-	-
X15	0.275	0.727	0.248	0	1	-	-
X16	0.006	0.001	0	0	0	1	-
X17	0.500	0.028	0.456	0.019	0	0	1

Table 4 shows the systematic analysis procedure. Under the existing conditions, independent variables had to be lesser than > 0.5. From Table 4, it is shown that independent variables X2 and X15 had high correlation coefficient, so the relationship between these two variables were quite strong. Based on the terms, only one variable can be used in the model, so the researchers chose variable X15 since this variable had higher correlation coefficient than variable X2 on Y as dependent variable. Similarly, all variable with correlation coefficient lesser than > 0.5 were chosen in this model, which were X6, X13, X16 and X15.

Table 5 shows the systematic analysis procedure. Similar to the test previously performed, independent variables had to be lesser than > 0.5. From Table 5, independent variables X2 and X13 had high correlation coefficient, so the relationship between these two variables were quite strong. Variable X2 was chosen since this variable had higher correlation coefficient than variable X13 on Y as dependent variable. Variables used in the model were X2, X6, X15, X16 and X17.

3.5 Regression model analysis of market visitor attraction

All variables that passed correlation test were later analyzed by regression analysis in order to get the best model.

Table 6. Results of Modeling Visitor Movement Attraction on Holiday

Model		Unstandardized Coefficients		Standardized Coefficients	T	Sig.
		B	Std. Error	Beta		
1	(Constant)	2.076	.263		7.890	.000
	Income	.000	.031	.000	.004	.997
	Distance from house to market	-.054	.049	-.060	-1.087	.278

Time needed from house to market	.015	.029	.028	.517	.605
Shopping expense	.013	.028	.024	.464	.643

Before reading the results of each variable's effect, we have to make sure whether those variables had significant effect on Y or not, by checking the t value and significance value of each variable. If we use t value, it is said to be significant if t value is greater than t table, while if we use significance value, it is said that the variable is significant if sig value is lesser than a (in this case, the value = 5%). We can see in the table that no variables had significant effect on dependent variable.

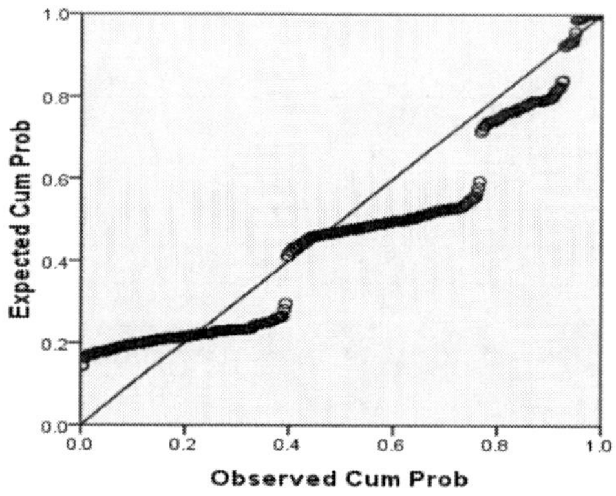

Fig 3. Graph of Normality Test for Holiday

Figure 3 shows the graph model for normality test performed with SPSS, where the existing points spread around diagonal line despite the extreme point, and its distribution follows the direction of diagonal line. Based on the independent variables, this regression model can then be used to predict the amount of trip attraction to the market on workday.

Table 7. Results of Modeling Visitor Movement Attraction on Workday

Model		Unstandardized Coefficients		Standardized Coefficients	T	Sig.
		B	Std. Error	Beta		
1	(Constant)	3.090	.420		7.351	.000
	Number of people at home	.087	.075	.060	1.152	.250
	Income	.084	.060	.074	1.416	.158
	Time needed from house to market	.026	.053	.027	.485	.628
	Shopping expense	.082	.042	.113	1.968	.050
	Shopping time	-.039	.088	-.025	-.437	.662

Similar to the previous procedure, we have to make sure whether those variables had significant effect on Y or not, by checking the t value and significance value of each variable. If we use t value, it is said to be significant if t value is greater than t table, while if we use significance value, it is said that the variable is significant if sig value is lesser than a (in this case, the a value = 5%). We can see in the table that variable X16 had significant effect on dependent variable.

Figure 4 shows the graph model for normality test performed with SPSS, where the existing points spread around diagonal line despite the extreme point, and its distribution follows the direction of diagonal line. Based on the independent variables, this regression model can then be used to predict the amount of trip attraction to the market on holiday.

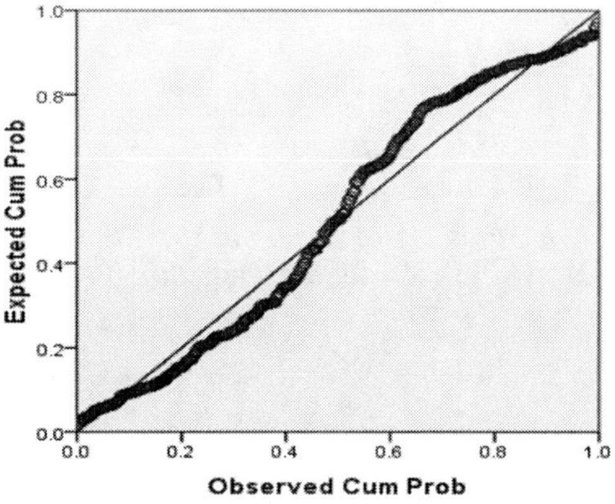

Fig 4. Graph of Normality Test for Workday

4 Conclusions

Based on the analysis results, no factors had significant effect on the trip attraction of Poso Central Market visitors on workday, while shopping expense (X16) had significant effect on the trip attraction on holiday. From the regression equation, model of trip attraction for visitor movement to Poso Central Market on holiday was Y = 2.076 + 0.997 X6 +0.276 X13 + 0.605 X15 + 0.643 X16, where R^2 = 0.004 below 5% or tends to close to 0, therefore it is concluded that the ability of independent variables in explaining variable variation is very limited. Meanwhile, based on the regression equation, trip attraction for visitor movement to Poso Central Market on workday was Y = 3.090 + 0.250 X2 +0.158 X6 + 0.628 X15 +0.050 X16 +0.662 X17, where R^2 = 0.030 below 5% or tend to close to 0. Therefore, it can be concluded that the ability of independent variables in explaining variable variation is very limited as well.

References

1. Ramli, M.I. et al. *Accommodating flexible daily temporal constraint on a continuous choice model of departure time for urban shopping travel*. International Journal of Urban Sciences Vol.15, No 3 .London. (2011)

2. Setiowati, B. dkk. *Studi Komparasi Pemilihan Moda Kereta Api Eksekutif dan Pesawat Terbang (Studi Kasus: Asal Semarang Tujuan Jakarta)*. Simposium FSTPT 6 UNHAS. No. Jurnal 77. Makassar. (2003)
3. Kurniadi, A. dkk 2000: *Studi Pemilihan Metoda Trip Assignment Untuk Pemodelan Kebutuhan Transportasi Kotamadya Surabaya*. Simposium FSTPT 3. Yogyakarta. (2000)
4. Asri, A dan Hamid Aly, S. *Pemilihan Moda Bus dan MPU Untuk Trayek Makassar-Majene*. Simposium FSTPT XIII. No Jurnal 74. Semarang. (2010)
5. Magribi, L. *Model Pemilihan Moda Angkutan Laut dan Penyebrangan Suatu Aplikasi Stated Preference Untuk Studi Kasus Rute Kendari-Raha di Sultra*. (2003)
6. Wicaksono, S.: *Analisa Faktor–Faktor yang Mempengaruhi Pemilihan Moda Transportasi Penduduk Kerja di Kecamatan Sukmajaya Depok Menuju Tempat Kerja Dengan Menggunakan Metode Analitic Hierarchy Process*. Skripsi, Jurusan Teknik Sipil Fakultas Teknik Sipil dan Perencanaan, Universitas Gunadarma, Jakarta. (2008)
7. Iskandar, dkk *Pemilihan Moda Antar Bus dan MPU untuk Trayek Makassar-Palopo*. Simposium FSTPT XIII. No Jurnal 14. Semarang. (2010)

Estimation urban railway demand in Yogyakarta using Bivariate Ordered Probit Model

Wida Yuliar Rezika [1,*], *Sapto* Priyanto[1], *Muhammad* Zudhy Irawan[1]

[1] Universitas Gadjah Mada, Depart. of Civil and Environmental Engineering, Yogyakarta, Indonesia

Abstract. The urban railway system is believed to solve transportation problems caused by the high growth of private vehicles and urbanization. This study is going to analyze the potential demand for the urban railway in Yogyakarta, Indonesia based on bivariate ordered probit model. The survey of preference stated with 120 samples conducted in Yogyakarta. The model of train demand is distinguished between public transport users and private vehicle users using seven scenarios. In-train travel time, waiting time, tariff, and ticketing discount for students are four factors considered in the model. The demand model shows that in-train travel time is the most important factor influence for train demand. Meanwhile, the scenario result reveals that respondents except student are willing to pay more to obtain shorter travel time, students who use private vehicle are reluctant to shift into the train, and ticketing discount brings no effect to stimulate them to use the train.

1 Introduction

High urbanization in developing countries, triggered by the discrepancy of job opportunity between urban and rural area, and followed by significant growth in the number of private vehicles, has caused an increase in urban transportation problem (Vuchic,2007). In solving those problems, implementing an urban railway system becomes an effective solution to mitigate congestion on a regional level (Creemers et al., 2015). The railway system is worthy of the urban transport backbone of its superior features such as high occupancy, environmentally friendly, and energy saving (Profillidis, 2006).

Several major cities in Indonesia are facing rapid growth in urbanization and vehicle ownership, such as Yogyakarta. The Ministry of National Development Planning shows the population growth rate in urban areas is 2.75% higher than the national average growth which is 1.7%. Moreover, the number of the motorcycle has doubled in the last five years (Bureau of National Statistics, 2014). Regarding to the problems, the Ministry of Transportation arranges regulation to improve the transportation services and solve the transportation problems. It is necessary to provide a urban railway system besides the existing bus rapid transit in Yogyakarta. Then, the master plan of the national railway has

* Corresponding author: widayuliar@gmail.com

been decided that Yogyakarta urban railway must be constructed in 2016 and operated in 2020 (Ministry of Transportation and Communication, 2010).

Yogyakarta has a great prospective for urban railway development because of its high-population-density, around 134 persons per hectare (Bureau of Yogyakarta Regional Statistics, 2014). Bus transportation becomes the main public transport in Yogyakarta. Yet, the failure to present the bus arrival time information and the uncertainty of departure time impact to the reluctance to using public transportation. The presence of urban railway service, with segregated tracks, is expected to shift private vehicle users to train users. Figure 1 shows the transportation network and public transport facilities in Yogyakarta.

This study is aimed to estimate the potential demand for Yogyakarta urban railway related to the affecting factors such as fare and ticketing discount, in-train travel time, and waiting time. It also simulate how the changing of these factors influnces the demand in a scenario-based analysis. Bivariate ordered probit model is used because the ability to capture the ordinal response (strongly disagree to strongly agree) compare to the other choice of decision-making models (disagree or agree). Meanwhile, factor analysis is engaged to identify and arrange the service factors which impact the willingness of the train users candidate and then classified it into bivariate ordered model.

Usually, most studies on demand used discrete choice analysis. Those studies have focused on examining the behavior of the decision-making process such as choice of car type (Tanaka et al., 2014; Hackbarth and Madlener, 2013; Ito et al., 2013) and choice of travel mode (Dissanayke and Morikawa, 2005; Habib, 2012; Irawan and Sumi, 2012). Based on the author's knowledge, only few studies use bivariate/multivariate ordered probit model for estimating the potential demand. Most of them are using multinomial logit, multinomial probit and nested multinomial logit. The implementation of ordered probit in this research becomes crucial because the model able to predict the level of certainty as a train user (i.e. strongly won't use; won't use; neither won't use nor will use; will use; and strongly will use).

Fig 1. Transportation network and public transport facilities in Yogyakarta

The rest of this paper is devided into five sections. In the second section, factor analysis is describe as a preliminary survey to obtain a simplification of factors influences the decision of the user to choose the train. In the third section, a bivariate ordered probit model is explained as a research method. The fourth section describes the selection of variable included in the model. Finally, the last two section are the result and conclusion sections. Both of the sections discuss and draw conclusions on how the demand for urban railway can be increased by the implementation of several scenarios regarding the performance of the urban railway.

2 Bivariate Ordered Probit Model

The estimation demand of urban railway as a travel mode choice process can be described by the utility function. McFadden (1974) explains the functions consists of two components; the observed utility of the alternative and a random term. The first component consists of specific attributes of the travel mode and user characteristics. The second component is unobserved attributes and characteristics of the utility. The bivariate ordered probit model is derived by defining the ordinal data y for each observation. For instance with two products (Greene and Hensher, 2009):

$$y_{i,1} = \beta'_1 X_{i,1} + \varepsilon_{i,1} \ , \quad y_{i,1} = j \ \dots \ if \ \mu_{j-1} < y_{i,1} < \mu_j \ , \ j = 0, \dots, J_1 \tag{1}$$

$$y_{i,2} = \beta'_2 X_{i,2} + \varepsilon_{i,2} \ , \quad y_{i,2} = j \ \dots \ if \ \theta_{j-1} < y_{i,2} < \theta_j \ , \ j = 0, \dots, J_2 \tag{2}$$

where,
Y : integer ordering level of certainty as a train user
β : estimable parameters
X : explanatory variables effect to use urban railway
μ,θ : estimable threshold parameters that define y, j integer ordered certainty levels
ε : random error terms, assumed to be normally distributed (*N*) with zero mean and variance of one
The cross-equation correlated error terms are :

$$\begin{pmatrix}\varepsilon_{i,1}\\ \varepsilon_{i,2}\end{pmatrix} \sim N\left[\begin{pmatrix}0\\0\end{pmatrix}, \begin{pmatrix}1 & \rho\\ \rho & 1\end{pmatrix}\right] \tag{3}$$

Where ρ cross-equation correlation coefficient of the error terms.
The bivariate ordered probit model with ordered selection joint probability for $y_{i,1}$ = j and $y_{i,2}$ = k is defined as follows:

$$P\left(y_{i,1} = j, y_{i,2} = k \middle| X_{i,1}, X_{i,2}\right) = \begin{pmatrix}\Phi_2\left[(\mu_j - \beta'_1 X_{i,1}), (\theta_k - \beta'_2 X_{i,2}), \rho\right] \\ -\Phi_2\left[(\mu_{j-1} - \beta'_1 X_{i,1}), (\theta_k - \beta'_2 X_{i,2}), \rho\right]\end{pmatrix} - \begin{pmatrix}\Phi_2\left[(\mu_j - \beta'_1 X_{i,1}), (\theta_{k-1} - \beta'_2 X_{i,2}), \rho\right] \\ -\Phi_2\left[(\mu_{j-1} - \beta'_1 X_{i,1}), (\theta_k - \beta'_2 X_{i,2}), \rho\right]\end{pmatrix} \tag{4}$$

where Φ stands for standard normal cumulative distribution function.

Variable Specification
Dependent Variables
The dependent variable is discrete variable represents the level of certainty as a train user, which can take one of five discrete values, ranging from very unlikely user to very potential user of Yogyakarta urban railway.

Independent Variables

Based on the comprehensive review of studies with respect to modal choice by De Witte et al. (2013), independent variables are divided into four factors: (i) socio-demographic, (ii) spatial characteristics, (iii) journey characteristics, and (iv) socio-psychological factors. However, the four factors above can be classified again into two main categories related to modes attributes and travelers' characteristics. There are several studies use mode attributes or traveler characteristics only (e.g. Recker and Golob, 1979; Lerman and Ben-Akiva, 1976; Irawan and Sumi, 2012). Most of the recent studies use both categories in the analysis. (e.g., Can, 2013; Masiero and Zoltan, 2013; Yarlagadda and Srinivasan, 2008; Hensher and Rose, 2007; Ewing et al., 2004). Related to mode choice decision process, Can (2013) mentions travel time, travel cost, income, demography, private vehicle ownership, comfort, convenience, and safety are commonly used as independent variables.

This research focuses on user preferences for particular attributes of urban railway which are affected their mode choice decision. The independent variables included in our study are in-vehicle travel time, waiting time, travel cost, and ticketing discount for the student. Apparently, major time component of travel time is in-vehicle travel time and waiting time. These two factors have significant contribution on mode choice decision (Arentze and Molin, 2013; Creemers et al., 2012; Wardman, 2004; Murray, 2001). Waiting time of public transport is also crucial factor in determining travel mode, especially for short trip. Walle and Steenberghen (2006) states Dutch travelers feel 2.3 minutes delay on in-vehicle travel time is equal to 1 minute delay on waiting time.

Looking forward to the travel cost factors, several previous studies show travel cost becomes the main factor in determining the public transport demand (Souche, 2010; Arentze, 2013; Creemers et al., 2015). Meanwhile, according to Paulley et al. (2006), fare and ticket discount for student could be an indicator that influences the travelers' decision to use public transport also. For Example, regional public transport in Germany offers fare discount both for children and university students and it shows significant effect for its demand (Buehler and Pucher, 2012).

Survey And Design

The data of the research are collected from October to December 2014 in Yogyakarta region. The survey is conducted by face to face interview with the respondents. There are 120 respondents compiled the survey. The survey is sectionalized based on several aspects. Section 1 explores information about the respondent characteristics such as age, gender, occupation, and existing travel mode. Section 2 focuses on familiarity with Yogyakarta urban railway system, an introduction to the railway route, location of the stations, and integration with existing public transport and the stated preferences discrete choice experiment. Respondents are also asked about the importance of a wide range of modes attributes, including those used in the choice experiment, in their mode choice decision.

The research uses stated preference survey method to measure traveler's preferences. As mentioned before, the urban railway attributes are described as in-vehicle travel time, waiting time, travel cost, and ticket discount for student. Table 1 shows the attribute levels of discrete choice experiment used in our study. Since the survey focuses on preferences for particular attributes of train, the researcher generates complete fractional factorial design according to independent variables shown in Table 1. By using this method, there are 81 choice sets for the respondents. Those choices cause difficulty for the respondent to answer the questions. In order to solve the problem, we implement an orthogonal planning method (Louveire et al., 2000). It produces 7 choice sets confront to 120 respondents as presented in Table 2.

. **Table 1**. Attribute levels

Variable	Indicator	Level	Remark
Independent Variables	Tariff (flat tariff)	0	Rp. 4,000 (0.3 USD)
		1	Rp. 5,000 (0.38 USD)
		2	Rp. 6,000 (0.45 USD)
	In-train travel time	0	More than existing travel time
		1	Similar to existing travel time
		2	Less than existing travel time
	Waiting time	0	20 minutes
		1	15 minutes
		2	10 minutes
	Ticketing discount for student	0	15%
		1	25%
		2	50%
Dependent Variable	Response	0	Strongly won't use
		1	Won't use
		2	Neither won't use nor will use
		3	Will use
		4	Strongly will use

Table 2. Choice sets of questionnaire

Scenario	Tariff	In-Train Travel Time	Waiting Time	Fare Discount *	Response (5 scales)
1	Rp. 6,000	Less than existing	15 min.	25%	
2	Rp. 4,000	Similar to existing	20 min.	15%	
3	Rp. 5,000	Less than existing	20 min	15%	
4	Rp. 6,000	Similar to existing	10 min.	50%	
5	Rp. 5,000	Similar to existing	10 min.	15%	
6	Rp. 5,000	More than existing	15 min.	25%	
7	Rp. 4,000	More than existing	15 min.	15%	

3 Data Description and Estimation Results

Respondents' Characteristic

Table 3 shows descriptive statistics on the demographic profile of the respondents. The result model categories of the research are divided into four groups: (i) public transport users except for student, (ii) students as public transport user, (iii) private vehicle users except for student, and (iv) students as private vehicle user. It makes the percentage of respondents within each category is approximately equal to 20.23%, 19.3%, 30.95%, 29.52%, respectively. The entire collected data can be analyzed supported by the direct interview survey.

Model Results

The estimated parameters of bivariate ordered probit model are shown in Table 4 and Table 5, for public transport users and private vehicle users respectively. The estimated parameters are statistically significant at the 5% level for both public transport and private vehicle users. It means every independent variables influence significantly for the choice decision. Each parameter also has expected sign. Tariff of train has a negative sign, indicated that high fares are disfavored by travelers. However, tariff becomes the most unconcerned about factor by travelers, except students. Other factors have a positive sign, indicated that high speed, short waiting time, and high train fare discount will increase the probability of travelers to use the urban railway service.

In-train travel time is the most significant factor influencing train choice decision both for bus users and private vehicle users. It can be emphasized that all of the parameters are greater in private vehicle users than those in public transport users, except for the factor of tariff for non-student users. It represents those private vehicle users are more sensitive to train attributes. Even for (non-student) travelers, they are willing to pay more to obtain the better performances of the train.

Simulations

The main focus of the scenario analysis is on the impacts of modes attributes on the probability of travelers to use train mode. The best scenario can also be achieved by using this simulation. As explained before, there are seven scenarios as seen in Table 2. The simulation result regarding the potential demand in each scenario is displayed in Table 6 and Figure 2. Generally, train demand generated from student-private vehicle users is the lowest compared to the other users. This situation describes the disinterest of students who use private vehicles to shift into using train mode. They believe private vehicle is faster than the train. They also cannot obtain the advantages offered by using the train.

Table 3. Respondents' characteristics

Demographic Variables	Category	Frequency	Percentage (%)
Age	11 – 20 years old	54	45.00
	21 – 30 years old	39	32.50
	31 – 40 years old	20	16.67
	Over 40 years old	7	5.83
Gender	Male	66	55.00
	Female	54	45.00
Occupation	Student	63	52.50
	Business sector staff	22	18.33
	State sector staff	26	21.67
	Other sector staff	9	7.50
Travel mode	Public transport	47	39.17
	Motorcycle	49	40.83
	Motorcycle (escorted)	7	5.83
	Car	11	9.17
	Car (escorted)	6	5.00

Table 4. Parameter estimates for public transport users

Variable	Student		Non-Student	
	Parameter	**Std. error**	**Parameter**	**Std. error**
Constant	`	0.231	-0.654	0.205
Tariff	-2.138	0.419	-0.905	0.198
In-train travel time	2.46	0.292	1.901	0.186
Ticketing discount for student	1.365	0.29	-	-
Waiting time	0.922	0.192	1.201	0.189
μ_1	0.102		0.851	
μ_2	0.505		0.967	
μ_3	2.169		1.818	

Table 5. Parameter estimates for private vehicle users

Variable	Student		Non-Student	
	Parameter	**Std. error**	**Parameter**	**Std. error**
Constant	-0.942	0.242	-0.729	0.202
Tariff	-4.200	0.440	-0.693	0.186
In-train travel time	4.398	0.308	2.125	0.184
Ticketing discount for student	2.736	0.314	-	-
Waiting time	1.766	0.202	1.280	0.176
μ_1	2.183		0.927	
μ_2	3.221		1.428	
μ_3	5.955		2.748	

Scenario 1 results are the highest mode shifting (in average) by 51.59%. Scenario 1 also produces the highest demand for train service for students who use the bus and non-student private vehicle users. In scenario 2, the probability of using train mode is significantly decreased compared to scenario 1 for bus users and private vehicle users. It occurs because travel time (in-vehicle and waiting time) in this scenario is longer than travel time in scenario 1, although it offers very much cheaper fare. Shorter in-vehicle travel time, yet followed by an increased tariff, is able to increase the demand for train mode as shown in scenario 3. Therefore, it can be remarked that travel time is considered more than the cost. According to the fourth scenario, the scenario that offers the cheapest fare for students and highest ticket discount shows an insignificant impact on train demand. Yet, the result of highest demand is generated from student private vehicle users. Then, the scenario 5 gives the highest train demand produced from non-student bus users. Similar travel time to current travel mode yet offering shorter waiting time is indicated attracting high demand for urban railway. It means when train speed is unable to pass other vehicles' speed, especially private vehicles (i.e. motorcycles) and creates shorter waiting time; it can be alternative way to increase the train demand.

The scenario 6 is the worst scenario. The poor demand for train modes produced is more influenced by in-vehicle travel time rather than other factors. It can be understood because in-vehicle travel time offered in the sixth scenario is much longer than the 5 scenarios before. The impact of scenario 6, an unexpected demand occurs in the seventh scenario. Reducing tariff and maintaining other variables have no significant impact on the increase of train service demand in this scenario. Moreover, the similar number is shown in train service demand generated from students who use private vehicles.

Table 6. Demand for urban railway service in the scenarios (in %)

Scenario	PT Users		Private Vehicle Users	
	Non-Student	**Student**	**Non-Student**	**Student**
1	76.42	52.79	74.86	2.28
2	28.43	34.09	8.85	0.62
3	66.28	46.81	53.19	1.07
4	50.80	45.62	42.86	2.94
5	82.38	24.20	69.85	0.08
6	1.46	0.33	0.19	0.01
7	10.20	2.62	1.39	0.01

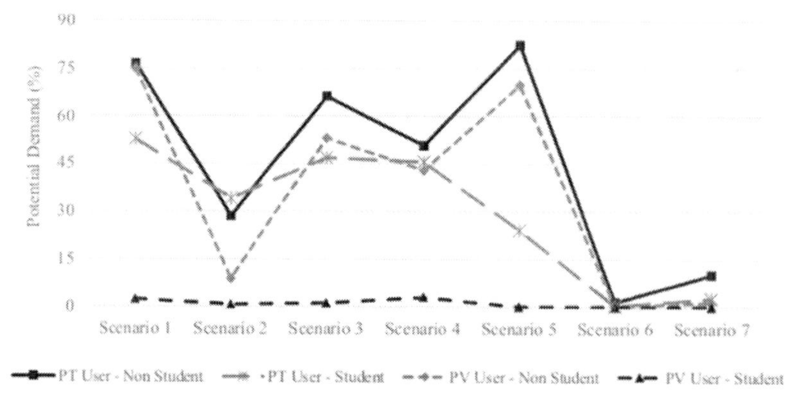

Fig. 1. Simulated potential demand for Yogyakarta urban railway

4 Conclusions

The research studies the potential demand for Yogyakarta urban railway service by using bivariate ordered probit model. The result of the research shows students who use private vehicles are very reluctant toward the train. It could be great barrier in terms of shifting to train use. However, the railway service as an alternative mode can be accepted for private vehicle users for work. In-train travel time and waiting time become the major factors influenced Yogyakarta urban railway demand for bus users and private vehicle users. Yet, ticket and fare discount for students have no significant outcome for train service demand. Travelers except the students are willing to pay more to obtain an optimal travel time. From the 7 scenarios simulated, scenario 1 is the best scenario in generating demand for train mode. However, scenario 6 and scenario 7 should not be considered as policy making options, since both of the scenarios produce poor demand for urban railway.

5 References

1. Arentze, T. "Personalized travel information systems: a Bayesian method to learn users' personal preferences in multimodal transport networks." *IEEE Transactions on Intelligent Transportation System*, Vol. 14, No. 4, pp. 1957–1966, DOI: 10.1109/TITS.2013.2270358. (2013).
2. Arentze, T., Molin, E.J.E. "Travelers' preferences in multimodal networks: design and results of a comprehensive series of choice experiments." *Transportation Research Part A*, Vol. 58, pp. 15–28, DOI: 10.1016/j.tra.2013.10.005. (2013).
3. Buehler, R., and Pucher, J. "Demand for public transport in Germany and the USA: an analysis of rider characteristics." *Transport Reviews*, Vol. 32, No. 5, pp. 541–567, DOI: 10.1080/01441647.2012.707695. (2012).
4. Bureau of Yogyakarta Regional Statistics (2014). Yogyakarta Statistic Annual Report, Yogyakarta, Indonesia.
5. Can, V.V. "Estimation of travel mode choice for domestic tourists to Nha Trang using the multinomial probit model." *Transportation Research Part A*, Vol. 49, pp. 149–159, DOI: 10.1016/j.tra.2013.01.025. (2013).

Mode choice analysis using discrete choice model from transport user (Case study: Jakarta LRT, Indonesia)

Dwi Novi Wulansari [1,*], and *Milla* Dwi Astari [2]

[1]17 Agustus 1945 Jakarta University, Civil Engineering Department, Jakarta, Indonesia
[2]Yapis Papua University, Civil Engineering Department, Jayapura, Indonesia

Abstract. Jakarta Light Rail Transit (Jakarta LRT) has been planned to be built as one of mass rail-based public transportation system in DKI Jakarta. The objective of this paper is to obtain a mode choice models that can explain the probability of choosing Jakarta LRT, and to estimate the sensitivity of mode choice if the attribute changes. Analysis of the research conducted by using discrete choice models approach to the behavior of individuals. Choice modes were observed between 1) Jakarta LRT and TransJakarta Bus, 2) Jakarta LRT and KRL-Commuter Jabodetabek. Mode choice model used is the Binomial Logit Model. The research data obtained through Stated Preference (SP) techniques. The model using the attribute influences such as tariff, travel time, headway and walking time. The models obtained are reliable and validated. Based on the results of the analysis shows that the most sensitive attributes affect the mode choice model is the tariff.

1 Introduction

One of the drivers of economic activity is the availability of good transportation services. However, the current use of public transportation in Indonesia is still not able to facilitate the movement that occurs efficiently. Improper vehicles, accident risks, long waiting times, over capacity, crime and pollution are some of the most common issues facing most of the users of public transport, especially during peak hours. To overcome these problems, the active role of some parties involved in public transportation is required. In the Law of the Republic of Indonesia number 22 [1] regarding traffic and road transport, it is mentioned that the government is responsible for public transport in order to meet the needs of safe, safe, convenient and affordable transportation.

Through the Jakarta Governor's decree number 84 year 2004 regarding determination of macro transportation pattern in DKI Jakarta, stipulating the direction of development of transportation system, one of them is to add primary road network, bus priority, light rail transit and mass rapid transit (MRT). Currently, the government of DKI Jakarta is developing a transportation system, one of which is to build Jakarta Light Rail Transit (Jakarta LRT). Jakarta LRT as a transport system using a light rail is planned to connect

* Corresponding author: dwi.novi@uta45jakarta.ac.id

Jakarta with the surrounding cities such as Bekasi, Depok and Bogor. The development of Jakarta LRT will be integrated with other modes, such as TransJakarta Bus, KRL-Commuter Jabodetabek and MRT.

The objective of this study is to obtain a mode choice model that can explain the probability of choosing Jakarta LRT based on travel attributes, ie tariff, travel time, headway and walking time. In addition, this study also aims to determine the sensitivity of mode choice when changes to travel attributes are made. Changes in travel attributes will affect the probability of choosing between Jakarta LRT and Bus Transjakarta or KRL-Commuter Jabodetabek. Thus the attribute value of Jakarta LRT can be determined in such a way as to avoid overloading on both public transport.

2 Literature

2.1 Discrete choice models

According to Tamin [2], generally, discrete choice model expressed as probability of each individual to choose an option that is a function of socio-economic characteristics and the attractiveness of such options. To declare the appeal of an alternative, used the concept of utility (defined as something that is maximized by each individual). Domencich and McFadden, and Williams, as quoted from Tamin [2], suggests that each set of utility options U_{in} for each individual n. Modelers who is also an observer of the system does not have complete information about all of the elements considered by individuals who make choices. So, in making the model assumed that U_{in} can be expressed in two components, namely:

- V_{in} is measured as a function of the measured attributes.
- Random choice εin, reflecting the particulars of each individual, including mistakes made by modelers.

In general, these effects can be expressed:

$$U_{in} = V_{in} + \varepsilon_{in} \tag{1}$$

where:

U_{in} = utility of alternative i for decision makers n

V_{in} = deterministic utility function of modes i for the individual n

ε_{in} = random error (random error) or stochastic component and a particular distribution function

2.2 Binominal logit model

Decision-making on the binomial logit model is determined on a pair of discrete alternatives, which will be selected alternative is to have the largest utility, the utility in this case is seen as random utility. With two alternate modes of the equation can be written as follows.

$$P_2 = \frac{exp^{U_2}}{exp^{U_2 + exp^{U_1}}} \tag{2}$$

$$P_1 = 1 - P_2 \tag{3}$$

where:

P1 is the probability for mode 1

P2 is the probability for mode 2

By using linear regression method, the probability that the individual chose mode 2 is a utility function of the difference between the two modes. Assuming a linear utility function, then the difference can be expressed in the form of utility difference in the number of n relevant attributes between the two modes, formulated as follows:

$$U_2 - U_1 = \beta_0 + \beta_1(X_{12} - X_{11}) + \beta_2(X_{22} - X_{21}) + \cdots + \beta_n(X_{n2} - X_{n1}) \quad (4)$$

Where U_2-U_1 is the individual response to the statement of choice, β_0 is a constanta, β_1, β_2 and β_n are the coefficients of each attribute are determined by multiple linear regression. Thus, the probability value which is reviewed both modes can be written in the form of the following equation.

$$P_2 = \frac{exp^{U_2}}{exp^{U_2} + exp^{U_1}} = \frac{exp^{(U_2 - U_1)}}{1 + exp^{(U_2 - U_1)}} \quad (5)$$

$$P_1 = 1 - P_2 = \frac{1}{1 + exp^{(U_2 - U_1)}} \quad (6)$$

Utility value as the individual response can also be expressed in terms of the probability of selecting a particular mode, known as transformation Berkson-Theil, the equation is as follows.

$$Ln\left[\frac{P_2}{1 - P_2}\right] = \beta_1 + \beta_1.(X_{12} - X_{11}) + \beta_2.(X_{22} - X_{21}) + \cdots + \beta_n.(X_{n2} - X_{n1}) \quad (7)$$

2.3 Survey techniques

According to Nobuhiro Sanko [3], Revealed Preference (RP) data is observation data / questions based on what people actually do. In these data, because the behavior of individuals who are actually known, which is usually assumed that the information is reliable and can be obtained from retrospective questionnaire. While Stated Preference (SP) data is data base on reaction given by the respondent is not the actual behavior, but only a statement of preference. SP technique is characterized by the use of experimental design to build an alternative hypothesis to the situation, which is then presented to the respondents. Several factors need to be considered in the design of the SP experiment is as follows:

1. Response Form (Ranking/ Rating/ Choice/ Degree of Preference)
2. Analytical Method
   Available analytical method is related to response form. Pearmain et al. [4] introduces four types of analytical methods, i.e.,
   - Naive or graphical methods
   - Non-metric scaling
   - Regression
   - Logit and probit models
3. Number of samples
   Data collection needs huge cost. After the analytical method has been determined, we need to decide the necessery number of samples.
4. Attributes (Measurement)
   How many levels should be treated and how to set attributes (absolute value, percentage and so on) should be considered.
5. Attribute Levels

This factor considers how many levels should be treated and how to set attributes (absolute value, percentage, etc.). Attribute levels in experimental design usually is 'orthogonal', namely to ensure that the attributes presented to respondents varied independently of one another. The result is that the effect of each attribute on the response level more easily isolated. This is to avoid a 'multi-colinearity' between attributes, which is a common problem with the data RP.

3 Research methodology

The analytical approach in this research is using discrete choice analysis. The discrete choice function is solved by binomial logit model, where the research data is obtained by using the survey technique Stated Preference (SP) so that the respondent preference of the mode model attribute is obtained. Juan de Dios Oruzar and Luis G. Willumsen 2011, stated that in determining the amount of sample stated preference can be done 75-100 respondents for each segment can already give the right results. In this study used a sample of 277 respondents for both models of mode choice. Research questionnaires were distributed around the location of the Jakarta LRT station plan. Choice modes were observed between 1) Jakarta LRT and TransJakarta Bus, 2) Jakarta LRT and KRL-Commuter Jabodetabek. Assuming that utility functions are linear, the utility differences can be expressed in terms of differences in the number of n relevant attributes between the two modes, formulated as follows:

$$U_{LRT}U_{Mode} = \beta_0 + \beta_1.(X_{1LRT} - X_{1Mode}) + \beta_2.(X_{2LRT} - X_{2Mode}) + \cdots + \beta_n.(X_{nLRT} - X_{nMode}) \tag{8}$$

Where U_{LRT} - U_{Mode} is the individual response to the mode choice statement, β_0 is the constant, β_1, β_2 and β_n are the coefficients of each attribute determined through multiple linear regression. Thus, the probability values of both modes are reviewed as follows.

$$P_{LRT} = \frac{exp^{(U_{LRT}-U_{Mode})}}{1+exp^{(U_{LRT}-U_{Mode})}} \tag{9}$$

$$P_{Mode} = 1 - P_{LRT}\frac{1}{1+exp^{(U_{LRT}-U_{Mode})}} \tag{10}$$

Where P_{LRT} is the probability for Jakarta LRT and P_{Mode} is the probability for TransJakarta Bus or KRL-Commuter Jabodetabek. The utility value is an individual response expressed in terms of the probability of choice of the two modes offered. Choice responses are analyzed using point ratings presented on the probability scale. The probability scale is then transformed into a numerical scale, called the Berkson-Theil transform, the equation is as follows.

$$Ln\frac{P_{LRT}}{1-P_{LRT}} = \beta_0 + \beta_1.(X_{1LRT} - X_{1Mode}) + \beta_2.(X_{2LRT} - X_{2Mode}) + \cdots + \beta_n.(X_{nLRT} - X_{nMode}) \tag{11}$$

The numerical scale value is the dependent variable on the regression analysis whereas the independent variable is the difference of the attribute value between the two modes (tariff, travel time, headway and walking time). After the transformation, then the process of estimating the equation of utility function is done using least square regression with the assumption that the equation of utility difference function is linear. Validation of the model is done to determine the level of confidence obtained model. Furthermore, model sensitivity analysis was conducted to find out the attributes that most influence the probability of choosing Jakarta LRT.

Table 1. Transformation value

Point Rating	Transformation value	
	Probability Scale	Numerical Scale
Definitely choose Jakarta LRT	0,9	2,197
Maybe choose Jakarta LRT	0,7	0,847
Point Rating	Transformation value	
	Probability Scale	Probability Scale
Do not know	0,5	0,000
Maybe choose existing mode	0,3	-0,847
Definitely choose existing mode	0,1	-2,197

4 Analysis

Based on 300 questionnaires that have been distributed, obtained 277 eligible answers and the remaining 23 answers are not eligible. Furthermore the data is summarized and displayed in the following table.

Table 2. Data characteristics of respondents

Number	Data		Percentage (%)	
			Bus	KRL
1	Gender			
	a.	Male	43,33	55,12
	b.	Famale	56,67	44,88
2	Age			
	a.	< 17 years	3,33	1,57
	b.	17 - 20 years	22,67	27,56
	c.	21 - 30 years	30,67	35,43
	d.	31 - 40 years	30,00	18,90
	e.	41 - 50 years	6,00	8,66
	f.	> 50 years	7,33	7,87
3.	Profession			
	a.	Student	31,33	41,73
	b.	Goverment Employees	9,33	7,09
	c.	Employees of BUMN / Private Sector	29,33	36,22
	d.	Entrepreneur	19,33	8,66
	e.	Indonesian State Army (TNI/POLRI)	1,33	0,00
	f.	Pensionary	0,67	1,57
	g.	Housewife	8,67	4,72
4.	Income Level			
	a.	≤ Rp 1.000.000	28,00	32,28
	b.	Rp 1.000.001 - 2.000.000	20,00	17,32
	c.	Rp 2.000.001 - 3.000.000	21,33	14,96

Number	Data		Percentage (%)	
			Bus	Bus
	d.	Rp 3.000.001 - 4.000.000	18,00	18,11
	e.	Rp 4.000.001 - 5.000.000	5,33	7,09
	f.	Rp 5.000.001 - 6.000.000	4,00	5,51
	g.	Rp 6.000.001 - 7.000.000	2,00	1,57
	h.	Rp 7.000.001 - 8.000.000	0,67	1,57
	i.	> Rp 8.000.000	0,67	1,57
5.	Transportation Cost per Month			
	a.	≤ Rp 50.000	8,00	7,09
	b.	Rp 50.001 - 100.000	10,00	6,30
	c.	Rp 100.001 - 150.000	28,00	15,75
	d.	Rp 150.001 - 200.000	14,67	12,60
	e.	Rp 200.001 - 250.000	10,67	11,81
	f.	Rp 250.001 - 300.000	8,00	15,75
	g.	Rp 300.001 - 350.000	6,00	9,45
	h.	Rp 350.001 - 400.000	4,67	5,51
	i.	> Rp 400.000	10,00	15,75

Table 3. Travel characteristics of respondents

Number	Data		Percentage (%)	
			Bus	Bus
1.	Income Level		54,15	45,85
2.	Travel frequency			
	a.	1 - 14 times per month	34,67	28,35
	b.	15 - 28 times per month	5,33	7,09
	c.	29 - 42 times per month	40,00	40,94
	d.	43 - 56 times per month	6,00	7,09
	e.	> 56 times per month	14,00	16,54
3.	Travel purpose			
	a.	Work / Trade	48,67	45,67
	b.	School / College	30,67	41,73
	c.	Daily shopping	4,67	1,57
	d.	Recreation / Treatment	16,00	11,02
4.	Travel time			
	a.	1 - 15 minute	8,67	7,09
	b.	16 - 30 minute	24,00	17,32
	c.	31 - 45 minute	11,33	9,45
	d.	46 - 60 minute	18,67	18,90
	e.	61 - 75 minute	2,00	0,79
	f.	76 - 90 minute	14,00	17,32
	g.	> 90 minute	21,33	29,13
5.	Origin of travel area			
	a.	Central Jakarta	10,67	10,24
	b.	South Jakarta	15,33	3,94
	c.	East Jakarta	8,00	3,15
	d.	West Jakarta	15,33	11,02
	e.	North Jakarta	7,33	5,51
Number	Data		Percentage (%)	
			Bus	Bus
	f.	Others	43,33	66,14

4.1 Logit-binominal model analysis

With linear regression-appraisal using the logit-binomial model, the results of the utility function are shown in the following table:

Table 4. Estimation of utility function (LRT vs Bus)

Number	Model Variables	LRT vs Bus			
		Coefficients	*Standard Error*	*t Stat*	*P-Value*
1	Constants	0,967	0,176	5,478	5,E-08
2	Tariff (X_1)	-3,E-04	1,E-05	-22,999	6,E-99
3	Travel time (X_2)	-0,141	0,040	-3,480	0,001
4	Headway (X_3)	-0,009	0,007	-1,358	0,175
5	Walking time (X_4)	-0,018	0,004	-4,348	1,E-05
	F	140,460			
	R^2	0,295			

From the estimation results, the following information is obtained.

1. Considering the F statistic value in the table, it can be seen that the F value is greater than the F-critical (F-critical value = 2,45, where the level of significance (α) = 0,05 = 5% and the observation number> 120), Can be concluded in the mode choice model, there is a relation between the attribute (X) and the choice response (Y).
2. The equation of the choice model has a value of R^2 of 0.295, which means that all attributes affect the utility change in the model is 29.5%.
3. In the table note that the coefficient of equation of utility function for tariff, travel time, headway and walking time is negative (-), it can be concluded that the bigger difference in attribute value of both modes will minimize the choice of Jakarta LRT.
4. Considering the value of P-value in the table, it can be seen that the attribute of tariff, travel time and walking time has a lower P-value value than the interest rate of 0.05 which means the attribute is acceptable in the utility equation. The headway attribute has a P-value greater than its importance level (> 0.05), so the headway attribute will be removed and the equation will be estimated in stepwise regression (backward elimination).

Table 5. Estimation of utility function (LRT vs KRL)

Number	Variabel	LRT vs KRL			
		Coefficients	*Standard Error*	*t Stat*	*P-Value*
1	Constants	1,1972	0,2203	5,4344	7,E-08
2	Tariff (X_1)	-0,0003	1,E-05	-29,4022	8,E-142
3	Travel time (X_2)	-0,2690	0,0828	-3,2499	0,0012
4	Headway (X_3)	-0,0086	0,0069	-1,2422	0,2144
5	Walking time (X_4)	-0,0184	0,0041	-4,4420	1,E-05
	F	224,081			
	R^2	0,441			

From the estimation results, the following information is obtained.

1. Considering the F statistic value in the table, it can be seen that the F value is greater than the F-critical (F-critical value = 2,45, where the level of significance (α) = 0,05 = 5% and the observation number> 120), Can be concluded in the mode choice model, there is a relation between the attribute (X) and the choice response (Y).
2. The equation of the choice model has a value of R^2 of 0.441, which means that all attributes affect the utility change in the model is 44,1%.

3. In the table note that the coefficient of equation of utility function for tariff, travel time, headway and walking time is negative (-), it can be concluded that the bigger difference in attribute value of both modes will minimize the choice of Jakarta LRT.
4. Considering the value of P-value in the table, it can be seen that the attribute of tariff, travel time and walking time has a lower P-value value than the interest rate of 0.05 which means the attribute is acceptable in the utility equation. The headway attribute has a P-value greater than its importance level (> 0.05), so the headway attribute will be removed and the equation will be estimated in stepwise regression (backward elimination).

Using binomial logit model, the probability of choice of Jakarta LRT is known by the following equation.

$$P_{LRT} = \frac{exp^{(U_{LRT}-U_{Mode})}}{1+exp^{(U_{LRT}-U_{Mode})}} \quad (12)$$

And the probability of choice of existing mode is,

$$P_{Mode} = 1 - P_{LRT} \frac{1}{1+exp^{(U_{LRT}-U_{Mode})}} \quad (13)$$

With the utility function being used as follows.

- For utility difference function between Jakarta LRT and Bus
  U_{LRT}-U_{Bus} = $\beta_0 + \beta_1 X_1 + \beta_2 X_2 + \beta_3 X_3 + \beta_4 X_4$
  U_{LRT}-U_{Bus} = 1,031 + (-0,0003 X_1) + (-0,141 X_2) + (-0,018 X_4) (

- For utility difference function between Jakarta LRT and KRL
  U_{LRT}-U_{KRL} = $\beta_0 + \beta_1 X_1 + \beta_2 X_2 + \beta_3 X_3 + \beta_4 X_4$
  U_{LRT}-U_{KRL} = 1,257 + (-0,0003 X_1) + (-0,269 X_2) + (-0,018 X_4) (

4.2 Model sensitivy analysis

To know the sensitivity of the model, it is necessary to illustrate the model sensitivity graph of the attribute change. From the graph, it can be seen the change of probability of choice of Jakarta LRT if gradual attribute value change is done.

Based on the sensitivity graph for both model mode model, it can be seen that the slope of the line shows the negative direction, stating that the greater difference between attribute value between Jakarta LRT and the existing mode will minimize the probability of choosing Jakarta LRT. The graph also has a steep slope, this means that if there is a change in the value of the attribute in the selection model it will cause a change in the probability of mode selection.

4.3 Model elasticity analysis

From model elasticity analysis result, it can be seen that tariff attribute has the highest elasticity coefficient value compared to other attributes, it shows that tariff is the most sensitive attribute in influencing mode choice.

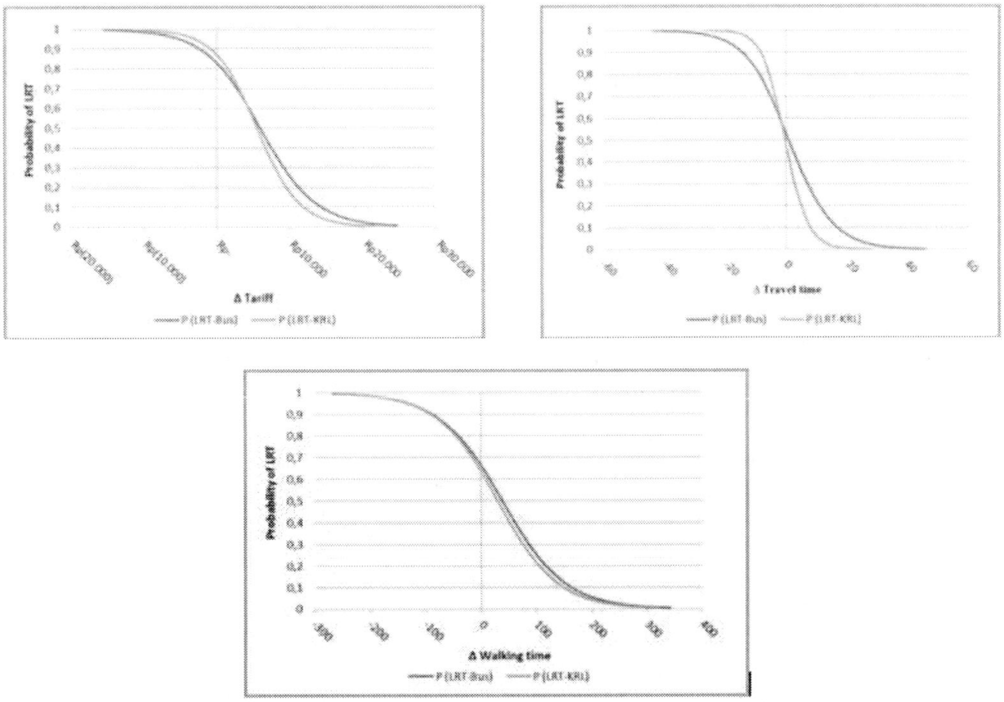

Figure 1. Attribute sensitivity graph

Table 6. Direct elasticity value

Number	Attribute	Coefficient of elasticity	
		Bus	KRL
1	Tariff (X_1)	-0,316	-0,512
2	Travel time (X_2)	0,191	0,248
3	Headway (X_3)	0	0
4	Walking time (X_4)	0	0

5 Conclusions

Based on the result of modeling study that influenced the Jakarta LRT mode choice in DKI Jakarta, the following conclusions are obtained:

1. Based on the individual response to the choice of travel scenarios, it can be seen that the choice attributes that are the consideration of respondents are tariff, travel time and walking time.
2. Based on the attribute sensitivity graph can be seen the slope of the line indicates the negative direction, stating that the greater value of attribute difference will reduce the probability of choosing Jakarta LRT.
3. Based on the results of elasticity analysis, it is known that the most sensitive attributes affect the probability of choosing Jakarta LRT is the tariff.
4. To obtain a probability value of 0.5 in the mode choice model, the difference in tariff attribute between Jakarta LRT and the existing mode (TransJakarta Bus or KRL-Commuter Jabodetabek) is Rp. 5,500 - 6,000.

Acknowlegment

Acknowledgments are addressed to the Directorate of Research and Community Service, Directorate General for Research and Development, Ministry of Research, Technology and Higher Education as a funding body.

References

1. *Laws Of The Republic Indonesia No. 22 Of On Road Traffic And Transportation.* (2009)
2. O.Z. Tamin, *Perencanaan, Pemodelan, & Rekayasa Transportasi: Teori, Contoh Soal, dan Aplikasi.* Penerbit ITB, page 388-433.(2009)
3. N. Sanko, *Guidelines for Stated Preference Experiment Design.* A dissertation submitted for the degree of Master of Business Administration, School of International Management. (2001)
4. D. Permain, and J. Swanson, *Stated Preference Techniques : A Guide to Practice, Steer Davies Gleave and Haque Consulting Group,* London, page 2-59. (1991)
5. D.N. Wulansari, *Analisis Pemilihan Moda Angkutan Penumpang Menuju Bandara (Studi Kasus : Bandar Udara Internasional Soekarno – Hatta)*, Journal of Kajian Teknik Sipil, page. 90-100. E-ISSN 2502-8456.(2016)
6. S. Sihite, and M.S. Surbakti, *Kajian Pemilihan Moda Transportasi Antara Angkutan Kota Dengan Monorel Menggunakan Metode Stated Preference (Studi Kasus: Rencana Pembangunan Monorel Kota Medan).* In Proceedings of the 18th FSTPT International Symposium (Bandar Lampung, August 28th, 2015). (2015)

Psycho-social attributes and quality of life: Overview and pilot study of commuters' travel behaviour

Taufik Tai[1,*], *Rohana* Ngah[2], and *Muhammad Zaly* Shah[1]

[1]Universiti Teknologi Malaysia, Department of Environmental Engineering, Johor, Malaysia
[2]Universiti Teknologi MARA Depart. of Business and Management, Shah Alam, Malaysia

Abstract. As metropolitan cities have improved and upgraded their public transport infrastructure to encourage higher public transport ridership, the next challenges understand factors that influence the individuals' travel behaviour. Psycho-Social Attributes and Quality of Life are important elements to study the public transport ridership in metropolitan cities. This paper offers a short overview of the role of psycho-social and quality of life in understanding the travel behaviour of public transport in cities. The paper explores attributes such as social norms, personal norms, freedom, safety and comfort that have an impact on travel behaviour. Twenty-nine participants took part in the pilot study. Partial Least Square was used to analyze data. Results showed psycho-social attributes has strong direct relationship to travel behaviour while quality of life can be a good support for travel behaviour. Recommendation and suggestion are discussed.

1 Introduction

Many metropolitan cities have built a good public transit infrastructure, however, the ridership is still low [1]. Kuala Lumpur has gone into massive public transportation facelift for the last few years. According to Chuen et al [2], almost 83% out of 7.23 million trips daily in Kuala Lumpur are made using private transport and single occupancy vehicles. Kuala Lumpur recorded only 20% of public transport compared to Singapore (67%) and Hong Kong (87%). A study by Nurdden, Rahmat and Ismail [3] found that age, gender, car ownership, travel time, travel cost, household size and income were significant factors in in influencing the individual's choice of transportation. Studies have shown that the accessibility distance to the rail transport system, transit time, and time spent using the service are the significant factors that will attract travelers, the accessible distance for the public bus service, as well as the bus network, buses transit time, and travelling time are the significant factors that will attract travelers however, as the income has improved significantly, psycho-social attributes will be useful in making the potential shift from private to public transport.

* Corresponding author: rohanangah@salam.uitm.edu.my

Travel behavior is to a large extent habitual [4]. Travel behavior can also explain by social expectations about behavior such as norms, values, beliefs, attitudes and finally lifestyle [5]. Attitude can be defined as positive or negative evaluations or beliefs held about something that in turn may affect one's behavior; attitudes are typically broken down into cognitive, affective, and behavior components [6]. In addition, traveler's attitude is becoming an important aspect in travel behavior research [7, 8]. According to attitude theory, attitude refers to the evaluation of a behavior, which disposes a person to behave in a certain way toward it [7]. Attitude is conceptualized as positive or negative evaluations or beliefs of something that affects one's behavior and can be broken down into cognitive, affective, and behavior components [6]. Social psychologists described attitudes as an element in the decision-making process [7] whereas transportation researchers described attitudes as part of the decision process [9]. Additionally, attitude along with intentions have a significant impact in understanding travel behavioral choices [10]. In the case of choosing public transportation, attitude was found to be more significant indicators than demographics and travel needs [7, 11, 12]. Traveler's attitude is becoming an important aspect of travel behavior research [7]. According to attitude theory, attitude refers to the evaluation of a behavior, which disposes a person to behave in a certain way toward it [13]. Attitude is conceptualized as positive or negative evaluations or beliefs of something that affects one's behavior and can be broken down into cognitive, affective, and behavior components [6].

Social factors include general factors at the level of human society concerned with social structure and social processes that impinge on the individual. Psychological factors include individual-level processes and meanings that influence mental states. Sometimes, these words are combined as "psychosocial." This is shorthand term for the combination of psychological and social, but it also implies that the effects of social processes are sometimes mediated through psychological understanding [14].

Lifestyle in urban cities can influence travel behavior [15]. Median monthly household income for Malaysian in 2014 increased to RM4,585 than RM3,626 in 2012 which is growing at the rate of 11.7 per cent annually [16]. Therefore, the lifestyle plays an important role in determining travel behavior in public transport in KL. In addition, various psycho-social factors appear to play a part in determining people's travel behaviors and how they perceive their travel choice [17, 18].

A smaller number of studies examined the influence of various psycho-social variables on travel behavior. Hiscock *et al* [19] studied the perceived psycho-social benefits of car usage and ownership. In particular, the authors studied the significance of the car as providing protection, autonomy and prestige compared with public transport where the respondents were car owners and non-car owners. The results found that there were some psycho-social benefits to car users. Car users felt that they gained protection, autonomy and prestige from their car and car ownership gave them 'street-cred'. Their car provided them with protection from 'undesirable' people, provided autonomy, convenience and greater access to a greater range of destinations than public transport. Socially desirable attributes such as competence, skill and 'masculinity' were also perceived to be derived from car ownership. People who didn't own cars were felt to be eccentric particularly those who chose to travel by bicycle [17].

Cullinane [20] found similar psycho-social perceptions amongst students attending five universities in Hong Kong. The findings are interesting in that car ownership was extremely low amongst the participants, less than 1% owned a car at the time of the study with the overall Hong Kong population having car ownership levels of 49 cars per 1000 population in 1999. Forty percent of participants felt that public transport was plentiful and low-cost in Hong Kong and suppressed their demand for a car. Few participants felt that they would own a car within the next ten years. However, latent demand was high particularly.

A range of psycho-social variables has been identified as playing some part in influencing travel behavior many of which appear to be dominant in car-owning younger males. Those papers which identified these psycho-social variables tended to use subjective and qualitative measures and discussed the following [17]:

- Feelings of power
- Feelings of achieving 'street cred'
- Safety
- Protection from socially undesirable groups
- Feelings of prestige within peer group
- Identification with selected peer group
- Feelings of greater autonomy
- Perceptions of greater skill and competence through car ownership
- Perceptions of greater masculinity among male car-owners – interestingly no mention of greater femininity amongst female car-owners and users
- Non car-owners/users deemed to be 'eccentric' and hence undesirable

A study done by Verhoeven et al. [21] shows that psychosocial variables seemed to be more important than environmental variables where social norm, social modelling and social support are the most consistent psychosocial. Simons, De Bourdeaudhuij, Clarys,De Cocker , de Geus, Vandelanotte [21] adopt Attitude-Social influence-self- Efficacy (ASE) model to measure the travel behavior of working adults in Belgium focusing psycho-social attributes such as self-efficacy, social norm, modeling, social support, perceived benefits and perceived barriers. Meanwhile, Bourdeaudui, Teixeira, Cardon and Deforche [22] employs social variables, self-efficacy, perceived benefits and perceived barriers as psycho-social variables in their study on adults in Portugal and Belgium towards their perception of active public transport modes such as cycling and walking. Few other studies on psycho-social have been carried out to identify in western countries to measure their inclination towards usage of public transport.

Quality of Life (QoL) is perceived to be a reflection of how well individual needs and values are fulfilled in various fields of life as cited by Groot and Steg [23]. The Quality of Life (QoL) can be assessed from either individual or society level which further divided into objective and subjective indicators [23]. Time spent during congestion also affected drivers' well-being and quality of life as well as productivity where time is wasted during traffic congestion. The quality of life has been largely affected by urban transportation problems like traffic congestion. Studies on sustainable transport highlight the importance of QoL especially towards pro-environment and travel behavior [24].

Recently, factors like psycho-social, attitude, beliefs, personality and lifestylethat could influence travel behavior have been given more attention [25, 26]. Psycho-social attributes or factors are seldom used in travel behavior study especially in public transport. Therefore, this study is attempts to explore commuters' travel behavior to increase public transit mode share in metropolitan cities through psycho-social variables and service quality. The study would provide clear issues in low public transit ridership in metropolitan cities. In addition, the authority would be able to identify factors of psycho-social and service quality that influence public transport usage thus would be able to encourage higher public transit ridership. The aim of this pilot study was to demonstrate the impact of psycho-social attributes and quality of life on travel behavior. It is hoped this information will contribute to the next generation of travel behavior studies, and help elucidate the links between the travel behavior and travel demand management.

2 Discussion

2.1 Methodology

The pilot study was carried out to find out the validity of questionnaire and to amend it for a bigger scale survey. A total of 29 participants took part in the pilot test. Table 1 presents the overall sample structure for this study.

Table 1. Respondent profile

Demographic Traits		Frequency	%
Gender	Male	10	34.5
	Female	19	65.5
Marital Status	Single	23	79.3
	Married	9	20.7
Age	Below 20	2	6.9
	21 – 30	23	79.3
	31 – 40	3	10.3
	41 – 50	1	3.4
Education	High School	2	6.9
	Diploma	23	79.3
	Degree	3	10.3
	Post-graduate	1	3.4
Occupation	Unemployed	1	3.4
	Housewife	1	3.4
	Student	11	37.9
	Public Sector Employee	1	3.4
	Private Sector Employee	11	37.9
	Others	4	13.8

Based on gender, female constituted 65.5% while male represented 34.5% of the sample population. Most of the respondents were from the age group of 21 - 30 years old. The majority of respondents have a college diploma (79.3%). Most of the respondents are students and private sector employees.

2.2 Constructs development

The psycho social attributes (PSA) comprises of personal norm and social status. Quality of life (QoL) comprises of safety and comfort. PSA measurement was adopted from Moktarian et al [27], QoL from Steg and Giifford, de Groot [23] and travel behaviour was from Javid et al. [28]. Items statements in the variables sections are measured as subjective estimates using a five-point Likert scale (with 1 = strongly disagree and 5 = strongly agree).

2.3 Data analysis

Partial Least Squares (PLS) analysis technique using the SmartPLS 3.0 software [29] was employed to analyze the research model. Following the recommended two-stage analytical procedures by Anderson and Gerbing (1988), we tested the measurement model (validity and reliability of the measures) followed by an examination of the structural model (testing the hypothesized relationships) [29, 30]. To test the significance of the path coefficients and the loadings, a bootstrapping method was used [29].

2.4 Measurement model

The measurement model is assessedby two types of validity which are convergent validity and discriminant validity. Convergent validity of the measurement model is usually ascertained by examining the loadings average variance extracted (AVE) and also the composite reliability [30]. The loadings were all higher than 0.7, the composite reliabilities were all higher than 0.7 and the AVE values were also higher than 0.5 as suggested by Hair et al. [29] as shown in Table 2.

Table 2. Measurement model

1st Order Constructs	2nd Order Constructs	Item	Loadings	CR	AVE
Personal Norms		PN5	0.805	0.912	0.722
		PN6	0.825		
		PN7	0.881		
		PN8	0.884		
Social Status		SS1	0.783	0.861	0.755
		SS2	0.877		
		SS3	0.802		
	Psycho-Social Attributes	Personal Norms		0.873	0.501
		Social Status			
Safety		SAF1	0.875	0.861	0.755
		SAF2	0.863		
Comfort		C2	0.757		
		C3	0.832		
		C7	0.775		
		C8	0.790		
		C9	0.778		
		C13	0.740		
		C14	0.786		
	Quality of Life	Safety		0.899	0.506
		Comfort			
Travel Behavior		TB1	0.714	0.857	0.547
		TB2	0.724		
		TB3	0.777		
		TB7	0.807		

Table 3 verified the interrelation of three latent constructs. Discriminant validity of the measures (the degree to which items differentiate among constructs or measure distinct concepts) was examined by following the Fornell and Larcker [31] criterion of comparing the correlations between constructs and the square root of the AVE for that construct as presented in Table 3, where the square root of the AVE (in diagonals) were greater than the correlation values indicating the measures were discriminant. In sum, both convergent and discriminant validity of the measures in this study were established.

Table 3. Discrimant analysis

Constructs	Mean	S.D.	PSA	QoL	TB
PSA	3.40	0.466	***0.707***		
QoL	3.03	0.491	0.742*	***0.711***	
TB	3.39	0.475	0.686**	0.564	***0.740***

2.5 Structural model

Structural equationmodeling is used to test the causal relationships among psycho-social, quality of life and travel behavior. In PLS, the path coefficients (t-values) and *R*-square are used to measure the strengths of latent variables. As shown in Figure 1, the model's predictive explanatory power for travel behavior refers to *R*-square to explain how much of the variance in the dependent variable can be explained by the independent variable. Psycho-social attributes explain 55% of quality of life while 61% of travel behavior can be explained by psycho-social attributes and quality of life. In this model as shown in Table 5, Psycho-Social has direct (0.121) and indirect impact (0.054) on Travel behavior with total effect of 0.175. This shows that Psycho-Social positively influences travel behavior while quality of life can additionally strengthen travel behavior.

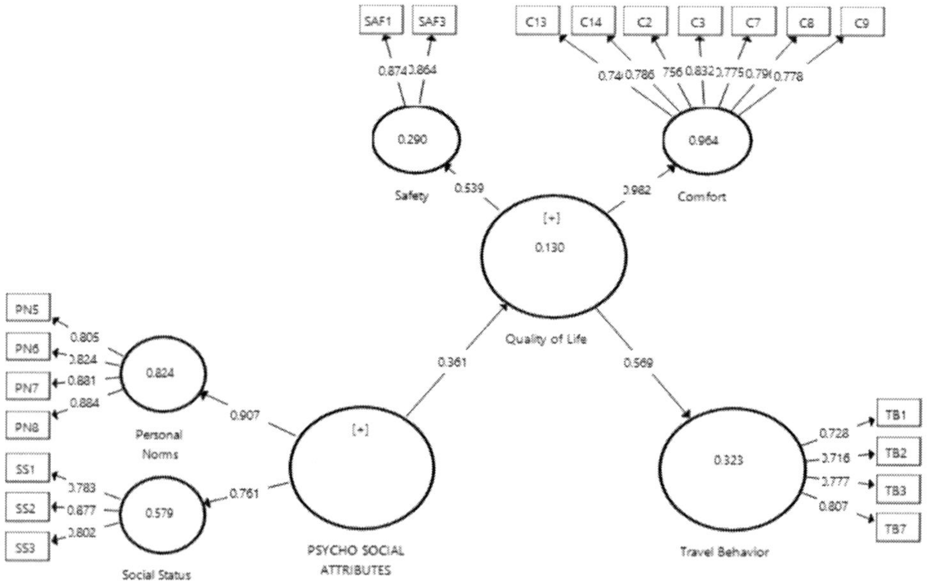

Fig. 1. Structural model

Table 4. Direct, indirect and total effects

	Direct effect	Indirect Effect	Total effects
Psyc-TB	0.121	0.054	0.175
Psyc-QoL	0.741		0.741
QoL-TB Travel Behavior	0.687		0.687

3 Conclusion

This paper has represented structural equation model of travel behavior that emphasize on psycho-social and quality of life on travel behavior. This study used pilot test to explore the relationship of latent constructs. The results demonstrate that psycho-social has direct and positive relationship to travel behavior while quality of life can be supporting elements to the relationship. This result is similar to study by Etminani-Ghasrodashti and Ardeshiri [5] where they found the strong relationship between psycho-social and travel behavior.

Psycho-social attributes are the important elements for the metropolitan cities' residents where social norms and social status play strong role in influence individual's behavior especially in choosing their daily mobility pattern. In addition, quality of life is also another keyelement that should be included in a study of travel behavior as the awareness of environmental issues has been widespread. Understanding these two elements would help the relevant authorities to plan their strategies especially in travel demand management in increasing public transport ridership to 48% by 2030 in line with Transformation Plan of Greater Kuala Lumpur under 11th Malaysian Plan. This study is not without limitation. First of all, the data was small as this is meant for pilot test. Thus, some of the empirical results were not encouraging. Secondly, most of the participants are students who might not have good exposure in public transport. However, this is a good preliminary study as students are the second largest public transport commuters. For future research, it is suggested that bigger sample to be collected to have more robust analysis and results. Beside, other elements such as service quality and trip chain attributes are less explored in travel behavior.

References

1. Choi, CC,and Loh, N., *Transport Policies and Patterns: A comparison of five Asian Cities. Journeys* .(2013).
2. Chuen, OC, Karim, MR, Yusoff, S, 2014, *Mode Choice between Private and Public Transport in Klang Valley, Malaysia*, The Scientific World Journal, Article ID 394587, http://dx.doi.org/10.1155/2014/394587. (2014).
3. Nurdden, A., Rahmat, R. A. O. K., and Ismail, A., *Effect of transportation policies on modal shift from private car to public transport in Malaysia*, Journal of Applied Sciences, 7(7), 1013-1018. (2007).
4. Verplanken, B., Aarts, H., van Knippenberg, A., and van Knippenberg, C., *Attitude versus general habit: Antecedents of travel mode choice*. Journal of Applied Social Psychology, 24,285-300. (1994).
5. Etminani-Ghasrodashti, R. and Ardeshiri, M., *Modeling travel behavior by the structural relationships between lifestyle, built environment and non-working trips*. Transportation Research Part A 78, 506–51. (2015).
6. Nairne, J.S., *Psychology: The Adaptive Mind* (3rd ed), Thomson Learning Inc. London. (2003).
7. Parkany,E., Gallagher, R., Viveiros, P. (20005). *Are Attitudes Important in Travel Choices?* Transportation Research Board Annual Meeting. (2005)
8. Thorpe, N., Hills, P., and Jaensirisa, S., P*ublic attitudes to TDM measures: a comparative study*. Transport Policy. 7, 243-257. (2000).
9. Sunkanapalli, S., Pendyala, R. and Kuppam, A. *Dynamic Analysis of Traveler Attitudes and Perceptions Using Panel Data. In* Transportation *Research Record:* Journal of the Transportation Research Board, 1718, -1, 2000, 52-60. (2000).
10. Outwater, M., S. Castleberry, Y. Shiftan, M. Ben Akiva, Y. S. Zhou and A. *Kuppam, Use of Structural Equation Modeling for an Attitudinal Market Segmentation Approach to Mode Choice and Ridership Forecasting*. Paper presented at the 10th International Conference on Travel Behaviour Research, Lucerne. (2003).
11. Garling, T., Eek, D., Loukopoulos, P., Fujii, S., Johansson-Stenman, O., Kitamura, R., Pendyala and Vilhelmson, B., *A conceptual analysis of the impact of travel demand management on private car use*. Transport Policy, 9, 59-70. (2002).
12. Fujii, S. and Garling, T., *Application of Attitude Theory for Improved Predictive Accuracy of Stated Preference Methods in Travel Demand Analysis*. Transportation Research Part A , 37(4), 389-402. (2003).

13. Ajzen, I. and Fishbein, M., *Understanding attitudes and predicting social behavior.* Prentice-Hall,Inc., New Jersey.(1980).
14. Stansfeld, S., and Rasul, F. , *Psychosocial factors, depression and illness. In A. Steptoe (Ed.), Depression and physical illness (pp. 19–52).* Cambridge: Cambridge University Press. (2007)
15. Krizek, K. J., *Household Lifestyles and Their Relationship to Land-Use and Transportation Planning. Minneapolis: Humphrey Institute of Public Affairs. http://www.cura.umn.edu/sites/cura.advantagelabs.com/files/publications/35-4-Krizek.pdf.* (2005).
16. Department of Statistic, *https://www.dosm.gov.my/v1/.* (2014).
17. Curtis,C., and Perkins, T., *Travel Behaviour: A review of recent literature. Impacts of Transit Led Development In A New Rail Corridor Working Paper No. 3.* Urbanet Department of Urban and Regional Planning Curtin University. (2006).
18. Lyons, G., Farag, S. and Haddad, H, *The substitution of communications for travel? In: Ison, S. and Rye, T., eds., The Implementation and Effectiveness of Transport Demand Management Measures*: An International Perspective. Ashgate, 211-232. ISBN 9780754649533 Available from: http://eprints.uwe.ac.uk/8692. (2008)
19. Hiscock, S. Macintyre, A. Kearns, and A. Ellaway, *Means of transport and ontological security: do cars provide psycho-social benefits to their users? Transportation Research* Part D, 7 (2), 119–135. (2002)
20. Cullinane, S, *The relationship between car ownership and public transport provision: a case study of Hong Kong,* Transport Policy, 9(1), 29-39. (2002).
21. Verhoeven H, Simons D, Van Dyck D, Van Cauwenberg J, Clarys P, and de Bourdeaudhuij I., *Psychosocial and Environmental Correlates of Walking, Cycling, Public Transport and Passive Transport to Various Destinations in Flemish Older Adolescents.* PLoS ONE 11(1): e0147128. https://doi.org/10.1371/journal.pone.0147128. (2016)
22. De Bourdeaudhuij, I, Teixeira, PJ, Cardon, G, and Deforche, B, *Environmental and psychosocial correlates of physical activity in Portuguese and Belgian adults,* Public Health Nutrition: 8(7), 886–895, DOI: 10.1079/PHN2005735. (2005).
23. Groot, JIM and Steg, L., *The role of value orientations in evaluating quality of life consequences of a transport pricing policy*, Transportation Research Part D 11 (2006) 160–165. (2016).
24. Steg, L., and Gifford, R., *Building Blocks for Sustainable Transport, Volume 1, Emerald Group Publishing Limited.* DOI:10.1016/B978-0-08044-709-4.00011-2. (2008).
25. Ory, David.,Mokhtarian, P., *Modeling the structural relationships among short-distance travel amounts, perceptions, affections, and desires*. Transp. Res. Part A 43, 26–43. (2009).
26. Van Acker, V., Mokhtarian, P., Witlox, F., *Car availability explained by the structural relationships between lifestyles, residential location, and underlying residential and travel attitudes*. Transp. Policy 35, 88–99. (2014)
27. Mokhtarian PL, I Salomon and ME Singer,, *What moves us? an interdisciplinary exploration of reasons for traveling*. Transport Reviews. (2015).
28. Javid, M.A., Okamura, T. Nakamura, F., Tanaka, S., and Wang, R., *Factors Influencing the Acceptability of Travel Demand Management (TDM) Measures in Lahore*: Application of Behavioral Theories. Proceedings of the Eastern Asia Society for Transportation Studies, 9. (2013).
29. Hair, J. F., Hult, G. T.M., Ringle, C. M., &Sarstedt,M., *A primeron partial least squares structural equation modeling (PLS-SEM).Los Angeles*: Sage Publication. (2014).

30. Gholami, R., Sulaiman, A. B., Ramayah, T., and Molla, A., *SeniorManagers' Perception on Green Information Systems (IS) Adoptionand Environmental Performance: Results from a Field Survey*.Information Management, 50(7), 431–438. (2013).
31. Fornell, C., andLarcker, D. F., *Evaluating Structural EquationModels with Unobservable Variables and Measurement Error.Journal of Marketing Research*, 18(1), 39–50. (1981).

Analysis of mode transportation performance and satisfaction level of Jenggala Commuter Line (Sidoarjo-Mojokerto)

Hera Widyastuti[1,*], *Asep* Yayat Nurhidayat[2] , *Ahmad* Soimun[2] , *Cahyani* Setyarini[2] , *Nafilah* El Hafizh[2] , and *Arinda* Leliana[2]

[1]Civil Engineering, Faculty of Civil Environmental and Geo Engineering, 6011 Surabaya, Indonesia
[2]Civil Engineering, Faculty of Civil Environmental and Geo Engineering, 6011 Surabaya, Indonesia

Abstract. The increasing traffic density in Sidoarjo-Mojokerto is caused by the inadequancy of road capacity available to handle the growth of vehicle every years, and as a result, traffic jams arise in the highway, travel time will be extended, and require more cost. Therefore to handle this situation a mode of transportation which can carry a mass of passengger in the same time or a mass rapid transit, is necessary. Every customer demand the best service they can get from the transport provider and as a result that make an important indicator to measure the quality of performance or service. Analysis results reveal that travel in 45 minutes, boarding-alighting ranges from 1-3 minutes, Load factor 0.88. Based on the results of the analysis of the customer satisfaction level of conformity obtained on Jenggala commuter line IPA analysis with an average of 93.5% which shows that the customer are satisfied with the service and with a CSI value of 0.827.

1 Introduction

Jenggala commuter line is one of the train line operated by PT Kereta Api Indonesia which serves 6 times route between Sidoarjo-Mojokerto (passengers service). The route between Sidoarjo and Mojokerto is estimated to take 45 minute from departure to arrival. This train will serve the line from Sidoarjo Station, Tulangan Station, Tarik Station, and to end its journey at Mojokerto Station. Increased traffic density in Sidoarjo and Mojokerto arise because the existing road capacity is not proportional to the increase in traffic volume which caused by the increasing number of private vehicle users such as cars and motorcycles. To overcome these problems, it is necessary to provide a mode of transportation that can move a massive number of passenger at the same time, or often referred to as mass rapid transport. The mass rapid transport available between Sidoarjo-Mojokerto is Jenggala commuter line or commonly called KA Komuter SIMO.

The existence of Jenggala commuter line is in an effort to improve its performance requires assessment of perceptions and expectations of service users. Because service users

* Corresponding author: hera.widyastuti@yahoo.co.uk

are the most eligible parties to evaluate and measure the extent to which service quality they receive.

One of the indicator for assesing performance is to assess the minimum service standards provided by the railway service providers both in terms of in terms of infrastructure (stations) and the means (railway). The Government has issued several regulations on service standards at the station that is Minister of Transportation Regulation No.48 of 2015 on Minimum Service Standards For Railway Trainers and Station Standardization Guidelines of 2012 [1]. Based on this rule it is necessary to evaluate the extent to which the application of minimum service standards And transportation infrastructure Jenggala commuter line.

2 Methodology

This research was conducted by using interview system to the respondents to obtain their satisfaction of Jenggala comuuter line and direct survey in field to know the performance of Jenggala commuter line. To measure respondent satisfaction of various factors relating to the operation of commuter trains used a questionnaire with the question format adapted to the Importance Performance Analysis (IPA). Data collection was done by distributing questionnaires to 51 respondents on weekdays. Preliminary survey conducted to evaluate the questionnaire and to test the validity and reliability to test whether each question is valid and reliable. In addition to interviews and questionnaires, this study also conducted a survey directly in the field to obtain primary data including the time of departure and arrival between trains (headway), the travel time between stations (travel time), comfort seating and standing capacity of carriages, the total capacity of the train fire and load factor.

The importance of the quality of service is how important a variable of service for customers to service performance [2]. The 5-level Likert scale is used to measure the level of importance that is very important, important, quite important, less important and not important. These five levels are scored as follows Very important answer given a score of 5, Important answer is given a score of 4, Quite important answer is given a score of 3, Less important answers are given a score of 2, An unimportant answer was given a score of 1.

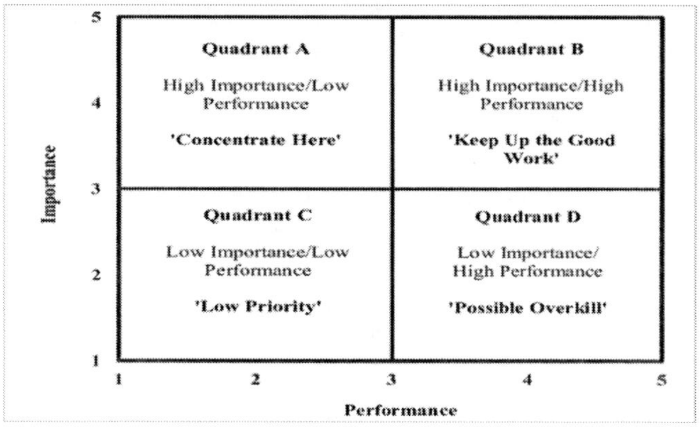

Fig. 1. Importance-Performance Grid Cartesian Diagram

3 Analysis and Discussion

Equations should be centred and should be numbered with the number on the right-hand side.

3.1 Facilities of Jenggala Commuter Line

The Jenggala commuter line owned by the Indonesian Ministry of Transportation and operated by PT Kereta Api Indonesia in Java serving the Mojokerto-Tulangan-Sidoarjo route. The train was inaugurated on November 12, 2014 to serve pioneer routes on the line between Sidoarjo Station to Tulangan Station, Tarik Station and continued to Mojokerto Station. Jenggala commuter line facility consists of two series of the latest air conditioning KRDI made by PT Industri Kereta Api (INKA) Madiun, each consisting of four units of diesel economy class railway rail.

3.2 Jenggala Commuter Line Performances

3.2.1 Train boarding-alighting

Data that can also be obtained from this survey is a train waiting time from the difference between actual departure time and actual arrival time. Furthermore this study can show the value of time delay from the difference between actual departure and arrival time with scheduled departure and arrival.

Table 1. Boarding-alighting on Sidoarjo Station-Mojokerto Station Line (PM).

Station	KA 289		KA 295		KA 294		KA 296	
	Schedule	Survey	Schedule	Survey	Schedule	Survey	Schedule	Survey
Sidoarjo	2.35	2.34	4.45	4.44	4.45	4.45	3.05	3.03
Tulangan	2.35	2.31	4.45	4.45	4.45	4.45	3.05	3.05
Tarik	2.49	2.49	4.45	4.45	4.45	4.45	3.05	3.05
Mojokerto	2.49	2.48	4.45	4.41	4.45	4.41	3.05	3.04

3.2.2 Travel time

As a survey conducted on May 17, 2017 and May 19, 2017 in conjunction with the survey of the number of passengers up and down are also carried out during peak hours (peak hours).

Table 2. Travel Time Route Station Sidoarjo-Station Mojokerto.

Station	Survey		Travel Time	
	Departure	Arrived	Schedule	Survey
Sidoarjo	4.04	4.06		
Tulangan	4.17	4.18	00.13	00.12
Tarik	4.38	4.40	00.21	00.22
Mojokerto	4.45	4.51	00.11	00.11

Travel Time = 45 minutes = 2,700 second. According to the field survey and according to PT KAI schedule, the average travel time for the Jenggala Sidoarjo-Mojokerto railway in the afternoon is 45 minutes.

3.2.3 Car carriage capacity

The capacity of the carriage can be known from the measurement of the seating dimension and the place to stand on the train. Jenggala train uses inter-city economic railway train which has seat type 2 seating. The distance between the side facing seat and the floor of the vehicle used for the standing passenger. The area provided for the standing area is the floor of the vehicle squeezed by the two adjacent seats as well as the area that is on the train bordes as the train is running.

3.2.4 Seating comfortability

R = Ad/m (1)
= ((0.9 x 0.5)x 18)+(0,9x0,5)x18))/ 72
R = 0.255 m2/space.
Then the comfortability of Jenggala commuter line seating would not meet the comfort standard of 0,3-0,5 m^2 /space.

3.2.5 Standing place comfortability

Empirically the standing place in Jenggala commuter line have meet its standart with standing space more than 0.25 m2 / space. And there are handles provided.
M' = Ab/Alfa (2)
= (16 x 0.55) + (1.4 x 2.76) + (1.4 x 2,76)/ 0,25= (8.8 + 3.9 + 3.9)/ 0.25 = 66 passengers

3.2.6 Train totalcCapacity

Train capacity according to real conditions can accomodate sitting and standing passengers with a capacity of 20% from the total seating capacity. The capacity of one car as mentioned earlier is 72 for sitting and 22 for standing passengers, in this way the total capacity (Cv1) for each train consisting of 4 car each are:
Cv 1 = (m + m ') x number of cars = (72 + 66) x 4 = 552 passengers
Capacity in accordance with the statement of the Director of PT KAI that the capacity of the rail transport economy is 100% of seated passengers. So the total capacity (Cv2) for 1 (one) series of Jenggala commuter line consisting of 4 units of carriages are:
Cv2 = m x number of cars = 72 x 4 = 288 passengers

3.2.7 Load factor

Load factor analysis on Jenggala commuter line is the calculation of the terms of each station. The stages in calculating the load factor on each station is follows:
- Conducting on train survey to obtain the number of passenger boarding and leaving the train at each station.
- Calculate the summary of passenger in the following way: The number of passenger on the first station is the summary of passenger boarding and the amount of passenger of the next station.
- The Load factor in accordance with the statement of Director of PT KAI that economic train transport capacity is 100% of passengers seated. Number of seats per carriage is 72 seats, and a series of trains consisting of four (4) carriage. So that the capacity of the train is 288 seats

$$LF = \frac{((\textit{Boarded Pass}+\textit{Boarding Pass})-\textit{Leaving pass}}{\textit{capacity}} \quad (3)$$

$$LF = \frac{((0+254)-0}{288} = 0.88$$

And for Load factor in full condition and passenger all standing:

$$LF = \frac{((0+254)-0}{532} = 0.48$$

Table 3. Load Factor Jenggala Commuter Line Sidoarjo Station-Mojokerto Station.

Station	Passanger			Load Factor
	Boarding	Alighting	Total	
Sidoarjo	254	0	254	0.88
Tulangan	13	0	267	0.93
Tarik	0	13	254	0.88
Mojokerto	0	254	0	0.00

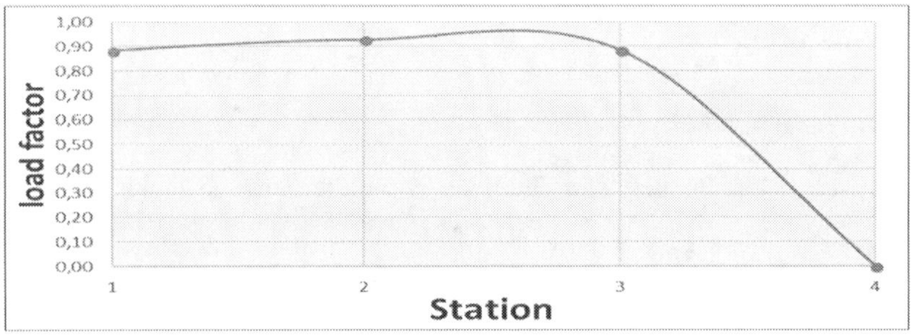

Fig. 2. Load Factor Jenggala Commuter Line Sidoarjo Station-Mojokerto Station

Table 4. Variable and Percentage Commuter Line Sidoarjo Station-Mojokerto Station.

Variable	Percentage
Gender	Male : 39.22%, Female : 60.78%
Age	At the age of 18 – 25 are 39,22%, At the age of 36 - 45 are 25,49%, At the age of 26 – 35 are 23,53% And at the age of 46 – 55 are 11,76%
Occupation	Student 15,69%, Housewive 19,61%, Employee 35,29 %, Businessman 15,69%, And Teacher 3,92%.
Income rate	Rp.2.500.000-Rp.5.000.000,- = 39,22%. >Rp.5.000.000 =5,88%. Rp.1.500.000,- Rp.2.500.000,- = 19,61%, Rp.500.000,-Rp.1.500.000,- =13,73%, Rp.500.000 = 21.57%
Travel Destination	Private bussiness 13,73%, Shopping and recreation 13,73%, work 39,22% and study 17,65%.
Transportation mode used before Jenggala commuter line	Motorcycle = 49.02%, Private car = 27.45%, Public Transport= 13.73 %.
Reasons to switch to using Jenggala commuter line	Faster travel = 47.07%, Cheaper cost = 21.57%, Safe and Comfortable = 23.53%, Distance from station to destination = 7.84%
Mode used to reach station	Motorcycle = 70.59%, Private car = 11.76%, Public Transport = 7.84 %, walking = 5.88% and other = 3.92%
Mode used to reach destination from station	Motorcycle = 47.06%, Private car, Public Transport = 17.65%, walking, others= 7.84% dan pedicab (rickshaw) = 1.96%

Table 5. Validity and Reliability Test of Jenggala Commuter Line Performance.

Station	Passanger			Load Factor
	Board	Leave	Total	
Mojokerto	94	0	94	0.33
Tarik	9	1	102	0.35
Tulangan	4	14	92	0.32
Sidoarjo	0	92	0	0.00

3.2.8 Validity and reliability test

The validity test is used to determine the feasibility of the questionnaire question items in defining a variable. Questionnaire question items should in principle support certain variables that are used as research variables. The criteria used to judge the validity of the test results is the value of the correlation (r), which is called the coefficient of validity. R value calculation results compared with the value of r tables where if the value of r table is smaller than the value of r count then the questions are considered valid.[3]

The Reliability Test is a measure of stability and consistency of respondents in answering questions items related to the dimensions of research variables. Reliability provides an illustration of the extent to which a measurement result can be reliable that means the extent to which the score of measurement results free from measurement error.

Table 6. Load Factor Jenggala Commuter Line Sidoarjo Station-Mojokerto Station.

Att.	**R Valuae**		**r table**	**Remark**	**R Valuae**		**r table**	**Remark**
	Satisfaction	**Expectation**			**Satisfaction**	**Expectation**		
P1	0,4174	0,4867	0,2759	**Valid**	0,6790	0,2868	0,2759	**Reliabel**
P2	0,3776	0,7648	0,2759	**Valid**	0,4829	0,3560	0,2759	**Reliabel**
P3	0,4970	0,6032	0,2759	**Valid**	0,2845	0,5106	0,2759	**Reliabel**
P4	0,4838	0,5073	0,2759	**Valid**	0,4083	0,3260	0,2759	**Reliabel**
P5	0,4558	0,6953	0,2759	**Valid**	0,5329	0,6990	0,2759	**Reliabel**
P6	0,2788	0,6866	0,2759	**Valid**	0,3122	0,4898	0,2759	**Reliabel**
P7	0,5116	0,5620	0,2759	**Valid**	0,5244	0,5037	0,2759	**Reliabel**
P8	0,4361	0,7510	0,2759	**Valid**	0,3783	0,4022	0,2759	**Reliabel**
P9	0,6516	0,7313	0,2759	**Valid**	1,0204	0,4575	0,2759	**Reliabel**
P10	0,5895	0,8081	0,2759	**Valid**	0,5352	0,3952	0,2759	**Reliabel**
P11	0,6870	0,6825	0,2759	**Valid**	0,6813	0,4022	0,2759	**Reliabel**
P12	0,5572	0,7477	0,2759	**Valid**	0,5744	0,6651	0,2759	**Reliabel**
P13	0,7471	0,8091	0,2759	**Valid**	0,5613	0,6005	0,2759	**Reliabel**
P14	0,5765	0,6778	0,2759	**Valid**	0,2730	0,6398	0,2759	**Reliabel**
P15	0,6055	0,6984	0,2759	**Valid**	0,6997	0,3652	0,2759	**Reliabel**
P16	0,7385	0,8788	0,2759	**Valid**	0,6275	0,5198	0,2759	**Reliabel**
P17	0,4976	0,8321	0,2759	**Valid**	0,6997	0,4068	0,2759	**Reliabel**
P18	0,6738	0,8399	0,2759	**Valid**	0,5429	0,3237	0,2759	**Reliabel**
P19	0,7052	0,7320	0,2759	**Valid**	0,6159	0,3260	0,2759	**Reliabel**
P20	0,4976	0,7872	0,2759	**Valid**	0,6997	0,3676	0,2759	**Reliabel**

3.2.9 Customer satisfaction index (CSI) on jenggala commuter line

The benefits of Customer Satisfaction Index (CSI) is to determine the level of user satisfaction of land transport services, especially trains [4]. Determining or measuring the level of user satisfaction of land transportation services can be determined by the CSI value indicator that considers the service user's expectation level of the factor to be defined.

Table 7. CSI Recommendations [5].

No	Number of Indeks	Interpretasi CSI
1	0,81 – 1,00	Very satisfied
2	0,66 – 0,80	Satisfied
3	0,51 – 0,65	Quite satisfied
4	0,36 – 0,50	Less satisfied
5	0,00 – 0,34	Not satisfied

Table 8. Results of Customer Satisfaction Index Calculation

Attribute	Expectation Level Median Socre	*Importance Weighting Factor (WF)*	Level of Statisfaction Median Score	*Weighted Score (WS)*
P1	4,45	0,051	4,22	0,21
P2	4,39	0,050	4,22	0,21
P3	4,20	0,048	4,10	0,20
P4	4,45	0,051	4,06	0,20
P5	4,35	0,049	4,24	0,21
P6	4,31	0,049	4,37	0,21
P7	4,25	0,048	4,16	0,20
P8	4,43	0,050	4,12	0,21
P9	4,33	0,049	4,14	0,20
P10	4,39	0,050	4,12	0,21
P11	4,39	0,050	4,16	0,21
P12	4,37	0,050	3,88	0,19
P13	4,45	0,051	4,22	0,21
P14	4,45	0,051	4,04	0,20
P15	4,45	0,051	4,08	0,21
P16	4,43	0,050	4,00	0,20
P17	4,51	0,051	4,08	0,21
P18	4,57	0,052	4,25	0,22
P19	4,45	0,051	4,18	0,21
P20	4,49	0,05	4,08	0,21
TOTAL	**88,14**	**1,00**	**82,69**	**4,13**
Customer satisfaction Index (CSI)				**0,827**

With CSI value of 0.827 identified that users of Sidoarjo-Mojokerto commuter are at a very satisfied level (the values between of 0.81 -1.00). This is considered as quite good with the improvements made by PT. (Persero) Kereta Api Indonesia continuously to improve services toward consumers.

3.2.10 Important performance analysis (IPA)

Importance-Performance Analysis (IPA) is a tool to analyze or to compare the extent to which performance / service can be perceived by service users compared to the expected level of satisfaction. The Importance Performance Analysis (IPA) method was first introduced by Marty and James with the aim to measure the relationship between consumer perceptions and the priority of product / service quality improvement, also known as quadrant analysis [4]. IPA has been generally accepted and used in various fields of study because of its ease application and display of analytical results that facilitate performance improvement proposals [6].

Table 9. Importance Performance Analysis (IPA) Jenggala Commuter Line.

Attribute	Satisfaction		Expectations		Comformity Rate (%)
	Xi	($\bar{X}$)	Yi	($\bar{Y}$)	
The accuracy of the train travel schedule	215	4,45	227	4,22	94,71
Service schedule of current train awareness	215	4,39	224	4,22	95,98
Officers always show attitude ready to serve in helping the customer	209	4,20	214	4,10	97,66
Safety and comfortability on board time	207	4,45	227	4,06	91,19
Available CCTV in train at least 1 piece in 1 series	216	4,35	222	4,24	97,30
The evacuation point notice sticker is easy to see and easy to read	210	4,31	220	4,37	95,45
Seating number facility	212	4,25	217	4,16	97,70
Toilet WC facilities	210	4,43	226	4,12	92,92
Ticket price offered	211	4,33	221	4,14	95,48
The availability of fire extinguishers in rail (APAR)	210	4,39	224	4,12	93,75
Hygiene in trains	212	4,39	226	4,16	93,81
Availability of tools to inform the travel route	198	4,37	223	3,88	88,79
In-train seating facilities	215	4,45	227	4,22	94,71
First Aid Kit	206	4,45	227	4,04	90,75
Security officer / conductor	208	4,45	227	4,08	91,63
Path guidance and evacuation procedures	204	4,43	226	4,00	90,27
There is an emergency telephone number	208	4,51	230	4,08	90,43
Station information to be visited / skipped sequentially	217	4,57	233	4,25	93,13
Information in a visual form placed in a strategic place, as well as audio must be sounds obvious	213	4,45	227	4,18	93,83
Length of trip from nearest station	208	4,49	229	4,08	90,83
Total		**88,14**		**82,69**	**1870,32**
	$\bar{\bar{X}}$ =	**4,4**	$\bar{\bar{Y}}$ =	**4,1**	**93,5**

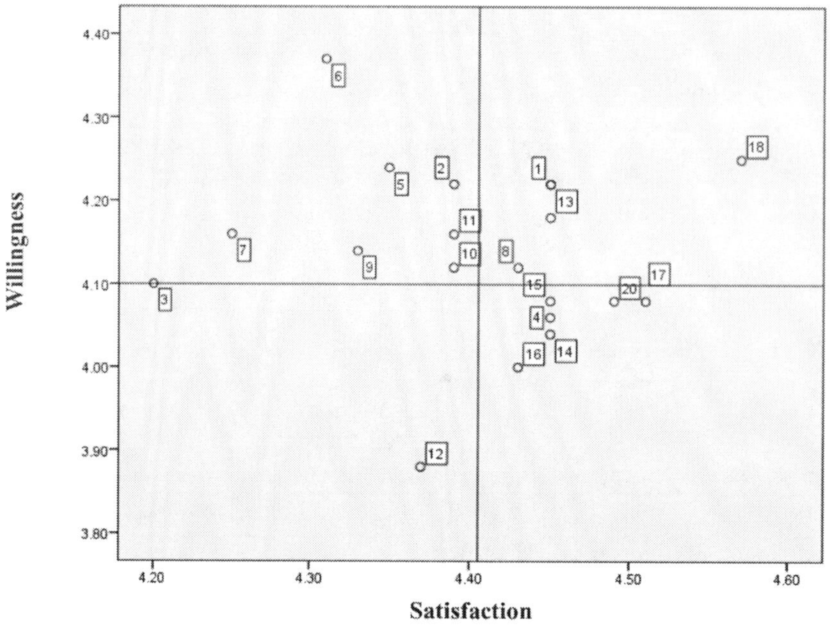

Fig. 3. Cartesian Diagram Importance Performance Analysis (IPA) Jenggala Commuter Line

From the IPA graphic results shows which attributes are included in the quadrant:
Quadrant I "Maintain Performance":
- The accuracy of the train travel schedule (attribute 1)
- Toilet WC facilities Functioning in accordance with the technical standards of operation and odorless clean coming from the toilet (attribute 8)
- In-train seating facilities (attribute 13)
- Station information to be visited / skipped sequentially (attribute 18)
- Information in a visual form placed in a strategic place, as well as audio must be sounds obvious (attribute 19)

Quadrant II "tends to be excessive":
- Safety and comfortability on board time (attribute 4)\
- First Aid Kit (attribute 14)
- Security officer / conductor (attribute 15)
- Path guidance and evacuation procedures (attribute 16)
- There is an emergency telephone number (attribute 17)
- Length of trip from nearest station (attribute 20)

Quadrant III "Low Priority":
- Availability of tools to inform the travel route (attribute 12)

Quadrant IV "Improve Performance":
- Service schedule of current train awareness (attribute 2)
- Officers always show attitude ready to serve in helping the customer (attribute 3)
- Available CCTV in train at least 1 piece in 1 series (attribute 5)
- The evacuation point notice sticker is easy to see and easy to read (attribute 6)
- Seating number facility (attribute 7)
- Ticket price offered (attribute 9)
- The availability of fire extinguishers in rail (APAR) (attribute 10)
- Hygiene in trains (attribute 11)

4 Conclusions

In general, users of Jenggala commuter line are satisfied with the quality of service on the train.

- The condition of Jenggala commuter line has been complied the Minister of Transportation Regulation No. 48 of 2015 on Minimum Service Standards of Railway Transportation where all the criteria contained in the regulations are already exist and applied properly.
- Customer Satisfaction Index (CSI) determine the level of user satisfaction on land transportation services, especially railway. With a CSI value of 0.827 identified that the users of the Jenggala commuter line are at a very satisfied level (the values of 0.81-1.00).
- Based on the respondents' answers obtained the level of conformity on Jenggala commuter line with an average IPA analysis of 93.5% which states the respondents are very satisfied with the services provided. The lowest level of conformity is found in attribute to (12) Availability of tools to inform the travel route (88.79%), and highest in attribute to (7) seating number facility (97.7%).

References

1. Department of Transportation of Indonesia, *Regulation of the Minister of Communications No. 48 of 2015 on The National Transportation System*. Jakarta (2012).
2. Hardiansyah. *Quality of Public Service Concepts, Dimensions, Indicators, and Implementation*. Yogyakarta (2011).
3. A, Parasuraman. *The Behaviorial Consequence of Service Quality*, Journal of Marketing, Vol 60. (2001).
4. Febriyanti, R. *Analysis of Commuter Train Service Surabaya-Sidoarjo From The Side Of User Satisfaction*. Surabaya: Institute of Technology Sepuluh Nopember (2009).
5. Tjiptono, F., and Chandra, Gregorious. *Service, Quality, and Satisfaction*. Ed 3, Yogyakarta: Andi Ofset (2011).
6. John A. Martilla and John C. James. *Important Performance Analysis*. (1997).

Analysis of the visual safety perception and the clarity of traffic signs and road markings in the presence of road lighting in straight and curved road

Endah Setyaningsih[1*], *Leksmono* Suryo Putranto[1], *Soegijanto*[3], and *F.X. Nugroho* Soelami[3]

[1] Tarumanagara University, Doctoral Program of Civil Engineering, 11440 Jakarta, Indonesia
[2] Tarumanagara University, Department of Civil Engineering, 11440 Jakarta, Indonesia
[3] Bandung Institute of Technology, Fac. of Civil and Environmental Engineering, Bandung, Indonesia

Abstract. The presence of road lighting helps a car driver to identify nearby cars. The availability of road lighting can also reduce the occurrence of security disturbances however it has not been considered as the cause of accident. A study was conducted to analyze the visual perception of car drivers toward the road lighting. The research was conducted on Cikampek and Cipularang freeways which include straight and curved road sections. Data collection was conducted using questionnaires as the measurements instrument by asking the participants to watch the video footage of the freeway. The results show that the car drivers feel safer and more clarity in identifying the traffic signs and road marking in the presence of road lighting. Furthermore, yellowish road lighting provides more visual safety perception than white lighting which opposed with other previous studies. This result was obtained from the existing condition in the questionnaires without any control variable.

1 Introduction

Road lighting has not been considered as an important issue for road safety, especially in Indonesia. There were many road lightings being installed which has not met the requirements. According to Eloholma, et al [1], the main function of road lighting is to maintain human safety and provide good visibility conditions and reduce potential hazards by providing illumination to the objects on the road and around the cars. While Li [2], mentioned that good road lighting is needed to estimate the car speed, monitoring harmful objects beside the car, and keep the distance between the cars. Therefore, the main function of road lighting is to facilitate road users in recognizing objects around them quickly, accurately and comfortably.

* Corresponding author : endahs@ft.untar.ac.id

Several studies have been conducted in Jakarta, Bandung and other cities. The results indicate the presence of uneven illuminance on the road surface, and less or excessive illumination according to the road lighting standards [3-5]. This study provides an overview of car drivers' perceptions to the road lighting, in relation to safety and clarity in identifying traffic signs and road markings. The research was conducted on Cikampek and Cipularang freeways. The freeways have various geometric elements of the road, including uphill slopes, decline, downhill slopes, straight sections and curved sections. A road with such complex geometric elements must consider the visibility of the road users that meet the safety standards [6]. The limitation of visibility may cause an accident. An action to overcome the risk is by installing or improving the road lighting [7].

The Cikampek freeway connects the Jakarta freeway to Cikampek, from the Cawang interchanges (Km 0) to the Cikampek interchange (Km 73). While the Cipularang freeway connects Cikampek to Padalarang (KM 121). Currently the Cikampek freeway has become the densest segment of the Trans Java freeways network, connected and integrated with the inner freeway City of Jakarta, Jakarta Outer Ring Road (JORR). Most of Cikampek freeway have 4 lanes in each directions. The busiest freeway at present is the Cikampek freeway with 552,322 cars pass through each day and were dominated by 87% of vehicle category 1 (light vehicles). The rest were heavy vehicle s ranging from vehicle category II (2 axles trucks/ buses) to vehicle caterogy V (5 axles truck trailers) [8]. The Cipularang freeway is crosses many hills and valleys with various road geometrics in the form of successive curved roads, as well as long uphill slopes from Jakarta direction and its reverse downhill slopes direction. The Cipularang freeway is connected directly to Cikampek freeway which is the shortest road from Jakarta to Bandung and vice versa, whiles also the busiest freeway.

The study used the questionnaires as a measuring instrument. The data collection was conducted by asking the participants to watch the video footage of the freeway with straight and curved road section. The videos were captured at night along Cikampek and Cipularang freeway in Jakarta to Bandung direction and its opposite direction using latest technology action camera. The detail locations of the video recording are shown in Table 1. The video recording was used with the assumption that the freeway condition and circumstances can be represented by the video footage which has the advantages of time saving, cost-effectiveness, reduction of accident risk, and controlled numbers of the respondents. However, it also has the disadvantages of the fact that the condition of the roads represented in the video recording will greatly depend on the resolution of the video recordings. In addition, the setting of the experimental room that resembles the conditions in a car is very difficult to obtain and the experience of respondents in a real condition might vary from the testing environment which needs several adjustments. The participants filled out the questionnaires after watching the video recordings with a duration limited between 40 and 60 seconds to avoid the participants from getting exhausted. The conditions in the video recording were in line with the condition in the questionnaires.

The questionnaire consists of 6 conditions for Jakarta to Bandung direction (JB) and Bandung to Jakarta direction (BJ). The definitions for freeway road lighting conditions are:

1. The straight and curved freeway road of condition 1 is the straight and curved freeway roads without road lighting which is intended as a freeway road without installed road lights.
2. The straight and curved freeway road of condition 2 is the straight and curved freeway road with white light road lighting, and the illuminance according to existing condition.

3. The straight and curved freeway road of condition 2 is the straight and curved freeway road with yellowish light road lighting, and the illuminance according to existing condition.

Table 1 The Locations, road lighting conditions, luminaires arrangements and types of lights which were used in the questionnaires

No	Section of freeway	Condi tion	Loca tion	With or without light	Lumminaire arrangements	Type, wattage lamp and mounting height
1	condition 1 straight section	JBS1	Km 46	without light	-	-
2	condition 2 straight section	JBS2	Km 35	with white light	single-sided	LED 90W 9 m
3	condition 3 straight section	JBS3	Km 47	with yellowish light	central	HPS 150 W 11 m
4	condition 1 curved section	JBC1	Km 76	without light	-	-
5	condition 2 curved section	JBC2	Km 70	with white light	central	LED 50 W 7.5 m
6	condition 3 curved section	JBC3	Km 84	with yellowish light	single-sided	HPS 150 W 11 m
7	condition 1 straight section	BJS1	Km 107	without light	-	-
8	condition 2 straight section	BJS2	Km 68	with white light	central	LED 50 W 7.5 m
9	condition 3 straight section	BJS3	Km 120	with yellowish light	single-sided	HPS 150 W 11 m
10	condition 1 curved section	BJC1	Km 88	without light	-	-
11	condition 2 curved section	BJC2	Km 111	with white light	single-sided	LED 50 W 7.5 m
12	condition 3 curved section	BJC3	Km 97	with yellowish light	single-sided	HPS 150 W 11 m

The questionnaires in this study accommodate 6 variables, including road lighting quality, traffic signs and road markings, which are the road facilities that must be observed by the road users. The variables are Illuminance, Luminance, Glare, Correlated Color Temperature (CCT), traffic signs and road markings. Those variables were compiled as the research instruments and actualized in the 16 questions/statements about the visual perception of the road lighting as shown in Table 2.

Table 2 Statement/Question items in the Questionnaire

No	Statement/Question item	Perceived Rate				
1	Light from the road surface	1	2	3	4	5
2	Light from a light source (road light)	1	2	3	4	5
3	Light of the surrounding light (another object)	1	2	3	4	5
4	Glare from the road surface	1	2	3	4	5
5	Glare from the light source (road light)	1	2	3	4	5
6	Glare from the ambient light (another object)	1	2	3	4	5
7	Glare cause by the car's light from opposite direction	1	2	3	4	5
8	Secure	1	2	3	4	5
9	Comfortable	1	2	3	4	5
10	Evenness of light on the road surface	1	2	3	4	5
11	Visibility of road marking	1	2	3	4	5
12	Visibility of any potholes	1	2	3	4	5
13	Visibility of road median	1	2	3	4	5
14	Visibility of other cars	1	2	3	4	5
15	Visibility colors of traffic signs	1	2	3	4	5
16	Visibility of sign symbol and text	1	2	3	4	5

Note: 1 = very negative, 5 = very positive

The design concept used in the research is referred to similar previous studies [9, 10, 11]. The questionnaire used the Likert measurement scale with 5 stages of measurement from numbers 1 to 5. Number 1 indicates a very negative condition and number 5 is very positive. For example, a 'light' statement marked with number 1 has the meaning of very dim and number 5 means very 'bright'. However, the statement for 'glare' and 'contrast' is in opposite, which means that number 1 is very "not glare" and very "not contrast" and number 5 is very glare and very contrast. Therefore, the statements of glare and contrast with number 1 means very positive and verynumber 5 means very negative. The participants were asked to provide a circle on the numbers corresponding to their judgement. Questionnaires were filled in a dark room (without lighting from the lights or the sun) which was intended to simulate the condition in a car at a dark night because it is clearer to watch the video in the dark room. The participants used the light from a mobile phone or a flashlight when they were filling the questionnaire in the dark room. Before the completion of the questionnaire, the participants were divided into a group of 3 each, and were explained about each item in the questionnaire and the proper procedure to watch the video and how to play the video. In addition, the introduction to road lighting was also given to the participants to obtain equal judgement on road lighting condition from each participant before the completion of the task. Each group required 30 to 45 minutes to complete the questionnaire including the 10 minutes explanation time. When the participants already finished watching the video, guidance on how to fill out the questionnaire were then given. Food and beverages were provided for the participant after the completion of the data collection.

The population of the study is the car drivers who has been driving through the Cikampek and Cipularang freeways or at least has the experience passing through the freeway as a passenger. The population is categorized as an infinite population as it has a large numbers of population members and not all of them are recognized. The sampling was not entirely random; therefore to obtain the diversity of the data, the randomness of the sample was accomplished by taking various data ranges, such as various age, gender, education, occupation, origin and residency. By taking such approach, it is expected that the characteristics of the population become objectively retrieved. The sample was selected to meet the following criteria: a person with a least at least 1 (one) year driving experience, minimum age 20 years, normal vision or has been corrected with glasses [12]. The respondents should pursue at least Senior High School. The research used the reliability test using item-total correlation method and Alpha Cronbach's cut-off minimal = 0.80 which is considered as good [13]. Discrete variable is used for the measured data because measurement of road lighting has 2 conditions: with or without road lighting, whereas the interval of the measurement is scaled between 1 and 5 [14]. The data processing used parametric statistic method, namely Paired Sample t-test and Pearson correlation test.

2 Data and result

2.1 Analysis of visual safety perception with and without road lighting

The statistical analysis of the test results for the visual safety perception with or without road lighting on the straight and curved section on the freeway from Jakarta to Bandung (JB) and Bandung to Jakarta (BJ) direction are summarized in Table 3 from 58 participants. The test results of visual safety perception using white lighting versus yellowish lighting are shown in Table 4.

Table 3 The test results of visual safety perception with or without road lighting on the straight and curved section on the freeway in the Jakarta-Bandung (JB) and Bandung-Jakarta (BJ) direction

Pair	Condition	Mean	Mean Difference	Significant Level	Significant? (Yes/No)
1	JBS1	2.54	-	-	-
	JBS2	3.76	-1.22	<0,001	Yes
2	JBS1	2.54	-	-	-
	JBS3	3.43	-0,89	<0,001	Yes
3	JBC1	2.21	-	-	-
	JBC2	3.48	-1.28	<0,001	Yes
4	JBC1	2.21	-	-	-
	JBC3	3.74	-1,53	<0,001	Yes
5	BJS1	2.25	-	-	-
	BJS2	3.43	-1,18	<0,001	Yes
6	BJS1	2.25	-	-	-
	BJS3	4.00	-1,75	<0,001	Yes
7	JBC1	2.21	-	-	-
	JBC2	3.23	-1,02	<0,001	Yes
8	JBC1	2.21	-	-	-
	JBC3	3.83	-1,62	<0,001	Yes

Note: S = straight and C = curved

The results in Table 3 can be divided into 2 directions: Jakarta – Bandung (JB) and Bandung – Jakarta (BJ) with each can be further classified into 2 different types of road, that is: straight and curved road. The analysis of the results will be discussed based on those available outcomes in the following section.

2.2 Analysis of Jakarta to Bandung direction

The result of paired sample t-test for Jakarta-Bandung direction (JB) shows that there is significant difference of visual safety perception on the straight and curved section of the freeway. The similar results were also obtained for the road section with or without road lighting, either for white light or yellowish light as shown in Table 3.

2.2.1 Straight road with and without road lighting

On the straight road section, the comparison of the visual safety perception was divided into 2 categories: without road lighting versus road lighting with white light and without road lighting versus road lighting with yellowish light.

The visual safety perception in the straight road without road lighting compared to road lighting with white light shows significant difference with $p < 0.001$. The car drivers prefer more to the road using road lighting with white light. The mean of the visual safety perception is 3.76 for road with white light road lighting while the mean of visual safety perception for the road without lighting is 2.54.

Similar trend also appears in the comparison of the visual safety perception on a straight road without road lighting compared to the road with yellowish road lighting ($p < 0.001$). The mean of visual safety perception for the road without road lighting is 2.54 which is smaller than the mean of visual safety perception for the road having road lighting with yellowish light (3.43).

2.2.2 Straight road with road lighting using white light versus yellowish light

Furthermore, the comparison of the visual safety perception between the white light road lighting and yellowish light road lighting shows that the white light road lighting outperforms with mean = 3.76, while the mean for the yellowish road lighting is 3.43.

Table 4 shows that the there is no difference of visual safety perception between the lighting with white light and the yellowish light on the Jakarta – Bandung highway ($p= 0.055$). The visual safety perception of the car drivers with the white lighting is 3.76 and 3.43 with the yellowish lighting.

2.2.3 Curved road with and without road lighting

On the curved road, the comparison of the visual safety perception was also divided into 2 categories as on straight road. The analysis is as follows.

The comparison of curved road without road lighting versus curved road with white light road lighting shows significant difference of visual safety perception between the road without road lighting and the road with white lighting ($p < 0.001$). Visual safety perception are greater when there is road lighting with white light (mean = 3.48) compared to the road without lighting (mean = 2.21).

There is also a significant difference in safety and security perception in the curved road when the road without road lighting is compared to the road lighting with yellowish light ($p < 0.001$). The visual safety perception is greater when there is road lighting with a yellowish light (Mean = 3.74) compared to the road without road lighting (mean = 2.21).

2.2.4 Curved road with road lighting using white light versus yellowish light

The visual safety perception is greater when there is road lighting with a yellowish light (Mean = 3.74) compared to the road lighting with white light (Mean = 3.48) which supported by the p value = 0.017.

Table 4 The test results of feel visual safety for road lighting using white light versus yellowish light on the straight and curved section on the freeway in the Jakarta-Bandung (JB) and Bandung-Jakarta (BJ) direction.

Pair	Condition	Mean	Mean Difference	Significant Level	Significant? (Yes/No)
1	JBS2	3.76	-	-	-
	JBS3	3.43	0,33	0,055	Yes, using white and yellowish light
2	JBC2	3.48	-	-	-
	JBC3	3.74	-0,25	0,017	Yes, using yellowish light
3	BJS2	3.43	-	-	-
	BJS3	4.00	-0,57	<0,001	Yes, using yellowish light
4	BJC2	3.23	-	-	-
	BJC3	3.82	-0,59	<0,001	Yes, using yellowish light

Note: S = straight and C = curved

2.3 Analysis of Bandung - Jakarta direction

The analysis of Bandung – Jakarta direction applied the same comparison method in Jakarta – Bandung direction. The results indicate that the paired sample t-test for Bandung-Jakarta (BJ) also demonstrate significant difference of visual safety perception between the road without road lighting and the road having road lighting using either white or yellowish light as seen in Table 3.

2.3.1 Straight road with and without road lighting

Significant difference of visual safety perception on the straight road without road lighting and road lighting with white light has p value less than 0.001. The car drivers selected the road with white light road lighting more than the road without road lighting which supported by the mean for the road with white lighting equal to 3.43 which is higher than the mean of the road without road lighting (2.25).

The similar result for the visual safety perception on a straight road without road lighting compared to the road with road lighting using yellowish light with $p < 0.001$, mean for yellowish light = 4,00 and mean for the white light = 2.25.

2.3.2 Straight road with road lighting using white light versus yellowish light

In this straight road section from Bandung to Jakarta, the car drivers prefer the yellowish light with mean = 4.0 which is higher than the white light road lighting with, mean = 3.43 and the p value is less than 0.001.

2.3.3 Curved road with and without road lighting

On the curved road of the freeway from Bandung to Jakarta, the car drivers chose the road with lighting better than the road without lighting (mean = 2.22) either with white lighting (mean = 3.23) or with yellowish lighting (mean = 3.82), p value < 0.001.

2.3.4 Curved road with road lighting using white light versus yellowish light

When comparing between the yellowish lighting and white lighting, the car drivers prefer the yellowish lighting (mean = 3.83) than the white lighting (Mean = 3.23) with the p value less than 0.001.

2.4 Analysis of traffic signs and road markings clarity with and without road lighting

The result of paired sample t-test for the clarity of traffic signs and road markings on the straight and curved road from Jakarta to Bandung and from Bandung to Jakarta is presented in Table 5.

2.5 Analysis of Jakarta to Bandung direction

Table 5 indicates significant preference to the road with road lighting than the road without road lighting on the freeway from Jakarta to Bandung.

Table 5. The test results in clarity of the traffic signs and road markings in the direction of Jakarta-Bandung (J) and Bandung - Jakarta (B) freeway with and without road lighting for the straight and curved road (N = 58)

Pair	Road Marking condition	Mean	Mean difference	Significant Level	Significant? (Yes/No)
1	TR_JS1	2.95	-	-	-
	TR_JS2	3.94	-0,99	<0,001	Yes
2	TR_JS1	2.95	-	-	-
	TR_JS3	3.35	-0,40	<0,004	Yes
3	TR_JC1	2.77	-	-	-
	TR_JC2	3.65	-0,88	<0,001	Yes
4	TR_JC1	2.77	-	-	-
	TR_JC3	3.83	-1,07	<0,001	Yes
5	TR_BS1	2.49	-	-	-
	TR_BS2	3.42	-0.93	<0,001	Yes
6	TR_BS1	2.49	-	-	-
	TR_BS3	3.97	-1,49	<0,001	Yes
7	TR_BC1	2.61	-	-	-
	TR_BC2	3.40	-0,79	<0,001	Yes
8	TR_BC1	2.61	-	-	-
	TR_BC3	3.84	-1,23	<0,001	Yes

Note: TR= traffic signs and road marking, S = straight and C = curved

2.5.1 Straight road

At Jakarta to Bandung direction on the straight roads, the results are significantly support the road with road lighting, either with white lighting ($p < 0.001$) or yellowish lighting ($p < 0.004$). The car drivers are confident that the traffic signs or road markings are clearer on the road that have road lighting with white light (Mean = 3.94) and yellowish light (M = 3.35) compared to the roads without road lighting (Mean = 2.95).

2.5.2 Curved road

For the curved roads on the freeway from Jakarta to Bandung the results also support the road with road lighting better than the road without road lighting, either white lighting ($p < 0.001$) or yellowish lighting ($p < 0.001$). The car drivers believe that traffic signs and road markings are more visible on the road that have road lighting with white light (mean = 3.65) and yellowish light (mean = 3.83), whereas the traffic signs and road markings on the roads without road lighting is less clear (mean = 2.77).

2.6 Analysis of Bandung - Jakarta direction

Result of paired sample t-test test on Bandung-Jakarta freeway shows that there is also difference of clarity level of the traffic signs or road markings between the road without road lighting and the road with road lighting.

2.6.1 Straight road

For the straight roads, the difference supports the road with road lighting using either white light ($p < 0.001$, mean = 3.42) or yellowish light ($p < 0.001$, mean = 3.97). On the road without road lighting, the traffic signs and road markings are less visible (mean = 2.49).

2.6.2 Curved road

The identical results are obtained for the curved roads, there is also an indication of less visible traffic signs and road markings on the road without road lighting. The difference has the p value < 0.001 for the comparison between the road without road lighting and the road that has road lighting with white or yellowish light. The traffic signs or road markings are more visible on the road with white lighting (mean = 3.40) and yellowish light (mean = 3.84) compared to the roads without road lighting (mean = 2.61).

Based on above analysis and the statement 'yes' in the statement column that has been shown in Table 5, it is obvious that the car drivers can clearly recognize the traffic signs and road markings on Cikampek and Cipularang freeway that equipped with road lighting either on the straight or curved roads.

3 Conclusions

Build upon the research results it can be concluded that the car drivers on the Cikampek and Cipularang freeway feel safer with the availability of the road lighting either with white light or yellowish light on the straight and curved road for both Jakarta-Bandung and Bandung-Jakarta directions. The results are corresponding to the research of Fotios et al [9] which mentions that the pedestrians feel more reassured when walking after dark around the residential area.

The visual safety perception is greater with a yellowish light road lighting compared to white light. This result is diverse from other researches, for example: Fotios [9] stated that road lighting with white light provide safer feeling than yellowish light; while Knight [15] expressed that people perceive white light to be more comfortable and safer than the yellowish light; and Morante [11], reported that road lighting with white light provide higher perceived safety than yellowish light.

The different of the results might be related to the facts that the research in Cikampek and Cipularang freeway used the existing lighting condition without any control to all the variables. The existing lighting condition at the time of the research used LED 50 W (white light) with the mounting height is 7.5 m and HPS150 W (yellowish light), with the mounting height is 11 m (see table 2) with an exception for the straight road lighting which utilized the LED 90 W white light and HPS150 W (yellowish light). To match with other researchers the study must applied the lighting condition using lamps with the same flux luminous and in the same road conditions.

The presence of road lighting provides clarity in recognizing traffic signs and road markings for car drivers on the straight and curved roads.

Acknowledgment

Thanks to Ministry of Education and Culture for the research funding of the doctoral program. Additional thanks to PT Jasa Marga, Tbk, Purbaleunyi branch who has provided the secondary data on Cipularang freeway, which is very useful in this research.

References

1. Eloholma, M., Ketomäki, J., Halonen, L., Meas. and An, *Luminances and Visibility in Road Lighting – Conditions* (2004)
2. Li, F., Chen, Y., Liu, Y., and Chen, D., Leukos **VIII-3,** *Comparative In Situ of LEDs and HPS in Road Lighting*, 205 – 214, (2012)
3. Armanto, R. Soelami, N., Soegijanto, Proc. SENVAR/2nd ISESEE **9th**, *Evaluation on Lighting Condition and Visual Legibility of Road Surface and Traffic sign in Bandung City* (2008).
4. Setyaningsih, E., Fat, J., Zureidar, I. And Wardhani, L., Proc. EECCIS, **7** th, *Performance Of LED Lights Installed on DKI Jakarta Streets,* (UB, Malang, 2014)
5. Santoso, W.J., *Study of Public Road Lighting Supply in Jakarta East Area* (Perpustakaan ITB, Bandung, 2009)
6. Ministry of Public Works, *Guideline of Inter-city Road Geometric Plan* (1997).
7. Putranto, L.S., *Rekayasa Lalu-lintas Edisi 2* (PT Indeks, Jakarta 2013)
8. PT. Jasa Marga Tbk, *Rekapitulasi Data Kecelakaan Ruas Dawuan-Padalarang Barat, Jawa Barat* (2011-2015)
9. Fotios, S, Proc. CIE **2014.** Lighting Quality and Energy Efficiency, *Road Lighting And Pedestrian Reassurance After Dark:A Review Of The Evidence* (2014)

10. Haans, A. and de Kort, Y.A.W., Env. Psy. **32**, *Light distribution in dynamic street lighting: Two experimental studies on its effects on perceived safety, prospect, concealment, and escape* (The Netherlands, 2012)
11. Morante, P, Fnl. Rep. for Groton Utl, Groton, Connecticut, *Mesopic Street Lighting Demonstration and Evaluation*. (USA, Troy, 2008)
12. Lewis, A.L., Illum. Eng. Soc., **Wtr.99**, *Visual Performance as a Function of Spectral Power Distribution of Light Sources at Luminances Used for General Outdoor Lighting*, (1999).
13. Murti, *Validitas dan Realibilitas Pengukuran* (UNS, Solo. 2011)
14. Privitera, G.J., *Statistic for the Behavioural Science 2nd Ed.* (Sage Publication Inc,, Canada, 2015)
15. Knight, C, Light. Res. Tech., **42-3**, 313-330, *Field Surveys investigating the effect lamp spectrum on the Perception of Safety and Comfort at Night* (2014)

MATEC Web of Conferences **181**, 04002 (2018) https://doi.org/10.1051/matecconf/201818104002
ISTSDC 2017

Effect of type of road humps on vehicular speeds on residential roads

Leksmono Suryo Putranto[1*], and *Kevin* Kurniawan[1]

[1] University of Tarumanagara, Dept. Of Civil Engineering, 11440 Jakarta, Indonesia

Abstract. To reduce potential pedestrian fatalities, in residential road maximum speed is 30km/hour. Apart from installing maximum speed signs, installing road humps may ensure speed reduction. This paper is intended to compare light vehicles and motorcycles speeds and speeds reductions due to road humps. Data collection was done in Modernland, Tanggerang City. Two road segments were observed, i.e. segments with standard and non-standard road humps. Observation was made in morning, noon and afternoon periods. A portable speed gun was used to measure the speed at about 50m and 25m before the road humps. Some mean difference statistical analyses were conducted for both speeds and speeds differences between different type of road humps and between pairs of observation periods. The 0.05 significant level was used. Surprisingly, mean of speeds differences (50m vs 25m) in standard road hump is significantly higher than in non-standard road hump.

1 Introduction

In residential roads, there are many daily domestic activities, such as, children crosing the roads to reach or return from school, senior citizens walking in the road (where pedestrian path is not available) to get fresh air in the morning, housewives crossing the roads to reach or return from grocery shops etc. To reduce potential pedestrian fatalities due to crash between motorized vehicles, in residential road maximum allowable speed is 30km/hour. Therefore, apart from installing maximum speed signs, installing road humps may ensure speed reduction. This paper is intended to compare light vehicles and motorcycles speeds and speeds reductions due to road humps. Data collection was done in Modernland, Tanggerang City. Two road segments were observed, i.e. segments with standard and non-standard road humps.

2 Literature reviews

Road fatalities are strongly related with speeds. As understood, kinetic energy when a pedestrian was crashed by a vehicle is half the mass of the vehcile multiplied by the square

* Corresponding author : lexy_putranto@yahoo.co.id

of the speed. Therefore part of kinetic energy from the speed is incredibly significant. So speed reduction will significantly help fatalities reduction [1]. Figure 1 shows the impact of speed on probability of pedestrian fatalities (accident causing death of road users). It can be seen that due to involvement of speeds in the magnitude of kinetic energy, there are huge increase of fatalities probability from almost zero to beyond 80% if speed is increased from 30 km/hour to 50 km/hour.

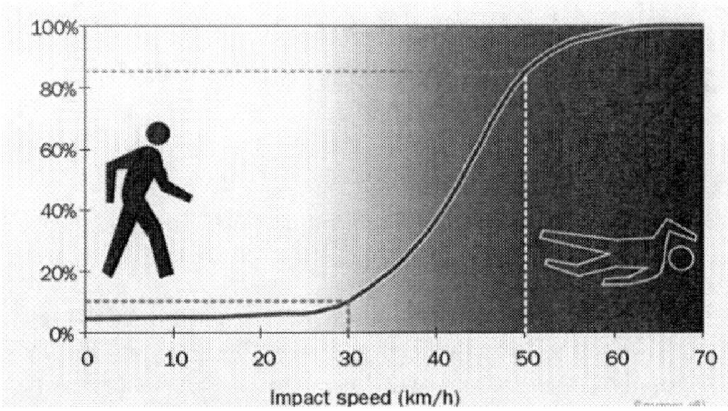

Figure 1 Impact of motorized vehicle speed to probability of pedestrian fatalities
Source: OECD/ECMT Transport Research Centre [2]

Speed also determines stopping distance. Figure 2 shows and adaptation of Australian Transport Safety Bureau (ATSB) calculation of stopping distance [2]. Again we find benefit or speed reduction, i.e. required stopping distance reduction. For safely stop during driving at 30 km/hour, ones only required less than 15m stopping distance (includes both reaction and braking distance). According to [3], speed limit is 30 km/hour.

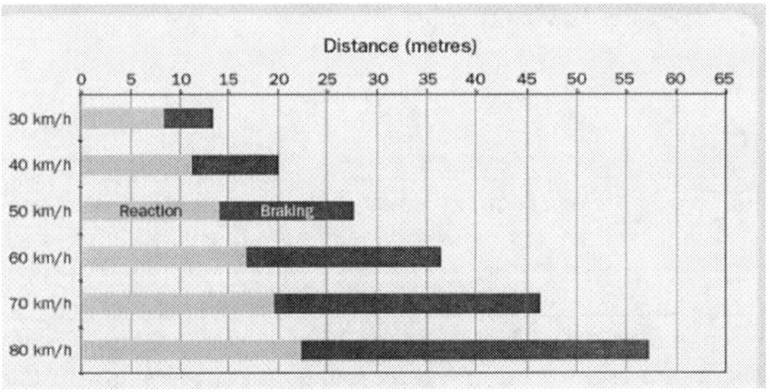

Figure 2 Effect of vehicular speeds on stopping distance
Source: OECD/ECMT Transport Research Centre [2]

In Indonesia, road hump is regulated in [4]. In Chapter 2 Verse 1, road hump is defined as additional equipment on road which installed to force road user to reduce speed. Road hump should be installed perpendicularly to the road axis with certain width, thickness and gradient. The choice of material for road hump should consider road user safety.

In the same decree Chapter 4 Verse 1 regulate that road hump should be installed in

- Residential road
- Local road (class III C)

• Roads with construction works

According to Chapter 5 Verse 1 of the same Decree, a road sign as shown in Figure 3 should be installed at certain distance before the road hump. This road sign is regulated in [5] on appendix 1 table 1 no. 6b.

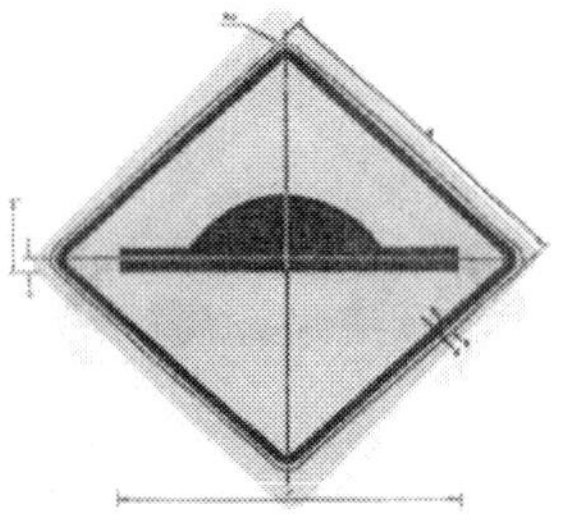

Figure 3 Road sign to indicate road hump ahead

According to Chapter 5 Verse 2 of [3], the road hump should be painted with incline white line road markings. Example and dimension of the marking is presented in Figure 4.

Figure 4 Example and dimension of road hump markings

According to Chapter 6 Verse 4 of [3], the dimension of Road Hump can be persented as Figure 5 and regulated as follow:

- The shape of road hump is trapezoid
- The maximum height is 12cm
- The maximum gradient of incline part is 15%
- The minimum width of flat part is 15cm

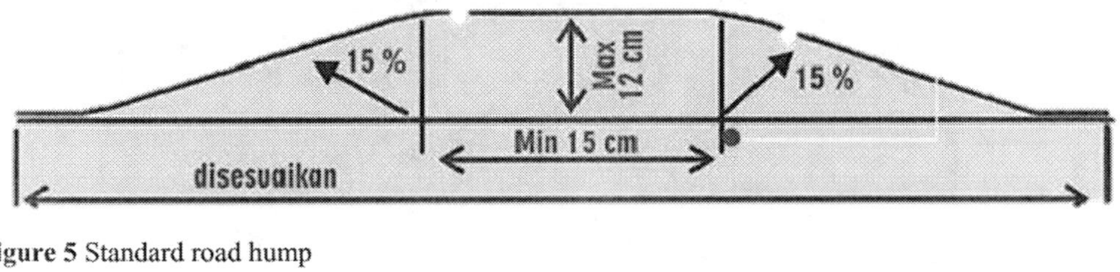

Figure 5 Standard road hump

Chapter 7 of [3] regulate the material of road hump can be the same with material of the road, can be from rubber or any other materials with similar effect and should consider road user safety.

3 Data collection method

Data collection was done in Modernland, Tanggerang City. Two road segments were observed (Figure 6), i.e. segments with standard road hump (Figure 7) and segment with non-standard road humps (Figure 8). Observation was made in morning, noon and afternoon periods. Data Collection for speeds related to standard hump was done twice (Monday, 17 April 2017 and Tuesday 25 April 2017). Data Collection for speeds related to non-standard hump was also done twice (Tuesday, 18 April 2017 and Monday 24 April 2017). A portable speed gun was used to measure the speed at about 50m and 25m before the road humps.

Figure 6 Observed segments in modern land indicated with red circles

Figure 7. Observed segment with standard road hump

Figure 8. Observed segment with non-standard road hump

The speed gun was operated by a surveyor assisted by another surveyor who recorded the speed data manually. Two types of vehicle were observed, i.e. light vehicle and motorcycle. In each observation period the the length of observation was 60 minutes. Every one minute a pair of speed data (at 50m and at 25m before the speed hump) was observed. It can be either a light vehicle or a motorcycle, whichever measured first at that particular minute. Therefore number of observed light vehicles and motorcycles in each observation period is not equal.

4 Method of data analysys

All analyses were made seperately for light vehicles and motorcycles. Several mean difference statistical analyses were conducted at significant level of 0.05 to evaluate the followings:

1. Mean difference of speed reduction between vehicles travelling toward standar road hump and non-standard road hump.
2. Mean difference of speed reduction between pairs of observation period (morning-noon, morning-afternoon and noon-afternoon) within sthe same type of road hump.
3. Mean difference of speed at 50m before standard road hump and non-standard road hump.
4. Mean difference of speed at 25m before standard road hump and non-standard road hump.

5 Results

Table 1 summarizes various speeds related to standard road hump.Table 2 summarizes the same things related to non-standard hump.

Table 1. Summary of various speeds related to standard road hump

	Speeds of Light Vehicles (km/hour)			Speed of Motorcycles (km/hour)		
	50m before	25m before	Reduction	50m before	25m before	Reduction
N	179	179	179	181	181	181
Mean	39.09	14.77	24.32	45.14	15.93	29.20
Minimum	31.00	12.00	16.00	35.00	13.00	19.00
Maximum	51.00	19.00	34.00	53.00	20.00	37.00

Table 2 Summary of Various Speeds Related to Non-Standard Road Hump

	Speeds of Light Vehicles (km/hour)			Speed of Motorcycles (km/hour)		
	50m before	25m before	Reduction	50m before	25m before	Reduction
N	202	202	202	158	158	158
Mean	34.93	14.79	20.13	37.27	15.16	22.11
Minimum	30.00	12.00	14.00	30.00	3.00	16.00
Maximum	43.00	19.00	26.00	49.00	20.00	30.00

It can be seen that in general speeds 50m before the standard road hump were higher than speeds 50m before to non-standard road hump both for light vehicles and motorcycles. As the speeds 25m before both type of road humps were about the same then the speeds reductions before standard road hump were higher than speed reductions before the non-standard road hump for both light vehicles and motorcycles. There is no logical nor theoretical explanation regarding these consistent results. It is also found that motorcycles speeds were higher than light vehicles. It should be noted that all mean speeds (and off-course maximum speeds) at 50m before any types of road hump were higher than 30 km/hours (beyond the regulation for residential road speed limit). Even the minimum speeds at 50m before any types of road hump were as high as 30 km/hour. Therefore additional road humps might be justified along with additional speed limit sign and public awareness program to reduce speed there.

For further tables on this paper RH stands for Road Hump, SRH stands for Standard Road Hump and NSRH standsfor Non-Standar Road Hump. Table 3 summarizes the mean difference test of speed reduction between standard road hump and non-standard road hump for morning, noon and afternoon observation periods for both light vehicles and motorcycles. It can be seen that all test results were significant at 0.05 and with higher mean of speed reduction for standard road hump meaning that mean of speed reduction for standard road hump is significantly higher than mean of speed reduction for non-standard road hump for all type and vehicles and for all observation periods.Table 4 summarized similar things with Table 3 except that it discuss speeds 50m before the road hump. Table 5 summarized similar things with Table 3 except that it discuss speeds 25m before the road hump. All analyses in Table 4 were significant at 0.05 and although not all analyses in Table 5 were significat at 0.05 but the mean differences were relatively low. The test results in Table 4 and Table 5 correspond with the results in Table 3, i.e. the significant speeds reductions were result of relatively high mean value of speeds at 50m before road hump and relatively low speeds at 25m before road hump.

Table 3 Mean Difference of Speed Reduction between SRH and NSRH

	Mean of Speed Reduction (km/hour)					
	Light Vehicles			Motorcycles		
	Morning	Noon	Afternoon	Morning	Noon	Afternoon
N SRH	58	65	56	62	55	64
N NSRH	74	63	65	46	57	55
Mean SRH	24.0466	24.8153	24.0446	28.8952	29.6636	29.1094
Mean NSRH	20.0541	20.5556	19.8308	22.4130	22.1316	21.8545
Mean Difference	3.9925	4.2598	4.2138	6.4821	7.5320	7.2548
Significant Level	<0.001	0.001	<0.001	<0.001	<0.001	<0.001
Significant? (Yes/No)	Yes	Yes	Yes	Yes	Yes	Yes

Table 4 Mean difference of speed at 50m before RH between SRH and NSRH

	Mean of Speed Reduction (km/hour)					
	Light Vehicles			Motorcycles		
	Morning	Noon	Afternoon	Morning	Noon	Afternoon
N SRH	58	65	56	62	55	64
N NSRH	74	63	65	46	57	55
Mean SRH	39.2190	39.4000	38.6250	39.2190	44.5091	44.8125
Mean NSRH	35.6216	35.4444	33.6462	35.6216	37.7193	36.8727
Mean Difference	3.5973	3.9556	4.9788	3.5973	6.7897	7.9397
Significant Level	<0.001	<0.001	<0.001	<0.001	<0.001	<0.001
Significant? (Yes/No)	Yes	Yes	Yes	Yes	Yes	Yes

Table 5. Mean difference of speed at 25m before RH between SRH and NSRH

	Mean of Speed Reduction (km/hour)					
	Light Vehicles			Motorcycles		
	Morning	Noon	Afternoon	Morning	Noon	Afternoon
N SRH	58	65	56	62	55	64
N NSRH	74	63	65	46	57	55
Mean SRH	15.1724	14.5846	14.5804	16.3306	15.7545	15.7031
Mean NSRH	15.5676	14.8889	13.8154	14.8043	15.5877	15.0182
Mean Difference	- 0.3951	0.2440	0.7649	0.049	0.1668	0.6849
Significant Level	0.065	0.244	<0.001	0.001	0.573	0.03
Significant? (Yes/No)	No	No	Yes	Yes	No	Ya

Tables 6 through 8 shows that there were almost no significant mean difference between speed reduction between pair of observation periods (morning-noon, noon-afternoon and morning-afternoon). The original plan of morning period was 06.00-07.00, noon period was 12.00-13.00 and afternoon period was 17.00-18.00 to describe morning peak hour, noon off-peak hour and afternoon peak hour. However, the survey team misunderstood the instruction and instead conduct the morning period at 10.00-11.00 and afternoon period at 15.00-16.00. These periods of the day might be considered to belong to off-peak hours and this may cause no significant difference of speed reduction.

Table 6. Mean difference of speed reduction between morning and noon observation

	Mean of Speed Reduction (km/hours)			
	Light Vehicles		Motorcycles	
	SRH	NSRH	SRH	NSRH
N Morning	58	74	62	46
N Noon	65	63	55	57
Mean Morning	24.0466	20.0541	28.8952	22.4130
Mean Noon	24.8154	20.5556	29.6636	22.1316
Mean Difference	-0.7688	-0.5015	0.7684	0.2814
Significant Level	0.248	0.166	0.248	0.598
Significant? (Yes/No)	No	No	No	No

Table 7. Mean difference of s'peed reduction between noon and afternoon observation

	Mean of Speed Reduction (km/hours)			
	Light Vehicles		Motorcycles	
	SRH	NSRH	SRH	NSRH
N Noon	65	63	55	57
N Afternoon	56	65	64	55
Mean Noon	24.8154	20.5556	29.6636	22.1316
Mean Afternoon	24.0446	19.8308	29.1094	21.8545
Mean Difference	0.7707	0.7247	0.5542	0.2770
Significant Level	0.220	0.024	0.384	0.598
Significant? (Yes/No)	No	Yes	No	No

Table 8. Mean difference of speed reduction between morning and afternoon observation

	Mean of Speed Reduction (km/hours)			
	Light Vehicles		Motorcycles	
	SRH	NSRH	SRH	NSRH
N Morning	58	74	62	46
N Afternoon	56	65	64	55
Mean Morning	24.0466	20.0541	28.8952	22.4130
Mean Afternoon	24.0446	19.8308	29.1094	21.8545
Mean Difference	0.0019	0.2232	-0.2142	0.5585
Significant Level	0.998	0.564	0.730	0.313
Significant? (Yes/No)	No	No	No	No

6 Conclusions and recommendations

From the results of the analyses, it can be concluded that:

1. Mean of speed reduction for standard road hump is significantly higher than mean of speed reduction for non-standard road hump for all type and vehicles and for all observation periods.
2. There were no significant difference of mean of speed reduction between pairs of observation period because of incorrect schedule of morning and afternoon observation (ending up with all observation periods belong to off-peak hours).

From the results of the analyses that showing relatively high speed exceeding speed limit for residential are, in Modern Land it is recommended to add more road humps, install speed limit signs and conduct public awaraness to reduce speed of motorized vehicles. For further research, it is important to conduct the survey at morning and afternoon peak hours to understand possible difference in the result compare to the off-peak hour. It is also recommended to measure the speed after the road hump to determine the suggested distance between adjacent road humps.

References

1. ____, Global Road Safety Partnership., *Speed management: a road safety manual for decision-makers and practitioners* (2008)
2. ____, OECD/ECMT Transport Research Center., *Speed Management report* (2006).
3. ____, Minister of Transportation Decree No. 111 Year 2015 on *Procedure of Speed Limit Determinaton for residential road* (2015)
4. ____, Minister of Transportation Decree No.3 Year 1994 on *Road User Control Equipment*, (1994)
5. ____, Minister of Transportation Decree No. 61 Year 1993 on *Traffic Signs on Street,* (1993)

Prediction model of cyclist's accident probability in the City of Malang

Imma Widyawati Agustin[1,*], *Muhammad* Zainul Arifin[2]

[1] Brawijaya University, Depart. of Urban and Regional Planning, Malang, East Java, Indonesia
[2] Brawijaya University, Depart. of Civil Engineering, Malang, East Java, Indonesia

Abstract. The development of transport is currently very rapid. This is because the increasing needs of the community, is directly proportional to the needs of public transport modes need to be able to support their activities. Malang is the second largest city in East Java province that has transportation problems such as traffic jams and high accident rate. The main purpose of the research is to create prediction model of cyclists' accident probability in the City of Malang. The research used frequency analysis and logistic regression analysis. The results showed that The prediction model of cyclists' accidents involving bicycle users in the City of Malang was formulated with $P = \frac{1}{1+e^{-(-1.047+1.103x_3+1,617x_6-1.965x_{15}+2.029x_{16}+1.057x_{24})}}$ with X_3 = income, X_6 = ownership of bicycle, X_{15} = behavior of check the brakes, X_{16} = behavior of check the tires, and X_{24} = cycling in groups.

1 Introduction

Traffic accidents according to the Law Number 22 [1] on The Road Traffic and Transportation is an unexpected and accidental incident on roads involving vehicles with or without other road users that result in human casualties and or loss of property. Accidents are influenced by several factors, including human factors, vehicle factors, road factors and environmental factors. Accidents due to human factors are influenced by the individual characteristics of the rider. Vehicle factors are affected by the reliability conditions of the components in the vehicle to maneuver when the vehicle is moving. Road factors are influenced by road infrastructure conditions that support comfort, safety and security in driving. While environmental factors are influenced by the bad weather around the crash site that can affect the driver's visibility in driving [2]. Accidents involving cyclists in Malang are 4 cases in 2014 - 2015 and in 2016 decreased to 3 cases of accidents. Generally cases of bicycle accidents in the city of Malang is happening on the road that does not have a special lane bicycle. Based on accident data from the police at least almost all road functions such as arterial road, collector, and environment there are cases of bicycle accidents [3].

The socio-economic characteristics of cyclist have an influence on bicycle ownership, movement characteristics, and behavior of cyclist. In this study the discussion of socio-

* Corresponding author: immasaitama@ub.ac.id

economic characteristics of cyclist includes age, gender, education, occupation, and income [4]. The movement was formed as a result of the activities carried out not in residence. This means that the relationship between the regions is instrumental in creating a space travel and distribution patterns of land use affect the travel patterns of people [5]. Characteristics of movement in this study focus on Purpose of movement, Distance, Travel time, Intensity of use, Start time activity [6]. Sulistio et al. [7] explains that the characteristics of riders based on their behavior are differentiated into driving preparation, driving attitude, driving experience, driving care, and knowledge in driving.

There are several problems related to cyclist accident in the City of Malang, traffic condition mixed between motorized and non-motorized vehicles so that all types of vehicles move together in the same lane or not separate. In addition there are many cyclists who do not use cycling safety equipment such as bicycle helmets (Figure 1a). Roads that currently operate are many that are incompatible with the hierarchy of road functions, especially for bike lanes where often these lanes are used by motor vehicles and used for other activities (Figure 1b). Cyclist behavior that could endanger the safety of themselves and motorists in the vicinity such as carrying cargo in bulk (Figure 1c).

(a) (b) (c)

Fig. 1. Some problems related to bicycle accident
Sources: Field survey, 2017

2 Methods

Model predictions accident opportunities in this study is a model developed to predict the chances of an accident that might occur in the future, with the variables used in this study is the characteristics of cyclist such as socio-economic characteristics, the characteristics of the movement, and behavioral characteristics. The characteristics of cyclists' accidents in this study were based on the experience of accidents during cycling that had been experienced by cyclist in the study area.

This research consists of several stages, firstly: the preparation stage. In this preparation stage, literature study is used as reference in the implementation of the study. It also determined the purpose of the study undertaken, as well as determine the related objects to be used for research. The second stage is the collection of data in the form of primary data and secondary data that support the study. Primary data were obtained from distributing questionnaires to respondents, interviews to interviewees and field surveys. While the secondary data obtained from related agencies such as the Regional Development Planning

Board in Malang, Transportation Department, Public Works Department, East Java Regional Police, Central Bureau of Statistics, and others. The third stage is the data analysis required in accordance with the purpose of the study that has been determined. Data analysis was done in the form of descriptive analysis of the variables obtained from the primary data and secondary data, and logistic regression analysis to create an accident probability model that can occur. The last stage is the preparation of the discussion and conclusions on the results of studies that have been done (Figure 2).

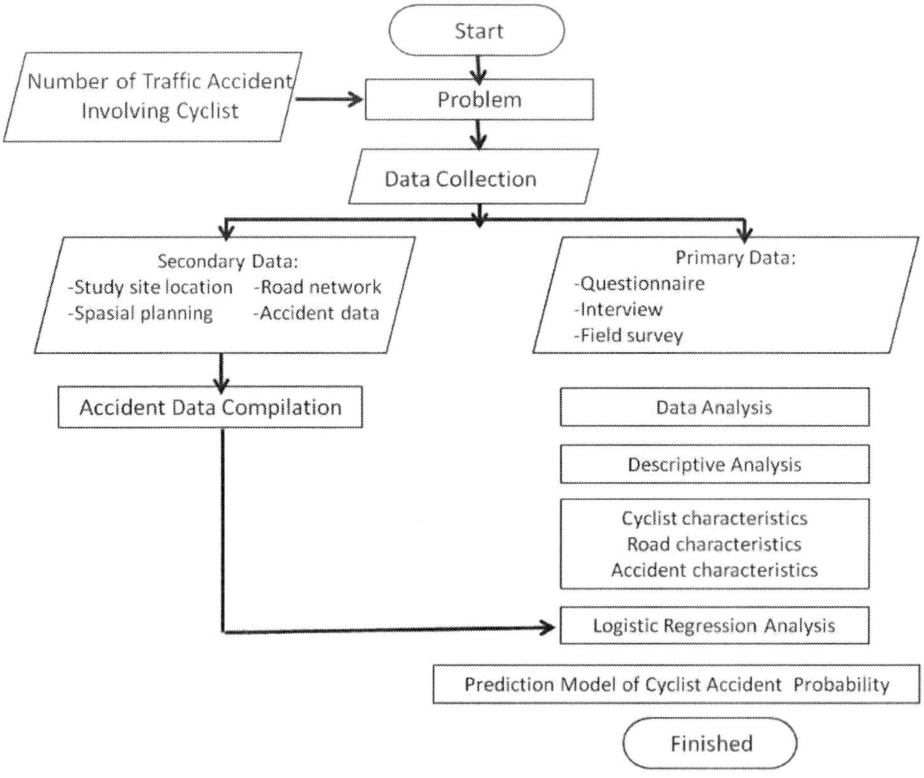

Fig. 2. Research framework

The stages taken to create a chance model of accidents are the determination of the respondent, in case the respondent is a cyclist in the study area who has been actively cycling once in the past 6 months. From the distribution of questionnaires in the study area then the data input into the SPSS program to then be tested between each explanatory variables with the variable response (univariate correlation) in this case is conducted by enter method in SPSS program. This identification is conducted by Chi Square Test, where each variable will be tested one by one to know whether the variable has an effect on the chance of having an accident.

The independent variables that have been shown to have significant effect will be reprocessed by the Wald Test. The variables in the study will be tested as a whole for the next estimation of a logistic model that aims to select the variables that have a significant effect on the chance of having an accident (Figure 3). This method works by issuing one by one variable that proved not significant (significance more than 0.05).

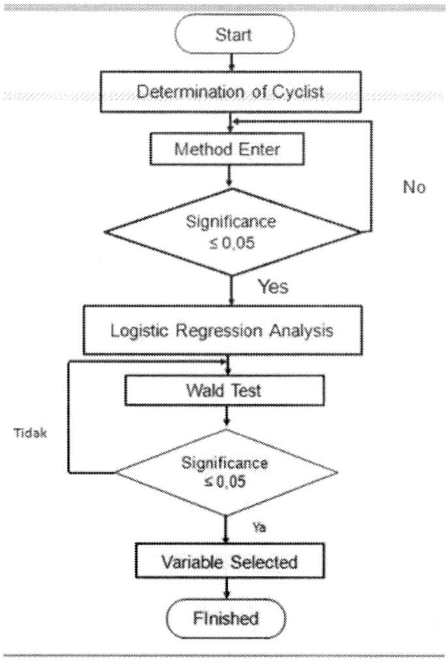

Fig. 3. Logistic regression analysis

According to Sulistio et al [7] the establishment of a logit model is based on the function of logistical opportunity that is specified as follows:

$$P_i = F(\beta_0 + \beta_1 X_{1i}) = \frac{1}{1+e^{-z}} = \frac{1}{1+e^{-(\beta_0+\beta_1 X_{1i})}} \ldots\ldots\ldots\ldots\ldots(1)$$

Furthermore, based on the formation of logit model according to Sulistio et al [7], the model structure used in this study is as follows:

$$P_{(BA)} = \frac{1}{1+e^{-(\beta_0+\beta_1 x_1+\beta_2 x_2+\beta_3 x_3+\cdots)}} \ldots\ldots\ldots\ldots\ldots(2)$$

Information:
$P_{(BA)}$ = chance of bicycle accident occurrence
e = natural number (2,71828)
β = explanatory variable coefficient (predictor)
X= explanatory variable (predictor)

3 Results and discussion

3.1 Characteristic of cyclist

Characteristics of cyclists based on socio-economic in the study area of Malang City conducted to 100 cyclists, obtained the following results:

- The age of majority of cyclists is 15-25 years old in the study area and the majority is male.
- The majority of cyclists have a recent education level of Senior High School / Vocational High School / Madrasah Aliyah, working as a student.
- Income of cyclists is around Rp.1.000.000 to Rp.2.000.000.

The majority of bicycle ownership is owned by themselves with 1 bicycle ownership. As for other vehicles, the majority of cyclists have motorcycles.

Table 1. Recapitulation of socio economic characteristic of cyclist in the City of Malang

Characteristics of Socio Economic	Results	Percentage (%)
Gender	Male	73
Age	15-25 years old	48
Education	SMU/SMK/MA	48
Occupation	Student	47
Income	Rp. 1.000.000 - Rp. 2.000.000	29
Ownership of bicycles	Owned by themselves	90
Ownership of bike	1	63
Amount of ownership	Motorcycle	57

Characteristics of cyclists based on movement are as follows:

- The majority of cyclists in Malang City use bicycles for sport purpose.
- The distance traveled by the majority of cyclists in the city of Malang is less than 5 km with travel time less than 30 minutes.
- The intensity of bicycle use by respondents for the study area of Malang City is uncertain.

Table 2. Recapitulation of movement characteristic of cyclist in the City of Malang

Characteristics of Movement	Results	Percentage (%)
Purpose and objective	Sport	28
Distance	< 5 km	66
Travel time	< 30 minutes	53
Intensity of use	uncertain	32
Start time activity	06.01-08.00 am	61

Based on behavioral aspects, the majority of cyclists in the City of Malang never check the bicycle lights, always check the condition of the brakes and bicycle tires and sometimes check the condition of the bicycle chain. While on driving, the majority of cyclists in Malang never wear helmets, never wear gloves, never wear elbow protector and never wear a knee protector. In addition, when driving, the majority of cyclists in the city of Malang also never rode (**"berboncengan"**), never cycling in groups, never frolic, do not break through the red light and never carry more goods / payload. From the analysis results also found that cyclists in Malang City sometimes wear brightly colored clothing while driving, sometimes preceding ("**menyiap**") another vehicle from the right, sometimes traveling when it rains and always give a sign when going to turn right (Table 3).

Table 3. Recapitulation of behavior characteristic of cyclist in the City of Malang

Focus	Results	Percentage (%)
Behavior before cycling		
Check the condition of the bike lights	never	52
Check the condition of the bicycle brake	always	45
Check the condition of the bicycle tire	always	60
Check the condition of the bicycle chain	occasionally	41
Behavior while cycling		
Wear a helmet	never	61
Wearing gloves	never	62
Wear a knee protector	never	78
Wear protective elbows	never	81
Wear bright / bright clothing	occasionally	51
Rode (**"Berboncengan"**)	never	81
Cycling in groups	never	53
Frolic	never	56
Break through the red light	never	79
Bringing goods / cargo in large quantities	never	79
Preceding (**"menyiap"**) another vehicle from the right	occasionally	47
Marking when it will turn right	never	42
Traveling by bicycle in the rain	occasionally	49

The characteristics of cyclist accident in the study area of Malang City are as follows:

- The majority of cyclist in the City of Malang had experienced accidents while cycling that is equal to 61 people from 100 cyclists.
- The disadvantages of accident costs in this study are differentiated from cost loss due to injury treatment and cost loss due to bicycle repair. The majority of cyclist in the City of Malang suffered losses due to costs of injury and loss costs due to bicycle repair for less than Rp. 50,000.
- The majority of bicycle accidents that happened to cyclists in the study area of Malang occurred in the afternoon with the majority of causes of accidents due to slippery or damaged roads.

3.2 Prediction model of cyclist accident probability

Prediction model of cyclist accident probability based on the resulting socio-economic aspects are:

$$P_{(BA)} = \frac{1}{1 + e^{-(0,844+1,128x_3-1,491x_4)}}$$

Prediction model of cyclists' accidents probability based on socio-economic characteristics in the City of Malang is influenced by X_3 (education) and X_4 (occupation).

Example calculation:

$$P_{(BA)} = \frac{1}{1 + e^{-(0,844+1,128(1)-1,491(0))}} = 0,783$$

Based on the example can be interpreted if the chance of accidents bike users in Malang based on socio-economic characteristics influenced by the level of education and work.

Bicycle users with education other than Senior High School/Madrasah Aliyah (MA)/Vocational High School and work as a student / students have an accident of 78.3 per cent.

Prediction model of cyclist accident probability based on movement aspects are:

$$P_{(BA)} = \frac{1}{1 + e^{-(1,386-1,272x_{12})}}$$

Prediction model of cyclists' accidents probability based on movement characteristics in the City of Malang is influenced by X_{12} (intensity of bicycle use).

Example calculation:

$$P_{(BA)} = \frac{1}{1 + e^{-(1,386-1,272(0))}} = 0,800$$

Based on the example can be interpreted if the chance of cyclists' accidents in Malang City based on the characteristics of the movement is influenced by the intensity of use. Cylist with the use of bicycles that are not necessarily a chance to crash by 80 per cent.

Based on the correlation test that has been done on the response variable and explanatory variable in the movement characteristics it is found that X_{15} (checking the brake condition), and X_{18} (wearing helmet) have an effect on the chance of the accident. There are two variables that proved significant that X_{15} and X_{18} then made a selection of variables that proved significant with the Backward Wald method. The value of regression coefficient (B) variable using helmet is -0.872. This negative regression coefficient (B) means that the less wearing of the helmet the more likely the accident. The significance value of these variables less than 0.05 indicates that these variables significantly influence the response variable. As for the establishment of accident opportunities utility is formulated by:

$$Z = 1,253 - 1,060\,X_{15}$$

Prediction model of cyclist accident probability based on behavior aspects are:

$$P_{(BA)} = \frac{1}{1 + e^{-(1,253-1,060x_{15})}}$$

Prediction model of cyclists' accidents probability based on behavior characteristics in the City of Malang is influenced by X_{15} (checking the bicycle brake).

Example calculation:

$$P_{(BA)} = \frac{1}{1 + 1 + e^{-(1,253-1,060(0))}} = 0,548$$

Based on the example can be interpreted if the chance of cyclists' accident in Malang based on behavioral characteristics influenced by the behavior when driving is checking the condition of the bicycle brake. Cyclist who drive while not checking the brake condition have the chance to crash by 54.8 per cent.

4 Conclusion

The results showed that the majority of cyclist in the city of Malang had an accident while cycling. The prediction model of cyclists' accidents involving cyclist in the City of Malang was formulated with $P_{(BA)} = \frac{1}{1+e^{-(-1,047+1,103x_3+1,617x_6-1,965x_{15}+2,029x_{16}+1,057x_{24})}}$ with X_3 = income, X_6 = ownership of bicycle, X_{15} = behavior of check the brakes, X_{16} = behavior of check the tires, and X_{24} = cycling in groups.

Acknowledgements

Special thanks to all member of Transportation Laboratory Department of Civil Engineering, EIIS Laboratory of Department of Urban and Regional Planning, Engineering Faculty, Brawijaya University for their support and kindly help to this research, especially for Debi and Rizaldy and all of Bicycle Research Team that we can not mention one by one, Thank you very much.

References

1. Republic of Indonesia. *Law of the Republic of Indonesia No.22 of 2009 on Traffic and Transportation.* (2009)
2. Warpani. *Management of Traffic and Road Transport*. Bandung: ITB. (1999)
3. East Java Police. *Traffic Accident Data*: http://korlantas.info
4. Patriya, Rukma Nur. *Characteristics of Motorcycle Users in Malang City. Thesis Not Published.* Malang: Universitas Brawijaya. (2008)
5. Tamin, Ofyar Z. *Transport Planning and Modeling.* Bandung: ITB Bandung. (2000)
6. Khisty, C. Jhotin, and Lall, B. Kent. *Fundamentals of Transport Engineering.* Bandung: Erland. (2003)
7. Sulistio et al, Harnen. Motorcycle *Accident Model On A Road.* Malang: Universitas Brawijaya. (2010)
8. Central Bureau of Statistics of Malang. *Malang City In Figures 2016*. Malang: Central Bureau of Statistics. (2016)

Ro – Ro Ferry safety: do the passengers aware the safety aspects during voyage?

Antoni Arif Priadi[1,*], *Lili* Purnama Sita[1], *Vidia* Selasdini[1], *Retno* Sawitri Wulandari[1]

[1] Sekolah Tinggi Ilmu Pelayaran Jakarta, Jakarta, Indonesia

Abstract. Beginning of 2017, the society had been shocked by the ship accident which occurred at Kepulauan Seribu Jakarta which had caused the death of dozens passengers on board who died are likely to be caused by panic and do not know what should to do. The objective of research is to gather the information related to the safety knowledge of Ro-Ro ferry passengers during voyage at sea. The gap analysis method is applied to find the differences between perception and expectation of Ro-Ro ferry passenger towards safety aspects. The result indicated that there is gap for all variables which can be categorized into three groups. The results give the important point for developing further approach to avoid similar incident in the. Further research on the detail approach need to be carried out which focus on the effective way in increasing the safety awareness of Ro- Ro ferry passenger.

1 Introduction

1.1 Background of the research

The first ship incident happened in Januari 2017, the KM Zahro Express ferry. She was carried over 200 passengers, caught fire during her way to Tidung Island, Thousand Islands. The National Disaster Mitigation Agency (BNPB) reported 23 passengers were killed while 17 people suffered injuries. It further said 194 passengers survived the incident while 17 others were reported still missing [1]. The Police already stated that the Master of KM Zahro Express as a suspect being considered to meet two elements of evidence, such as the ship's safety equipment and the responsibility on safety, security and the property of vessel, seafarer & cargo. The second vessel is KM. Mutiara Sentosa I which also caught fire in Masalembu Island in May 2017, caused 5 peoples died [2]. Result of the investigation from the KNKT said that the fire came from the lower deck from the truck. There were also differences from the numbers on the manifest to the number of peoples evacuated [3]. The similar accident occurred in April 2014 at South Korean vessel. The Sewol carried 476 persons when it took a capsized off the southwestern island of Jindo [4]. The victims is 304 loss of lives and of those, 250 were

* Corresponding author: puket1.stip.jkt@gmail.com

student. The question raised in these incidents is why there are a lot of passengers missing? Are they aware of the safety aspect on board the ship? Do they have sufficient knowledge on how to abandon the ship? Those questions lead to this research aimed to analyze the safety knowledge of the passenger during voyage by using gap analysis.

1.2 Previous research

Safety is a human perceived quality that determines to what extend the management, engineering and operation of a system is free of danger to life, property and the environment [5]. Another definition, safety is "the state of not being exposed to danger, a protective device (as on a weapon) to prevent accidental operation" [6]. Based on this definition, safety is really related to the perception of human towards a thing. For example, related to the voyage using a ship, the human will behave a safety matter regarding their perception towards safety on board ship.

A research related safety perception showed that eventhough the overall safety knowledge of the passengers can be considered good, but there is some differences between groups of passengers [7]. Younger passengers and passengers on shorter trips generally have less safety knowledge than older passengers and passengers on longer trips. Their study also addressed the effects of two different formats of presenting safety relevant information on passengers' perception of trust, safety and risk on board. Another research identified five safety perception themes. [8]. The results specify that passengers perceive safety through the architecture of the passenger ship, the life-saving appliances, communication, emotions and other people. The article gives to safety research on passenger ship design, where human perceptions and reactions to the surrounding environment significantly affect behavior.

1.3 Research methodology

The methodology applied for this research is gap analysis. The gap analysis is the comparison of the perception to the expectation on safety related issues. Gap analysis consists of several steps. The first step is the data collection method, followed by validity test, reliability test and at the end the gap analysis is applied.

Gap analysis is used to determine the steps which is need to be taken for further movement from recently condition to the expected condition in the future. The gap analysis is also a means of ccomparasion between actual performance and future performance. This gap analysis also identifies the actions plan to achieve the future performance [9].

Sample of data collection was done through random sampling for the ship passenger who travelling from Port of Merak to Bakauheni. To decide the number of sample from the population, the table and guidance is employed. In general, the minimal number of sample for correlation research is 30 [10]. The primary data used in this research is obtained from survey with questionnaires distributed in June 2017.

Having primary data, the validity and reliability test are conducted. Validity test is used to measure validity of a questionnaire. A questionnaire is stated valid if the question of the questionnaire could explore things which will be measured by such questionnaire. The validity test is applied through the comparison of the calculated r with r from table. If the calculated r for each r questionnaire items values a positive and has higher values then it could be stated that the questionnaire items is valid [11].

The reliability is the instrument to measure a questionnaire which is an indicator of variable [12]. High reliability measurement means that the measurement is able to provide reliable result. A questionnaire items is said to be reliable if the answer of respondent to the questionnaire is consistent. To determine the coefficient of reliability, the Cronbach's Alpha is applied. If the value of Cronbach's Alpha is more than 0.60, it means that the variable is reliable. For this purpose, the SPSS version 18 is used.

To do the gap analysis between perception and expectation, the difference average values of each perception and expectation items is calculate which expectation values – perception values [13].

$$Q = P(\text{Perception}) - E(\text{Expectation}) \quad (1)$$

The safety matters for this research consists of 16 variables. The variables name is started from passenger ticket checking (V1), passengers luggage checking (V2), passengers room direction (V3), muster station direction (V4), location of muster station (V5), Alarm sign (V6), Public addressor volume (V7), abandon instruction (V8), information for evacuation (V9), Crew identity (V10), Lifejackets location (V11), Lifejackets instruction (V12), Instruction to use lifejackets by crew (V13), Firefighting apparatus location identification (V14), Firefighting apparatus instruction (V15), and Instruction to use firefighting instruction (V16).

2 Discussion

2.1 Port of Merak

The Port of Merak is a seaport located in front of the Pulau Merak Besar sub district of Cilegon, Banten, on the Nnorthwestern tip of Java, Indonesia. The port is a key transport link between Java and Sumatra and is a major service provider for the heavy passenger and commercial ferry traffic from Merak to Bakauheni across the Sunda Strait on the Ssouthern tip of Sumatra. The overview of Port of Merak is illustrated in figure 1.

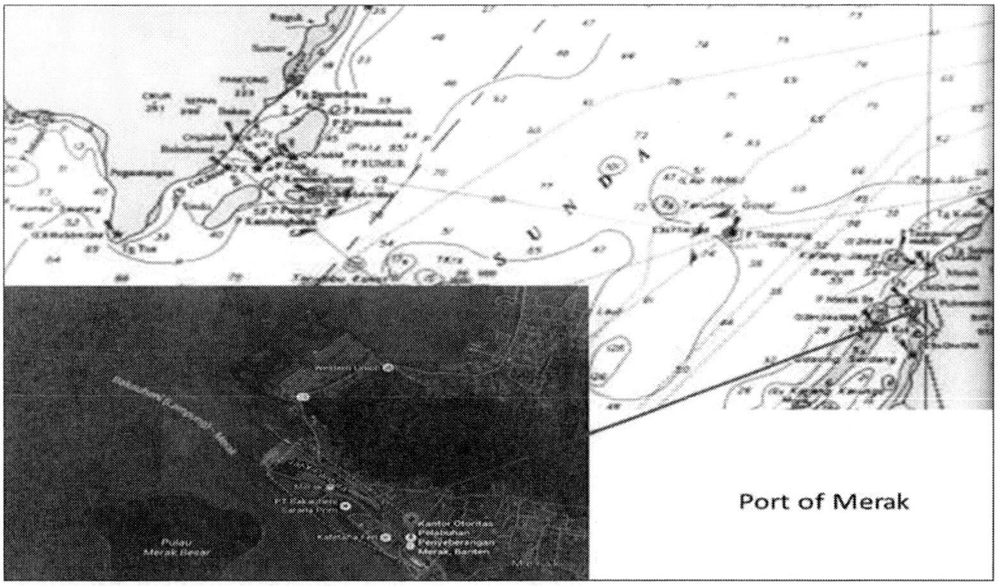

Fig. 1. Overview of Merak Port

Port of Merak consists of five jetty. The nearest one to the entrance gate is jetty no. 1. The furthest jetty is located at jetty no. 5. The layout of Port of Merak is illustrated in figure 2.

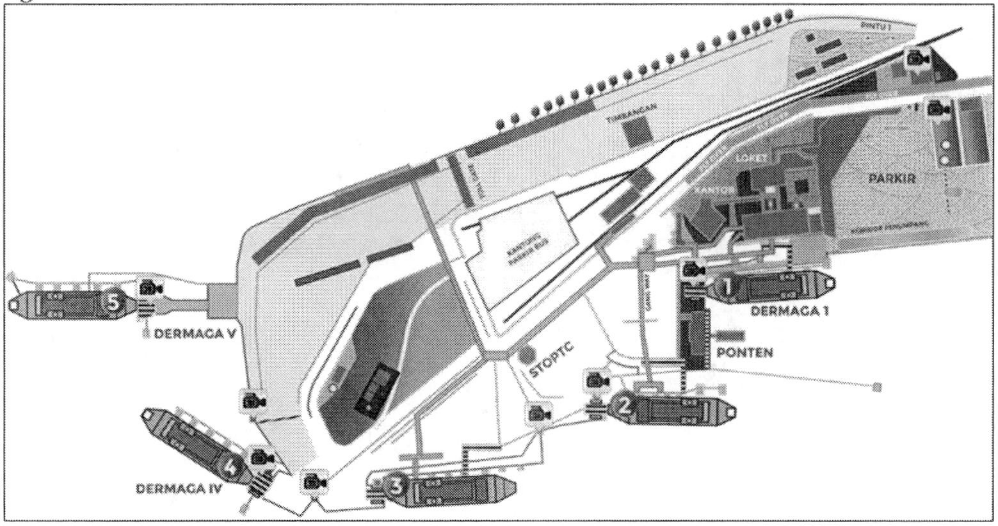

Fig. 2. Layout of jetty at Merak Port
Source :http://bestananda.blogspot.co.id/2015/07/pelabuhan-merak-banten.html

Growth in demand for ferry services has been rapid in recent years. For instance during holiday such as Eid Fitr 2017, there are more than 200 thousands passenger travelling by ship from Merak to Bakaheuni and vice versa. This number of passenger is counted for several days during Eid Fitr 2017. Additionally, more passengers travel together with their car or motorcycle. The detail of port production is figured out in table 1.

Table 1. Number of cargo during eid fitr 2017

Review of Tranportation Production During Eid Fitr Holiday for Merak Port, 2017		
No	**Type of Cargo**	**2017**
1	Passanger	256.638
2	Passenger with car	1.180.912
3	Motorcycle	105.601
4	Cars	186.182
5	Bus	9.323
6	Truck	27.724

Source : modified from http://oppmerak.dephub.go.id/#link

2.2 Respondent profile

The respondent profile includes the respondent age, education background, employment status and travelling number by ship. The total number of respondent is 115 passengers. For the respondent age, 66 % of passengers have age less than 31 years old. The educational background, the 51 passenger graduated from senior high school and 33% graduated from higher education. Further, regarding the employment status, 60% of passenger worked for private sector and 14% worked as civil servant. To the last respondent profile, the travelling number on the same route, the 65% of passenger have previous experience between once and three times. For four times to six times, there are only 22% passengers accounted.

2.3 References

Based on the methodology, the first step of analysis is the validation of data. The validation of data consists of the perception data and expectation data. The result of Pearson Correlation is used to determine the validation of data. The result of validation test is figured out in table 2. The 16 variables are valid for this research.

Table 1. The validation test result

No.	Variables	Corr Perception	Corr Expectation
1	Check ticket	,539**	,670**
2	Check luggage	,427**	,655**
3	Direction to passenger space	,574**	,669**
4	Direction to muster	,652**	,686**
5	Identification muster	,626**	,714**
6	Identification muster	,704**	,767**
7	Public addressor	,464**	,675**
8	Emergency guidance	,701**	,767**
9	Evacuation information	,631**	,736**
10	Crew identity	,704**	,682**
11	Lifejacket location	,761**	,538**
12	Lifejacket instruction	,718**	,722**
13	Crew inform use lifejacket	,658**	,700**
14	Fire fighting app	,571**	,643**
15	Fire fighting instruction	,592**	,724**
16	Crew inform use ff app	,696**	,637**

The next test is the reliability test. The first test is applied for the perception of respondent as the result is shown in table 3. Based on the value of Cronbach's Alpha which is 0.896, it means the respondent perception is reliable for this research. Number of respondent are 115 and there are 16 questions in the questionnare, total number data collected are 1.840 to analyze. This quistionarre is arranged to identify the requirements of passenger's safety on board.

Table 2. Reliability test for perception

Cronbach's Alpha	Cronbach's Alpha Based on Standardized Items	N of Items
,896	,900	16

The similar value is obtained for expectation of respondent as it is shown in table 4. The value of Cronbach's Alpha is 0.926. Therefore, it could be stated that the data is reliable for this research.

Table 3 Reliability test for expectation

Cronbach's Alpha	Cronbach's Alpha Based on Standardized Items	N of Items
,926	,927	16

The The gap analysis result is illustrated in figure 3. Three areas of variable is identified having highest value such as the checked ticket before entering ship (V1), checked baggage before entering ship(V2) and evacuation information (V9). The passenger expects that luggage/baggage need to be checked for safety reason as well as passenger ticket. The evacuation information to the passenger is also a matter that the passenger expects have right information if there is signal to head muster station.

The second group consists of five areas which are fire-fighting apparatus instruction (V15), Crew identity (V10), Lifejackets location (V11), Instruction to use lifejackets by crew (V13), and muster station direction (V4). The third group has lower value of gap which are lifejackets instruction (V12), instruction to use fire-fighting instruction (V16), passengers room direction (V3), location of muster station (V5), alarm sign (V6), public-addressor volume (V7), and fire-fighting apparatus location identification (V14). For the third group, most of passenger aware with it, yet some passenger feel need for more information.

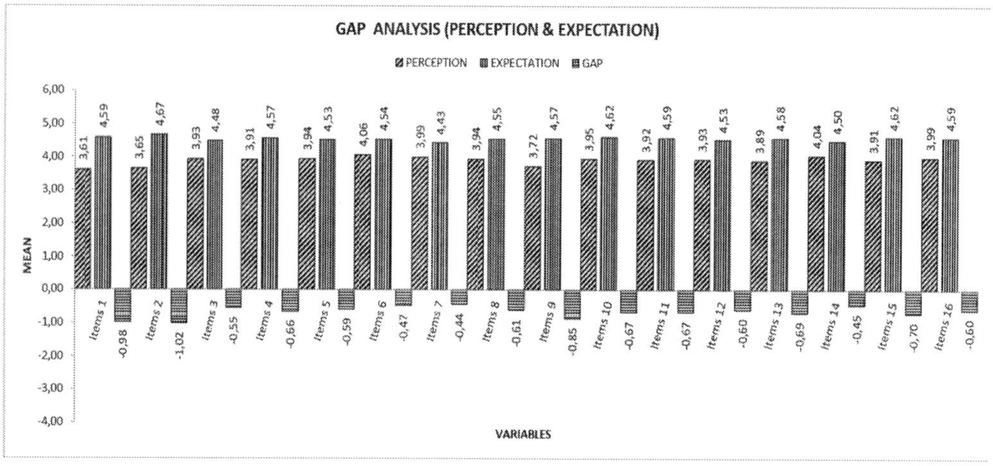

Fig. 1 Gap Analysis Result

3 Conclusion

Refer to the discussion above, the overall result of the gap analysis shows that for the safety related matter, there is gap for all variables. It means that there is need to give the more information to the passenger as actually they have high expectation to the safety related matter during voyage at Ro- Ro ferry. It could be stated that the passenger have high awareness level which is shown through the expectation level. It also could be stated that method to give more information of related safety need to be carried out not only by crew but also by the port authority through brochure, ticket, etc. While the crew member need to aware this matter to focus more on the safety knowledge of the passengers before departing port. The risk of accident of the ship during voyage is always there. When this situation occurs, the readiness of the passengers to counter the situation really depends on the passenger's knowledge of the safety related matter. Otherwise, the passengers will not know what to do and later just jump to the sea. The shipping companies also need to encourage and to establish the procedure of safety related matter to the passengers through the social engineering for instance, the life jacket is available at the passenger's seat and should be put on or the passengers is not allowed to sit. The shipping companies need to give more efforts to inform the passenger about emergency guidance, evacuation information and lifejacket instruction as the result of analysis. Further research on the different perception of safety by education parameter need to be carried out for finding relationship between safety related matter and education level.

References

1. The Jakarta Post. "ministry-to-standardize-thousand-islands-service-after-zahro-express-incident." thejakartapost.com. Retrieved June 10, 2017 (www.thejakartapost.com/news/2017/01/02/ministry-to-standardize-thousand-islands-service-after-zahro-express-incident.html). (2017)
2. CNN Indonesia. "penumpang-km-mutiara-sentosa-1-sudah-dievkuasi." www.cnnindonesia.com. Retrieved June 10, 2017 (www.cnnindonesia.com/nasional/ 20170520094349-20-216076/penumpang-km-mutiara-sentosa-1-sudah-dievkuasi/). (2017)
3. Nasional.tempo.co. "www.nasional.tempo.co." km-mutiara-sentosa-i-terbakar-knkt-selidiki-truck-sumber-api. Retrieved June 10, 2017 (www.nasional.tempo.co /read/news/2017/05/21/058877324/km-mutiara-sentosa-i-terbakar-knkt-selidiki-truck-sumber-api). (2017)
4. Dailymail.co.uk. "http://www.dailymail.co.uk." Eerie-pictures-untouched-classroom - belonging-students-teachers-killed-Sewol-ferry-disaster-South-Korea. Retrieved June 10, 2017 (http://www.dailymail.co.uk/news/article-3605976/Eerie-pictures-untouched-classroom-belonging-students-teachers-killed-Sewol-ferry-disaster-South-Korea.html). (2016)
5. Kuo, C. *Safety Management and Its Maritime Application*. London: The Nautical Institute. (2007)
6. Merriam-webster.com. n.d. "safety." www.merriam-webster.com/thesaurus. Retrieved June 10, (http://www.merriam-webster.com/thesaurus/safety). (2017)
7. Hystad, S.W., Oyeniyi S.O., and Jarle, E. *Safe travel: Passenger assessment of trust and safety during seafaring*. Transportation Research Part F 38:29–36. (2016)
8. Ahola, Markus, Pekka, M., Pentti, K., and Juhani, P. *Perceiving safety in passenger ships – User studies in an authentic environment*. Safety Science 70 (2014) 70 :222–232. (2014)
9. Suroto, A. *Gap Analysis*. http://sis.binus.ac.id. Retrieved June 10, 2017 (http://sis.binus.ac.id/2015/07/28/gap-analysis-analisa-kesenjangan/). (2015)
10. Sekaran, U. *Metodologi Penelitian untuk Bisnis*. Jakarta: Salemba Empat. (2006)

11. Sujarweni, V. W. *SPSS Untuk Penelitian*. Yogyakarta: Pustaka Baru. (2015)
12. Sunyoto, D. *Analisis Regresi dan Uji Hipotesis*. Yogyakarta: Media Pressindo. (2009)
13. Wijaya, T. *Manajemen Kualitas Jasa*. Jakarta: PT. Indeks. (2011)

Safety inspection on level crossing JPL 727 km 537+453 Patak-Pathukan Road, Sleman, Yogyakarta

Dian Setiawan M[1,*], *Mayang* Sari[1], and *Noor* Mahmudah[1]

[1]Muhammadiyah University of Yogyakarta, Department of Civil Engineering, Yogyakarta, Indonesia

Abstract. Level crossing (LC) safety inspection between a highway and a railroad on Pirak-Pathukan Road, Sleman, Yogyakarta is necessary because this LC is located near various community centers, has high traffic volume, intersects with railway double tracks, and intersecting angle that is not perpendicular. This study aims to evaluate LC technical condition, analyse delay time and vehicles queue length, and evaluate pavement structure condition. The research results indicate: 1) LC technical conditions do not meet the requirements of Regulation of Director General of Land Transport No. 770 Year 2005; 2) The longest duration of LC gate closure occurred on Sunday at 15:05, that is 360 seconds, the highest traffic flow occurred on Monday (from the South side) that is equal to 1443 skr/day, and the longest delay time occurs on Sunday at 15:05, that is 498 seconds; 3) The value of pavement condition index (PCI) is 82% very good.

1 Introduction

According to Undang-Undang No. 22 year 2009, Traffic safety is a circumstance to avoid a person from accident risk during traffic that can be caused by humans, vehicles, and/or environment. Trains become one of transportation modes that provides several benefits to the community because the travel cost is relatively cheap, can transport many people or goods related its capacity, efficient in time, safe and comfortable [1]. There are some factors that affecting accidents at highway and railroad crossings, which are vehicles, drivers, natural and weather conditions, signs and markings, level crossing geometric design, and road pavement conditions [2]. The safety inspection on a level crossing is a systematic examination of road and railway track on a level crossing to identify hazards, errors and deficiencies that can cause accidents. Inspection activities are usually conducted by PT. Kereta Api Indonesia or Ministry of Transportation.

The city of Yogyakarta has a number of level crossings, one of which is a level crossing that located at Pirak-Pathukan Road, Sleman, Yogyakarta. This level crossing is located in the area that close to various community activity centers, such as Gamping market, Gamping sub-district office, Godean sub-district office, PKU (Public Health Center) of Muhammadiyah Gamping, Tamantirto village hall, Mejing II Primary School, State Junior

* Corresponding author: diansetiawanm@ft.umy.ac.id

High School 1 Kasihan, High School State 1 Godean, and Universitas Muhammadiyah Yogyakarta. In addition, this location is also adjacent to the national road of Wates Road in the South side and Godean Road in the North side. The purpose of this study is to inspect

the safety of level crossing in Pirak-Pathukan Road, Sleman, Yogyakarta on JPL 727 KM 537 + 453. This study aims to evaluate technical condition of level crossing, analyse delay time and length of vehicles queue that occurring as the impact of level level crossing gateclosure, and to evaluate pavement structure condition using Pavement Condition Index (PCI) method. Secondary data in the form of train departure and arrival schedule information and list of level crossing in the Special Region of Yogyakarta obtained from PT. KAI DAOP VI Yogyakarta.understanding that surface run-off has to discharge immediately through drainage channel. If something occurs on the channel, it is the beginning of the continuous destruction of the road.

One solution offered is what people called as Sustainable Urban Drainage Systems (SUDS), i.e. a system which consists of one or more structures developed to manage surface water run-off [3]. The basic concept of this system is to improve water usefulness, to minimize the loss, and also to improve environmental conservation. One of the efforts is to detain or reduce the speed of the surface water run-off when rainfall and immediately infiltrate the water into soil. As a whole, SUDS can be categories into several rainfall retention facilities, as seen in Figure 1.

2 Sustainable road drainage system concept

2.1 Level crossing

Based on the Regulation of Director General of Land Transport No. 770 Year 2005 about technical guidelines for level crossing between road and railway track, the requirements for the construction of a level crossing includes [4]:

1) Road surface is not higher or lower than rail head (0.5 cm tolerance).
2) There is a 60 cm long surface measured from outer side of rail track.
3) The maximum gradient for vehicle passing (calculated from the highest point on the rail head) is 2% measured from outer side of the flat surface and 10% for the next 10 meters.
4) The maximum width of level crossing for one line is 7 meters.
5) The angle of intersection between rail track and road is at least 90 degrees and the length of the straight road must be at least 150 meters from rail track.
6) Must be equipped with an opponent rail or other construction to ensure a groove for the train wheel.
7) Road segment that can be crossed between a road and a railroad has the following requirements: Class III, at least has 2 lanes and 2 directions, not on the curve, vertical alignment less than 5% from the rail track outer point, fulfill the visibility requirement, and in accordance with the General Plan of Spatial Planning (RUTR).

2.2 Traffic flow

Traffic flow (Q) is expressed in units of light vehicles (skr) per hour for one or more periods, for example during peak periods of morning, noon, or afternoon. Traffic flow is converted from vehicle per hour to skr per hour by using the equivalent value of light vehicles (ekr). The ekr values for each vehicle type are described in Table 1 below.

Table 1 Value of light vehicle equivalent (EKR)

Vehicle Type	EKR Value
Light Vehicles (LV)	1.0
Heavy vehicles (HV)	1.3
Motorcycle (MC)	0.4

Source: Indonesian Road Capacity Guidance (PKJI) (2014)

$$Q = Q_{LV} \times ekr_{LV} + Q_{HV} \times ekr_{HV} + Q_{MC} \times ekr_{MC} \quad (1)$$

Where:
Q : Traffic flow (skr/jam)
QLV : Light vehicles flow
QHV : Heavy vehicles flow
QMC : Motorcycles flow
ekr : Equivalent value of light vehicles

2.3 Delays

According to the Indonesian Road Capacity Guidance [5], delay calculations include traffic delays and geometric delays. At a level crossing, the delay is affected by a *stopped delay* as a geometric delay and congestion delay as a traffic delay. Systematically can be expressed as follows:

$$T = T_L + T_G \quad (2)$$

Where:
T_L : Traffic delay
T_G : Geometric delay

2.4 Queue length

According to the Indonesian Road Capacity, the queue length is a vehicle queuing along the level crossing approach and expressed in meters. The queue length starts to be measured when the level crossing gateis closed until the level crossing gateis opened.

2.5 Pavement condition index (PCI) method

According to Shahin [6], the assessment of pavement structures condition by Pavement Condotion Index (PCI) method is assessed based on the type of damage and the level of damage, and it can be used as a reference in maintenance efforts. Stages of calculation of pavement condition assessment as follows:

Density

Density is the percentage of damage to the total area of a type of damage to the area in a research unit measured in square meters or long meters. To calculate the value of density used the following formula:

$$Density = \frac{Ad}{As} \times 100\% \quad (3)$$

or

$$Density = \frac{Ld}{As} \times 100\% \quad (4)$$

with:
Ad = Total of damage area (m^2)
Ld = Total of damage length (m)
As = Total area of segment unit (m^2)

2.6 Deduct value

Deduct Value is a reduction value for each type of damage obtained from the graph that shows density and deduct value relationship.

2.7 Total deduct value (TDV)

Total Deduct Value is the total of deduct value for each type of damage and level of damage in a segment unit.

2.8 Corrected deduct value (CDV)

Corrected Deduct Value obtained from the graph that shows relationship between TDV and CDV by choosing the suitable graph with number of *Deduct Value* that more than score 2.

2.9 Pavement condition index (PCI)

Once the CDV is known, then the PCI value for each research unit or segment can be calculated by the following equation:
PCIs=100-CDV (5)
Where:
PCIs = PCI for each segment unit
CDV = CDV for each segment unit

2.10 Rating

The quality of pavement structures can be determined by inputting PCI values that have been calculated into PCI levels, i.e Failed (0-10%), Very Poor (11-25%), Poor 26-40%), Fair (41-55%), Good (56-70%), Very Good (71-85%), and Excellent (86-100%) [6].

3 Research stages

This research was conducted at level crossing JPL 727 STA. 537 + 453 in Pirak-Pathukan Road, Sleman, Yogyakarta (Figure 1).

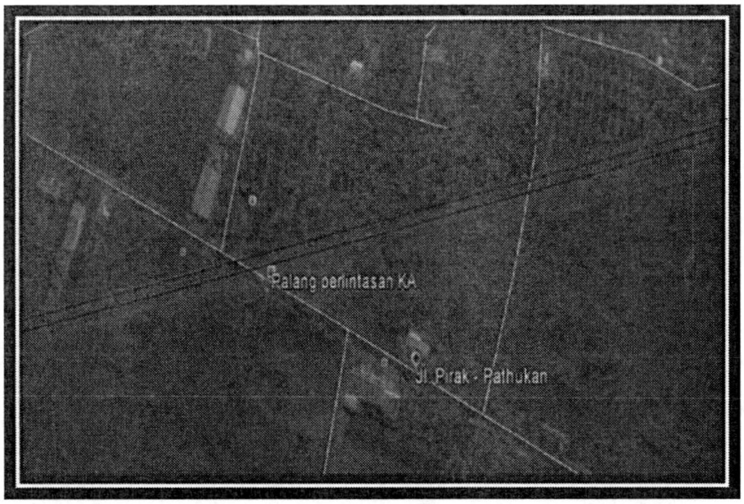

Fig. 1 Research sites

The primary data were obtained by conducting the following survey:

3.1 Infrastructure condition

The survey was conducted by direct observation along the segment of highway and railway line related to the signs, markings and traffic signal lights.

3.2 Level crossing geometric

The geometric survey was carried out by using the Garmin 76csx GPS tool to obtain the intersection angle of level crossing.

3.3 Crossing door closure duration

The survey was conducted to determine the time length of level crossing gateclosure.

3.4 Vehicle time-delay

The vehicle time-delay survey is conducted to obtain the time it takes for the vehicle to pass through a nuisance and obstacle at a level crossing.

3.5 Vehicle queue length

The queue length survey is intended to obtain the queue length that occurs at one time closure of the crossing gate.

3.6 Pavement structure deterioration

The survey of road pavement structural damage is done by direct observation along the highway under review.

3.7 Analysis of level crossing technical conditions

A comparative stage between the parameters in the Regulation of Director General of Land Transport No. 770 Year 2005 about technical guidelines for level crossing between Road and Railway Track with the existing technical condition of level crossing on Pirak-Pathukan Road, Sleman, Yogyakarta (JPL 727 KM 537 + 453) can be seen in Table 2 below.

Table 2. Result of technical condition analysis

Parameters	Technical Standards SK Dirjen No. 770 (2005)	Results in the Field	Information
Train Schedule	Must fit Gapeka	Often late	Does not meet the requirement
Intermittent time between trains during peak hour	Minimum 6 minutes	Less than 6 minutes	Does not meet the requirement
Number of passing trains	25 – 50 trains/day	89 trains/day	Does not meet the requirement
Level crossing location	Avoid curve in the horizontal alignment	Not in curve	Meet the requirement
Road class	Class III	Class III	Meet the requirement
Rail track elevation	Rail head and road surface at the same height level	Rail head&road surface has different height level	Does not meet the requirement
Level crossing width	One line must be ≤ 7 m	One line 5 m	Meet the requirement
The angle of intersection between rail track & highway	90^0	33^0	Does not meet the requirement
Distance between consecutive level crossings	> 800 meter	500 meter	Does not meet the requirement
Average Daily Traffic	300-500 vehicles/day	2956 vehicles/day	Does not meet the requirement
Signs	Must be equipped with warning or prohibition traffic signs	There are no warning signs No. 23 and 24 as well as prohibition signs No. 5c	Does not meet the requirement

Table 2. Result of Technical Condition Analysis (Continued)

Parameters	Technical Standards SK Dirjen No. 770 (2005)	Results in the Field	Information
Markings	Must be equipped with road markings	No transverse and symbol markers, rapping ribbons, and medians	Does not meet the requirement
Traffic light	Two red lights flashing and sound	Two red lights flashing and sound	Meet the requirement
Physical and non physical facilities at level crossings	Guard posts, JPL officers, semboyan lists, gapeka lists	Everything is available	Meet the requirement

3.8 Impact analysis of level crossing closure

The steps that required to obtaining delay duration and vehicle queue length due to the closure of level crossing gate are as follows:

3.9 Closure duration

A survey of level crossing gate closure duration is done to find variations of level crossing gate closure duration caused by the passing train. The data of level crossing gate closure duration for two days of observation are shown in Figure 2 below.

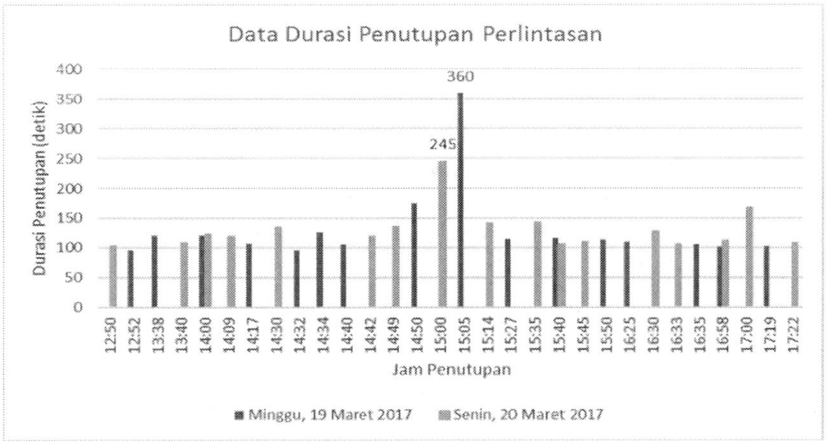

Fig. 2 Level crossing gate closure duration

3.10 Traffic flow

The data of traffic flow survey result at level crossing is shown in Figure 3.

3.11 Delay Time and vehicle queue length

The queue length varies on each approach lane of level crossing for each level crossing gate closure time. The data of delay time can be seen in Figure 4.

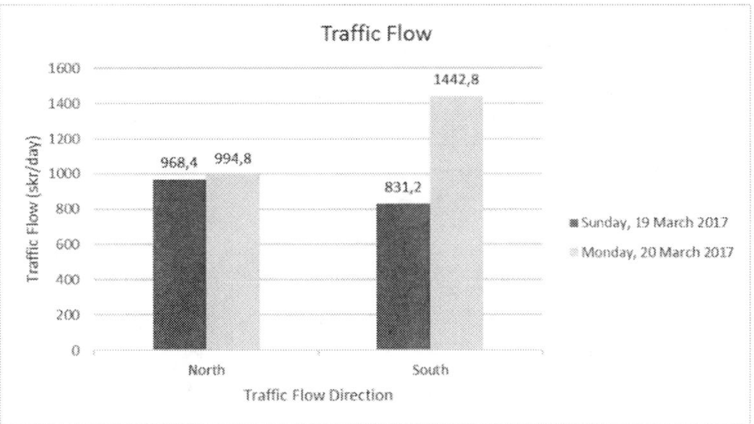

Fig. 3 Traffic flow

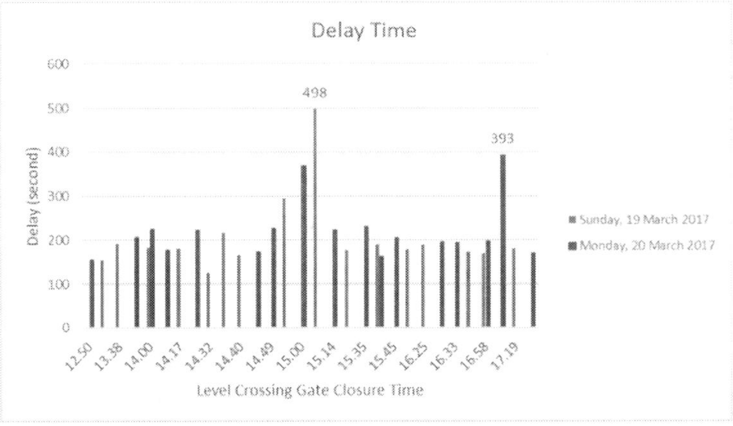

Fig. 4 Delay time

From the survey of delay time and vehicles queue length that conducted for 6 hours from 12:00 to 18:00 on Sunday, 19 March 2017, it can be obtained the result that the longest delay time occurred at 15:05 for 498 seconds. The delay time is influenced by the duration of the level crossing gate closure due to the passage of two trains at almost the same time, causing the vehicle queue length is 65 meters in the North side and 87 meters in the South side of level crossing. Furthermore, from the survey of delay time and vehicles queue length that conducted for 6 hours from 12:00 to 18:00 on Monday, 20 March 2017, it can be obtained the results that the longest delay time occurred at 17:00 for 393 seconds which is causing the vehicle queue length at that hour is 31 meters in the North side and 125 meters in the South side of level crossing.

3.12 Analysis of pavement structure condition

Table 3 PCI survey form

	ASPHALT PAVEMENT SKETCH: CONDITION SURVEI DATA SHEET FOR SAMPLE UNIT				SKETCH:	
1.	Alligator cracking	(m^2)	7. Weathering / Raveling	(m^2)	14. Railroad Crossing	(m^2)
2.	Block Cracking	(m^2)	8. Polished Agregat	(m)	15. Rutting	(m^2)
3.	Edge Cracking	(m)	9. Potholes 10. Patching and Utility Cut Patching	(m^2) (m^2)	16. Corrugation	(m^2)
4.	Joint Reflection Cracking	(m)	11. Bleeding	(m^2)	17. Shoving	(m^2)
5.	Slippage Cracking	(m^2)	12. Depression	(m)	18. Swell	(m^2)
6.	Long and Trans Cracking	(m^2)	13. Lane/Shoulder Dropp-Off	(m)	19. Bumps and Sag	(m^2)

STA	*Distress* *Severity*	*Quantity*					Total	*Density* (%)	*Deduct Value*	Total (DV)
0+125	10 (L)	35					35	9,33	17	35
s/d	11 (M)	0,36	0,48				0,84	0,22	2	
0+200	12 (L)	3					3	0,8	8	
	14 (L)	9	9				18	4,8	8	

Based on a survey conducted on a road segment (200 meters to the North and 200 meters to the South) of level crossing JPL 727 STA. 537 + 453, it was observed that there were road damages at STA. 0 + 125 up to STA. 0 + 200. This damage has been analysed using PCI method. The result of pavement condition analysis by PCI method can be seen in Table 3.

Based on calculations, Total Deduct Value (TDV) is 35 with Q value is 3, resulting Corrected Deduct Value (CDV) is 18 from the CDV chart. Furthermore, the pavement condition value is obtained by subtracting 100 (one hundred) with the value of CDV that has been obtained. If the CDV value is known, then the PCI value of each segment unit can be calculated by the formula (5).

$$PCI_{(S)} = 100 - CDV \tag{6}$$

= 100 – 18 = 82

From the PCI value of each segment unit, it can be concluded that the amount of PCI value of this road segment as in the following figure.

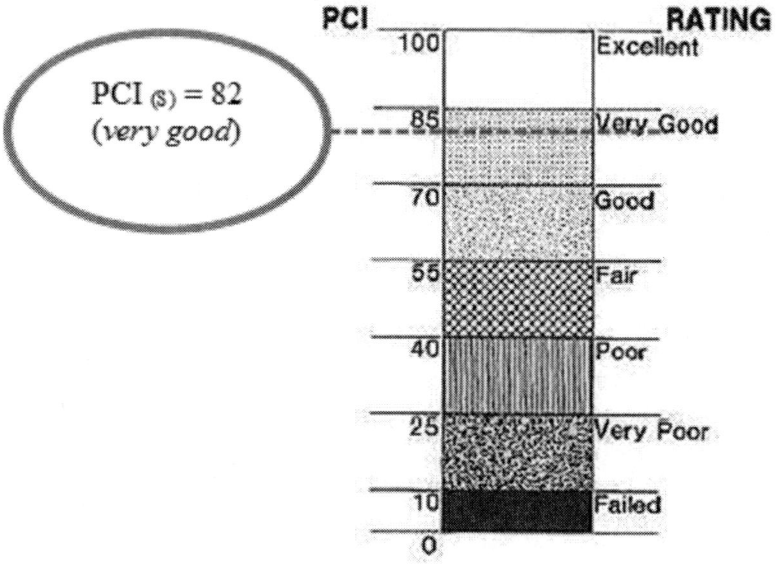

Fig. 5 PCI value chart

4 Conclusion

Based on the research that has been done, obtained the following conclusions:

1. The level crossing at Pirak-Pathukan Road, Sleman, Yogyakarta on JPL 727 KM 537 + 453 according to the Regulation of Director General of Land Transport No. 770/2005 about technical guidelines for level crossing between road and railway track does not meet the requirements due to the passing train journey is often late, time intervals between trains during peak hours is less than six minutes, number of trains passing is more than 50 train/day, rail head and road pavement is in different elevation, distance between the consecutive level crossings is less than 800 meter, average daily traffic is 2956 vehicles /day, signs and markings are incomplete, and intersection angles between road and rail track is not perpendicular that could provide negative impact both for vehicle and train driver that passing this level crossing.

2. Some parameters are analyzed as the impact of level crossing gate clossing, including as follows:
   a. The longest closure duration of level crossing gate occurred on Sunday, March 19, 2017 at 15:05 for 360 seconds, while on Monday, March 20, 2017 occurred at 15:00 for 245 seconds.
   b. The highest traffic flow that occurred on Sunday, March 19, 2017 comes from the North side that is equal to 1932 vehicles/day or equal to 969 skr/day. Furthermore, the highest traffic flow that occurred on Monday, March 20, 2017 comes from the South side that is equal to 2956 vehicles/day or equal to 1443 skr/day.
   c. The longest delay time on Sunday, March 19, 2016 occurred at 15:05 for 498 seconds with a queue length is 65 meters from the North side and 78 meters from the South side. Moreover, the longest delay time on Monday, March 20, 2017 occurred at 17:00 for 393 seconds with a queue length of 31 meters from the North Side and 125 meters from the South Side.
3. There are 4 types of damage on the road segment JPL 727 KM 537 + 453 along the 200 meters North-South direction, which are : patching (9.33%), bleeding (0.22%), depression (0.8%), and railroad crossing (4.8%). The value of the pavement conditions index (PCI) is 82% therefor it can be concluded that the road pavement is in very good condition.

Reference

1. Utomo, S.H.T. *Jalan Rel.* Beta Offset: Yogyakarta. (2013).
2. Hasan, B. *Evaluasi Kelayakan Perlintasan Sebidang Studi Kasus Perlintasan Sebidang Patukan Gamping Sleman Yogyakarta.* Tugas Akhir S-1 Jurusan Teknik Sipil, Fakultas Teknik UGM: Yogyakarta. (2009).
3. Estonia Latvia Programme, *Handbook on Sustainable Urban Drainage Systems, European Union*. (2013).
4. Direktorat Jenderal Perhubungan Darat. *Pedoman Teknis Perlintasan Sebidang Antara Jalan Dengan Jalur Kereta Api*. (2005)
5. Direktorat Jenderal Bina Marga. *Pedoman Kapasitas Jalan Indonesia (PKJI).* Departemen Pekerjaan Umum. Jakarta. (2014).
6. Hadiyatmo, H.C. *Pemeliharaan Jalan Raya*. Gadjah Mada University Press, Yogyakarta. (2007).

Characterization of Reclaimed Asphalt Pavement (RAP) as a road pavement material (National Road Waru, Sidoarjo)

Ari Widayanti[1,*], *Ria* A. A. Soemitro[1], *Januarti* Jaya Ekaputri1, and *Hitapria* Suprayitno[1]

[1]Institut Teknologi Sepuluh Nopember, Depart. Civil Enginering, Surabaya, Indonesia

Abstract. Reclaimed Asphalt Pavement (RAP) is a dredging pavement material using Cold Milling Machine. The application of Reclaimed Asphalt Pavement is increased year by year. Due to the increasing application of RAP year by year which implicates environment condition, especially in damaging natural resources, the research on material used in RAP needs to be conducted, so RAP can be optimally utilized. To achieve optimal performance, data of RAP characteristics reviewed from microstructural analysis is necessary. The objective of this research is to obtain the characteristics of Reclaimed Asphalt Pavement. The method used was literary study based on previous research. Material tests used were XRF, SEM and FTIR. The object of study was RAP material taken from national road Waru Sidoarjo. The major compositions of Reclaimed Asphalt Pavement obtained were Kaolin, Lithium, Tetraborate, Dextrin. SEM graphics showed the morphological and surface texture of RAP. FTIR graphic presented the functional group of Reclaimed Asphalt Pavement showing O-H C-O acid in the peak of graphic. From XRD result, the major compounds of Reclaimed Asphalt Pavement obtained were Calcium, Sodium, Aluminum, Silicate.

1 Introduction

Road is one of the supporting infrastructure to support community activities. Reclaimed Asphalt Pavement is a dredging road pavement material using Cold Milling Machine. The accumulation of Reclaimed Asphalt Pavement materials is averages 50000 ton/annual, in order to optimize RAP utilization efforts [1].

Thought to process and utilize Reclaimed Asphalt Pavement in other countries has been widely studied. Many countries such as America, India, Taiwan and Japan used 10-40% of RAP as road pavement. Even in Singapore has managed to utilize RAP of 100% for pavement layer material. Based on research [2] in the FHA/Federal Highway Administration and NCHRP/National Highway Cooperative Research Program (provide input to AASHTO to update existing standards for use of Reclaimed Asphalt Pavement). Use of RAP is limited to the base and usage. RAP in road pavement that requires new material should also. Reclaimed Asphalt Pavement percentage is used 25-30% in 2010,

* Corresponding author: ariwidayanti1973@gmail.com

even close to 50%. RAP usage application can be up to 30% for surface layer on Asphalt Concrete, 100% RAP in hot mix, warm and cold mix on low traffic road, parking area, bicycle and pedestrian area. Use of 30% RAP on base and sub base on primary road, and use 100% RAP on the shoulder of the road [3]. Use of RAP must be cost effective, environmental friendly and produce optimum performance of pavement layers.

Some research results that have been done include testing the nature of RAP by mixing RAP with lime added materials. From this mixing result, obtained asphalt pavement extracted asphalt pavement equal to 4.16%. The addition of lime is 4.5%. From the test results obtained increase in mixed properties can be improved by gradation engineering and lime addition [4].

The other research used RAP with 20%, the addition of filler fly ash by 4%, 5%, 6%. From the results of this research, the use of RAP 20% with the most optimal mixed performance result is 5% filler can produce flow, stability Marshall in accordance with standard [5]. The use of 25% reclaimed asphalt pavement can meet the requirements of grain AC-WC gradation resulting in 6% Optimum Bitumen Content with 60-70% and 5.9% asphalt Pen with asphalt TRS 55 type modification [6]. The composition of 20% RAP from Pilang - Probolinggo road, 6.5% coarse aggregate, 25% medium aggregate, 45% fine aggregate and 1% cement in hot asphalt mixture with modified asphalt can meet the requirements of asphalt concrete layer with Optimum Bitument Content 5.6% with synthetic Elastomeric asphalt and 6.2% with asphalt processed [7].

Initial mixing with 40% RAP and 60% new material is had consideration of aging effect of pavement layer. Optimum composition is obtained 30% RAP Gemekan – Jombang road and 70% new material with Optimum Bitumen Content of 5.70%. While 20% RAP and 80% new material with Optimum Bitumen Content 5.9% [8]. Composition 25% RAP Pandaan - Malang road and 75% new material with abrasion value of 19.73% meet the requirements of specification with Optimum Bitumen Content of 6.35% [9].

Replacement of 100% RAP granular material yields the highest score on Indirect Tensile Strength and Modulus Resilient in both dry and wet conditions. The comparable condition is to use Pen asphalt 60/70 and 80/100, and the better one is to use asphalt Pen 80/100 [10]. Comparison of Reclaimed Asphalt Pavement between 0%, 10%, 20%, 30%, 40%, 100% and natural materials for hot mix asphalt using 60/70. The result is that the percentage of Reclaimed Asphalt Pavement of 30% yields high stability value, and on the other hand increases the possibility of brittle effect on highway pavement mix [11].

Addition of Reclaimed Asphalt Pavement by 20% improves all properties of the asphalt mixture. The addition of Reclaimed Asphalt Pavement produces better performance than using only natural aggregates [12]. Use of 30-48% Reclaimed Asphalt Pavement exhibits better properties (volumetric properties, mechanical and performance properties) than using natural materials [13].

Research in the laboratory used the percentage of RAP by 10%, 20%, 30%, 40% RAP. There is no consistency of change in addition to physical properties (penetration, ductility, softening point) on the mixture. RAP usage ranges within 10-40% can be adopted for use on new roads using RAP [14]. Increasing the percentage of RAP on a mixture is containing tire waste, indicating a higher stability value. The addition of RAP with tire wastes results in a good workability score. The addition of tire waste can increase the resistance to be able to withstand deformation due to traffic load [15].

Addition of RAP by 25% with tire waste variation can add resilient modulus value to temperature variation. This addition is also recommended for areas with high temperature tropical climate. Addition of RAP and tire waste influence the technical life of road and its resistance to fatigue [16].

Chemical composition was obtained by XRF test. The minerals and morphology of material are obtained by XRD and SEM test. The results of the XRF test obtained Calcium

Oxide and Iron Oxide are the two major chemical formers of EAF and BOF slag. The results of XRD show both BOF and EAF slag samples are very complex, with many peak points, of many types of minerals in the sample. XRD analysis showed the presence of MgO and CaO in BOF and EAF sample slag. The SEM micro graph shows the majority of the sand sizes of steel slags are sub angular to the angular. It shows very rough surface texture with different crystal structures. The characteristics of steel slag samples are considered in more detail for review of steel slag properties [17].

Many countries have used RAP, especially to improve the performance of road pavement quality. Compared to the use of natural materials, using RAP has more benefits, such as to save natural materials in which it affects to the sustainability of the environment. These studies review the application of RAP in some countries with a percentage of use from 0% to 100% depending on the source of the RAP and part of the road pavement layer. However, testing of RAP characteristic has not been done in detail. Therefore, it is necessary to test of the RAP material microstructural to support better material properties results, in this case the focus of the review on the performance of composite materials and the attachment between combining material with each other.

2 Discussion

2.1 Method

RAP material can be obtained from milling in Waru National Road - Sidoarjo. Dredging process of road pavement used Cold Milling Machine. The RAP sample was broken and ground to fine powder. The figure of RAP material is given in Figure 1.

Figure 1 Reclaimed asphalt pavement material

Materials test consist of XRF (X-Ray Fluorescence), XRD (X-Ray Diffraction), SEM (Scanning Electron Microscopy) and FTIR (Fourier Transform Infra Red). XRD, FTIR and SEM test did in Material and Metallurgy Laboratory, Institut Teknologi Sepuluh Nopember Surabaya. XRF test did in Energy and Environmental Laboratory, Institut Teknologi Sepuluh Nopember Surabaya.

2.2 Analysis

2.2.1 XRF analysis

RAP material can be obtained from road national Waru, Sidoarjo Indonesia. Material test used XRF (X-Ray Fluorescence). The chemical composition of Reclaimed Asphalt Pavement can be presented in Table 1.

Table 1. Element Compound of Reclaimed Asphalt Pavement

Element Compound	Test result (% by weight)
SiO_2	38
Fe_2O_3	26.8
CaO	16.3
Al_2O_3	11
SO_3	2.9
TiO_2	1.8
K_2O	1.73
MnO	0.585
SrO	0.37
CuO	0.13
V_2O_5	0.11
BaO	0.2
Re_2O_7	0.06
ZrO_2	0.055
ZnO	0.045

Based on XRF test in Table 1 can be obtained major element compound of RAP consist of SiO_2, Fe_2O_3, CaO by weight percentage. Material with high SiO_2 content will produce hard material, commonly found in nature as sand or quartz. For road pavement, this material is water-absorbing or hygroscopic. Relating to its capacity as a refractory material, it can be used as high temperatures protection. SiO_2 has more different crystalline forms (polymorphs) than amorphous forms.

Table 2. Oxide compound of reclaimed asphalt pavement

Oxide Compound	Test result (% by weight)
Fe	36.6
Si	26.7
Ca	19.8
Al	8.3
K	2.36
Ti	1.92
S	1.9
Mn	0.868
Sr	0.65
Ba	0.4
Cu	0.22
V	0.11
Zn	0.096
Zr	0.083
Re	0.06

Based on XRF test in Table 2 can be obtained major oxide compound of RAP consist of Si, Fe and Ca by weight percentage. Silica is hard mineral, chemical inert (no reaction to any chemicals), and has low melting point. Its condition shows the strong bond among thatoms.

Figure 2. SEM micrograph of reclaimed asphalt pavement (1000x, 5000x, 10000x)

2.2.2 SEM analysis

The SEM image for the RAP sample is presented in Figure 2. It shows morphology and surface texture of RAP material.

The SEM micrographs shows that the majority of Reclaimed Asphalt Pavement materials were sub-angular and angular, very rough surface textures and good to be used in road pavement as it improved the interlocking aggregate in road pavement material as well as the asphalt concrete mixtures quality.

2.2.3 FTIR analysis

An FTIR analysis was performed on Spectrum RAP. FTIR spectrometer in the range of 4000 – 400 cm^{-1} at room temperature. The result of FTIR analysis can be presented in Table 3.

Table 3. FTIR Test Result of RAP

Index	Match	Compound Name
18452	35.30	Kaolin
18201	34.95	Lithium tetraborate
1151	34.55	Dextrin
18371	33.76	Potassium heptafluoroborate
18206	33.71	Ammonium tetrafluoroborate
49	33.63	TRANS-PIPERYLENE
18162	33.31	Metaphosphoric acid
1579	33.17	Glyoxal trimeric dihydrate
18280	32.28	Ammonium-d4 dideuteriumphosphate
18292	32.13	Sodium bismuthate

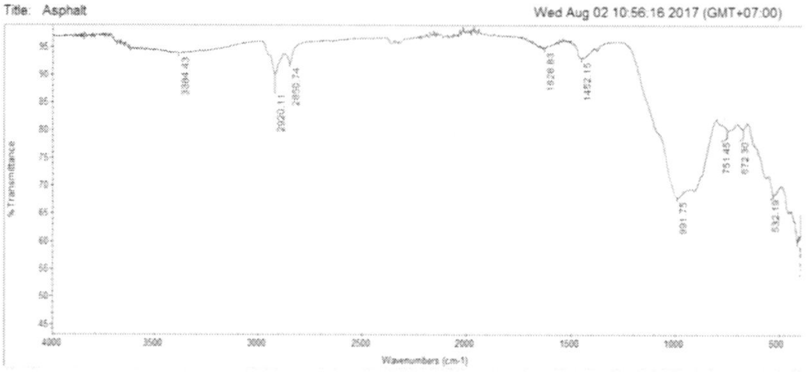

Figure 3 FTIR spectra of reclaimed asphalt pavement sample

Figure 3 shows that FTIR spectrum of the RAP sample. The strong peak at ~1000 cm^{-1} was related to Al-O and Si-O asymmetric stretching vibrations. The bands seen bond molecule of RAP. It shows O-H C-O acid.

Table 4 Wavenumber of FTIR result

Wavenumber (cm^{-1})	Functional Group	Peak Description
3384.43	O-H (alcohol)	Strong and broad
2920.11	C-H (sp^3 carbon)	Strong, broad and multi banded
2850.74	C-H (sp^3 carbon)	Strong, broad and multi banded
1628.83	C=C (alkene, aromatic ring)	
1452.15		
991.75		
751.45	Phenyl group	Strong
672.30	Phenyl group	Strong
532.19		

2.2.4 XRD analysis

XRD analysis was presented in Figure 4 and Table 4.

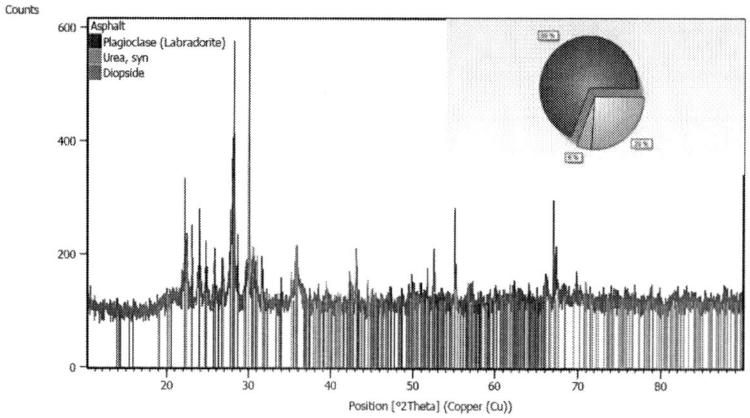

Figure 4. XRD of reclaimed asphalt pavement sample

Table 5. Result of XRD test

Ref. Code	Score	Compound Name	Displacement (^{0}2Th)	Scale Factor	Chemical Formula
01-083-1370	45	Calcium Sodium Aluminum Silicate	0.213	0.484	Ca 0.65 Na 0.35 (Al 1.65 Si 2.35 O 8)
01-089-2835	21	Urea	0.203	0.124	$C O (NH_2)_2$
01-083-1392	20	Magnesium Iron Calcium Sodium Silicate	0.180	0.404	(Mg 0.964 Fe 0.036) (Ca 0.94 Na 0.06) (Si_2O6)

3 Conclusions

In conclusions, the results of this research were:

1. Major compositions of Reclaimed Asphalt Pavement in object taken were Kaolin, Lithium Tetraborate and Dextrin
2. SEM graphic presented morphology and surface texture of Reclaimed Asphalt Pavement. The SEM micrographs showed majority of Reclaimed Asphalt Pavement material were sub-angular and angular and very rough surface texture which were good to be road pavement structure because it improved the interlocking aggregate in road pavement material
3. Reclaimed Asphalt Pavement molecule banding at peak graphic showed O-H (alcohol) and C-H (carbon).
4. XRD result showed that the major components of RAP were Calcium, Sodium, Aluminum, Silicate.

4 Acknowledgements

The research in degree scholarship is provided by LPDP/Fund Managers - Ministry of Finance, Indonesia.

References

1. Budianto, Herry, *Menuju Jalan yang Andal* (2009)
2. Cooper, S.J, *'Asphalt Pavement Recycling with Reclaimed Asphalt Pavement (RAP)'* (2011)
3. Widger, A., Skilnick, F, *Utilization of Recycled Asphalt in Cold Mixes and Cold in Place Recycling Processes Guidelines.* Clifton Associates Ltd.(2012)
4. Astuti, W.W., Sunarjono, S., Sutanto, M.H, *'Analisis Pengaruh Bahan Tambah Kapur terhadap Karakterisrik RAP (Reclaimed Asphalt Pavement)'*. (2015)
5. Handayani, R, *'Analisa Penggunaan Reclaimed Asphalt Pavement (RAP) sebagai Bahan Campuran Beraspal Panas Tipe Asphalt Concrete-Binder Course (ACBC) dengan Menggunakan Fly Ash (Studi Kasus Ruas Jalan Taman Waru)'*(2016)

6. Harahab, S, Soemitro, R.A., Budianto, H, *'Optimalisasi Penggunaan Reclaimed Asphalt Pavement (RAP) sebagai Bahan Campuran Beraspal Panas (Asphaltic Concrete) Tipe AC-Wearing Course (AC-WC) Gradasi Kasar dengan Aspal Pen 60-70 dan Aspal Modifikasi Jenis TRS 55 (Studi Kasus Jalan Nasional Pandaan-Malang dan Jalan Nasional Pilang-Probolinggo)'*. (2013)
7. Herawati, N. Soemitro, R.A.A., Budianto, H, *'Analisis Penentuan Komposisi Optimal Penggunaan Reclaimed Asphalt Pavement (RAP) sebagai Bahan Campuran Beraspal Panas (Asphaltic Concrete) Menggunakan Aspal Modifikasi (Studi Kasus Jalan Pilang – Probolinggo)'*. (2012)
8. Kusmarini, E.P., Soemitro, R.A.A., Budianto, H, *'Analisis Penggunaan Reclaimed Asphalt Pavement (RAP) dan Aspal Pen 60 – 70 sebagai Bahan Campuran Beraspal Panas (Asphaltic Concrete) (Studi Kasus Ruas Jalan Gemekan – Jombang dan Pandaan – Malang)'*(2012)
9. Falevi, R, *'Optimalisasi Penggunaan Reclaimed Asphalt Pavement (RAP) sebagai Bahan Campuran Beraspal Panas (Asphaltic Concrete) Tipe AC-Wearing Course (AC-WC) Gradasi Halus dengan Menggunakan Aspal Pen 60-70 Variasi Abrasi Agregat Baru (Studi Kasus Jalan Nasional Pandaan – Malang)'*(2012)
10. Ortiz, O. R., Berardinelli, A., Carvajal-Munoz, Fuentes, L.G, *Evaluation of 'Hot Mix Asphalt Mixtures with Replacement of Aggregates by Reclaimed Asphalt Pavement (RAP) Material'*. (2012)
11. Parveez, P., Srikanta, Y., Konnur, D, *'Study on the Effect of Reclaimed Asphalt Pavement (RAP) on the Mechanical Behaviour of Hot Mix Asphalt'*(2008)
12. Pradyumna, T.A., Mittal, A., Jain, P.K *'Characterization of Reclaimed Asphalt Pavement (RAP) for Use in Bituminous Road Construction'*(2013)
13. Shen, J., Amirkhanian, S., Aune, J.M,*Effects Of Rejuvenatin Agents Of Superpave Mixtures Containing Reclaimed Asphalt Pavement'*. (2007)
14. Sunil, S. Mallesh, K.M., Chandrasekaraiah 'Experimental Investigation on The Perfomance of Bituminous Mixes wth Reclaimed Asphalt Pavement (RAP) Materials (Case Study Tumkur to Chidradurga-NH4)'. IJRET: International Journal of Research in Engineering Technology **03** 06 (2014)
15. Xiao, F., Amirkhanian, S.N., Shen, J., Putman, B, *'Influences of Crumb Rubber Size and Type on Reclaimed Asphalt Pavement (RAP) Mixtures'*. (2009)
16. Xiao, F., Amirkhanian, S., Juang, C. H, *Rutting Resistance of Rubberized Asphalt Concrete Pavements Containing Reclaimed Asphalt Pavement Mixtures'*. Journal of Materials in Civil Engineering (2007)
17. Yildirim, I.Z., Prezzi, M. 2011. '*Chemical, Mineralogical, and Morphological Properties of Steel Slag'*. **2011** (2011)

MATEC Web of Conferences **181**, 06001 (2018) https://doi.org/10.1051/matecconf/201818106001
ISTSDC 2017

Analysis of congestion cost at signalized intersection using Vissim 9 (Case study at Demak Ijo Intersection, Sleman)

Noor Mahmudah [1,*], *Rizkie* Akbar[1] , and *Muchlisin*[1]

[1]Muhammadiyah University of Yogyakarta, Depart of Civil Engineering, Jl. Brawijaya, Indonesia

Abstract. Due to imbalance of road capacity and traffic volume, so traffic congestion will be occurred either along the road and intersection. Demak Ijo intersection is one of signalized junction located in the western part of Yogyakarta with high traffic volume so traffic congestion is frequently occurred. The aim of this study is to analyze the performance of existing traffic condition and then estimate the congestion cost at signalized intersection by modeling (simulation) using Vissim 9. The analysis results show that existing traffic condition is in very bad condition (level F), average delay of 80 seconds, average queue length of 48.73 meters with congestion cost is about Rp. 2,830,336 per hour.

1 Introduction

Congestion is one of significant transportation problems in many urban roads in Indonesia. Traffic congestion will be occurred either along the road and intersection due to imbalance the road capacity and traffic volume. In addition, mostly congestion will generate problems such as the reducing of vehicle speed that lead to the increasing of travel time, higher operating cost, and negative impact to the environment like air pollution produced by motorized vehicle.

The congestion will increase the transportation cost. As stated by Stubs in Sugiyanto [1], the cost of congestion is a relationship between speed and flow on the road and the relationship of speed and vehicle operating cost. It means when the existing traffic flow is increased hence the average traffic speed will decrease. Since the speed of vehicle is slower so the travel time is longer and then lead to the higher vehicle operating costs.

In general, the transportation in Yogyakarta is mixed traffic and overloaded on some road segments. It is due to the average growth of private vehicle in Yogyakarta city is 4.04% per year [2]. Demak Ijo intersection is one of the signalized junction which is located in the western part of Yogyakarta with high traffic volume where traffic congestion is frequently occurred. Therefore, this study attempts to analyze the congestion cost as a result of the low traffic performance at signalized intersection.

* Corresponding author: noor.mahmudah@umy.ac.id.

2 Literature review

Congestion is defined as the impedance vehicles impose on each other, due to the speed-flow relationship, in conditions where the use of a transport system approaches its capacity [3]. Tzedakis [4] stated that the theory of congestion cost was not based on the concept of homogenous traffic and the relationship of speed and flow. The congestion cost is the quantification of delay cost and changes in vehicle operating cost caused by slow traffic. In addition, Nash in Basuki and Siswandi [5] highlighted that congestion cost is a travel expense due to traffic delay or additional vehicle volume approaching or exceeding road service capacity. Verhoef in Sugiyanto [2] defined the congestion cost as representation of the difference of marginal social cost (MSC) to marginal private cost (MPC). Some studies related to congestion cost have been conducted by Tzedakis [4], Basuki and Siswadi [5], and Sugiyanto [3, 6]. Mostly studies focused on congestion cost that were occurred in road segment.

Theory of Tzedakis [4] highlighted the congestion cost resulted by slow traffic by time period (i.e. per hour) that can be estimated by considering the followings:

1) Slow vehicle (traffic) due to congestion will cause the delay of fast vehicle;
2) As a result the queuing time will be occured on the road segment or intersection;
3) The expected value of the sum queuing time of fast vehicle impeded by the slow one is then estimated;
4) Based on the expected value of the sum queuing time, the expected value of delay costs and the change in vehicle operating costs are then quantified;
5) These costs per hour are then estimated by multiplying the costs per slow vehicle journey by the hourly number of slow vehicle journeys;
6) The formula to calculate the congestion cost is as written on equation 1

$$C = N * [G\,A + (1 - \frac{A}{B})\,V']\,T \tag{1}$$

where:
C = Congestion Cost (Rupiah),
N = Vehicle Volume (Vehicle),
G = Vehicle Operating Cost (Rp/veh.km),
A = Existing Speed (km/hour or kph),
B = Ideal Speed (km/hour or kph),
V' = Vehicle Time Value (Rp/veh.hour)
T = Delay Time (Hour).

The Vehicle Operating Cost (VOC) in Yogyakarta can be calculated using formula proposed by Sugiyanto [6] as written in equation 2 below.

$$Voc = 0.4937v^2 - 60.218v + 29 \tag{2}$$

where
V = vehicle speed (Km/Hour)

Vehicle Time Value is estimated by using Table provided by Indonesian Highway Capacity Manual (1995) as described on Table 1 below.

Table 1. Vehicle time value
(Source: Indonesian Highway Capacity Manual, 1995)

Type of Vehicle	Vehicle Time Value (Rp/veh.hour)	
	Gross Domestic Product	Welfare maximation
Motor Cycle	315,00	736,00
Passenger Car	1.925,00	3.281,00
Small Bus	7.385,00	12.572,00
Big Bus	9.800,00	18.211,00
Small Truck	4.970,00	5.605,00
Medium Truck	4.970,00	5.605,00
Big Truck	4.970,00	736,00

3 Methods

The method applied in this study is as illustrated on Figure 1. Primary and secondary data collection have been conducted in order to obtain traffic data, cycle time (of traffic light), geometric data, road condition, vehicle speed, traffic volume, etc. These data are then utilized in traffic modeling/simulation by using software Vissim 9. The considerations are taken into account to the minimum congestion cost, queue length and delay, but the highest Level of Service (LOS).

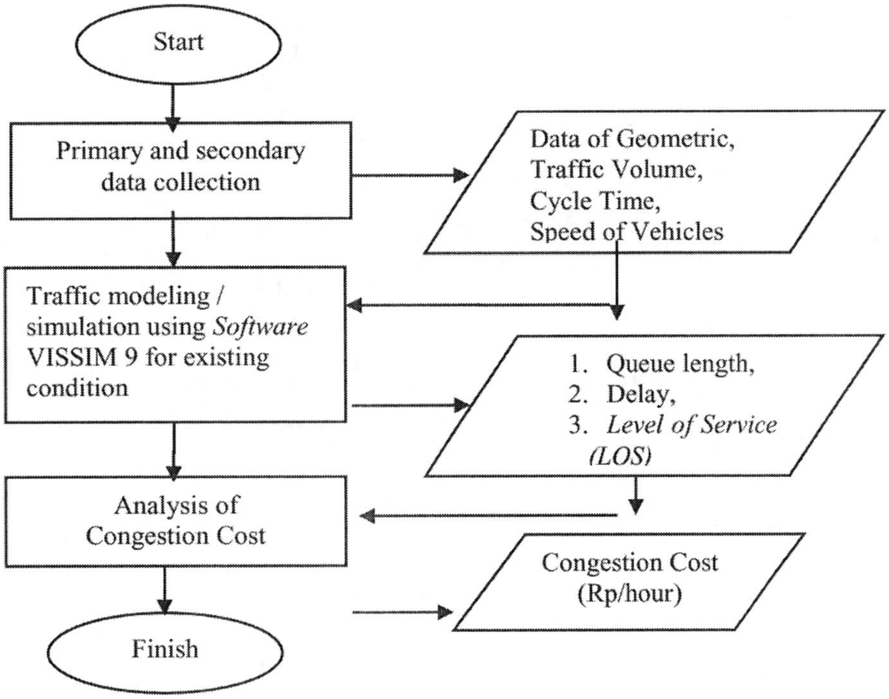

Figure 1. Research method

The study was conducted at signalized intersection of Demak Ijo, Godean Street, Sleman, Yogyakarta as shown in Figure 2 and 3. Traffic data utilizing in this study was collected by Umar [7], which was carried out for 12 hours on weekday at 06.00 up to 18.00 WIB. Moreover, in order to determine the geometric data at intersection, cycle time, road and environmental condition, the observation survey were also carried-out. The vehicle speed (spot speed) was conducted on three time: 06.00-08.00; 12.00-14.00, and 16.00-18.00 which represents the weekday traffic condition for every type of vehicle such as HV (Heavy Vehicle), LV (Light Vehicle), and MC (Motor Cycle).

Figure 2. Demak Ijo intersection, Godean, Sleman

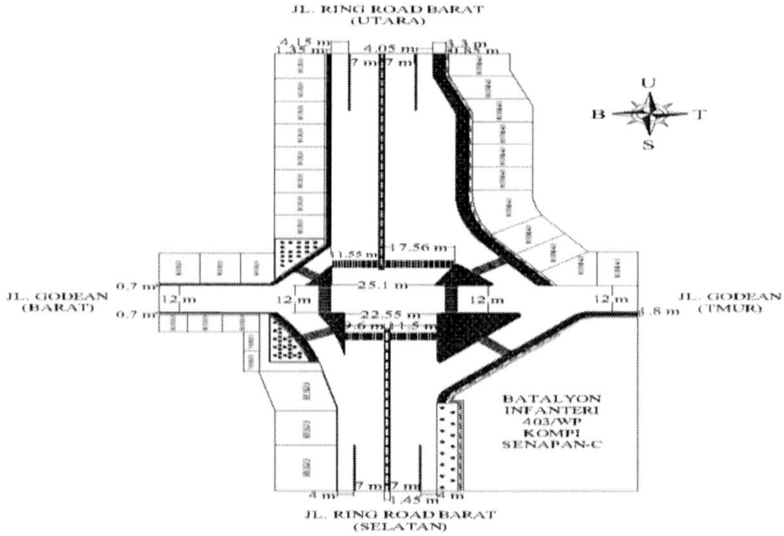

Figure 3. Geometric condition at Demak Ijo intersection, Godean, Sleman

4 Analysis result and discussion

4.1 The existing phase condition

The existing phase condition at Demak Ijo intersection can be seen in the Table 2 below. The total cycle time is 148 second for all arms.

Table 2. Existing phase condition

Signal	Arm	Type of Arm	Time (second)			
			Red	Green	Amber	All Red
Phase 1	N	Protected (P)	138	35	3	4
Phase 2	E	Protected (P)	148	25	3	4
Phase 3	S	Protected (P)	138	35	3	4
Phase 4	W	Protected (P)	148	25	3	4
Cycle Time (second)			148			

4.2 The traffic volume at peak hour

The survey of traffic counting was conducted by Umar [7] in the period of 12 hours starting from 06.00 up to 18.00. However, the data used for modelling the traffic at intersection is selected only during the peak hour (06.15.- 07.15). The traffic volume at peak hour obtained from the survey is as illustrated in Tabel 3.

Table 3 Data of traffic volume at peak hour [7]

Time	Arm Junction		HV	LV	MC	NMV	Total
6.15 s.d 7.15 WIB	North (A)	T. Left	1	68	540	16	625
		Straight	26	332	1135	2	1495
		T. Right	5	118	452	0	575
	Total		32	518	2127	18	2695
	East (B)	T. Left	1	87	268	2	358
		Straight	6	91	749	7	853
		T. Right	2	50	482	19	553
	Total		9	228	1499	28	1764
	South (C)	T. Left	5	21	87	6	119
		Straight	28	689	3118	8	3843
		T. Right	5	67	434	2	529
	Total		38	777	3639	16	4491
	West (D)	T. Left	1	210	650	9	870
		Straight	6	169	3838	34	4047
		T. Right	0	62	703	1	766
	Total		7	441	5191	44	5683

4.3 Stages of traffic modeling by using Vissim 9

In general, the stages of traffic modeling using software Vissim 9 as illustrated in Figure 4, which consist of input data, analysis, and obtain the output.

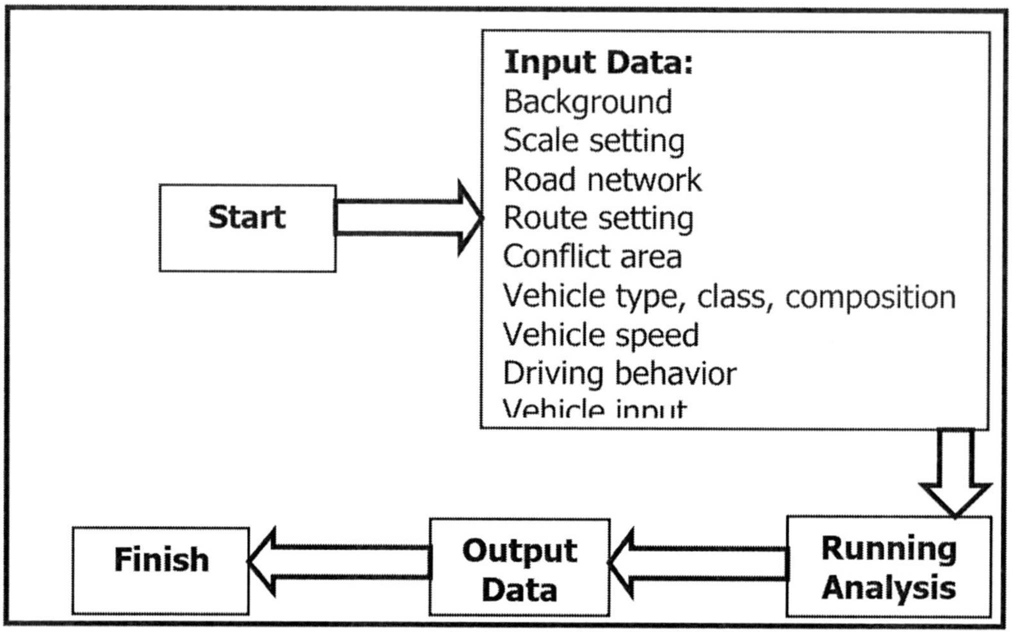

Figure 4. Stages of modeling using software Vissim 9

4.4 Output data of Vissim 9

The output resulted from simulation using software Vissim 9 is as shown in Table 4. It can be seen from the table that the maximum vehicle delay occured is 232 second and the longest queing is 214 meter at west arm with level of service F (very bad).

Table 4. The output data of Vissim 9 for existing condition

Time Interval	Arm	Direction	QLen (m)	QLen Max (m)	Vehs (All) (veh)	LOS (All)	LOS Val (All)	Veh Delay (All) (second)	Stop Delay (All) (second)	Stops (All) (veh.)
0-3600	North	(N) - (E)	01.24	20.51	96	LOS_A	1	08.36	0,088889	0,052083
0-3600		(N) - (S)	103.21.00	201.34.00	143	LOS_F	6	123.42.00	106.14.00	02.43
0-3600		(N) - (W)	0,884722	67.07.00	73	LOS_E	04.05	60.685	49.48.00	01.02
0-3600	East	(E) - (S)	01.12	18.24	64	LOS_A	1	3.055	2.235	00.14
0-3600		(E) - (W)	72.49.00	150.05.00	86	LOS_F	6	110.25.00	85.92	02.13
0-3600		(E) - (N)	1,014583	58.31.00	62	LOS_E	04.05	58.975	52.845	0,065278
0-3600	South	(S) - (W)	00.38	13.03	39	LOS_A	3	37.31.00	25.445	01.04
0-3600		(S) - (N)	128.18.00	186.33.00	143	LOS_F	6	135.61	104.975	04.07
0-3600		(S) - (E)	13.39	69.96	76	LOS_D	4	48.42.00	37.26.00	01.21
0-3600	West	(W) - (N)	0	12.07	145	LOS_A	1	1.195	00.52	00.18
0-3600		(W) - (E)	173.92	214.42.00	87	LOS_F	6	232.63	196.73	0,224306
0-3600		(W) - (S)	46.225	122.37.00	87	LOS_D	5	139.07.00	72.035	0,068056
0-3600	Average		48.73	214.42.00	1101	LOS_F	6	80	61.27.00	0,088889

4.5 Vehicle operating cost (VOC)

By employing equation 2 and considering the speed of vehicle, type of vehicle, and arm of intersection, so the vehicle operating cost (VOC) in existing condition at Demak Ijo intersection as shown in Table 5. The highest VOC is at west arm which is Rp. 1594/veh.km.

Table 5. Vehicle operational cost for existing condition

No	Arm	Vehicle Type	VOC (Rp/Veh.km)
1	NORTH	HV	1354
		LV	1221
		MC	353
2	EAST	HV	1478
		LV	1365
		MC	351
3	SOUTH	HV	1247
		LV	1158
		MC	378
4	WEST	HV	1582
		LV	1594
		MC	351

4.6 Travel time value (V)

The time value is calculated using the study of Indonesian Highway Capacity Manual (1995) as illustrated in Table 1. Based on the vehicle type, the travel time value for Heavy Vehicle is estimated Rp. 4,970, Light Vehicle Rp. 1,925, and Motor Cycle Rp. 315.

4.7 Congestion cost analysis

The total congestion cost at Demak Ijo intersection in existing condition can be seen on Table 6 below. The total congestion cost at the junction is Rp. 2,830,336 per hour.
Table 6. Total Congestion Cost for Existing Condition at Demak Ijo Intersection

5 Conclusions

By considering the analysis results of simulation using software VISSIM 9 for signalized intersection of Demak Ijo, Godean, Sleman, Yogyakarta, it can be concluded as follows.

1. The traffic performance of existing condition of signalized intersection Demak Ijo, Godean, Sleman, Yogyakarta is very bad (level F) with the highest traffic volume (peak hour) occurs in the morning at 06.15 - 07.15 WIB with traffic volume 14,633 vehicle, average delay of 80 seconds, and average queue length of 48.73 meters.
2. The congestion costs at signalized intersection of Demak Ijo, Godean, Sleman, Yogyakarta, which is occurred at peak hour, is Rp. 2,830,336 per hour.

Acknowledgements

My highly appreciations are addressed to my colleagues Muchlisin, M.Sc. and Dian Setiwan, M.Sc., Sc. and my student Muhammad Ade Alvianto Umar, who had been collaborating and contributing to this study in Civil Engineering Department, Universitas Muhammadiyah Yogyakarta.

Referencers

1. Sugiyanto, G, *Biaya Kemacetan (Congestion Cost) Mobil Pribadi di Central Business District* (2008)
2. Sugiyanto, G, et al, Pengembangan Model Biaya Kemacetan Bagi Pengguna Mobil Pribadi di Daerah Pusat Perkantoran Yogyakarta. Jurnal Transportasi **11**, 2, 87-94 (2011)
3. Sugiyanto, G, et al, Modeling the Effect of Congestion Pricing on Mode Choice in Yogyakarta, Indonesia. International Journal of Engineering & Technology IJET – IJENS **11**, 01. 85-94 (2011)
4. Tzedakis, A, Different Vehicles Speeds and Congestion Costs. Journal of Transport Economics and Policy **XIV,** 1, 81-103 (1980)
5. Basuki, Siswandi, Biaya Kemacetan Ruas Jalan Kota Yogyakarta. Jurnal Teknik Sipil **9**, 1, 71-80 (2008)
6. Sugiyanto, G, et al, Pemodelan Biaya Kemacetan Pengguna Mobil Pribadi Dengan Variasi Nilai Aktual Kecepatan Kendaraan. Jurnal Transportasi **12**, 2, 123-132 (2012)
7. Umar, M.A.A, *Analisis Kinerja Simpang Bersinyal pada Persimpangan Demak Ijo*, (2016)
8. Directorate General of Highway, *Ministry of Public Works the Republic of Indonesia* (2010)
9. Khisty dan Lall, *Dasar- Dasar Rekayasa Transportasi* (2005)
10. Mahmudah. N, Bayunagoro. D.H, Pemodelan Lalu Lintas pada Simpang Bersinyal di Kota Yogyakarta (Studi Kasus Simpang Pingit). Proceedings of SNTT IV. UMP. Purwokerto. 26 November 2016. ISSN: 2339-028X (2016)
11. Ministry of Transportation Republic Indonesia, *Peraturan Menteri No. 96 Tahun 2015. Tentang Pedoman Pelaksanaan Kegiatan Manajemen dan Rekayasa Lalu Lintas* (2015)Munawar, A, Dasar-dasar Rekayasa Transportasi (2005)
12. Putranto, L.S, *Rekayasa Lalu Lintas (Edisi 3)* (2016)
13. PTV VISSION, *PTV VISSIM 9 User Manual*. PTV AG (2016)
14. The Goverment of Republic Indonesia, *Undang - Undang Republik Indonesia No 22 Tahun 2009 Tentang Lalu Lintas dan Angkutan Jalan* (2009)

Weaving length and lane-changing behavior at two-sided weaving section along federal road FT050: jalan kluang – ayer hitam

Joewono Prasetijo[1,*], *Ishak* Baba[1], *Nurul* Elena A. Arifin[1], and *Alvin* John L. M. Siang[1]

[1] Universiti Tun Hussein Onn Malaysia, Faculty of Engineering Technology, Department of Transportation Engineering Technology, Johor, Malaysia
[2] Universiti Tun Hussein Onn Malaysia, Faculty of Civil and Environmental Engineering, Johor, Malaysia

Abstract. Weaving is defined as a movement of vehicles that crossing over the direct traffic from on-ramp to off-ramp. Apparently, weaving maneuver lead to collision risk and weaving turbulence especially at two-sided weaving section. The following study recognize the maximum weaving length and lane-changing rate on Federal Route 50 (FT050) segment. Visual recording technique was used at two different locations that are called as Site A (KM15) and Site B (KM16). Data of traffic volume was extracted by playback the video/personal computer. Analysis was made using spreadsheet and applied the methodology of the Highway Capacity Manual 2010. The study found that Site A has more tendencies to encounter longer weaving turbulence with maximum weaving length ranging from 1952 m to 2120 m, compared to site B that is 1866 m to 1882 m. Site A with a longer distance of weaving section 358 m has higher intense of total lane-changing rate that is 1142 lc/h compared to Site B with a distance of weaving section 316 m has total lane-changing rate of 812 lc/h. Shorter weaving distance may cause less lane-changing activity because drivers does not prefer to perform weaving at shorter weaving section as it is forced the drivers to perform weaving drastically in more crammed situation and risk to dealing with accident is huge. The findings shall help to understand more on causes of weaving turbulence at weaving section by referring HCM 2010 methodology.

1 Introduction

Weaving is an action that performed by merging traffic to cross the through lane over diverging movements [1]. Basically, weaving is considered as types of conflict that occurred in daily traffic and traffic streams conflicts are very common at any of weaving sections and intersections [2]. A weaving maneuver performed by drivers is normally depending on their own purpose. However, inefficiencies of driver when perform weaving

* Corresponding author: joewono@uthm.edu.my

will finally lead to risk of accident and also confusing maneuver [3, 4]. A weaving maneuver could be detected by the driver on direct traffic when the weaving's driver gives signal to change lane. Effect of the lane-changing of the weaving vehicles along a weaving segment could create weaving turbulence. As weaving creates conflict streams, therefore, the traffic streams behavior should be considered based on the current fundamental traffic theory [5].

Since collision risk at weaving section is large, human error could be one of the biggest factors. Somehow, by prevention this error, there might be a high probability to reduce the accident risk [6, 7]. In traffic engineering view, the collision risk at weaving section might be influenced by the length of the weaving segment itself. Previous study [8] proved that longer distance of weaving cause less collision risk because its influence the driver decision to change lane safely. If longer weaving section is more preferred by most of driver to perform weaving, then activity of lane changing is become higher as the volume of vehicles is increase. Affected by this is weaving turbulence that will be created by the increasing volume of vehicles and this could be measured by analyze the maximum weaving length of the section. Therefore, this study is carried out to investigate the maximum weaving length and lane-changing rate which is affected by volume ratio and weaving length at different sites. Methodology of chapter 12 in Highway Capacity Manual (HCM) 2010 is applied to assist the analysis.

1.1 Review on weaving behavior

A study on weaving section was conducted as early in 1965 by HCM which is the length of weaving section which used as approach. Since that, a lot of studies were conducted in order to analyze a weaving section in order to investigate the length, flow, lane-changing rate, capacity and speed. Basic purpose of investigating these is due to a problem regarding to weaving turbulence. This problem is often formed within the weaving section [9]. However, there is a lack of method used in HCM 1965 which is the volume and flows of non-weaving was not taken into consideration to analyze the rate of lane-changing. Year by year, HCM was upgraded in order to provide the best methodology of weaving analysis and finally the latest version of HCM was published in 2010. The methodology of freeway weaving analysis in HCM was adapted by the report of NCHRP which is a study was conducted in order to develop a new model or methodology for maximum weaving length and lane-changing rate. New version of HCM used one-sided and two-sided instead of using 'Type' to classify the weaving's configuration. Two-sided weaving section is classic and special as it is frequently occur on multilane highways and ramp to ramp flow is a main subject of analysis [10].

Due to weaving maneuver, lane-changing activity is become more aggressive in purpose of avoiding weaving vehicles. There were a lot of studies regarding to lane-changing in traffic but somehow lane-changing at weaving section is more complex compared to other situation [6]. Gap acceptance is a very famous approach in order to analyze lane-changing activity. Lane-changing at a roundabout also complex as a lot of confusing and weaving maneuver which was studied mostly by using gap acceptance approach. Speed adjustment is also one of approach used and studied by Uno as a speed adjustment model was developed to analyze speed adjustment processes of a lane-changing vehicle at weaving section. However, both approaches could not be applied as it does not measure a weaving turbulence impact. Approach of HCM 2010 is the best to use in order to analyze lane-changing rate at weaving section so that the impact of weaving turbulence could be known.

2 Discussion

It was earlier explained that data analysis was made using HCM 2010. Early stage of analysis is to find the short length L_S which will be as weaving length and also base length L_B. L_S is measured between the end point of any barrier marking that prohibit lane-changing while L_B is measured between right edge and left edge of the ramp meet. If there is no barrier found, $L_S = L_B$. Otherwise, if there is any presence of barrier, Eq. (1) is used.

$$L_s = 0.77 \times L_B \tag{1}$$

To determine at which distance the weaving section no longer has an impact of turbulence, maximum weaving length will be calculated using Eq. (2). It shows that the length of maximum weaving at certain weaving section is sensitive to volume ratio VR and number of weaving lanes N_{WL}. As stated in HCM 2010, N_{WL} is defined as zero for two-sided weaving. Therefore, only VR used as function and it use volume adjustment from volume data that calculated as in Eq. (3) through Eq. (7). Finding of L_{MAX} will be compared with value of L_S in order to specify if the section studied should be analyzed as weaving section or isolated merging and diverging. HCM 2010 specifies that a studied section only could be analyzed as weaving section if the value

$$VR = \frac{V_W}{V} \tag{2}$$

$$L_{MAX} = [5{,}728(1 + VR)1.6] - [1{,}566\ NWL] \tag{3}$$

where,
V_W = total weaving flow (pc/h)
V = total demand flow

$$V_p = \frac{V}{PHF \times f_{HV} \times f_P} \tag{4}$$

$$V_p = V_{NW} + V_W \tag{5}$$

where,
PHF = peak hour factor
f_p = 1.00 (assume drivers are familiar with the site)
f_{HV} = heavy-vehicle factor

$$PHF = \frac{V}{4 \times V_{15}} \tag{6}$$

where,
V = peak hourly volume
V_{15} = highest 15-minute volume

$$f_{HV} = \frac{1}{1 + P_T(E_T - 1)} \tag{7}$$

Where,
P_T = percentage of trucks and buses
E_T = passenger car equivalent for trucks and buses as in (Table 1)

Table 1 Passenger Car Equivalents for Extended General Highways

Factor	Type of Terrain		
	Level	Rolling	Mountainous
E_T (truck and buses) E_R (RVs)	1.5 1.2	2.5 2.0	4.5 4.0

Since lane-changing at weaving section consist of weaving and non-weaving maneuver, then analysis of lane-changing rate for both maneuver could be analyzed. This could be done by using Eq. (8) through Eq. (13). Calculation of non-weaving lane-changing rate requires index number in order to specify which formula between Eq. (11) to Eq. (13) should be use. To specify which equation should be use, the index number should fulfill the condition as stated below:

If $I_{NW} \leq 1{,}300$: $LC_{NW} = LC_{NW1}$
If $I_{NW} \geq 1{,}950$: $LC_{NW} = LC_{NW2}$
If $1{,}300 < I_{NW} < 1{,}300$: $LC_{NW} = LC_{NW3}$
If $LC_{NW1} \geq LC_{NW2}$: $LC_{NW} = LC_{NW2}$

$$LC_W = LC_{MIN} + 0.39[(L_S - 300)^{0.5} N^2 (1 + ID)^{0.8}] \tag{8}$$

where,
LC_w = Equivalent hourly rate at which weaving vehicles make lane changes within the weaving segment (lc/h)
LC_{min} = Minimum equivalent hourly rate at which weaving vehicles must make lane changes within the weaving segment to successfully complete all weaving maneuvers (lc/h)
L_s = Length of the weaving segment using the short length definition (ft) 300 ft is the minimum value
N = Number of lanes within the weaving segment; and
ID = Interchange Density (int/mi)

$$LC_{MIN} = LC_{RR} \times V_{RR} \quad \text{(for two sided)} \tag{9}$$

$$I_{NW} = \frac{L_S \times ID \times V_{NW}}{10{,}000} \tag{10}$$

$$LC_{NW1} = (0.206\, V_{NW}) + (0.542\, L_S) - (192.6\, N) \tag{11}$$

$$LC_{NW2} = 2{,}135 + 0.223(V_{NW} - 2000) \tag{12}$$

$$LC_{NW3} = LC_{NW1} + (LC_{NW2} - LC_{NW1}) \frac{I_{NW} - 1{,}200}{650} \tag{13}$$

Both equation of lane-changing rate are consisting of L_S function. This explained weaving length could influence the lane-changing rate. In turn, total lane changing rate could be determine by using Eq. (14) to find out the intensity of lane-changing activity at each site.

$$LC_{ALL} = LC_{W} + LC_{NW} \tag{14}$$

Findings

Measurement of short length L_S at both sites found that there is no barrier presence to prohibit the vehicles to weave at on-ramp. Therefore, L_S was assumed as equal to L_B. Measurement shows that weaving segment at Site A is longer than Site B. Table 2 shows the result of L_S at both sites. The following Figure 1 and Figure 2 show the current geometric and situations along the Federal Road FT050-Jalan Kluang.

Table 2.

Location	Short Length, L_S (m)
Site A	358
Site B	316

Figure 1 Site A; Federal Road FT050 KM15

Figure 2 Site B; Federal Road FT050 KM16

Besides, investigation of maximum weaving length was made by using Chapter 12 in HCM 2010 since the value of L_{MAX} for both sections was greater than L_S. Hence, both sites could be analyzed as weaving section. The analysis found that Site A has more tendencies to encounter weaving turbulence since the value of maximum weaving length is ranging from (1952 m – 2120 m) while Site B is ranging from (1866 m - 1882 m). This means that the impact of weaving turbulence at Site A could be felt as far as 168 m meanwhile impact of weaving turbulence at Site B could be felt as far as only 16 m. In term of operational, Site A is worse than Site B. Table 3 and Figure 3 summarize this finding for both sites.

Table 3. maximum weaving length

Location	Time Periods	L_{MAX} (m)	Range (m)	Impact of Weaving Turbulence Encountered(m)
Site A	10.00 - 11.00 am 11.00 - 12.00 am 4.00 - 5.00 pm	2018 1952 2082	(1952-2120)	168
	5.00 - 6.00 pm	2120		
Site B	10.00 - 11.00 am	1866	(1866-	16
	11.00 - 12.00 am 4.00 - 5.00 pm 5.00 - 6.00 pm	1867 1882 1875	1882)	

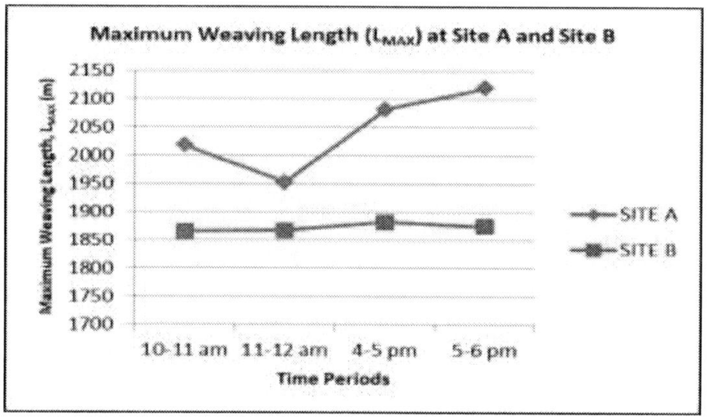

Figure 3. Graph of maximum weaving length at sites

In turn, analysis of lane-changing rate found that the most intense lane-changing activity of weaving occurred at Site A with a highest rate 555 lc/h. Meanwhile, a weaving lane-changing with a rate of 254 lc/h was recorded at Site B. These values were adapted during peak hours in order to obtain the absolute value due to highest volume of vehicles but even during off peak hour's period, Site A still encounter a higher value of LC_W compared to Site B. This difference explained that weaving maneuvers frequently happened at Site A with a longer weaving length because drivers more prefer to perform weaving in an unforced circumstance. Shorter length of weaving provides less comfortability to drivers to perform weaving because it happens in crammed situation. Table 4 and Figure 4 illustrate the finding of lane-changing rate of weaving vehicles at both sites.

Table 4. Finding of lane-changing rate of weaving vehicles

Location	Time Periods	LC_W (lc/h)	Absolute LC_W (lc/h)
Site A	10.00 - 11.00 am	391	555
	11.00 - 12.00 am	319	
	4.00 - 5.00 pm	451	
	5.00 - 6.00 pm	555	
Site B	10.00 - 11.00 am	190	254
	4.00 - 5.00 pm	254	
	5.00 - 6.00 pm	234	

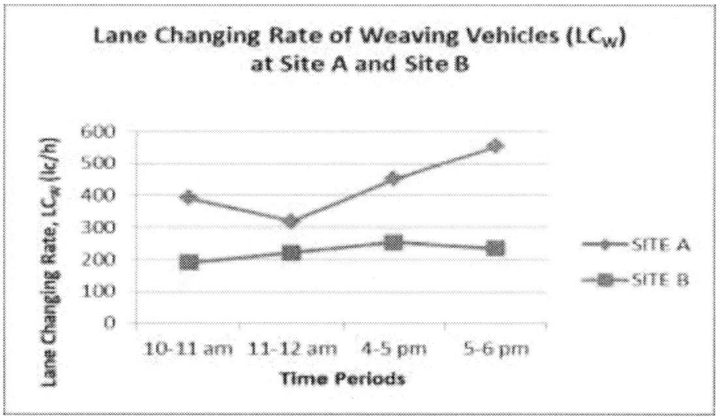

Figure 4. Graph of lane-changing rate of weaving vehicles at sites.

The final finding of this study found that the total lane-changing of both weaving and non-weaving vehicles were absolutely occurred during peak hours and the highest rate recorded at Site A with 1142 lc/h while Site B is 812 lc/h. This number shows in Table 5 and Figure 5. It also explained that Site A has more tendencies to encounter weaving turbulence due to high lane-changing activity. The high total lane-changing activity somehow affected by a lot of weaving vehicles crossing over the direct traffic. In term of operational, Site B is better than Site A because smooth traffic movement occurred at here. Meanwhile in term of safety, Site B is worse than Site A due to high risk of collision because of shorter weaving length.

Table 5. Finding of total lane-changing rate

Location	Time Periods	LC_{ALL} (lc/h)	Absolute LC_{ALL} (lc/h)
Site A	10.00 - 11.00 am	955	1142
	11.00 - 12.00 am	895	
	4.00 - 5.00 pm	999	
	5.00 - 6.00 pm	1142	
Site B	10.00 - 11.00 am	652	812
	11.00 - 12.00 am	754	
	4.00 - 5.00 pm	812	
	5.00 - 6.00 pm	773	

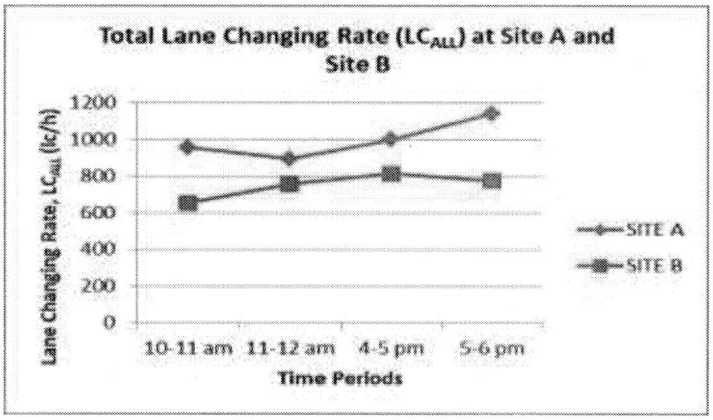

Figure 5. Graph of total lane-changing rate at site A and site B

3 Conclusions

Based on the finding, weaving turbulence tend to occur at Site A because of high number of vehicle using the weaving section at this site. In addition, it is also because of the development nearer to the site especially Taman Universiti residential. The drivers that tend to use the weaving section at this site are mostly consisting of the resident of Taman Universiti residential and also consist of students of UTHM. This factor not only leads to high impact of weaving turbulence but also increasing in lane-changing rate especially for weaving vehicles. Development available around the site is a factor that encourages the high intensity of weaving's lane-changing activity other than weaving length factor.

Acknowledgements

The authors would like to thank to the Universiti Tun Hussein Onn Malaysia for supporting the study through Contract research grant Vot. U395 and all individuals/organization that have made this study possible. Thank you is also extended to the Department of Transportation Engineering Technology, Faculty of Engineering Technology, Universiti Tun Hussein Onn Malaysia.

References

1. Cassidy, M. J., & May, A. D., Proposed Analytical Technique For Estimating Capacity And Level Of Service of Major Freeway Weaving Sections. Transportation Research Record 1320, pp. 99-109. (1991)
2. Prasetijo J., Pour M. H. & Ghadiri S. M. R., Capacity of Unsignalized Intersections under Mixed Traffic Conditions. Procedia – Social and Behavioral Sciences, Vol.**16**, pp. 676-685. (2011)
3. Prasetijo J. & Musa, W. Z., Modeling Zero – Inflated Regression of Road Accidents at Johor Federal Road F001. MATEC Web of Conferences, Vol. **47**. (2016)
4. Hosseinpour, M., Yahaya, A. S., Ghadiri, S. M., Prasetijo, J., Application of Adaptive Neuro-fuzzy Inference System for Road Accident. KSCE Journal of Civil Engineering. Vol.**17**(7), pp. 1761-1772 (2013).
5. Prasetijo J., Capacity and Traffic Performance of Unsignalized Intersections under Mixed Traffic Condition. Ph.D Thesis, Ruhr University Bochum, Germany (2007)
6. Uno N., Iida Y., Itsubo S. & Yasuhara S., A Microscopic Analysis of Traffic Conflict Caused by Lane-Changing Vehicle at Weaving Section, Kyoto University, Japan. (2002)
7. Halim, H., Adisasmita, S.A., Ramli, M.I., Aly, S.H., The pattern of severity of traffic accidents on traffic conditions heterogeneous. International Journal of Civil Engineering and Technology, Vol. **8**(4), pp. 1720-1729. (2017)
8. Crillo, & Anna, J., The Relationship of Accidents To Length of Speed-Change Laned And Weaving Areas On Interstate Highways. Highway Research Record, pp.17-32. (2002)
9. Shoraka M. & Che Puan O., Review of Evaluating Existing Capacity of Weaving Segments, Vol. **1**(3). (2010)
10. Highway Capacity Manual., Transportation Research Board, National Research Council , Washington, D.C. (2010)

MATEC Web of Conferences **181**, 06003 (2018) https://doi.org/10.1051/matecconf/201818106003
ISTSDC 2017

Congestion cost analysis of Condongcatur signalized intersection Sleman, D.I. Yogyakarta using PTV. Vissim 9

Muchlisin[1,*], *Muhamad* Yusup[2], and *Noor* Mahmudah[1]

[1]Muhammadiyah University of Yogyakarta, Depart of Civil Engineering, Jl. Brawijaya, Indonesia

Abstract. The congestion cost is one of the most important consideration when congestion happens. One of the most frequent congestion is in the intersection. And one of the intersections in Yogyakarta Special Region is Condongcatur signalized intersection, Sleman. This study aims to analyze the congestion cost of intersection by using PYV. Vissim 9 as a traffic microsimulation model. For the congestion costs, the author uses Tzedzakis, 1998 approach. The results show that in the existing condition, the average delay is 103.72 determining Vehicle Operating Cost (BOK), Existing Speed is 70 km/H for primary road and 60 km /H for secondary road, travel time value is KB : Rp. 4.970, KR: Rp. 1.925, and SM : Rp. 315. Therefore, the total congestion cost is Rp. 5.663.790, -/hour.

1 Introduction

The congestion cost is one of the most important consideration when congestion happens. One of the most frequent congestion is in the intersection. And one of the intersections in Yogyakarta Special Region is Condongcatur signalized intersection, Sleman. Condongcatur signalized intersection is a North Ringroad of Yogyakarta which is a connector between Provinces of Central Java and an access road to the city of Yogyakarta.

In this study, the author tries to approach how much the cost incurred in the signalized intersection. According to Stubs in Sugiyanto [1], he mentions that the cost of congestion is a relation between speed and flow on the road and the relation between speed and vehicle cost. If the existing traffic flow limit is exceeded, then the average traffic speed will decrease. When the speed starts to fall, then the vehicle operating costs (BOK) will increase in the range of 0-45 miles / hour and the time to travel wil increase. But in case at a signal intersection, not all vehicles have speed due to delay on traffic light effects.

Corresponding author: muchlisin@umy.co.id

Therefore, in this study the author uses simulation results from PTV. Vissim 9 student version as the diciding factor of the intersection delay. The intersection delay is used to predict how much it will cost. Vehicle operating costs (BOK) in this study refers to Sugiyanto [2], which has made an approach to determine the BOK from the speed of the vehicle.

2 Methodology

2.1 Framework analysis

The data used in this study is based on field survey and previous study which then modeled using PTV. VISSIM. 9. Flow chart that explain the methodology can be seen in Figure 1.

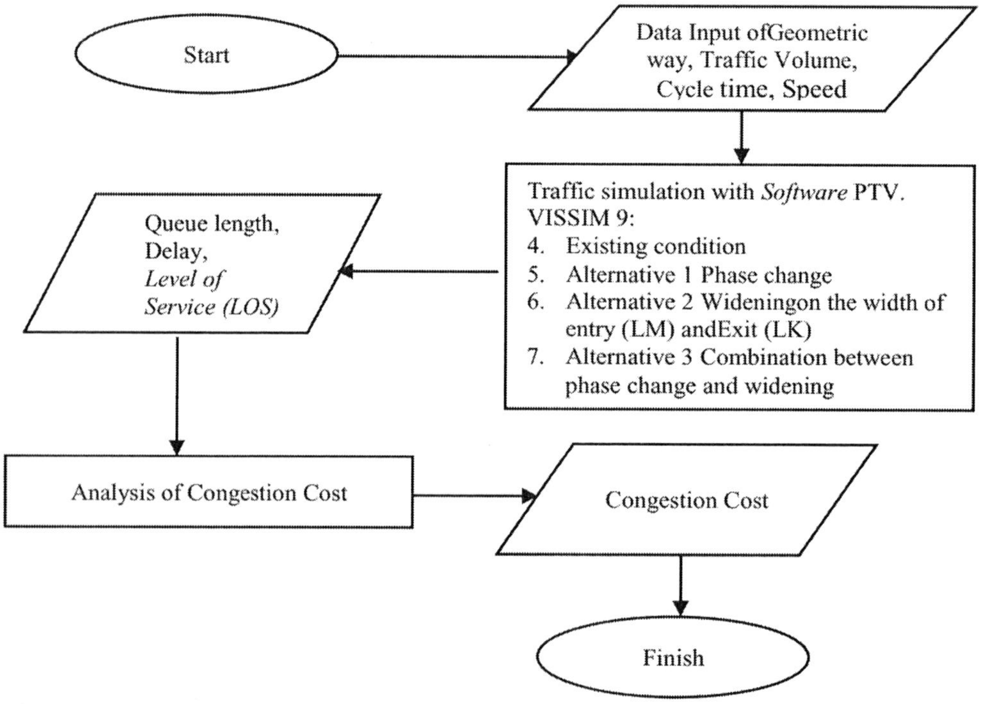

Figure 1. Framework analysis

2.2 Study location

This study is located at Condongcatur Signalized Intersection Sleman, Special Region of Yogyakarta which is a connecting the North Ringroad, Jl. Angga Jaya, and Jl. Affandi (Gejayan). The Location can be seen in Figure 2.

2.3 Study time

The traffic volume used in this study is secondary data on previous study by Maulidiah [3]. Data collection is done for 12 hours on weekdays at 06.00 – 18.00 WIB.meanwhile, in determining the geometric way, cycle time, and environmental conditions are carried out at the observation on Sundays, December 04, 2016. And to determine the speed of the vehicle

(Spot Speed) conducted on Wednesday, March 22, 2016 which represents the weekday through survey every type of vehicle KB (heavy Vehicle), KR (Light Vehicle), and SM (Motorcycle). The study was conducted with survey in the morning at 06.00 – 08.00 WIB, 11.00 – 13.00 WIB, and in the afternoon 16.00 to 18.00 WIB.

Figure 2. The study location (Source: Google Earth, 2016)

3 Results and analysis

3.1 Geometric condition

The geometric way in the intersection can be seen at the figure below.

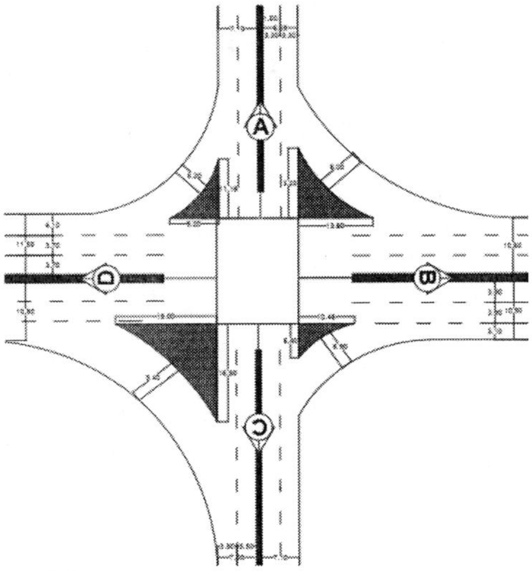

a. Wide of arm North (A) : 7,1 meter
b. Wide of arm East (B) : 10,9 meter
c. Wide of arm South (C) : 6,6 meter
d. Wide of arm West (D) : 11,5 meter

Figure 3. Intersection geometric

3.2 Type of road environment

Type of road environment can be seen at the table below.

Table 1. Type of road environment

Arm Code	Regional Condition	Type of Road Environment
Jl. Angga Jaya (N)	Shopping complex	Commercial
Jl. North Ringroad(E)	Shopping complex	Commercial
Jl. Gejayan (S)	Shopping complex, office complex	Commercial
Jl. North Ringroad(W)	Shopping complex	Commercial

3.3. Phase condition

Phase condition can be seen at the table below.

Table 2. Phase condition

Signal	Arm	Type of Arm	Time (second)			
			Red	Green	Amber	All Red
Phase 1	N	Protected (P)	162	34	3	4
Phase 2	E	Protected (P)	137	59	3	4
Phase 3	S	Protected (P)	167	29	3	4
Phase 4	W	Protected (P)	143	53	3	4
Cycle Time (second)			203			

3.4 Traffic volume

Traffic volume survey conducted by data Maulidiah, E [3]. The survey was taken on Monday, 12 january, 2016 which represent on weekdays condition. This is the traffic volume.

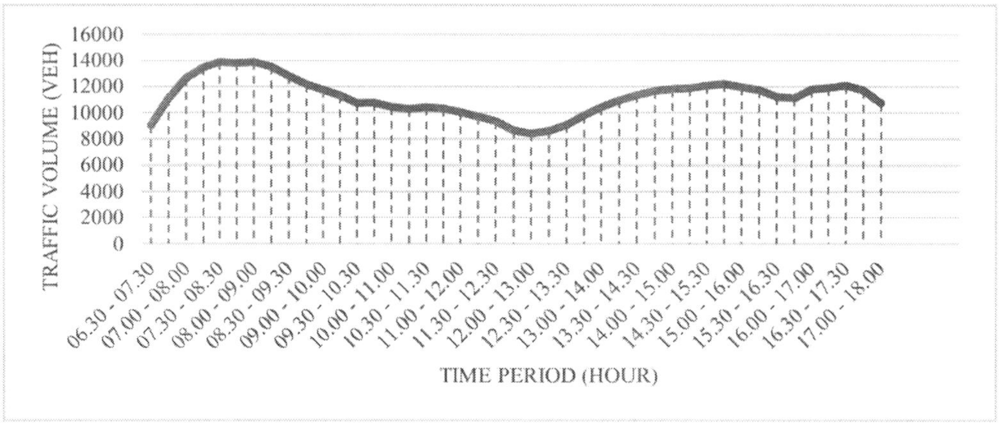

Figure 4. Traffic volume fluctuation

3.5 Traffic modeling by PTV. Vissim 9

This is the flowchart how PTV. Vissim used in this research.

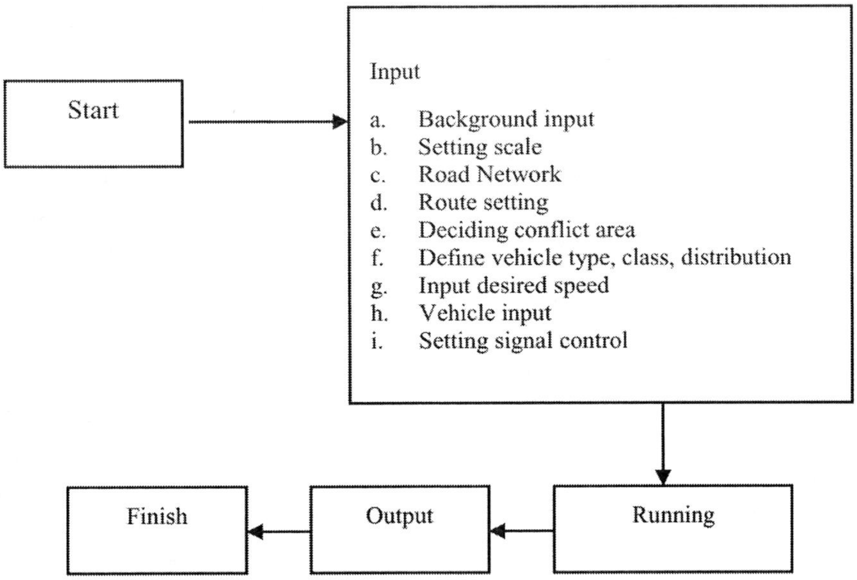

Figure 5. PTV.Vissim 9 process

And this is the output.

Table 2. PTV.Vissim 9 output on existing condition

NO	MOVEMENT	QLEN	QLENMAX	VEHS (ALL)	PERS (ALL)	LOS (ALL)	LOSVAL (ALL)	VEHDELAY (ALL)	PERSDELAY (ALL)	STOPDELAY (ALL)	STOPS (ALL)
1	Jl. Ringroad Utara (B) - Jl. Angga Jaya (U)	0.00	0.00	41.00	41.00	LOS_C	3.00	24.57	24.57	13.79	3.68
2	Jl. Ringroad Utara (B) - Jl. Ringroad Utara (T)	291.55	383.33	131.00	131.00	LOS_F	6.00	139.04	139.04	115.26	4.56
3	Jl. Ringroad Utara (B) - Jl. Affandi (S)	291.55	383.33	60.00	60.00	LOS_F	6.00	174.82	174.82	131.17	11.23
4	Jl. Angga Jaya (U) - Jl. Ringroad Utara (T)	0	0	48	48	LOS_A	1	6.47	6.47	3.01	0.17
5	Jl. Angga Jaya (U) - Jl. Affandi (S)	140.02	167.94	59	59	LOS_F	6	188.52	188.52	157.37	3.75
6	Jl. Angga Jaya (U) - Jl. Ringroad Utara (B)	140.02	167.94	13	13	LOS_F	6	115.25	115.25	94.15	2.46
7	Jl. Affandi (S) - Jl. Ringroad Utara (B)	24.66	148.29	19	19	LOS_D	4	36.57	36.57	21.43	1.68
8	Jl. Affandi (S) - Jl. Angga Jaya (U)	152.48	207.82	59	59	LOS_F	6	208.15	208.15	185.12	3.15
9	Jl. Affandi (S) - Jl. Ringroad Utara (T)	152.48	207.82	55	55	LOS_F	6	197.8	197.8	177.62	2.73
10	Jl. Ringroad Utara (T) - Jl. Affandi (S)	0.01	12.32	105	105	LOS_B	2	15.9	15.9	4.51	0.67
11	Jl. Ringroad Utara (T) - Jl. Ringroad Utara (B)	227.85	365.66	153	153	LOS_F	6	102.98	102.98	83.6	2.27
12	Jl. Ringroad Utara (T)- Jl. Angga Jaya (U)	227.85	365.66	47	47	LOS_F	6	122.59	122.59	99.35	3.55
	Average	**49.38**	**383.33**	**897**	**897**	**LOS_F**	**6**	**103.72**	**103.72**	**83.87**	**3.09**

3.6 Model validation

To test the validation of traffic modeling with Vissim, the author makes correlation with the number of real vehicle and from the model. Let's see the model validation from regression linier analysis with Microsoft Excel 2016.

Table 3. Comparison between the number vehicle from actual survey and from the model

Arm	Direction	Real (veh/h)	Model (veh/15 m)
	LT	671	105
E	ST	2513	153
	RT	600	47
	LT	866	19
S	ST	1059	59
	RT	979	55
	LT	112	41
W	ST	2935	131
	RT	1419	60
	LT	262	48
N	ST	1916	59
	RT	407	13

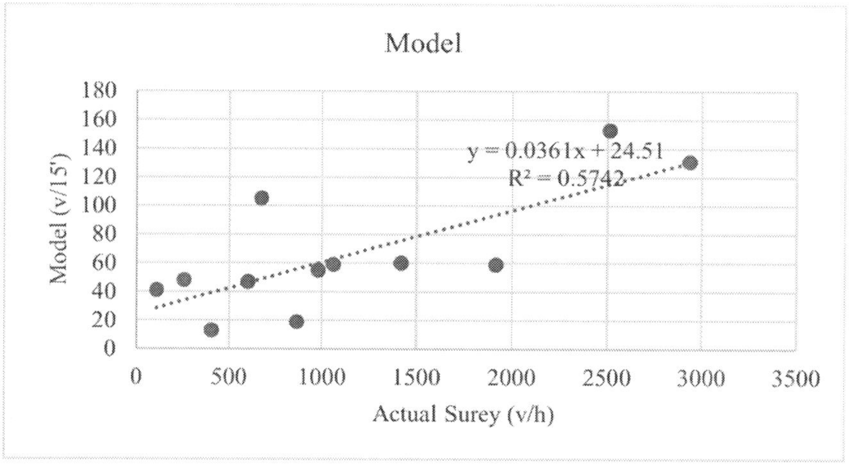

Figure 6. The regression analysis to validate the model

From the figure above, we can see that R square is 0,5742. The author argues that the the model is not too correlate with the real number of vehicle in Gejayan intersection. Because it is not close with 1. The other reason is about the software version. The author uses Student Version from this model. Student Version just only can run in 10 minute simulation.

3.7. Vehicle operating cost

The vehicle operating cost estimate by using Sugiyanto [2, 4] by using references prices at the end of September 2009.

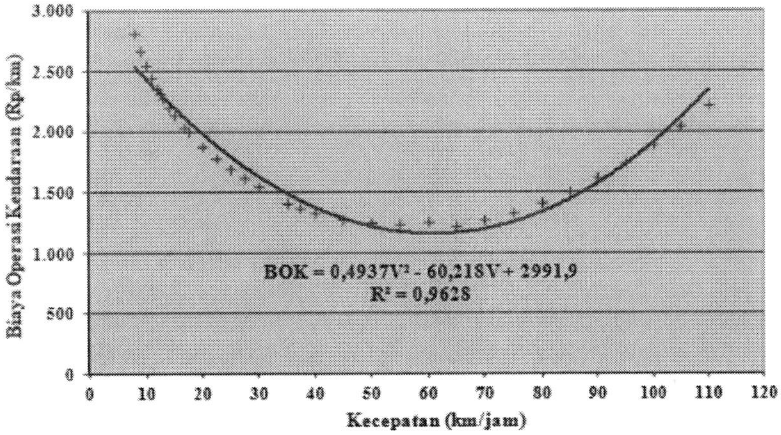

Figure 7. Speed relation and BOK personal car with LAPI ITB method [2]

From the graphic above, thus the BOK for light vehicle and heavy vehicle can be calculated as the equation below.

$$BOK = 0{,}4937V^2 - 60{,}218V + 2991{,}9 \quad (1)$$

With, V = Vehicle speed (Km/Hour)
Based on the vehicle type and arm, this is the result of BOK.

Table 3. Vehicle operational cost

No	Arm	Vehicle Type	BOK (Rp/Veh.km) (G)
1	NORTH	KB	1754.62
		KR	1621.97
		SM	375.07
2	EAST	KB	1207.84
		KR	1207.13
		SM	351.96
3	SOUTH	KB	1219.62
		KR	1177.69
		SM	365.70
4	WEST	KB	1440.24
		KR	1239.61
		SM	351.43

Note:
KB : Heavy Vehicle
KR : Light Vehicle
SM : Motorcycle

3.8 Existing speed (A) and ideal speed (B)

In this study, on the east and West arms which is the primary arterial road based on Government regulation No. 34 of 2006 on the Road is determined to be the lowest at 60 km/h, but based on the survey speed that is done at peak hours speed of 60 km / h is still exceeded, thus determined the ideal speed of 70 km/h to calculate the congestion charge.as for the North and South arms is determined at 60 km/h which is the primary collection road.

3.9 Travel time value (V)

The time value is calculated using the study of Indonesian Highway Capacity Manual [5] with approach of Gross Regional Domestic Product (GRDP). Based on the vehicle type, the time value for heavy vehicle is taken at Rp. M4.970, Light vehicle Rp. 1.925, and Motorcycle Rp. 315.

3.10 Queue and delay Time

The amount of queue or delay time is obtained from the output of VISSIM modeling.

Table 4. Queue and delay time (Vissim output)

No	Arm	Delay (second)	Queue Time (hour)
1	North	103.413	0.0287
2	East	80.49	0.0224
3	South	118.468	0.0329
4	West	112.81	0.0313

3.11 Congestion cost analysis

Congestion cost is a travel expense due to traffic delay or additional vehicle volume approaching or exceeding road service capacity [6]. Using the equation 1-2, then the congestion cost in the existing condition can be seen in Table 5.

$$C = N * \left[GA + \left(1 - \frac{A}{B}\right)V'\right]T \quad (2)$$

where:
C = Congestion cost (Rupiah),
N = Vehicle volume (Vehicle),
G = Vehicle operating cost (Rp/Veh.Km),
A = Existing speed (Km/Hour),
B = Ideal speed (Km/Hour),
V' = Vehicle time value fast (Rp/Veh.Hour),
T = Delay time (Hour).

Table 5. Total congestion cost analysis

No	Lengan	Congestion Cost (Rp/Hour)
1	North	IDR 774,589
2	East	IDR 1,573,246
3	South	IDR 1,351,843
4	West	IDR 1,964,113
Total of Congestion Cost		IDR 5,663,790

4 Conclusion

Based on the results of modeling and analysis using software VISSIM 9 at the signalized intersection Condongcatur Sleman Yogyakarta, it can be concluded as follows.

1. Existing condition of signalized intersection Condongcatur Sleman Yogyakarta shows the highest traffic volume (peakhour) occurred in the morning at 07.30 – 08.30 WIB with average delay rate of 103.72 seconds, average queue length of 49.38 meters, and Level of service level is F (very bad).
2. The costs incurred due to congestion that occurred at the signalized intersection CondongcaturSleman Yogyakarta is Rp. 5.663.790,-/hour at peak hour condition 07.30 – 08.30 WIB.

Acknowledgements

My great gratitude for Dr. Noor Mahmudah, ST. M.eng who had given me advise and support as colleague and mentor in UMY. I also should say thanks very much for Muhamad Yusuf and EstyMaulidiah who help to finish this study. My best tribute to my workplace in Civil Engineering Department, Universitas Muhammadiyah, Yogyakarta.

References

1. Sugiyanto, G.*Biaya Kemacetan (Congestion Cost) Mobil Pribadi di Central Business District. Media Teknik Sipil*(2008).
2. Sugiyanto, G.Pemodelan Biaya Kemacetan Pengguna Mobil Pribadi Dengan Variasi Nilai Aktual Kecepatan Kendaraan. Jurnal Transportasi **Vol.12 No.2**, 123-132 (2012).
3. Maulidiah, E.*Analisis Simpang Bersinyal Condoncatur, Sleman, DIY. UMY: Yogyakarta*(2012).
4. Sugiyanto, G. dkk.Pengembangan Model Biaya Kemacetan Bagi Pengguna Mobil Pribadi di Daerah Pusat Perkantoran Yogyakarta. Jurnal Transportasi**Vol.11 No.2**, 87-94 (2011).
5. Directorate General of Highway. *Highway Capacity Manual.Jakarta: Indonesia*(2010).
6. Basuki. Siswandi. Biaya Kemacetan Ruas Jalan Kota Yogyakarta. Jurnal Teknik Sipil***Vol.9 No.1***, 71-80 (2008).

7. Goverment of Republic Indonesia.. *Undang - Undang Republik Indonesia No 22 Tahun 2009 Tentang Lalu Lintas dan Angkutan Jalan. Pustaka Yustisia: Jakarta* (2009).
8. Ministry of Transportation Republic Indonesia. *Peraturan Menteri No 96 Tahun 2015. Tentang Pedoman Pelaksanaan Kegiatan Manajemen dan Rekayasa Lalu Lintas. Jakarta* (2015).
9. Khisty. Lall. *Dasar- Dasar Rekayasa Transportasi. Erlangga: Jakarta*(2005).
10. PTV VISSION.*PTV VISSIM 9 User Manual. PTV AG : Kalsrhure, Germany*(2016).
11. Sugiyanto, Gito. *Bahan Pembelajaran Mata Kuliah Ekonomi Transportasi. Universitas Jenderal Soedirman: Purwokerto*(2010).

Estimated travel time between the city (Case study: Waru Street-Mojokerto)

Budi S. Waloejo[1,*], *Dadang* Meru Utomo[1], and *Ninditarari* Salsabila W[1]

[1]Brawijaya University, Faculty of Engineering, Malang, Indonesia

Abstract. The area in Surabaya and Mojokerto is an industrial area that is increasing quite rapidly. Based on City Mojokerto in Figures Year 2016. From an industrial estate is a change of land use from agriculture or ponds into settlements and / or industries. The occurrence of congestion along this road because of the capacity that is not sufficient to accommodate the number of vehicles passing. The purpose of this research is to know the model of land use interaction, knowing the network performance and know the travel time to be passed. And analyze what components affect the calculation of travel time. Necessary data. The existing land layout along the road and the average traffic calculation data on the road in the study area. To know the travel time must be known in advance the capacity of the road that can be accommodated along the road. After the capacity is known the service performance becomes the main factor in determining road network performance. This research uses regression analysis for land use modeling.

1 Preliminary

Transportation is a tool used to move people or goods from the place of origin to the destination. Various types of transportation facilities that have developed rapidly raises various problems, one of which is the growth of the number of motor vehicles is not directly proportional to the growth of the road [1, 2]. The number of motor vehicles in Indonesia each year has increased. Based on data from the Traffic Police Corps, until July 2016 the population of vehicles in all parts of Indonesia reached 124,348,224 units. Each year vehicle growth reaches six million units per year and about 10-15 percent is the contribution rate of four-wheeled vehicles. According to the Chairman of the Association of Indonesian Automotive Industries (Gaikindo), road growth in Indonesia only reaches an average of 0.08 percent. Growth in the number of motor vehicles will result in traffic conditions that decline in quality of service, and cause discomfort such as congestion.

In addition to increasing the number of vehicles, the cause of congestion is the reduction of effective space of roads caused by parking of vehicles and / or activities of street hawkers on the shoulder of the road. Surabaya city development is very rapidly also shows the development towards the outskirts of the city directly adjacent to the city of Surabaya, such as Sidoarjo, Gresik, and Mojokerto. The border area of Surabaya and Mojokerto is an

* Corresponding author: budieswe@yahoo.co.id

industrial area that increases quite rapidly. Based on Mojokerto City In Figures 2016, the impact of increasing industrial area is a change of land use function from agriculture or pond into settlement and / or industry. Land use functions to influence urban congestion. The land will form the pattern of the resurrection zone, destination zone, and traffic volme. In an urban area it is characteristic that land use will be concentrated at one point which will result in over capacity. Reviewing that these 2 cities are the Center for Area Activity (PKW), which is filled with road users every day from and going to Mojokerto or vice versa. The occurrence of congestion along the road is due to insufficient road capacity to accommodate the number of vehicles passing by. If the capacity of the road network is almost saturated, then what will happen is a traffic jam that has adverse impacts on the social, economic, and environmental aspects. One of them is the loss of time for the wasted riders because it is spent on the road and the less productive time to work. So, with this research is expected to provide direction in the form of providing information in the form of traffic volume conditions for road users. Then the road user can choose and determine the route of the journey that will be passed.

2 Research methodology

This research is descriptive research where research which aim to describe about an event or situation. This study aims to produce a systematic, factual, and accurate information related to the characteristics of a particular population or region. In addition, this study is categorized as a correlation study to determine the level of variation in a bound factor or other factors based on the correlation coefficient. A qualitative approach is undertaken to determine the condition of the study area and the behaviour of the visitors of the land. Quantitative approach in this research is used to process linear regression model in order to get direction of problem handling strategy that happened in every field. Data collection methods used are: primary survey consists of; observations, interviews, and questionnaires. The secondary survey consists of; previous study surveys, agency surveys, and literature studies. Analytical methods used are qualitative analysis in the form of analysis of physical characteristics and road network and traffic characteristics. While quantitative analysis in the form of analysis of road network loading characteristics, free flow velocity analysis, and regression analysis. Additional analysis of the calculation of travel time calculation. To know the travel time must be known in advance the capacity of the road that can be accommodated along the road. Once the capacity is known to be obtained the level of service road that becomes the main factor in determining road network performance. The calculation of road capacity uses the following calculations [1]:

$$C = Co \ x \ FCW \ x \ FCSP \ x \ FCSF \ x \ FCCS \tag{1}$$

Where:

C	= Capacity (smp / hour)
Co	= Basic Capacity (smp / hour)
FCW	= Traffic Length adjustment factor
FCSP	= Median Adjustment Factor or Direction Separator
FCSF	= Side Constraint Adjustment Factor
FCCS	= City Size Adjustment Factor

Road network performance or service level using the following calculation [1].

$$VCR = V/C \tag{2}$$

$$VCR = \frac{\Sigma Vi + \Sigma V_{Eksternal}}{C} \tag{3}$$

Where,
VCR = Ratio Volume Capacity (service level indicator)
V = Traffic Volume (pcu / hour)
Vi = Amount of Vehicle Volume from Rise / Pull of Land
Vexternal = Amount of Ongoing Vehicle Volume On Main Street
C = Road Capacity (pcu / hour)

If the road capacity and road network performance is known then the calculation of travel time as follows [1].

$$t = \frac{\Sigma D}{V} \tag{4}$$

Where,
t = Travel Time (hours)
ΣD = Distance (km)
V = Velocity (km / h)

2.1 Research variable

The location of this research study is located at Jalan Waru which is the final boundary of Surabaya that will go to Mojokerto. The required data are land use type along the road segment and the average daily traffic data that can be obtained from the traffic counting result. The model of the rise and pull used to take the form of correlation between variables, land use, or land use characteristics from the study area as the response variable and the number of activities in the land use as explanatory variables [3, 4, 5]. Explanation of variables used are as follows:

Table 1 Research variable

No	Response variable (Dependent variable)	Explanatory Variables (Variable Free)
1	$Y_{Housing}$	X_1 = Number of Family Members X_2 = Number of Motor Vehicles X_3 = Number of Household Income
2	$Y_{PTN/PTS}$	X_4 = Building Area X_5 = Number of Students
3	$Y_{TK\text{-}SMK/SMU}$	X_6 = Number of Students X_7 = Number of Teachers / Staff X_8 = Number of Classes
4	$Y_{Pertokoan}$	X_9 = Number of Visitors
5	Y_{SPBU}	X_{10} = Number of Employees
6	Y_{Kantor}	X_{11} = Number of Employees X_{12} = Number of Visitors X_{13} = Building Area

No	Response variable (Dependent variable)	Explanatory Variables (Variable Free)
7	Y_{Apotik}	X_{14} = Number of Visitors X_{15} = Building Area
8	$Y_{Industri}$	X_{16} = Land Area X_{17} = Number of Employees X_{18} = ParkingArea
9	$Y_{Rumah\ Sakit}$	X_{19} = Daily Patient Amount X_{20} = Area of Inpatient Room

3 Result

3.1 Characteristics of Waru Street - Mojokerto

The type of road that connects Jalan Waru - Mojokerto is 4/2 UD or 2 undivided directions with 6 meters width of each line, hotmix asphalt pavement with good condition and has a median road of 0.5 meters. In accordance with the condition of the location of the study in the know Co = 6000 pcu/hour, FCw = 0.91, FCsp = 1, FCsf = 0.92, and FCcs = 1, as a result can be known the capacity of the road of 5023.2 pcu/hour.

3.2 Modeling of land use

The results of linear regression analysis resulted in land use modeling for all species contained along Waru Street - Mojokerto as follows:

Table 2. Results of modeling of origin and destination from land use

Land use		**Origin Modeling**	**Determination Coeffisient**
Housing	Residents	$Y_{Housing}$ = −0,386+0,002 X1+5,569 X2+0,00000000247 X3	1
Land use		**Destination Modeling**	**Determination Coeffisient**
Education	University	$Y_{University}$ = -0,355+0,008 X4+0,149 X5	0.994
	TK-SMK/SMU	$Y_{TK\text{-}SMK/SMU}$ = -0,254+0,505 X6+0,518 X7-0,018 X8	1
Trade and Services	Trade	Y_{Trade} = -0,431+0,075 X9	1
	SPBU	Y_{SPBU} = 0,447+142,318 X10	1
	Office	Y_{Office} = 0,108+0,662 X11+0,659 X12+0,00007194 X13	1
	Apotic	Y_{Apotic} = 25,323 + 0,322 X14 + 0,084 X15	1
Industry	Industry	$Y_{Industry}$ = -0,497-0,0000003505 X16+1,184 X17-0,000002801 X18	1
Medical	Hospital	$Y_{Hospital}$ = 13,715+0,291 X19+0,055 X20	0.999

3.3 Movement of volume of vehicles and volume of land use

By comparing the results of land use model with road capacity in getting the volume of vehicle movement in the following table. Peak volume of the largest movement is 12.253 smp / hour. The movement occurred at the peak hour at 17:00 to 18:00. the movement during peak hours is affected by the ratio of land use that reached 98%. The lowest volume of movement occurred at 11:00 to 12:00 of 7,640 smp / hour. Where the ratio of land use only reaches 51% effect on road capacity which only reaches 5,023 smp / hour.

3.4 Model of land use interaction with road network

The total movement of vehicles per hour in Jalan Waru - Mojokerto affects the level of service seen from its capacity. Calculation of the application of land use model in Waru-Mojokerto Street as a whole, then the result will interaksikan with Waru-Mojokerto road network system. The movement of vehicles from land use as the rise / pull and continuous current movement as well as the calculation of road service level by comparison of total volume with capacity can be seen in the following table.

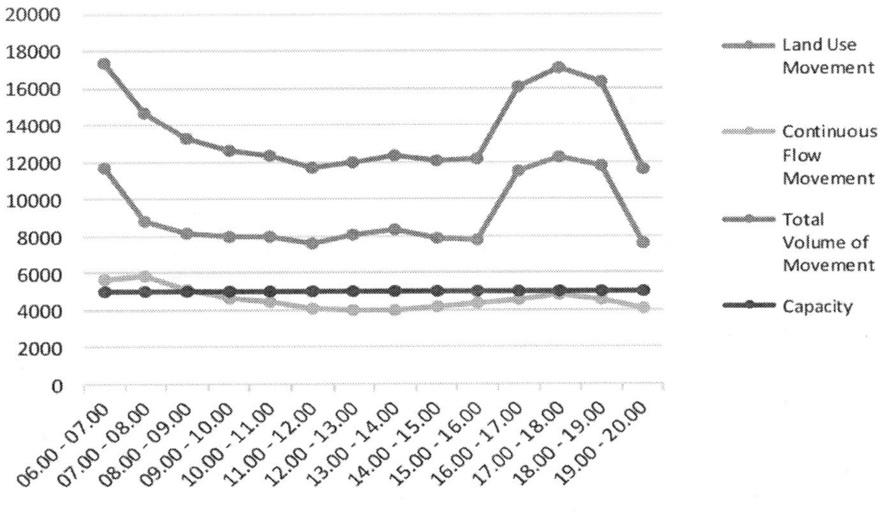

Fig. 1. Land use and road interaction

The interaction results of vehicle movement volumes from the rise / pull of land use with the road network system can show the performance of the road network or the level of road service. The calculation in the table above shows if at 6:00 to 20:00 the service level of the road is F. In the graph can also show the movement of the rise and / or pull along the road corridor has a very big influence compared with the continuous current. It affects the capacity of Jalan Waru-Mojokerto which is only 5023.2 smp / hour.

3.5 Estimated time travel

The calculation of travel time can be calculated with known average density divided by road length. Based on the calculation of the volume of movement with road capacity in Jalan Waru - Mojokerto, the service level is F, where the average of the density can be reached 30 km/hour. To determine the estimated travel time to be achieved is to compare the length of the road with its speed. The length of the road is known 34 km. Then get the time to travel is 68 minutes.

Table 3. Calculation of Travel Time

Time	Movement of Vehicles from Land Use (V_i)	Continuous Flow Movement ($V_{eksternal}$)	Total Volume Movement	Capacity	Ratio V/C	LOS	Travel Time (minutes)
06.00 - 07.00	11739	5663	17402.0	5023.2	3.46	F	68
07.00 - 08.00	8815	5887	14701.5	5023.2	2.93	F	68
08.00 - 09.00	8199	5093	13292.1	5023.2	2.65	F	68
09.00 - 10.00	7977	4663	12640.2	5023.2	2.52	F	68
10.00 - 11.00	7970	4404	12374.1	5023.2	2.46	F	68
11.00 - 12.00	7640	4037	11677.4	5023.2	2.32	F	68
12.00 - 13.00	8047	3978	12025.2	5023.2	2.39	F	68
13.00 - 14.00	8313	4008	12321.2	5023.2	2.45	F	68
14.00 - 15.00	7887	4183	12070.0	5023.2	2.40	F	68
15.00 - 16.00	7804	4386	12190.7	5023.2	2.43	F	68
16.00 - 17.00	11553	4498	16051.1	5023.2	3.20	F	68
17.00 - 18.00	12253	4866	17118.8	5023.2	3.41	F	68
18.00 - 19.00	11777	4579	16355.9	5023.2	3.26	F	68
19.00 - 20.00	7578	4040	11617.5	5023.2	2.31	F	68

Note: Length of road is 34 km.

4 Conclusion

4.1 Model of the relationship of traffic flow with travel time

On the Waru Street- Mojokerto road segment, the volume of vehicle movement from the raised / pulled generated land (127,551.35 pcu/hour) is smaller than the continuous movement volume on the main road (64,286 pcu/hour). For land that contributes the largest volume of movement is the use of industrial land. So the more land to build land for land on the main road will have an impact on increasing the level of congestion on the main road.

For the service level of Jalan Waru - Mojokerto and Interaction Model of Land Use - Road Network, it shows that at 06.00 - s / d - 20.00 the level of service = F. Meanwhile, the time required to take the road Waru Street - Mojokerto road with the length of the road 34 km is 68 minutes. Where the service level of the road = F, so the speed is 30 km/hour.

4.2 Components that affect the determination of travel time estimation

The amount of travel time on the road depends on the amount of current and capacity of the road segment.

1. Influential components in road capacity calculation is basic capacity, capacity correction factor for road width, capacity factor due to directional distribution, capacity correction factor due to side interference and capacity correction factor due to city size.
2. Traffic flow is divided into two types, namely internal currents (land use movement) and external currents (continuous currents).

References

1. Tamin, O. Z. *Perencanaan dan Permodelan Transportasi*, Bandung: ITB. (2000)
2. Sahrullah, Ramli, M.I., Ali, N., Rahim, R., A travel time estimation model of private cars in urban arterial roads based on heterogeneous traffic. International Journal of Civil Engineering and Technology, Vol. 8(7), pp. 676–685. (2017)

3. Damayanti, R. *Pengaruh Guna Lahan dan Pola Pergerakan Terhadap Tingkat Pelayanan Jalan di Sekitar Bandara Soekarno Hatta*. AGORA, Jurnal Arsitektur 2. (1997)
4. Kumalasari, D. *Pengaruh Guna Lahan Terhadap Tarikan Pergerakan, Biaya Kemacetan, dan Biaya Kecelakaan (Di Jalan KH. Abdul Fatah - Jalan Kapten Kasihin Tulungagung)*. Jurnal Rekayasa Sipil 4. (2011)
5. Sugiarto, Budi W. *Studi Pemasangan Indikator Estimasi Waktu Tempuh Perjalanan Antar Kota*. Dinas Perhubungan dan Lalu Lintas Jalan. Jawa Timur. (2015)

Evaluation of performance and service quality of Trans Bandung Raya Bus (case study: route of Elang Terminal – Jatinangor Terminal)

Antono Damayanto[1*], *Faizal* Kenedi[1], and *Bagea R.* Yogatama[1]

[1]Jenderal Achmad Yani University, Departement of Civil Engineering, Cimahi, Indonesia

Abstract. Land transportation needs are increasing. One of the problems is the uncontrolled number of private vehicles. Buses as public transportation become the right choice to overcome the problem. Bus rapid transit is a fast, convenient, secure and punctual. The research purpose is to know the performance of Trans Bandung Raya Bus (TBRB) based on the policy of Director General of Land Transportation Decision No.SK.687/AJ.206/DRJD/2002 and to analyze the service quality factors of TBRB. The method of service quality assessment is Importance Performance Analysis. The research showed that the total value of TBRB's performance is 23 and based on the performance standards of the Director General of Land Transportation Decision is the "Good" category. The value of TBRB service quality (2.63) is smaller than the expected value (3.31) which means that TBRB have to keep improving their performance. The service quality variables that need to be improved are Empathy and Assurance.

1 Introduction

Currently, the number of residents in West Java, especially the city of Bandung, continues to increase in line with the increase in the economy. This resulted in the transportation need, especially land transportation is increasing and this causes the movement of vehicular traffic increased rapidly, resulting in various problems of land transportation such as traffic congestion. One of the problem causes is the number of private vehicles increasing significantly. This is because people prefer to use private vehicles as a mode of transportation due to aspects of time schedule, comfort and security. The best solution to overcome this problem is the Government together with Private Parties should continue to improve the quantity and quality of public transportation, including Bus.

Bus rapid transit (BRT) is a bus management system with characters: fast, convenient and secure, and timely in terms of schedule. Bus rapid transit (BRT) is used as an alternative of transportation because it has better service quality than other buses system. In the BRT system, passengers up and down are done quickly at the bus stop, thus saving travel time, which ultimately improves the quality of bus services. Currently, Bandung is one of the cities that have Bus Rapid Transit (BRT), one of which is Trans Bandung Raya

* Corresponding author: antono@ymail.com

Bus (TBR). Bus TBR operation needs to be improved steadily, because it is not functioning optimally yet. Therefore, the researcher conducted a case study at route of Elang Terminal (Bandung City) to Jatinangor Terminal (Bandung District).

2 Performance

Performance is the ability or potential of public transportation to serve the needs of movement in an area, either in the form of transportation of goods and transportation of people (human) [1]. Performance is also the level of achievement or the work of the company of the goals to be achieved/tasks that must be implemented within a certain time [2].

2.1 Performance of public transport

The performance parameters of public transport are based on the Decision of the Director General of Land Transportation [3] is as follows:

Travel Time: Travel time is used to measure the travel time of a public transport for mileage per kilometer. Travel time can be calculated using equation as follows:

$$W = \frac{T}{J} \tag{1}$$

where, W is traveling time of public transport in minutes/km, T is the distance between segments in km and J is traveling time of public transport in minutes.

Travel speed: The speed of the urban public transport journey is the ratio of the operating distance to the travel time required by the transport in carrying out its service operation. The equations used in measuring travel speed as follows:

$$V = \frac{60J}{T} \tag{2}$$

where, V is traveling speed of public transport in km/hour, J is the distance of public transport routes in km and T is traveling time of public transport in minutes.

Load factor: Load factor is the ratio of the number of passengers with the capacity of the seat per unit of time. The limit of ideal load factor is < 70% (KM 35 of 2003). To determine the load factor used the following formula:

$$Lf = \frac{JP}{C} \times 100\% \tag{3}$$

where, Lf is the load factor (%), JP is the number of passengers / public vehicle and C is the passenger capacity / public vehicle.

Time of service or hours of operation: Service time is very influential on the acquisition of rit in one day, the operational cost, income and services of public transport provided to the community.

Circulation time: Circulation time is the time required by the city transport to undergo a round or two route rit from the original terminal back to the original terminal. The formula is used as follows:

$$CT_{ABA} = (T_{AB} + T_{BA}) + (\sigma_{AB} + \sigma_{BA}) + (T_{TA} + T_{TB}) \tag{4}$$

where, CT_{ABA} is the circulation time from A to B then return to A, T_{AB} and T_{BA} are average travel time from A to B and from B to A, σ_{AB} and σ_{BA} are the travel time deviation from terminal A to terminal B and from terminal B to terminal A, T_{TA} and T_{TB} are the stop time at Terminal A and Terminal B.

Frequency of service: Frequency is the number of vehicles that operate within one hour. Frequency calculation using the following formula:

$$f = \frac{N}{60} \tag{5}$$

where, f is the frequency of the number of vehicles per minute and N is the number of vehicles.

Time between vehicles (headway): Headway is the time interval between a public transport vehicle and the next public transport vehicle that must go through a certain point. The headway value is calculated by the following equation:

$$H = \frac{60}{f} \tag{6}$$

where, H is the time between vehicles in minutes and f is the frequency of service in vehicle per hour.

Waiting Time: Waiting time is the stop time of public transport at the origin or destination, the public transport waiting time can be calculated from half of Headway.

2.2 Standard appraisal of public transport performance

The performance appraisal of public transport uses the indicators and standard of value published by the Ministry of Transportation as follows [4]:

Table 1. Performance indicators of public transport

NO	INDICATOR	UNIT	ASSESMENT STANDARD		
			NOT GOOD (1)	MODERATE (2)	GOOD (3)
1	*Load factor*, busy hour	%	> 100	80 – 100	< 80
2	*Load factor*, no busy hour	%	> 100	70 – 100	< 70
3	Travel speed	Km/hour	< 5	5 – 10	> 10
4	*Headway*	Minutes	> 15	10 – 15	< 10
5	Time travel	Minutes/km	> 12	6 – 12	< 6
6	Service time	Hour	< 13	13 – 15	> 15
7	Frequency	Vehicle/hour	< 4	4 – 6	> 6
8	Number of vehicles operating	%	< 82	82 – 100	100
9	Waiting time	Minutes	> 30	20 – 30	< 20
10	Start and end time of trip	Time	05.00 - 18.00	05.00-20.00	05.00 >20.00

Source: Director General of Land Transportation 2002 in Marsudi [4]

Total performance value of public transport is categorized according to the standard in the following table [4]:

Table 2. Standard appraisal of public transport performance

CATEGORY	TOTAL VALUE
Good	18,00 - 24,00
Moderate	12,00 - 17,99
Not good	< 12

Source: Director General of Land Transportation 2002 in Marsudi [4]

2.3 TBR Bus Demand

The number of Bus TBR demand at any time of circulation can be estimated by the following equation:

$$K = \frac{CT_{ABA}}{H \times fA} \times 100\% \quad (7)$$

$$H = \frac{60 \times C \times Lf}{P_{maks}} \quad (8)$$

where, K is the number of vehicles per circulation time, CT_{ABA} is the circulation time in minutes, H is the time between vehicles, fA is the vehicle availability factor (100%), Lf is the load factor (used 70%), C is the passenger capacity of vehicle and P_{maks} is the maximum number of passengers

Number of TBR Bus needed in the busy time period is calculated by the following equation:

$$K' = K \times \frac{W}{CT_{ABA}} \quad (9)$$

where, K' is the number of vehicles busy time period, K is the number of vehicles per circulation time, W is busy time period in minutes and CT_{ABA} is circulation time in minutest.

3 Importance Performance Analysis (IPA)

Importance Performance Analysis (IPA) was formulated by John A. Martilla and John C. James [5] is a descriptive statistical method for measuring a person's level of satisfaction over the performance of others. Person satisfaction is measured by comparing the level of expectations / importance with perception / performance (Gap Analysis). If the expectation level is higher than perception, it means that the consumer has not reached satisfaction, or vice versa. The results of the analysis are set forth in a Cartesian diagram of Martilla [5], as follows:

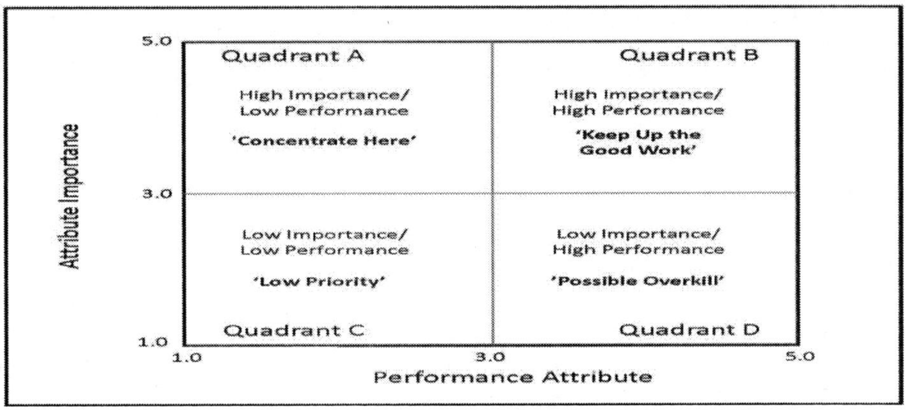

Fig. 1. Importance – Performance matrix (adapted from Martilla and James [5])

Quadrant A: aspects require immediate attention for improvement and are major weaknesses. Quadrant B: aspects indicate opportunities for achieving or maintaining competitive advantage and are major strengths. Quadrant C: aspects are minor weaknesses and do not require additional effort. Quadrant D: aspects indicate that business resources committed to these attributes would be overkill and should be deployed elsewhere".

4 Methodology and result

The structure of the research framework can be seen in Figure 2 below

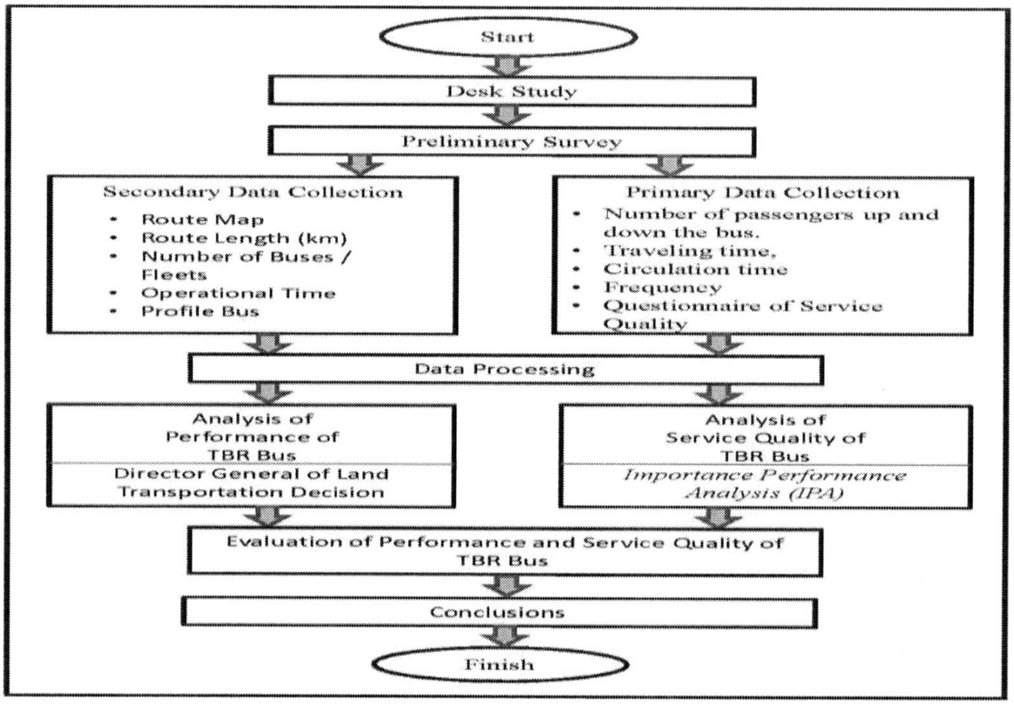

Fig. 2. Methodology of research

4.1 Data collection method

Data collection was done by two approaches, dynamic survey and static survey. Dynamic surveys are conducted on top of the vehicle to collect data, among others: the number of passengers up and down the vehicle, travel time including delays, stop time to take and down passengers, and vehicle speed. Static surveys are conducted outside the vehicle to record vehicles number in operation, travel time, circulation time and service frequency.

4.2 Variable selection of service quality

The service quality appraisal approach used is the ServQual method developed by Parasuraman, quoted from Muluk [6]. ServQual divides the service quality into five dimensions (main variable): Reliability, Responsiveness, Assurance, Empathy, and Tangibles (physical evidence).

The main variables of service quality are described in several sub variables, as follows: 1. Area and comfort of bus stops. 2. Cleanliness and neatness of bus stops. 3. Completeness of bus stop Facility. 4. Cleanliness and tidiness of officers. 5. Service time for customers. 6. Knowledge of officers of consumer needs. 7. Service is conducted professionally during the trip 8. Discipline and responsibility of officers. 9. Hospitality and courtesy of officers. 10. Service is done thoroughly. 11. The role of officer's responds to criticism and suggestions 12. Responsiveness of officer to handle consumer complaints. 13. Knowledge and ability of officers to handle problems. 14. Skills of officers to handle consumer needs. 15. Reliable and assurance security; 16. Officer's Attention to consumer. 17. The services provided do not view social status / gender. 18. The officer's care to the needs and the desire of consumers

4.3 Trans Bandung Raya Bus (TBRB)

Trans Bandung Raya Bus is a bus that is a mass transit mass transportation strived by the City Government of Bandung. The bus system is based on a service system that has characteristics such as: scheduled of bus departure, punctual, stop at bus stop, safe, convenient, reliable, affordable and friendly for the environment. TBRB has been operating in hopes of reducing the number of private vehicles and public transportation, as well as congestion solutions in Bandung and surrounding areas. TBRB is managed by Damri Public Company. The number of TBRBs available and has an operating license at the Elang Terminal - Jatinangor Terminal is 18 buses, but fully operational only 15 units.

The distance from Elang Terminal to Jatinangor Terminal is 26.7 Km. Location of research, routes and distances are described as follows:

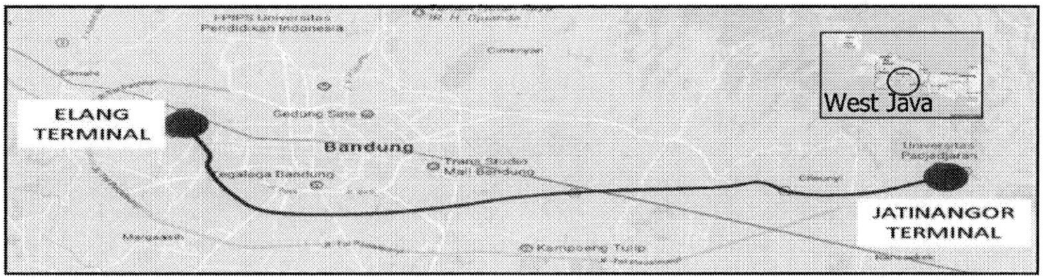

Fig. 3. Location of research

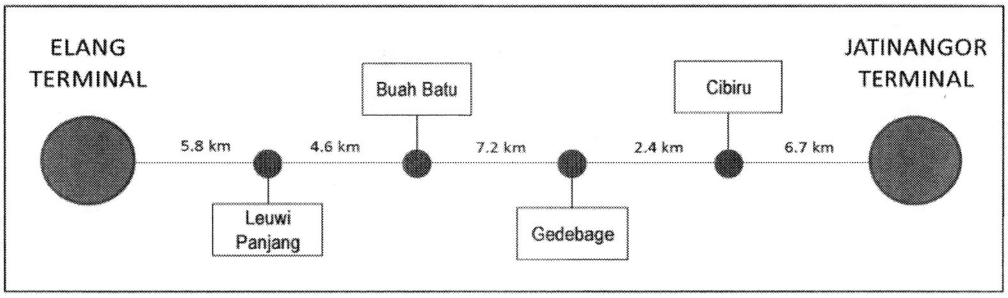

Fig. 4. Route and distances

TBR Bus Capacity based on interviews is 70 passengers, consisting of 30 seated passengers and 40 standing passengers. Passengers are allowed to stand because the distance and travel time is relatively short.

4.4 Performance analysis of TBR BUS

Based on the survey and analysis, the performance of Trans Bandung Raya Bus can be seen in the following table:

Table 3. TBR Bus performance for route of Elang Terminal – Jatinangor Terminal

NO	INDICATOR	UNIT	ASSESMENT STANDARD			RESULT	VALUE OF PERFOR MAN-CE
			NOT GOOD (1)	MODERATE (2)	GOOD (3)		
1	*Load factor*, busy hour	%	> 100	80 – 100	< 80	55.72	3
2	*Load factor*, no busy hour	%	> 100	70 – 100	< 70	42.86	3
3	Travel speed	Km/hour	< 5	5 – 10	> 10	23.21	3
4	*Headway*	Minutes	> 15	10 – 15	< 10	10	2
5	Time travel	Minutes/	> 12	6 – 12	< 6	69.02	1
6	Service time	Hour	< 13	13 - 15	> 15	15	2
7	Frequency	Vehicle/h	< 4	4 - 6	> 6	6	2
8	Number of operating veh.	%	< 82	82 - 100	100	83	2
9	Waiting time	Minutes	> 30	20 - 30	< 20	5	3
10	Start and end time of the trip	Time	05.00-18.00	05.00-20.00	05.00->20.00	05.00-20.00	2
Total Value of Performance							

Based on the performance assessment in the above table, it can be said that based on the standards of the Ministry of Transportation, the performance of Trans Bandung Raya Bus is included in the "Good" category.

4.5 Number of TBR Bus needs

Circulation time of TBR Bus on busy time for route of Elang Terminal - Jatinangor Terminal is 184.24 minutes. Circulation time of TBR Bus on not busy time is 165 minutes. The maximum number of passengers based on survey results is 331 people per hour. TBR Bus Needs per Circulation Time (K) on route of Elang Terminal - Jatinangor Terminal is calculated based on time between vehicles, is 21 vehicles / circulation. While in the busy hour period (K') requirement is 14 vehicles.

4.6 Service quality analysis of TBR Bus

The service quality of TBR Bus by users is measured based on the level of satisfaction. Service quality is divided into 5 main variables and 18 sub variables. Based on the analysis of main variables already mentioned above, the average value of service perception (Performance) is 2.63 and the average value of service expectation (Importance) is 3.30. The satisfaction level of TBR Bus's users is 79.39%. Means, at this time, service quality of TRB Bus has not satisfied users. The analysis results are described in the IPA quadrant, the result is as follows:

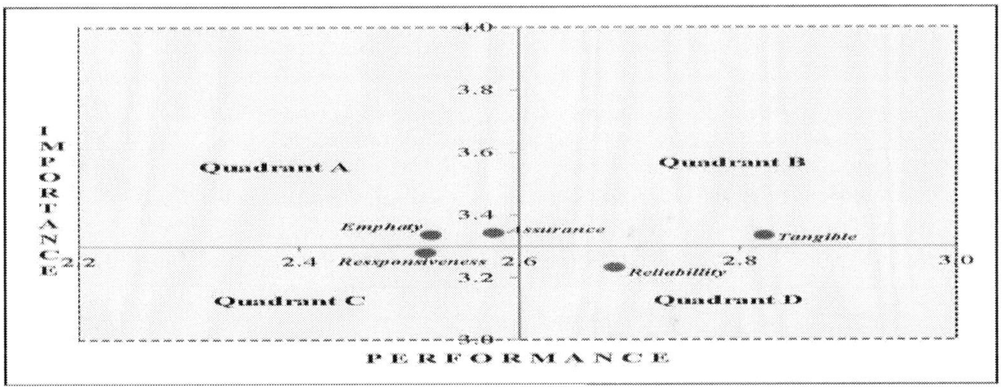

Fig. 5. IPA analysis of main variables of TBR Bus service quality

Based on the Cartesians diagram above, it can be explained: In the Cartesians diagram above, there are 2 main variables that are in quadrant A is Empathy and Assurance. This means that variables need to be improved / upgraded by the management of the TBR Bus. The implementation of TBR Bus service related to empathy and assurance still can't fulfill the expectation from TBR Bus users.

The responsive variable is in the quadrant C, which means that this variable needs to be increased steadily, although this is not a priority variable. The tangible variable is in the B quadrant, meaning the user judges this variable as a good category and needs to be maintained. While the variable reliability is in quadrant D, which means this variable is good, but felt more than is needed.

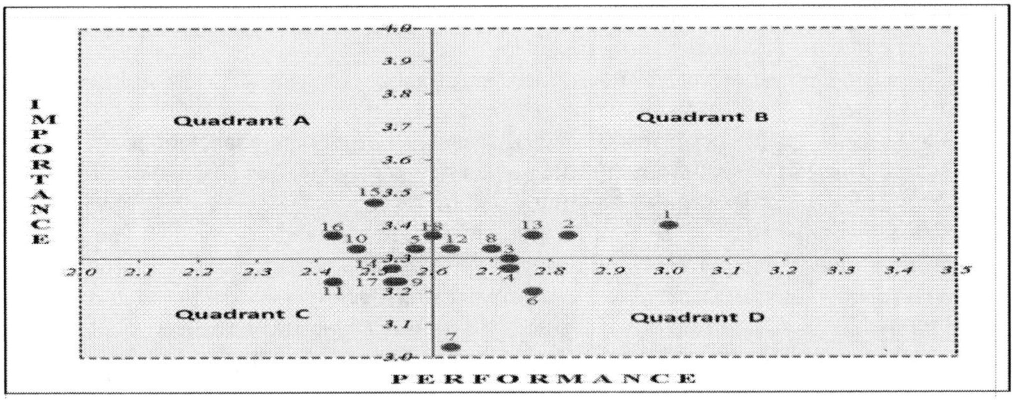

Fig. 6. IPA analysis of service quality sub variable of TBR Bus

Quadrant A (Concentrate here / Main Priority): In this quadrant, the user assesses, the service quality of TBRB is lower than the user's expectation, or it can be said that the service level of some aspects in quadrant A is still disappointing / unsatisfactory, so the management of TBRB should pay attention, and improve some aspects of it. Some aspects that need to be improved / improved quality are: service time for customers, service is done thoroughly, reliable and assurance security, officer's attention to consumer, and the officer's care to the needs and the desire of the consumers.

Quadrant B (Keep Up good Work): Shows some aspects that have been good or are appropriate between the implementation and expectations of users, so the factors need to be maintained and developed. These aspects are: area and comfort of bus stop, cleanliness and neatness of bus stops, completeness of bus stop facility, discipline and responsibility of officers, responsiveness of officers to handle consumer complaints, and knowledge and ability of officers to handle problems.

Quadrant C (Low Priority): Indicates some aspect of performance value is lower than the user's desire, so it is said to be less satisfactory so it needs to be improved, but that aspect includes low priority scale. These factors include: hospitality and courtesy of officers, the role of officer's responds to criticism and suggestions, skills of officers to handle consumer needs and the services provided do not view social / gender status.

Quadrant D (Possible Overkill): Factors in quadrant D are considered satisfactory and redundant in their implementation, although the user considers these factors less important. These factors are: Cleanliness and tidiness of officers, knowledge of officers of consumer needs, and service are conducted professionally during the trip. However, TBR Bus management should keep these factors performing well to maintain service quality for users, considering these factors will affect overall service quality.

5 Conclusion

Based on analysis results of performance and service quality of Trans Bandung Raya Bus on route Elang Terminal - Jatinangor Terminal, can be concluded:

Load factor value at busy hours is 55,72%, while load factor value at not busy hours is 42,86%, velocity 23,21 km/hour, 10 minutes headway, travel time 2,97 minutes/km, daily service time is 15 hours, frequency 6 vehicles/hour, total operating vehicles 83%, 5 minutes waiting time, starting and ending travel time from 05.00 WIB to 20.00 WIB. Based on the result above, the overall value calculated at 23 which categorized as good performances.

Number of TBR Bus of Elang Terminal – Jatinangor Terminal route needed is 21 units / circulation time, with 8.88 minutes headway. Currently, operational permits are available for 18 buses, but there are only 15 buses fully operational. This finding can be used by the government as a consideration to evaluate and to add more bus vehicles service on route of Elang Terminal – Jatinangor Terminal.

Based on calculations using Importance Performance Analysis (IPA) as an instrument to assess the service quality, results explained below:

a. The service quality performance of TBR Bus has not met the expectations of TBR Bus users. This can be seen from the value of service perception (performance) is lower than value of service expectation (importance): 2.63 <3.30.
b. Empathy and assurance have the lowest value based on user surveys. This indicates that aspects should be improved well (priority) by TBR Bus management.
c. The detail aspects that need to be improved are: speed of officers to serve customers, service is done thoroughly, reliable and assurance security, officer's attention to consumer, and the officer's care to the needs and the desire of the consumers.

References

1. Nur, N.K., Samang, L., Isran Ramli, M., Hamid, S., Study of modes transformation preferences of private transport based on travel characteristics and user behavior. International Journal of Applied Engineering Research. Vol. 11(22), pp. 10766-10771. (2016)
2. Hazian, M. *Analisis Kinerja Angkutan Kota di Kota Jambi*. Civil Engineering Department. University of North Sumatera. (2008).
3. ______, Decision of the Directorate General of Land Transportation no. SK.687 / AJ.206 / DRJD / 2002 concerning Technical Guidelines for the Implementation of Public Passenger Transportation in Urban Areas in Permanent and Regular Route.

4. Marsudi. *Analisis Kinerja Mobil Penumpang Umum (MPU) di Kota Salatiga*. Wahana Journal, Civil Engineering, **II** 3, December: 87-98. (2005).
5. Martilla, John A. and John C. James. *Importance-Performance Analysis*. Journal of Marketing. **41**(1). 77-79. (1977).

Evaluation passenger car unit for motorcycle in Indonesia Highway Capacity Manual (Case study: Bandung and Semarang)

Najid [1,*]

[1]Tarumanegara University, Department Civil Engineering, Jakarta, Indonesia

Abstract. Value of Passenger Car Unit or commonly known as PCU value is a value that is given to any vehicle that is classified into heavy vehicles, light vehicles (passenger car) and motorcycles. The value of passenger car unit on Indonesia Highway Capacity Manual (IHCM) set up in 1997 is based on a study conducted from 1980-1990 in several cities in Indonesia At the time of the study, the traffic conditions are very different to the current traffic conditions. That affects of difference traffic conditions are the composition of traffic, traffic regulations, traffic density, traffic discipline and the presence of mass transit, so that the results of traffic analysis do not always correspond to reality as there are anomalies in the determination of the level of road service (Najid, 2014). As well the incompatibility of the capacity value which is considered due to the incompatibility value of Passenger Car Units (PCU). Evaluation PCU become very important to get the value of traffic parameters into compliance with actually occur. In accordance with the traffic density is higher actually, then it is necessary to study for evaluation against PCU current value and the need to approach or to get the value of PCU more in line with current traffic conditions. Data collected at two cities, those are Bandung and Semarang. Based on analysis found PCU's value that got from survey have difference but not all significantly with PCU value in IHCM.

1 Background

As we know that actually the use of private transport is very high, especially motorcycles in almost cities in Indonesia, especially the cities in Java Island. Those phenomenon caused by rapidly growth of private vehicle usage especially motor cycle, while on some routes passengers of public transport decreased significantly. This shows the displacement of demand from public transport to private transport.

The high usage of private transport that cause in-efficiency of urban transportation, so the level of road service to be down [1]. The fall in the level of road service is an indicator that can be used by the Government in road development planning. The problem is in the analysis of the results obtained often not appropriate like traffic volume is greater than

* Corresponding author: najid2009@yahoo.co.id

capacity so that ratio of volume and capacity be more than one. Therefore, the determination of the value of passenger car unit which is the decisive factor in the calculation of traffic volume becomes very important to be evaluated.

The calculation of road capacity in IHCM using passenger car unit that strongly influenced by the condition or characteristics of the vehicle, traffic as well as by the characteristics of the road it self based on its function. So in this research need survey conducted in various road type.

This research undertaken Bandung city and Semarang city as case study, so the survey have conducted on those cities.

2 Research objectives

This research aims to :

- Determining the value of passenger car unit based on survey result.
- Identify relationships between types of road with the passenger car unit values.
- Formulate the magnitude of the mismatch between the value of passenger car unit based on survey and based IHCM.

3 Problem Identification and Limitation

- Value of PCU in IHCM look too optimistic and rude.
- Value of PCU in IHCM not consider the actual service conditions.
- PCU value that evaluated and analyzed is PCU value for urban roads.

4 Review of literature

Directorate General Bina Marga has recorded the total length of the existing road network in Indonesia, which reached 376176 km consisting toll road 741.97 km (0.20%), non-toll national roads 38 569 km (10.25%), road provinces 48 681 km (12.94%), district road 255 253 km (67.85%) and urban road 32 932 km (8.75%). In-efficiency traffic movement take place on urban roads in everyday [2].

As stated in the background the growth of private vehicle is cause declining the level of service in the mark with declined travel speed and increased the density of traffic on the road. The calculation of traffic volume is highly influenced by the value of PCU (Passenger Car Unit). Value of PCU is the value of equality of another vehicle to vehicle passenger cars were categorized as light vehicle (LV), while other vehicles are motorcycles (MC) and heavy vehicles (HV) including all medium bus and big bus as well as all the trucks.

The Directorate General of Highways, Ministry of Public Works in cooperation with the consultant at that time Sweroad of Sweden make traffic's handbook named Indonesian Highway Capacity Manual (IHCM) in year 1997. The manual has been made for determination of the PCU value of vehicles based on traffic conditions and the road conditions. PCU value of the vehicle is also distinguished in urban areas or in areas between cities. The following Table 1a-1c shows the value for the urban PCU based IHCM [3]:

Table 1a. Intersection

Vehicle Type	PCU Protected	PCU Opposed
Light Vehicle (LV)	1	1
Heavy Vehicle (HV)	1.3	1.3
Motor Cycle (MC)	0.2	0.4

Table 1b. Urban road (undivided road)

Road Type	Traffic Flow 2 ways (vehicle/hour)	PC		
		HV	MC	
			Road Width, Wc (m)	
			≤ 6	□ 6
2 lanes undivided (2/2 UD)	0-1799 ≥ 1800	1.3 1.2	0.50 0.35	0.40 0.25
4 lanes undivided (4/2 UD)	0-699 ≥ 3700	1.3 1.2	0.40 0.25	

Table 1c. Urban road (divided road)

Road Type	Traffic Flow/lane (Vehicl/hour)	PC	
		HV	MC
2 lanes 1 Way (2/1) 4 lanes 2 Ways (4/2 D)	0-1049 ≥ 1050	1.3 1.2	0.40 0.25
3 lanes 1 Way (3/1) 6 lanes 2 Ways (6/2 D)	0-1099 ≥ 1100	1.3 1.2	0.40 0.25

The magnitude of the dimensions of the vehicle will affect the value of PCU [4]. Research to determine the value of PCU been done by some researchers. either on the road section and at signalized intersection or intersection is not signalized. Past research in Surakarta by Rosma in determining PCU at signalized intersections. Value of motorcycle's PCU was determined with linear regression method, result of the research a value of motorcycle's PCU between 0.11 to 0.2. As for bus passengers's PCU between 1.02 to 1.72 [5].

As comparisson total population and density population in Bandung and Semarang showed on table 1d follow:

Table 1d. Population comparisson

No	City	Population (habitant)	Area (km^2)	Density habitant/km^2
1	Bandung	2.341.097	167,67	13.962,5
2	Semarang	1.622.520	373,78	4.340,8

Source: Central Statistic Bureau (2015) [6]

5 Methodology

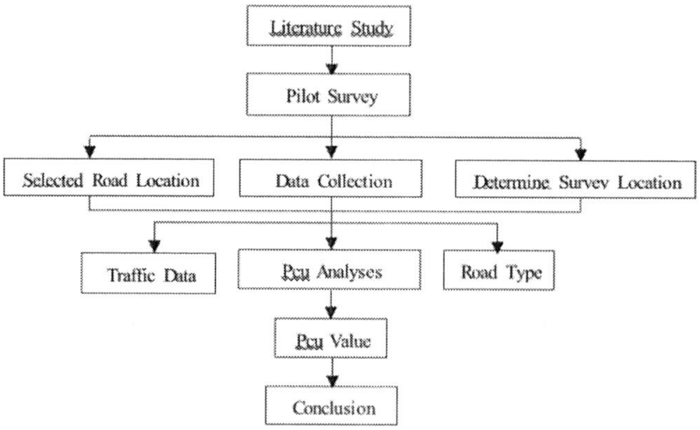

Figure 1 : Methodology

Flow chart showing the systematic study of the prose of this research that can be seen in Figure 1.

6 Determining factor passenger car unit

To determine PCU value, be used headway's method [7]. In this method, PCU various types of vehicles on the certain condition at certain road type calculated by headway's method. Estimated vehicles 1, 2, 3, ..., i, with the surveyed time of t1, t2, t3 ...., ti, is calculated as the vehicle passed the imaginer line. First Vehicle movement that passed the imaginer line be noted its time and the following vehicle be noted its time too. The difference of time is a headway. Do the same procedure for any following vehicle. The following vehicle consists of passenger car - motor cycle, motor cycle-passenger car, passenger car-passenger car and motor cycle-motor cycle.

$$PCU = ((HSM-Q) / NSM) / (HCC-Q / NCC) \quad (1)$$

Or

$$PCU.Sm = HAA / HCC \quad (2)$$

with:
HSM = Headway between motorcycle
HAA = Headway between motorcycle and car
HCC = Headway between car
Q = Correction Factor
NSM = total HSM
NCC = total HCC

7 Data collection

The flow of traffic on all roads are relatively high, especially at peak hour. The following road geometric data for the survey areas in Bandung:

- Pasteur road has a total road width of 23 metre with a median width of 0.5 metre and width of sidewalks on both sides of road each 2.0 metres, represents 6/2 D (Divided) and intensive business location
- Dipati Ukur road whole road width is 10 metres, with a median width of 0.5 metre and width of sidewalks on both sides of the road each 1.5 metres, represent 4/2 UD (Undivided) and solid business location
- Juanda road in whole width of the road is 26 metres, with a median width of 0.5 metre and width of sidewalks on both sides of road each 2.0 metres, represent 4/2 D and solid business location
- Tubagus Ismail road whole width of the road is 9 metres, without median and width of sidewalks on both sides of the road each 1.5 metres, represent 2/2 UD and residential location

The following road geometric data for the survey areas in Semarang :

- Thamrin road has a total width of 18,5 metres with a median width of 0.5 metre and width of sidewalks on both sides of road each 2.0 metres, represent 4/2 D and office location.
- Soegijopranata road has a total width of 25 metres without median and width of sidewalks on both sides of road each 2.0 metres, represent 6/2 UD and office location.

- Imam Bonjol road has a total width of 18 metres without median and width of sidewalks on both sides of road each 2.0 metres and 4/2 UD and business location
- Pandanaran road has total width 33 metres with a median width of 1.0 metre and width of sidewalks on both sides of road each 2.0 metres, represent 6/2 D and office location.

As comparison population in Bandung is 2,470,802 person and population in Semarang. From the data survey shows that the number of motorcycles is quite high compared to other type of vehicles.

7.1 Data collection for PCU of motorcycles

Table 2.1. Dipati Ukur Road 2/2 UD

Periode	Ta	Tc	Td	Tb	ta	tb
7.30-8.00	Mb-Mb	Mb-Mt	Mt-Mb	Mt-Mt	Xi-μ	Xi-μ
1	2.63	1.1	1.41	1.14	0.504	0.029
2	2.21	1.23	1.49	1	0.084	0.111
3	2.07	1.02	1.62	1.2	0.056	0.089
4	1.58	2.17	1.08	1.16	0.546	0.049
5	2.33	1.62	1.29	1.1	0.204	0.011
6	1.61	1.07	2.81	1.05	0.516	0.061
7	2.27	1.11	2.14	1.1	0.144	0.011
8	2.03	0.97	1.24	1.07	0.096	0.041
9	2.31	1.51	1.42	1.12	0.184	9E-03
10	2.22	1.07	3.8	1.17	0.094	0.059
Σ	21.26	12.87	18.3	11.11	1.394	0.289
μ	2.126	1.287	1.83	1.111		
					ta	tb
9.00-9.30	Ta	Tc	Td	Tb	Xi-μ	Xi-μ
1	2.47	1.38	3.02	1.04	0.90	0.13525
2	2.40	2.15	2.24	1.82	0.84	0.64475
3	2.60	2.61	1.26	1.35	1.04	0.17975
4	1.00	1.01	2.21	1.07	0.56	0.10025
5	1.12	1.10	1.16	1.14	0.44	0.03275
6	1.09	0.97	1.08	1.10	0.469	0.07525
7	1.09	3.06	1.18	1.16	0.469	0.01525
8	1.45	1.68	2.45	0.99	0.109	0.18525
9	0.99	2.85	2.90	1.05	0.569	0.12025
10	1.40	1.04	1.67	1.01	0.162	0.16025
Σ	15.62	17.83	19.13	11.70	3.785	1.09275
μ	1.562	1.783	1.91	1.17		

Data collection techniques on the road is use simple tools like the regular stopwatch, pen and note's board. Because many factors can affect the accuracy of the data and usually the human factor is the most decisive. The survey result show on the tables 2.1-2.4 (Bandung) and 2.5-2.8 (Semarang) as show on table.

Table 5.1 (road 2/2 UD) the headway car and motorcycle (Tc), motorcycle and car (Td) and motorcycle and motorcycle (Tb) in the morning (peak hour) is rather similar than during the day except Ta and Tc, the analysis shows the traffic volume in the morning is not significantly different with during the day.

Table 2.2. Juanda Road 4/2 D

Periode	Ta	Tc	Td	Tb	ta	tb
7.30-8.00	Mb-Mb	Mb-Mt	Mt-Mb	Mt-Mt	Xi-μ	Xi-μ
1	6.98	1.20	2.12	1.14	2.823	0.285
2	3.62	1.10	0.74	0.75	0.537	0.675
3	4.46	1.53	2.43	1.20	0.303	0.225
4	3.87	3.26	1.62	1.41	0.282	0.015
5	4.10	2.43	1.94	1.35	0.057	0.075
6	2.48	1.31	4.22	1.05	1.677	0.375
7	4.16	1.22	3.21	3.53	0.003	2.1
8	5.00	1.46	0.36	1.58	0.843	0.15
9	2.06	2.27	2.13	0.83	2.097	0.6
10	4.83	1.61	5.70	1.43	0.678	0.16
Σ	41.52	17.355	24.45	14.25	4.002	1.275
μ	4.152	1.7355	2.445	1.425		
9.00-9.30	Ta	Tc	Td	Tb	Xi-μ	Xi-μ
1	3.24	1.84	4.02	1.58	0.13	0.278
2	1.2	2.86	2.98	2.42	2.17	0.562
3	3.3	3.48	1.14	1.80	0.07	0.058
4	3.86	1.34	2.94	0.96	0.49	0.898
5	2.56	1.46	1.84	2.98	0.81	1.122
6	3.82	1.26	0.9	1.46	0.45	0.398
7	3.8	4.08	1	1.54	0.43	0.318
8	4.06	2.24	3.26	2.62	0.69	0.762
9	3.8	3.8	3.86	1.54	0.43	0.318
10	4.06	1.12	2.22	1.68	0.69	0.178
Σ	33.7	23.48	24.16	18.58	3.67	2.918
μ	3.37	2.348	2.416	1.858		

Table 2.2 (road 4/2 D) the headway between the car (Ta), car and motorcycle (Tc), motorcycle and car (Td) and motorcycle and motorcycle (Tb) in the morning (peak hour) is smaller than during the day but data deviation is higher, the analysis shows the traffic volume in the morning is rather similar with during the day.

Table 2.3. Tubagus Ismail Road 2/2 UD

Periode	Ta	tc	Td	Tb	ta	Tb
7.30-8.00	Mb-Mb	Mb-Mt	Mt-Mb	Mt-Mt	Xi-μ	Xi-μ
1	2.7	4.8	4.6	1.6	1.78	0.18
2	4.4	3.8	2.4	1.3	0.08	0.12
3	7.8	6.2	7.2	1.4	3.32	0.02
4	4.8	3.2	4.8	1.2	0.32	0.22
5	3.8	5.2	7.4	1.2	0.68	0.22
6	4.9	5.2	6.6	2.1	0.42	0.68
7	6.2	4.6	8.4	1.4	1.72	0.02
8	4.8	8.2	4.4	1.2	0.32	0.22
9	2.2	3.6	5.2	1.3	2.28	0.12
10	3.2	4.2	2.8	1.5	1.28	0.08
∑	44.80	49.00	53.80	14.20	6.18	0.76
μ	4.48	4.9	5.38	1.42		
9.00-9.30	Ta	tc	Td	Tb	Xi-μ	Xi-μ
1	3.24	1.84	4.02	0.58	0.13	1.358
2	1.20	2.86	2.98	2.42	2.17	0.482
3	3.30	3.48	1.14	1.80	0.07	0.138
4	3.86	1.34	2.94	0.76	0.49	1.178
5	2.56	1.46	0.84	2.98	0.81	1.042
6	3.82	1.26	0.90	1.46	0.45	0.478
7	3.80	4.08	1.00	1.54	0.43	0.398
8	4.06	2.24	3.26	2.62	0.69	0.682
9	3.80	3.80	3.86	0.54	0.43	1.398
10	4.06	1.12	2.22	4.68	0.69	2.742
∑	33.70	23.48	23.16	19.38	3.67	4.198
μ	3.37	2.348	2.316	1.938		

Table 2.3 (road 2/2 UD) the headway between the car (Ta), car and motorcycle (Tc), motorcycle and car (Td) and motorcycle and motorcycle (Tb) in the morning (peak hour) is higher than during the day, the analysis shows the traffic volume in the morning is not different significantly with during the day.

Table 2.4 (road 6/2 D) the headway between the car (Ta), car and motorcycle (Tc), motorcycle and car (Td) and motorcycle and motorcycle (Tb) in the morning (peak hour) is higher than during the day, the analysis shows the traffic volume in the morning is not different significantly with during the day.

From Table 2.1 - 2.8 we can see that headway time variation based on type of road and the location of road, those headway variation determine the value passenger car unit for motor cycle.

Table 2.4. Pasteur Road 6/2 D

Periode	Ta	Tc	Td	Tb	ta	tb
7.30-8.00	Mb-Mb	Mb-Mt	Mt-Mb	Mt-Mt	Xi-μ	Xi-μ
1	6.98	1.20	2.12	1.14	2.823	0.285
2	3.62	1.10	1.74	0.75	0.537	0.675
3	4.46	1.53	2.43	1.20	0.303	0.225
4	3.87	3.26	1.62	1.41	0.282	0.015
5	4.10	2.43	1.94	1.35	0.057	0.075
6	2.48	1.31	4.22	1.05	1.677	0.375
7	4.16	1.22	3.21	3.53	0.003	2.1
8	5.00	1.46	0.86	1.58	0.843	0.15
9	2.06	2.27	2.13	0.83	2.097	0.6
10	4.83	1.61	2.70	1.43	0.678	2.22E-16
∑	41.52	17.36	22.95	14.25	4.002	1.275
μ	4.152	1.7355	2.295	1.425		
					ta	tb
9.00-9.30	Ta	Tc	Td	Tb	Xi-μ	Xi-μ
1	4.2	1.84	4.02	1.58	0.92	0.658
2	3.8	2.86	2.98	2.42	0.52	0.182
3	4.4	3.48	1.14	1.8	1.12	0.438
4	2.2	1.34	2.94	1.76	1.08	0.478
5	3.8	1.46	0.84	2.98	0.52	0.742
6	3.8	1.26	0.9	1.46	0.52	0.778
7	2.4	4.08	1	1.54	0.88	0.698
8	3.2	2.24	3.26	2.62	0.08	0.382
9	1.4	3.8	3.86	1.54	1.88	0.698
10	3.6	1.12	2.22	4.68	0.32	2.442
∑	32.80	23.48	23.16	22.38	4.16	2.498
μ	3.28	2.348	2.316	2.238		

Table 2.5. Soegijopranoto Road 6/2 UD

Periode	Ta	Tc	Td	Tb	Ta	Tb
8.00-8.30	Mb-Mb	Mb-Mt	Mt-Mb	Mt-Mt	Xi-μ	Xi-μ
1	15	3	5	3	0.3	0.5
2	13.5	6	8	2	1.2	0.5
3	14	15	4	3	0.7	0.5
4	15	8	9	2.5	0.3	0
5	16	16	12	2	1.3	0.5
∑	73.5	48	38	12.5	3.8	2
μ	14.7	9.6	7.6	2.5		
					Ta	Tb
9.00-9.30	Ta	Tc	Td	Tb	Xi-μ	Xi-μ
1	25	15	11	4	3.8	0.6
2	22	14	15	4	0.8	0.6
3	20	18	13	3	1.2	0.4
4	22	17	14	3	0.8	0.4
5	17	19	13	3	4.2	0.4
∑	106	83	66	17	10.8	2.4
μ	21.2	16.6	13.2	3.4		

Table 2.6. Thamrin Road 4/2 D

Periode	Ta	Tc	Td	Tb	ta	Tb
7.30-8.00	Mb-Mb	Mb-Mt	Mt-Mb	Mt-Mt	Xi-μ	Xi-μ
1	1.43	1.22	1.71	1.41	3.56	0.284
2	1.77	0.34	3.1	2.21	3.22	1.084
3	0.52	0.37	1.11	1.24	4.47	0.114
4	0.59	0.35	0.31	2.12	4.4	0.994
5	4.3	2.52	1.12	0.34	0.69	0.786
6	1.74	1.16	1.25	1.15	3.25	0.024
7	4.06	2.9	1.61	1.32	0.93	0.194
8	2.11	1.74	0.31	0.21	2.88	0.916
9	2.78	1.72	0.46	0.23	2.21	0.896
10	30.6	0.47	3.12	1.03	25.61	0.096
Σ	49.90	12.79	14.10	11.26	16.34	3.262
μ	4.99	1.278833	1.41	1.126		
9.00-9.30	Ta	Tc	Td	Tb	Xi-μ	Xi-μ
1	1.46	1.71	0.52	0.5	0.733	0.254
2	1.7	1.8	1.45	0.39	0.493	0.364
3	1.88	1.42	0.89	0.67	0.313	0.084
4	1.43	0.32	0.72	0.46	0.763	0.294
5	1.32	1.88	0.52	0.89	0.873	0.136
6	1.77	0.31	0.5	0.63	0.423	0.124
7	1.8	0.98	1.51	0.35	0.393	0.404
8	2.11	1.24	1.46	1.44	0.083	0.686
9	4.27	0.45	1.34	1.62	2.077	0.866
10	4.19	0.8	1.15	0.59	1.997	0.164
Σ	21.93	10.91	10.06	7.54	3.175	1.132
μ	2.193	1.091	1.006	0.754		

Table 2.7. Pandanaran Road 8/2 D

Periode	Ta	tc	Td	Tb	ta	Tb
7.30-8.00	Mb-Mb	Mb-Mt	Mt-Mb	Mt-Mt	Xi-μ	Xi-μ
1	3.96	2.76	4.7	3.48	0.012	0.48
2	4.1	3.38	3.28	2.76	0.152	0.24
3	3.88	3.42	4.42	3.08	0.068	0.08
4	5.5	3.38	3.86	1.86	1.552	1.14
5	5.76	3.06	4.66	2.7	1.812	0.3
6	4.54	2.34	4.36	3.1	0.592	0.1
7	4.2	2.6	3.62	2.52	0.252	0.48
8	2.34	3.54	4.42	3.64	1.608	0.64
9	1.82	3.06	4.04	3.2	2.128	0.2
10	3.38	3.68	4.52	3.66	0.568	0.66
∑	39.48	31.22	41.88	30.00	3.596	2.24
μ	3.948	3.122	4.188	3		
					ta	Tb
9.00-9.30	ta	tc	td	Tb	Xi-μ	Xi-μ
1	3.16	0.84	4.32	2.14	0.882	0.498
2	3.38	5.66	1.02	2.68	0.662	0.042
3	7.2	1.9	2.28	2.8	3.158	0.162
4	2.24	1.96	1.54	2.8	1.802	0.162
5	4.64	2.24	1.04	2.32	0.598	0.318
6	3.14	2	2.28	2.44	0.902	0.198
7	4.38	2.34	2.14	3.4	0.338	0.762
8	4.88	1.2	1.32	2.6	0.838	0.038
9	4.14	1.34	2.82	3.4	0.098	0.762
10	3.26	1.56	2.46	1.8	0.782	0.838
∑	40.42	21.04	21.22	26.38	7.102	1.182
μ	4.042	2.104	2.122	2.638		

Table 2.8 Jalan Imam Bonjol 2/2 UD

Periode	ta	Tc	Td	Tb	ta	tb
7.30-8.00	Mb-Mb	Mb-Mt	Mt-Mb	Mt-Mt	Xi-μ	Xi-μ
1	4.65	0.8	1.41	0.76	1.882	0.19
2	2.41	0.73	0.49	0.5	0.358	0.45
3	2.97	1.02	1.62	0.8	0.202	0.15
4	2.58	2.17	1.08	0.94	0.188	0.01
5	2.73	1.62	1.29	0.9	0.038	0.05
6	1.65	0.87	2.81	0.7	1.118	0.25
7	2.77	0.81	2.14	2.35	0.002	1.4
8	3.33	0.97	0.24	1.05	0.562	0.1
9	1.37	1.51	1.42	0.55	1.398	0.4
10	3.22	1.07	3.8	0.95	0.452	1.16
Σ	27.68	11.57	16.30	9.50	2.668	0.85
μ	2.768	1.157	1.63	0.95		
9.00-9.30	ta	Tc	Td	Tb	Xi-μ	Xi-μ
1	1.62	0.92	2.01	0.29	0.065	0.679
2	0.6	1.43	1.49	1.21	1.085	0.241
3	1.65	1.74	0.57	0.9	0.035	0.069
4	1.93	0.67	1.47	0.38	0.245	0.589
5	1.28	0.73	0.42	1.49	0.405	0.521
6	1.91	0.63	0.45	0.73	0.225	0.239
7	1.9	2.04	0.5	0.77	0.215	0.199
8	2.03	1.12	1.63	1.31	0.345	0.341
9	1.9	1.9	1.93	0.27	0.215	0.699
10	2.03	0.56	1.11	2.34	0.345	1.371
Σ	16.85	11.74	11.58	9.69	1.835	2.099
μ	1.685	1.174	1.158	0.969		

8 Data analysis

8.1 PCU calculation

The results of calculation of the value PCU across the road at the morning peak hours and off-peak hours of daylight are presented in Table 3.1 and Table 3.2 below:

Table 3.1 Summary of value PCU at Bandung

Metode	Observation Location	PCU	
		Motor Cycle	
		Peak	Off Peak
Headway	Pasteur	0,28	0,33
	Dipati Ukur	0,22	0,26
	Juanda	0,24	0,33
	Tubagus Ismail	0.18	0,19

From the analysis of PCU obtained from four locations in Bandung area PCU Motorcycles values obtained varied from 0.18 to 0.28 during peak hours and the value of PCU Motorcycles at off peak hours vary from 0.19 to 0.33.

Table 3.2 Summary of value PCU at Semarang

Metode	Observation Location	PCU	
		Motor Cycle	
		Peak	Off Peak
Headway	Soegijopranoto	0,29	0,34
	Thamrin	0,24	0,33
	Padanaran	0,29	0,35
	Imam Bonjol	0.19	0,22

From the analysis of PCU obtained from four locations in Semarang area PCU Motorcycles values obtained varied from 0.19 to 0.29 during peak hours and the value of PCU Motorcycles at off peak hours vary from 0.22 to 0.35.

8.2. RMSE analysis of PCU value

Comparisson between data at road of Bandung and data at road of Semarang on off peak hour use Root Mean Suare Error (RMSE) Analysis describe at Table 6.3 below :

Table 3.3 RMSE on peak hour

ROAD TYPE	PCU VALUE			Difference (RMSE)		
	IHCM	BANDUNG	SEMARANG	2-3	2-4	4-3
1	2	3	4	5	6	7
6/2 D	0.4	0.33	0.35	53%	18%	42%
6/2 UD	-	-	0.34	-	-	-
4/2 D	0.4	0.33	0.33	58%	18%	48%
4/2 UD	0.4	0.26	-	60%	-	-
2/2 UD	0.4	0.19	0.22	45%	53%	16%
Average	-	-	-	0.54	0.30	0.35

Comparisson between data at road of Bandung and data at road of Semarang on peak hour describe at Table 6.4.

Table 3.4 RMSE off peak hour

ROAD TYPE	PCU VALUE			Difference (RMSE)		
	IHCM	BANDUNG	SEMARANG	2-3	2-4	4-3
1	2	3	4	5	6	7
6/2 D	0.25	0.28	0.29	12%	12%	21%
6/2 UD	-	-	0.29			
4/2 D	0.25	0.24	0.24	16%	4%	13%
4/2 UD	0.25	0.22	-	12%	12%	0%
2/2 UD	0.25	0.18	0.19	24%	28%	6%
Average	-	-	-	0.16	0.14	0.13

PCU value based on survey in Bandung and Semarang show that at the Off Peak Hour the PCU value of both cities more different with PCU based on IHCM compare at the Peak Hour. It means PCU value version IHCM is more not appropriate when off peak hour condition.

Based total population and density population shows the PCU of motor cycle looks influenced by total population or density population, which PCU of motor cycle in Bandung looks smaller than in Semarang (Bandung has more population than Semarang).

9 Conclusions and recommendations

9.1 Conclusion

1. Value PCU Motor cycles based on research results on average smaller than the value PCU based IHCM good at peak hour and off-peak hours.
2. At off peak hour the difference between PCU based on survey and PCU based on IHCM greater than at peak hour.
3. PCU value version IHCM is more not appropriate when off peak hour condition.
4. The density of traffic significantly influence the PCU value of motor cycle.
5. The traffic characteristic for same road type is different between Bandung and Semarang.
6. Total Population or Density Population in City influence PCU value of motorcycle, so more total population or density population shows bigger PCU of motorcycle.

9.2 Recomendation

1. Need to look at the relationship PCU value on the level of service roads on each road
2. The studies should be done on a smaller city or town has a population of fewer than Bandung and Semarang.

References

1. Nur, N.K., Samang, L., Isran Ramli, M., Hamid, S., *Study of modes transformation preferences of private transport based on travel characteristics and user behavior.* International Journal of Applied Engineering Research. Vol. 11(22), pp. 10766-10771. (2016)
2. Directorate General Bina Marga (2010), Road Classification in Indonesia, Directorate General of Highways, Ministry of Public Works Indonesia.
3. IHCM (1997), Indonesia *Highway Capacity Manual,* Directorate General of Highways, Ministry of Public Works Indonesia.
4. Izumi Okura, (2006), *Determination value of Passenger Car Equivalent (PCE)*
5. Rosma I (2003), *Determining PCU Value for Motorcycle at Intersection in Surakarta,* UNS University.
6. Central Statistic Bureau (2015)
7. Scraggs, D.A (1964), *determination of the passenger car eqivalent of a good vehicle in single lane flow at traffic signals,* U.K., Road Research laboratory, Harmondsworth and Langley.

Impact of the additional parking space on parallel parking maneuver time

Rudy Setiawan[1,*]

[1] Petra Christian University, Depart. of Civil Engineering, Surabaya, Indonesia

Abstract. The study reports the effect of additional parking space for the tandem parking upstream stall and downstream stall to reduce parking maneuver time compare with conventional parallel parking maneuver time. The experiments involving 295 students of Petra Christian University. Results indicate that, added one-meter additional parking space could reduce total maneuver time up to 29% for the upstream stall and 31% for the downstream stall, and also reduce 8% income from on-street parking fee, but also will reduce 35% loss due to the delay time caused by parallel parking maneuver.

1 Introduction

The availability of the car park is an inevitability for car user and will ease car user, especially when car park location is near by their destination. However, there are car park configurations that will affect the travel time of adjacent traffic flow causing by parking maneuver, specifically conventional parallel on-street parking.

Recently, car manufacturers have added intelligent vehicle parking assist systems to the car due to difficulties in parallel parking [1-3]. However, that feature takes a few years to become available for any cars [4]. Therefore, it is necessary to find an alternative to minimize parallel parking maneuver time.

One alternative to reduce the delay time of adjacent traffic flow due to parallel parking maneuvers is to use a tandem parking system, both upstream stall and downstream stall [5]. Although the drawbacks of tandem parking system are reduced efficiency the use of parking space and encourages on-street parking [6].

The study reports the effect of additional parking space for the tandem parking upstream stall and downstream stall to reduce parking maneuver time compare with conventional parallel parking maneuver time.

1.1 Parallel parking

Conventional parallel parking is a parking layout, in which car was parked parallel to the road in line with other parked cars (Figure 1). For few drivers, parallel parking maneuver is considered difficult and is required in test to get the car-driving license in Indonesia.

* Corresponding author: rudy@petra.ac.id

Conventional parallel parking enables the driver to park a car in a smaller space. Conventional parallel parking enables the driver to park a car in a smaller space. Reversing approach into the empty parallel parking space allows the driver to take advantage of an empty parking space a little bit longer than the car dimension.

One of the issues that arise from the on-street parallel parking is the maneuver time that affects travel time to the adjacent traffic. Tandem parallel parking both upstream stall and downstream stall, in which two cars were parked with additional parking space between them, the additional parking space could be used to ease entry and exit maneuver (Figure 2 and Figure 3), are an option to reduce delay time causing by maneuver time compare with conventional parallel parking [5].

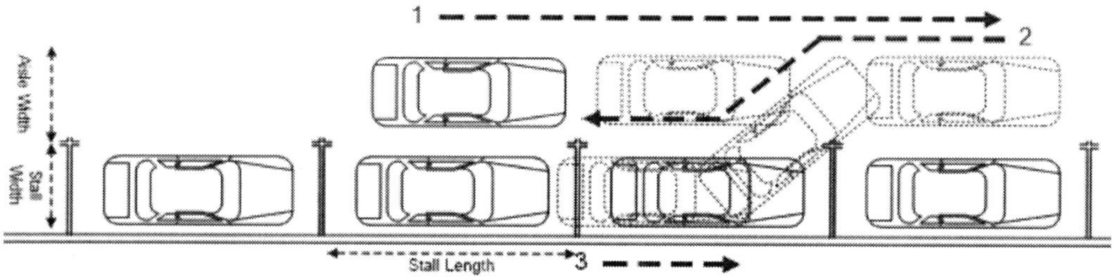

Fig. 1. Conventional parallel parking

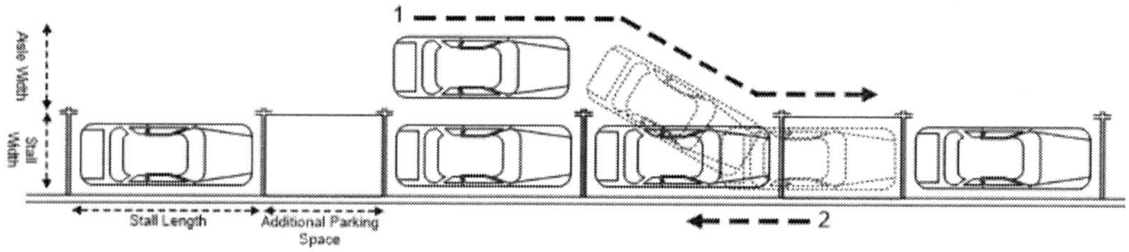

Fig. 2. Tandem parking upstream stall

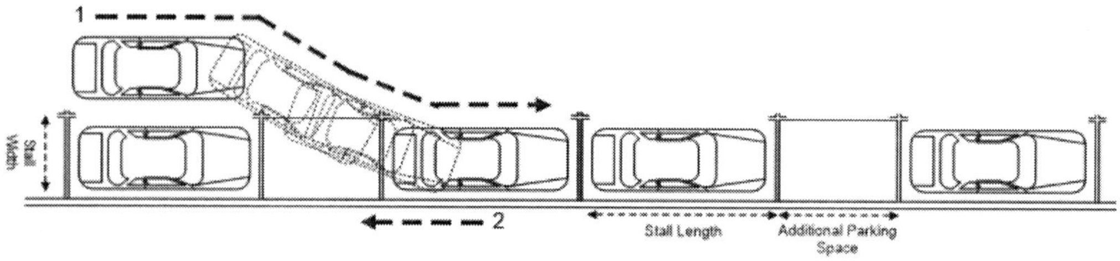

Fig. 3. Conventional parallel parking

2 Methods

The experiment involving 295 students of Petra Christian University, the car type varies from city car, sedan, Sport Utility Vehicle, Multi-Purpose Vehicle, and Double Cabin. The car dimension (length x width) participate in this experiment was varies from 3,410mm x 1,460mm to 5,255mm x 2,154mm, and the average is 4,148mm x 1,710mm.

In this experiment, the dimension of parking space for the conventional parallel parking is 5,700mm x 2,400mm (stall length x stall width), and aisle width is 2,700mm [7]. The dimension of parking space for the upstream stall and downstream stall is the same as the conventional parallel parking, with the additional parking space varies between 1.0m up to 4.5m with 0.5m interval.

The reduction of delay time to adjacent traffic causing by parallel maneuver time between conventional parallel parking and tandem parallel parking, compare using the Value of Time (VT). VT for Surabaya City at 2017 is IDR. 27,510 per car per hour, that obtain from the VT for Surabaya City at 2015 (IDR. 24,484) multiplied by interest rate 6% [8].

3 Results

The average entry and exit maneuver time for the conventional parallel parking are 22.3 secs (SD = 2.7) and 10.9 secs (SD = 1.3) respectively. The entry maneuver time are longer than the exit maneuver time because for the entry maneuver driver need to do it in three steps, first positioning the car, second reversing approach into the parking space, and third adjusting parking position inside the parking space.

Compare with conventional parallel parking, both tandem parallel upstream and downstream stall with additional parking space from 1.0m to 4.5m, have shorter maneuver time, both for entry maneuver (Figure 4 and Figure 5) and exit maneuver (Figure 6 and Figure 7). Additional parking space could reduce entry maneuver time about 39% to 72% (Figure 8), and reduce exit maneuver time about 2% to 40% (Figure 9) compare without additional parking space or conventional parallel parking. Overall, the difference of total maneuver time between conventional parallel parking and both tandem parking upstream stall and downstream stall about 29% to 60% (Figure 10)

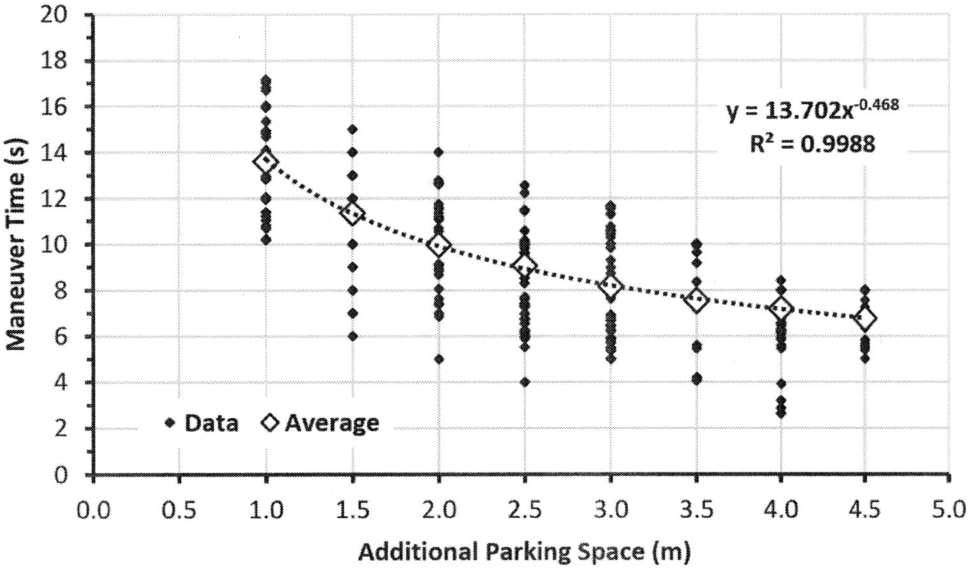

Fig. 4. Upstream stall entry maneuver time

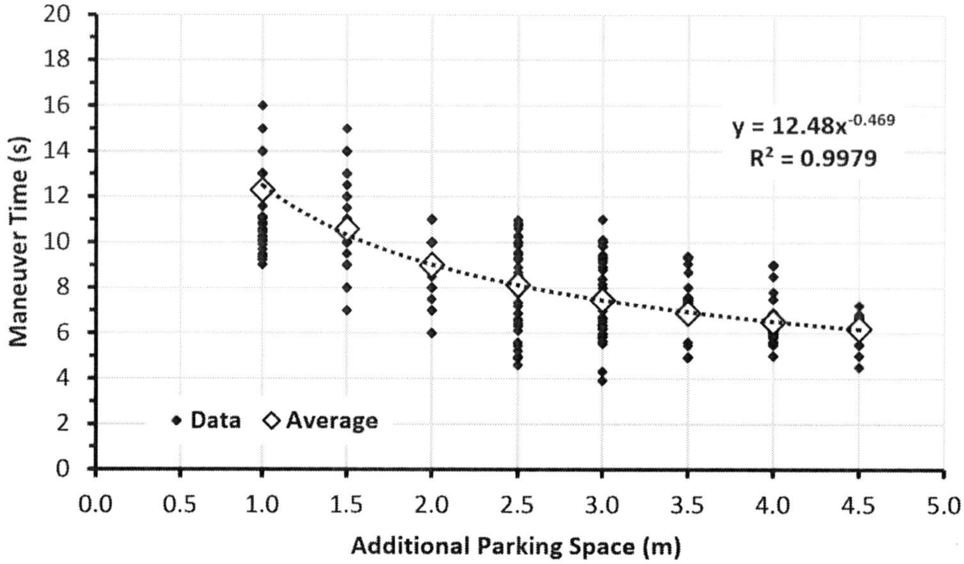

Fig. 5. Downstream stall entry maneuver time

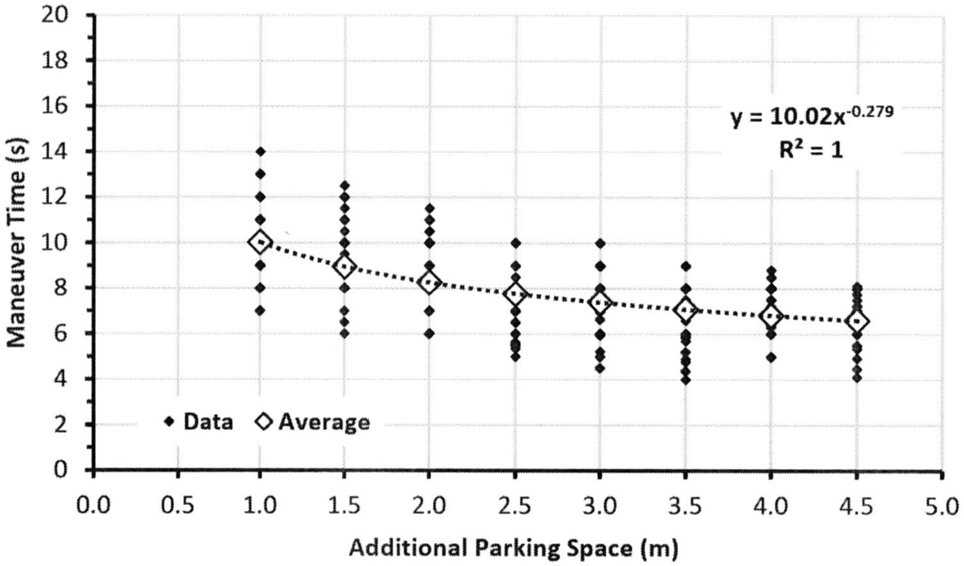

Fig. 6. Upstream stall exit maneuver time

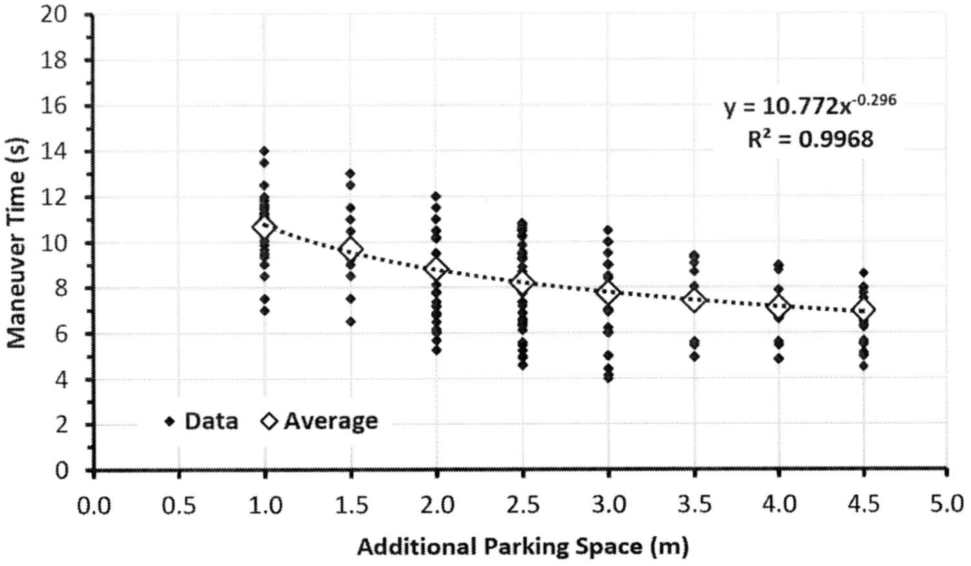

Fig. 7. Downstream stall exit maneuver time

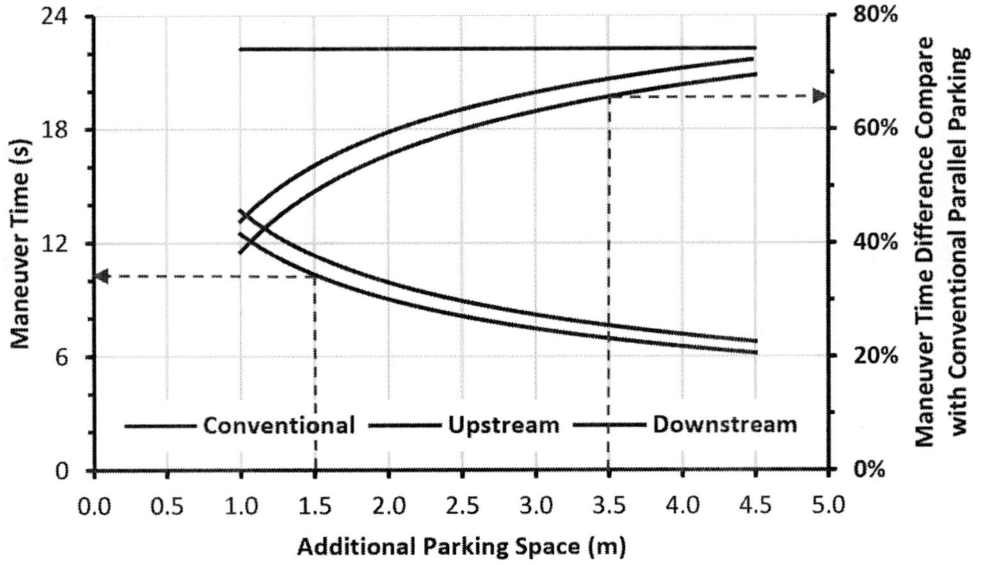

Fig. 8. Comparison of entry maneuver time

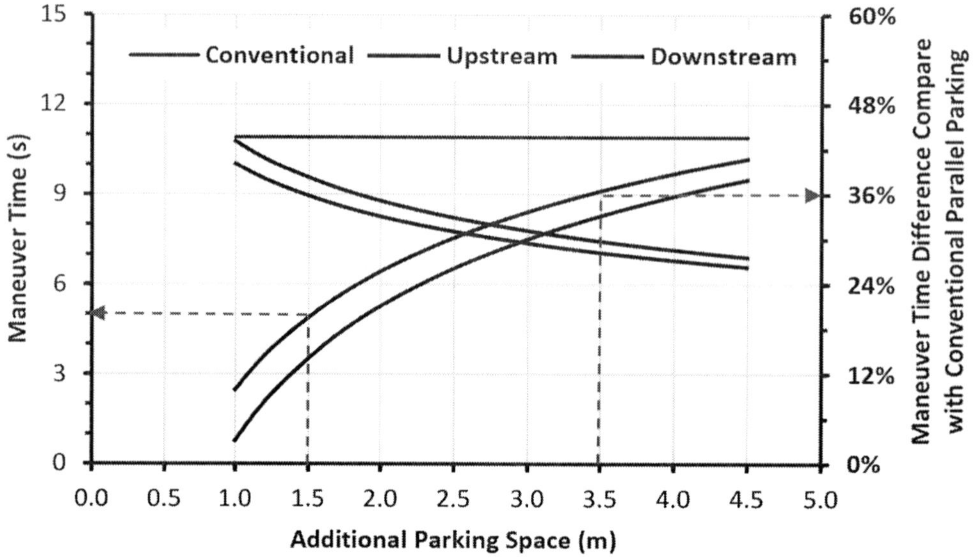

Fig. 9. Comparison of exit maneuver time

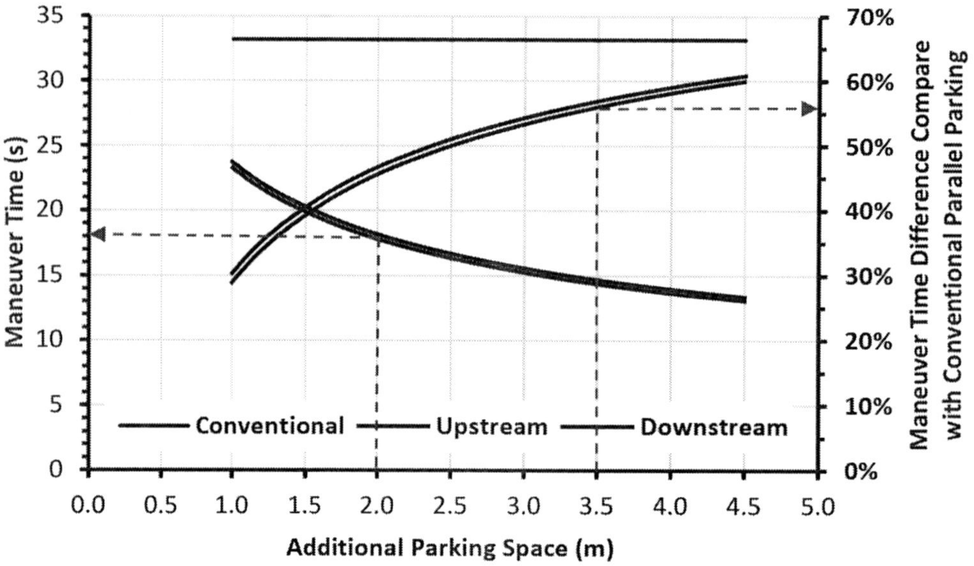

Fig. 10. Comparison of total maneuver time

Table 1 illustrated comparison between income from parking fee and loss due to delay time for stall length = 5.7m and varies dimension of additional parking space. Column 1 and column 2 are number of parking space per 100 meters segment of on-street parking, e.g. for additional parking space 1.0m, number of parking space for upstream or downstream = 100/(5.7 + (1.0/2)) = 16.1 parking space.

Table 1. Income from parking fee

Additional Parking Space (m)	Number of Parking Space		Income from Parking Fee (IDR/hour)		
	C	U and D	C	U and D	Difference
	(1)	(2)	(3)	(4)	(5)
1.0	17.5	16.1	175,439	161,290	-8%
1.5	17.5	15.5	175,439	155,039	-12%
2.0	17.5	14.9	175,439	149,254	-15%
2.5	17.5	14.4	175,439	143,885	-18%
3.0	17.5	13.9	175,439	138,889	-21%
3.5	17.5	13.4	175,439	134,228	-23%
4.0	17.5	13.0	175,439	129,870	-26%
4.5	17.5	12.6	175,439	125,786	-28%

C = Conventional, U and D = Upstream and Downstream
*average between upstream and downstream

Table 2. Loss due to delay time

Additional Parking Space (m)	Total Maneuver Time (secs)		Loss due to Delay Time (IDR/hour)		
	C	U and D	C	U and D	Difference
	(1)	(2)	(3)	(4)	(5)
1.0	33.2	23.3	8,892,381	5,741,143	-35%
1.5	33.2	20.3	8,892,381	4,805,548	-46%
2.0	33.2	18.0	8,892,381	4,110,719	-54%
2.5	33.2	16.6	8,892,381	3,645,685	-59%
3.0	33.2	15.4	8,892,381	3,260,912	-63%
3.5	33.2	14.4	8,892,381	2,964,268	-67%
4.0	33.2	13.8	8,892,381	2,736,908	-69%
4.5	33.2	13.3	8,892,381	2,547,517	-71%

C = Conventional, U and D = Upstream and Downstream
*average between upstream and downstream

Based on number of parking space on column 1 and column 2, if parking turn over = 2 and parking fee is fix IDR 5,000 per car, e.g. income from parking fee for upstream or downstream = 2 x 16.1 x 5.000 = IDR 161,290. Column 5 show the difference income from parking fee between conventional parallel parking and upstream/downstream stall for varies additional parking space.

Total maneuver time for conventional and upstream /downstream stall are shown in Table 2 column 1 and column 2 respectively. Column 2 is average total maneuver time between upstream stall and downstream stall, because the value for each maneuver almost the same (Figure 10).

Column 3 and column 4 show the loss due to delay time causing by maneuver time for VT = IDR 25.954/pcu/hour, traffic flow = 1.000 pcu/hour, e.g. loss due to maneuver time for additional parking space 1.0m = (16.1 x 2 x 23.3)/3,600 x 27,510 x 1,000 = IDR 5,741,143/hour. Column 5 show the difference loss due to delay time between conventional parallel parking and upstream/downstream stall for varies additional parking space.

4 Conclusions

Tandem parallel parking both upstream stall and downstream stall, are a good option to reduce delay time causing by maneuver time compare with conventional parallel parking. Results indicate that, added one-meter additional parking space could reduce total maneuver time up to 29% for the upstream stall and 31% for the downstream stall, and also reduce 8% income from on-street parking fee, but also will reduce 35% loss due to delay time causing by parallel parking maneuver.

References

1. Wolf, C. C., Ocklenburg, S., Ören, B., Becker, C., Hofstätter, A., Bös, C., … Güntürkün, O. *Sex differences in parking are affected by biological and social factors. Psychological Research*, *74*(4), 429–435. (2010)
2. Reimer, B., Mehler, B., and Coughlin, J. F. An *Evaluation of Driver Reactions to New Vehicle Parking Assist Technologies Developed to Reduce Driver Stress. New England University Transportation Center*, 1–26. (2010)
3. Su, D., and Chan, K. *Compensating for Operational Uncertainty in Man – Machine Systems : A Case Study on Intelligent Vehicle Parking Assist System, 1(September 2015), 1–13. http://doi.org/10.1115/1.4030438*. (2017)
4. Cherise Threewitt. *The 8 Best Self-Parking Cars*. (2017)
5. Iowa Department of Transportation Office of Design. *Parking on Urban Primary Highways Chapter 6C-8*. (2010)
6. Essex Works. *Parking Standards Design and Good Practice*. (2009)
7. Setiawan, R., Kurniawan, W., and Tomasoa, S. H. P. *Dampak Perubahan Dimensi Petak Parkir Terhadap Waktu Manuver Parkir Paralel*. In *Konferensi Nasional Teknik Sipil (KoNTekS) 4*. (Sanur-Bali, 2-3 Juni 2010). (2010)
8. Faridli, M. M., and Kartika, A. A. G. *Studi Kelayakan Pembangunan Jalan Lingkar Barat Dalam Kota Surabaya. Jurnal Teknik POMITS, 1*(1), 1–6. (2013)

On street parking and its impact on road performance

Sonya Sulistyono[1,*], *Harnen* Sulistio[1], *Ludfi* Djakfar[1], *Achmad* Wicaksono[1], and *Ririn* Endah Badriani[2]

[1]Brawijaya University, Department of Civil Engineering, Malang, East Java, Indonesia
[2]Jember University, Department of Civil Engineering, Jember, East Java, Indonesia

Abstract. Commercial areas have a high economic and parking activity. High parking activity that is not comparable to the availability of parking facilities and it can interfere with the performance of the road. This study based on congestion caused by activities on-street parking on Gajah Mada St., Rambipuji, Jember. This road broadness reduced because on street parking and disrupting the flow of traffic. From the survey on Thursday, March 5, 2015 obtained the following results: performance comparison roads with and without on-street parking shows capacity has increased 25%, DS decreased 25%, and the speed has increased 11%. Characteristics of the road parking for MC, LV and HV in order are: volume parking 257 vehicles, 62 vehicles, and 10 vehicles. Accumulated parking 46 veh/h, 20 veh/h, and 5 veh/h. Turn over parking 0,0997 veh/SRP/h, 0,4588 veh/SRP/h, and 1,0940 veh/SRP/h. Parking index of 0,27, 2,22, and 8,20.

1 Introduction

Transportation requires parking space to stop. If parking is not well served, it can disrupt traffic flow and cause congestion. Parking problems that often encountered is parking on the side of the road. This type of parking will incur a loss for both the driver and the public, if not properly managed. Road parking leads to road reduction and a disruption to traffic performance.

This study was conducted to review the effect of parking on the road body on it performance, especially on Gajah Mada St., Rambipuji, Jember. This road is the center of economic activity with parking facilities that do not meet, so the road is used as a parking lot. This road is a national road with a fairly solid traffic reaching 49,491 veh/ day. The type of road on Gajah Mada is 2/2 UD with total length of road 1.3 km. The effect of parking space usage on road performance is done along 100 m at the highest point of parking activity and traffic. The purpose of this research is to know the characteristics of parking on the road body that occurs, the comparison of road performance with and without the parking on the road body, and the influence of road parking on the road performance.

This study will approach the characteristics of parking and traffic characteristics such as research method [1] who have conducted research to obtain the results of road performance

* Corresponding author sonya.sulistyono@unej.ac.id

comparison with and without road parking. Characteristic is a basic trait that can provide an assessment of the services and problems that occur in the study area. Traffic characteristics include capacity, degree of saturation, and travel speed. While parking characteristics include peak parking hours, parking volume, parking accumulation, turn over parking, average parking time, parking capacity, and parking index.

2 Research methods

2.1 Location of study

Gajah Mada St. has a total length of 1.3 km. The object in this study is limited to 100 m, i.e. at STA 0 + 700 up to STA 0 + 800 or segment between Light shirt store up to Gajah Mada Sport (see figure 1). Restrictions on research sites are useful to facilitate surveys conducted in the field

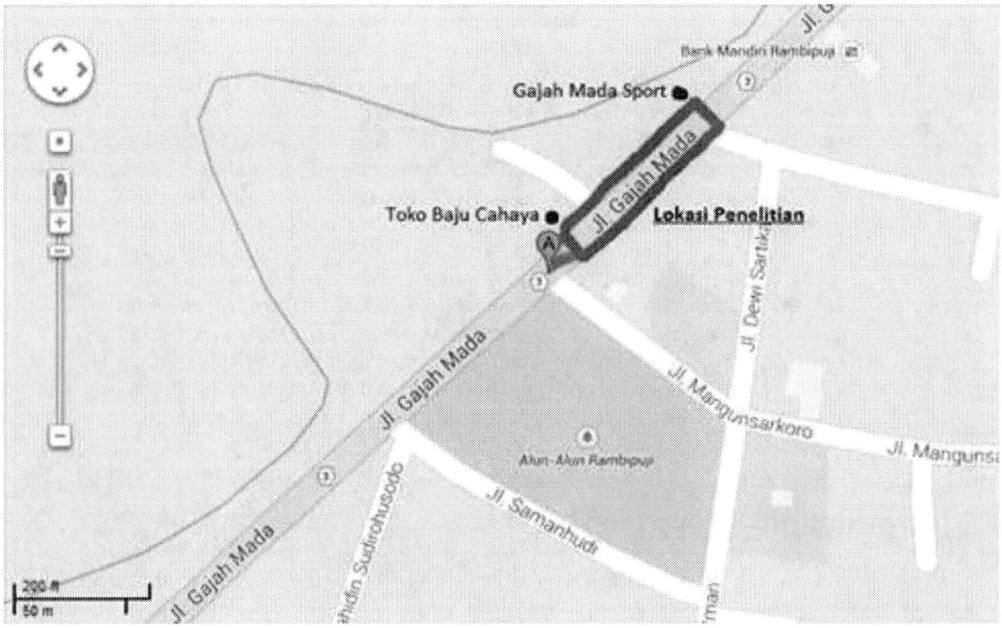

Figure 1. Location of study

2.2 Research Steps

The stages of the implementation of this study are as follows:

1. Initial stage. Identify the problems and analysis the existing conditions as needed. Based on the initial survey obtained parameters to be surveyed.
2. Planning phase. Planning the survey activities and run it systematically and not deviate from the problem. In the stage of determining the method and time of the survey.
3. Data collection phase. Conducted by collecting data by way of direct survey in the field. Surveys include: inventory surveys, traffic surveys, and parking characteristics surveys.

Data analysis phase. The analysis includes: analysis of traffic performance, analysis of parking characteristics, and analysis of the effect of parking on traffic performance.

3 Result and discussion

3.1 Road inventory

Inventory data is needed to support the data in the road performance analysis. Gajah Mada St. is the primary collector road. Type of Gajah Mada Road segment is 2/2 UD. The width of each lane is 3.75 m. The width of the road shoulder toward Jember is 2.2 m, while the shoulder width of the road towards Surabaya is 1.5 m.

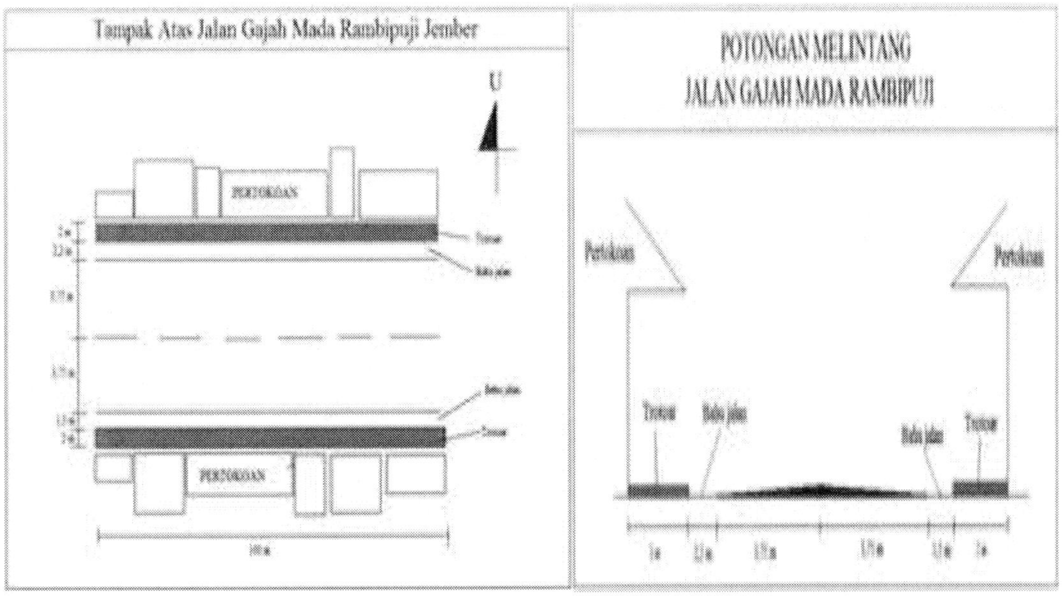

Figure 2. Road geometric of Gadjah Mada St.

3.2 Traffic volume

Traffic volume is the number of vehicles passing through a road segment over a period of time. From the survey conducted at 06.00-21.00 it can be seen that the composition of the use of road space, peak hour, and peak hour volume required in the calculation of road performance. Total traffic volume in two directions for 15 hours for MC of 36,679 vehicles, LV 10,449 vehicles, and HV 2,363 vehicles. Figure 1 presents the fluctuations in traffic flow that occurred on Gajah Mada St.

Based on figure 3 above, it can be obtained peak hour traffic. The path to town (towards Jember), peak hour morning at 06.15-07.15, peak hour noon 11.30-12.30, peak hour afternoon 15.15-16.15, and peak evening hours 18.30-19.30. While on the lane out of town (towards Surabaya), peak hour morning at 06.15-07.15, peak hour noon 12.15-13.15, peak hour of the afternoon 15.15-16.15, and peak hours of the night 17.30-18.30.

 https://doi.org/10.1051/matecconf/201818106008

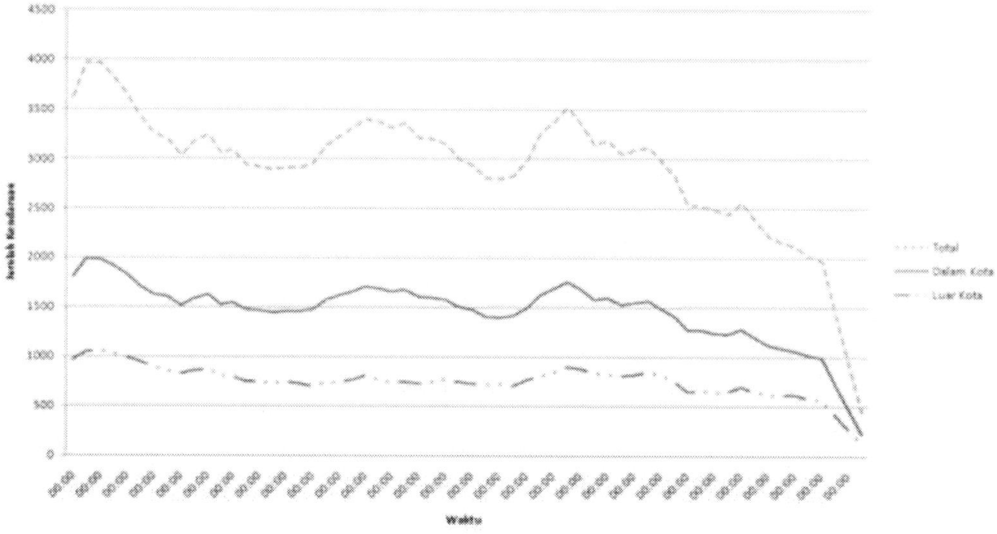

Figure 3. Traffic fluctuations (smp/h)

3.3 Road performance analysis

Roadway performance variables include free flow rate, road capacity, degree of saturation, travel speed and service level. The performance of the road segment analyzed with the data obtained in the field is the performance of the road with the street parking (parking on the road body). Performance of roads without the street parking is obtained by adding the effective lane width of the road body section used for parking, so parking vehicles are considered non-existent. The width of effective path on road parking condition is 7 m, and on condition without on street parking is 11,2 m. Comparison of road performance with and without on street parking is shown in table 1 below.

Table 1. Road performance analysis

No.	Variables of Analysis	Peak Hour							
		Morning		Afternoon		Evening		Night	
		With	Without	With	Without	With	Without	With	Without
1.	Side friction classification	Low	Very Low	Medium	Low	Medium	Low	Low	Very Low
2.	Capacity (C)	2890	3616	2832	3543	2832	3543	2890	3616
3.	Degree of Saturation (DS)	0.68	0.54	0.62	0.49	0.61	0.49	0.45	0.36
4.	Speed (V_{LV})	35	40	36	40	36	40	40	43
5.	Level of Service	D	C	D	C	D	C	D	C

Based on table 1, the performance with on street parking and without on street parking showed the capacity increased by 25%, the degree of saturation decreased by 25%, the travel speed increased by 11%, and the level of road service increased from the service level of road D to Level of service C. This condition occurs because of the reduction on the effective width of the road used for parking. The effective width reduction for a total of 2 lanes is 4.2 m, of 11.2 m width is effective only 7.5 m.

3.4 Parking characteristics

Parking characteristic is a basic trait that can provide an assessment of parking services and parking problems that occur in the study area [2]. Parking characteristics consist of parking accumulation, parking volume, parking turn over, parking index, parking duration, and parking capacity. Parking characteristics are needed to analyze the operational conditions and design the development of parking lots.

Parking accumulation is the number of vehicles being parked in a place at any given time. Parking volume is the number of vehicles parked in a certain place or parking area for a certain time. The time used is usually one day. The results of parking survey data analysis obtained peak parking hours, accumulation and volume of parking vehicles for motorcycles (MC), light vehicles (LV), and heavy vehicles (HV).

Table 2. Parking characteristics

No.	Factor of Analysis	Type of Vehicles		
		MC	LV	HV
1.	Peak hour of parking	10.00 - 11.00	11.00 - 12.00	13.30 - 14.30
2.	Parking volume (veh)	257	62	10
3.	Accumulation of parking (veh/h)	46	20	5
4.	Turn over parking (veh/SRP/h)	0.0997	0.45880	1.0940
5.	Duration of parking (minutes)	15	29	24
6.	Capacity of parking (SRP/h)	689	19	2
7.	Parking Index	0,27	2,22	8,20

Turnover parking is the rate of parking usage for a certain period of time. The parking turnover rate is obtained from the division between the total numbers of parked vehicles with the number of parking lots available during the observation time. The turn over parking value for MC is 0.0997, LV is 0.4588, and HV is 1.0940. The duration of the parking is the length of time the vehicle is parked on the parking lot. Parking duration is calculated by reducing parking time out with parking admission time. After getting the parking duration of each vehicle and each type of parked vehicle, it can be calculated that the average length of time parking. The average MC parking time was 15 minutes, LV was 29 minutes and HV was obtained 24 minutes.

Parking capacity is the number of vehicles that can be serviced by a parking lot during service time. Capacity can be calculated by dividing the number of available plots by the average length of parking time. Parking capacity for MC is 689 SRP / hour, LV is 19 SRP / hour, and HV is 2 SRP / hour. Parking index (IP) is calculated by dividing the number of parking vehicles with available parking lots. MC Parking Index obtained 0.27, MC of 2.22, and LV of 8.20. If IP> 1 then the parking requirement exceeds the existing capacity or problems occur. Conversely, if IP <1 then the parking needs are still under existing capacity or no problem. The parking for motorcycles has less than one IP, meaning parking space is no problem or still enough to accommodate vehicle parker. Parking for light and heavy vehicles experiences problems because of more than one IP value.

3.5 The effect of on-street parking on road performance

The use of parking space is the percentage level of parking space used in a certain time unit. Use of parking spaces can be used to review whether available parking spaces can or do not accommodate parked vehicles. Analysis on the use of parking spaces on the performance of roads in condition when there is on-street parking using correlation analysis to determine the strength of the relationship and the form of relationship between two variables. The strength of the relationship between two variables is called the correlation coefficient and is denoted by the symbol "R". The correlation coefficient value will always be between -1 to +1. The closer to 1 is the expected R value.

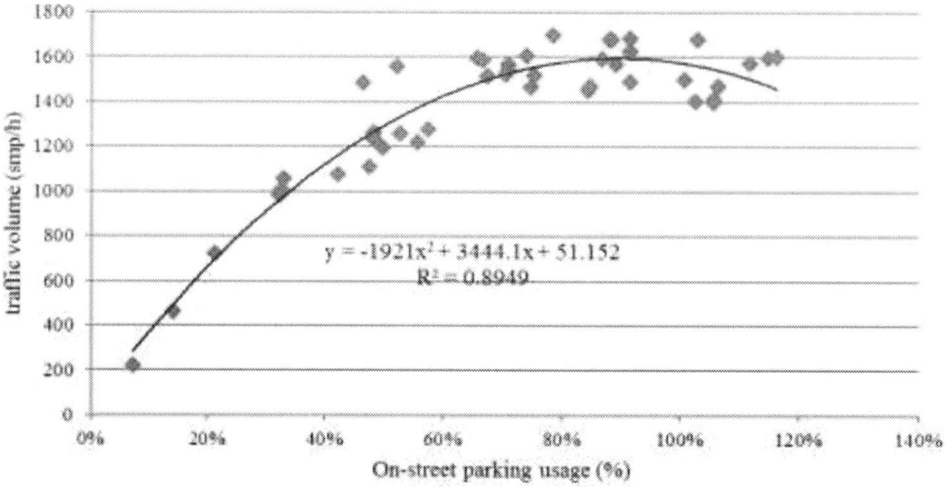

Figure 4. Relationship between parking usage and traffic volume

Based on figure 4, we got value of R^2 equal to 0,894 hence value of R is 0,946. With R value equal to 0,946 hence relation of use of parking space with traffic volume included in very strong correlation criterion.

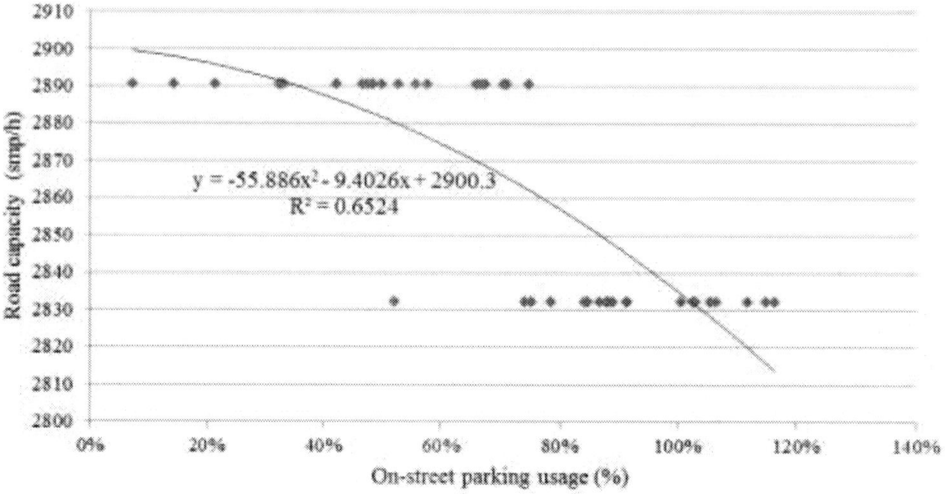

Figure 5. Relationship between parking usage and capacity

Based on figure 5 above got value of R^2 equal to 0,659 and value of R is 0,812. With R value equal to 0,812 hence relation of use of parking space with capacity included in very strong correlation criterion.

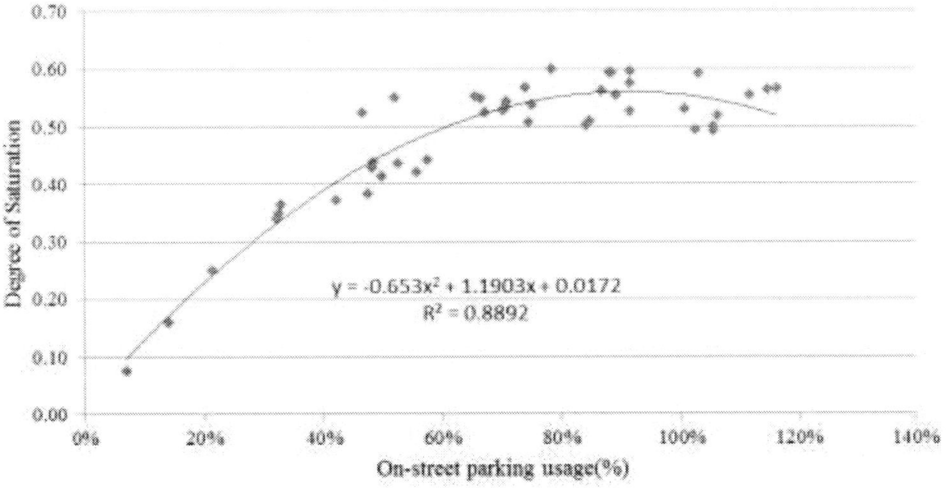

Figure 6. Relationship between parking usage and degree of saturation

Based on figure 6, we got value of R^2 equal to 0,889, then R value is 0,943. With the value of R the relationship of parking space usage with the degree of saturation included in the criterion is very strong correlation.

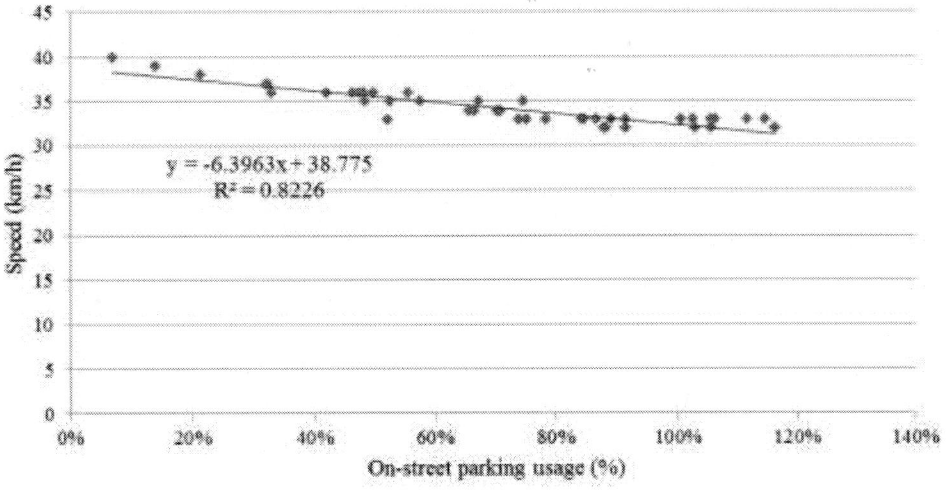

Figure 7. Relationship between parking usage and speed

Based on the figure 7, we obtained value of R^2 of 0.816 and obtained the value of R is 0.903. With R value of 0.903 then the relationship of parking space usage with traffic volume included in high correlation criteria with the form of relationship is negative linear.

Statistical approach

This statistical test is used to determine whether there is a difference between two pieces of data. This test uses two different test average is Test Z, because the amount of data to be tested is a large data which has 60 data (more than 30 data). Test Z is conducted to see whether there is a significant difference between road performance when there is on street parking and in the absence of on street parking. Road performance tested includes capacity, degree of saturation, and travel speed. Data X1 is a road performance with on street parking and X2 data is a road performance without on street parking.

Table 3. z-test value

Test variable	Reception area	Z-arithmetic
Kapasitas	-2,33> Z >2,33	-3,46
Derajat kejenuhan	-2,33> Z >2,33	5,52
Kecepatan tempuh	-2,33> Z >2,33	-8,12

The results of the above two-averaging test show that Z-arithmetic is not in the rejection area, so there is a real difference between the test variables (capacity, degree of saturation and speed) when there is on street parking and when there is no on street parking on the real ground 1 %. The on street parking has a great effect on road performance (capacity, degree of saturation and speed) that happened.

4 Conclusions

From the analysis of the relationship of parking space usage to the performance of the road segment, it can be seen that the use of parking space and road performance has a very strong relationship and strong influence to each other. It can be seen from the correlation coefficient between 0.8 - 1.00 which means the relationship is very strong.

References

1. Purbanto, I Gusti Raka. Karakteristik Parkir Pinggir Jalan (On Street Parking) dan Pengaruhnya terhadap Kinerja Ruas Jalan. *Jurnal Ilmiah Teknik Sipil.* Vol. 16, No. 2. Denpasar: Universitas Udayana. (2012).
2. Hobbs, F.D. *Perencanaan dan Teknik Lalu Lintas.* Yogyakarta: Gajah Mada University Press. (1995).
3. Direktorat Jenderal Bina Marga. *Manual Kapasitas Jalan Indonesia.* Jakarta: Departemen Pekerjaan Umum. (1996).
4. Direktur Jendral Perhubungan Darat. *Pedoman Teknis Penyelenggaraan Fasilitas Parkir.* Jakarta: Departemen Perhubungan Republik Indonesia. (1996).
5. Departemen Perhubungan. *Keputusan Menteri Perhubungan Nomor 14 Tahun 2006 Tentang Manajemen dan Rekayasa Lalu Lintas di Jalan.* Jakarta: Departemen Perhbungan Republik Indonesia. (2006).
6. Tamin, O.Z. *Perencanaan, Pemodelan, dan Rekayasa Transportasi.* Bandung: Penerbit ITB. (2000).

Study of efficiency pedestrian bridge crossing in the road of Pangerang Antasari, Banjarmasin

Hari Nukta Ramadani[1,*], *Hudan* Rahmani[1], and *Akhmad* Gazali[1]

[1]Universitas Islam Kalimantan Muhammad Arsyad Al-Banjary, Department of Civil Engineering, Faculty of Engineering, Banjarmasin, Indonesia

Abstract. You Pedestrians are one of the highest traffic accident objects. Several ways are used to reduce the number of accidents, among others by providing road crossing facilities, such as zebra cross and pedestrian bridge crossing. Unfortunately, the utilization of such facilities, especially pedestrian bridge crossing is still very less. This condition raises the need to identify efficiency use the pedestrian bridge crossing. The study location is a pedestrian bridge crossing in the road of Pangeran Antasari, Banjarmasin City. The studies method used are survey method and analysis method. The analysis shows that pedestrian percentage that doesn't use pedestrian bridge crossing is very big, that is 87% at busy time and 88% at not busy time. The solution to increase the use of pedestrian bridge crossing is by install a curb railing fence on both sides of the road (about 200 m).

1 Introduction

1.1 Background

Pedestrians are the elements of traffic that are often forgotten as part of the movement on the highway. From the number of traffic accidents that occur, pedestrians are the highest traffic object.

Several ways are used to reduce the number of accidents, including by providing road crossings, such as pelican zebra cross and pedestrian bridge crossing. Unfortunately, the utilization of such facilities, especially pedestrian bridge crossing is still very less.

1.2 Objectives

The main purpose of this study is to study the extent to which the efficiency of the use of pedestrian bridge crossing in the city of Banjarmasin on the road of Pangeran Antasari, In front of Mitra Plaza and looking for reasons why pedestrians use less and provide suggestions for improvement to increase the use of the pedestrian bridge crossing.

* Corresponding author: harinukta@gmail.com

1.3 Problem Definition

The main point of this study focused on measuring the benefits of pedestrian bridge crossing, both measuring the total flow of pedestrians in each bridge during the study, as well as the percentage of pedestrians crossing over the bridge. A field survey was conducted, aiming to obtain the above variables as well as factors affecting the ratio of total pedestrians crossing by using pedestrian bridge crossing.

Surveys were conducted: pedestrian survey that crossing over bridges and on highways, traffic flows, and geometric measurements of bridges and roads.

The most important survey was the pedestrian survey that crossed over the pedestrian bridge crossing and on the highway to get the number of pedestrians using pedestrian bridge crossing.

1.4 Influence Factors

The influence factors studied for planning a pedestrian bridge crossing include:
- Traffic volume
- Speed of vehicle
- Comparison of time crossing
- Comparison of travel time
- Pedestrian opinions on pedestrian bridge crossing

Expectations of this study may indicate which factors are important in increasing the use of pedestrian bridge crossing.

2 Literature Review

Marking an ordinary pedestrian crossing is associated with an increase in the number of accidents, both pedestrian accidents and accidents involving vehicles [1]. The reason for this is not very well known, but pedestrian feeling more safe might be one possible explanation [2].

Followed from the technical aspect, pedestrian bridge or "footbridge" as defined by the Dictionary of Scientific and Technical Term that are authored by Daniel N. Lapades refer pedestrian bridge as; Footbridge is a bridge for people on foot only. According to [3], this is the best solution fo rthe pedestrians to cross the road.

Pedestrian bridge is only one way to increase road network capacity. Pedestrians who are crossing are considered to be one source of congestion [4].

Ideally, road crossing facilities should be separated from vehicle streams bridge crossings (overpass/crossingbridge/footbridge), underground crossings (subway/underpass/tunnel), and skywalk so there is no intermediate conflict pedestrians with vehicles and no delays for vehicles [5].

Crossing bridge crossing is one of the safest crossing facilities. Indeed, the construction of pedestrian bridge crossing requires a high enough cost but this value is quite valuable considering for the interests and safety of pedestrians, but the facility is underutilized because pedestrians tend not to want to change the level of the path he passes [6].

The cause of lazy pedestrians using pedestrian bridges is due to tiring, as well as unpleasant bridge conditions such as, bridge elevation, steep stairs, dirty bridge condition, and the existence of beggars. More pedestrians choose to take the risk of getting hit by a vehicle because it feels faster and more practical. Another thing that encourages pedestrians prefer the route through the road is a median road that can be used as a refuge island when crossing [7].

This means that the crossing bridge will only be used if the route is shorter from the route through the road. To increase the use of bridges crossings need to be applied to the guardrail at the side of the road and / or in the middle of the road so if choosing the route through the road must take that route longer or even impossible [8].

Safety factor is a main factor for respondents who tend to utilize pedestrian bridges; while for respondents who tend not to utilize pedestrian bridges, ease factors (time and distance across) became the most influential factor. So it is necessary to reduce the time and distance across to be able increase the utilization of pedestrian bridges, one possibility is by installing a guardrail either on the sidewalk or in the median [9].

From a study of pedestrian bridge crossing conducted in the Bangladesh (Dhaka), It is found that 71% pedestrians prefer underpass rather than using foot over bridge. The reasons pedestrians are not using road crossing facilities are insufficient security, time consuming, poor entrance, hawker's problem, discomfort, takes a long walk etc [10].

3 Survey Method

The data were collected by limiting the survey to the number of variables considered important in determining the usefulness of a pedestrian bridge crossing.

These surveys were divided, four groups:

a. Pedestrian survey, consisting of counting the number of pedestrians, crossing time conferences, comparison of travel time and pedestrian interview.
b. The traffic survey, comprised of the calculation of a large volume of traffic and the composition and speed of the vehicle.
c. The observation and measurements of pedestrian bridge crossing, consisting of dimensions, building materials, maintenance conditions, types of pedestrian bridge crossing locations and billboards sometimes attached to the bridge.
d. Geometric road, consisting of the width of the road, the direction of the vehicle flow (one-way or two-way).

Surveys of pedestrian calculations, crossings, traffic volume, and vehicle speed are performed simultaneously at the same time.

3.1 Field Studies

Currently in the city of Banjarmasin there is only one pedestrian bridge crossing located on the road of Pangeran Antasari in front of Mitra Plaza, this bridge is located on a busy traffic vehicle and pedestrian.

The location of these crossing bridges can be seen on the attached city map. Photos of the crossing bridges are attached.

The field survey was conducted with the help of Civil Engineering Department students from the Faculty of Engineering, Islamic University of Borneo Muhammad Arsyad Al-Banjari Banjarmasin, from 1st June 2017 until 29th June 2017 for 30 days. At the appropriate time to conduct the survey, cultivated at a representative time (representative).

4 Method of Analysis

4.1 Variables

Table 1. Shows the variation in comparison of pedestrians using pedestrian bridge crossing. This is considered the most important variable to be explained in the analysis. In this analysis it is attempted to explain the comparison of (Pa / Pt), apart from other explanatory variables. The objective is to determine the individual relationships or relationships between variables.

Individual factors (independent variables) are taken into account where the 5 pedestrian requirements, traffic volume, vehicle speed, time of crossing, trip time comparison, pedestrian opinions and influence of billboards are also examined.

4.2 Statistic Analysis

In statistical analysis, Statistical and Parametic Testing is used in the analysis of this test data, in particular measuring the correlation or pooling of factors used to explain the usefulness of the pedestrian bridge crossing.

The Spearman Rank Correlation Coefficient Test is used to explore the relationship between all variables in the analysis that are not measured on a nominal scale (category).

The Eksak Fisher's probable test is used to measure the union between the variables although it is still somewhat crude, only such a way which is the most feasible alternative to this sample size and is widely used for similar problems.

5 Result and Discussion

Comparison of pedestrians using pedestrian bridge crossing.

Table 1. Number and Comparison of Pedestrian bridge crossing User

Situation	Average number of Pedestrians			% Under PB (Pb/Pt)	% Through PB (Pb/Pt)
	Under PB (Pb.o/j)	Through PB (Pa.o/j)	Total (Pt.o/j)		
Busy	142	22	164	87	13
Not Busy	92	12	104	88	12

Table 2. Time Crossing Comparison

Situation	Average Crossing Time In Seconds				Through/Under
	Through PB	Number of Samples	Under PB	Number of Samples	
Busy	57	29	07	48	2.5 : 1
Not Busy	56	16	21	35	2.7 : 1

Table 3. Travel Time Comparison

Situation	Average Travel Time In Seconds				Through/Under
	Through PB	Number of Samples	Under PB	Number of Samples	
Busy	215	17	145	28	1.5 : 1
Not Busy	211	9	126	36	1.7 : 1

Where,
Pb = The number of pedestrians passing under the pedestrian bridge crossing,
Pa = The number of pedestrians through the pedestrian bridge crossing,
o/j = people per hour,
Pt = Totals (Pa + Pb).

From Table 1. it can be seen that the percentage of pedestrians who do not use pedestrian bridge crossing (Pb / Pt) is very big, 87% at busy time and 88% at not busy, while using pedestrian bridge crossing (P / Pt) only 13% at Busy time and 12% at no busy time.

Table 1. also shows that if the actual flow of pedestrians using pedestrian bridge crossing is generally low (maximum of 22 people / h at busy time and 12 persons / h in non-busy time) this amount is much lower than the capacity of the bridge plan 860 persons / hour.

The relationship between the percentage of pedestrians using pedestrian bridge crossing (Pa / Pt) during busy times and pedestrian percentages using pedestrian bridge crossing at (Pa / Pt) unoccupied time can be searched with Spearman Rank Coefficient tests.

The result shows that both variables (Pa / Pt) at busy times and Pa / Pt at no busy time have significant correlations. In non-busy times the percentage of pedestrians using pedestrian bridge crossing tends to be smaller compared to busy times.

The relationship between Percentage of Pedestrians using pedestrian bridge crossing (Pa / Pt) and Total Pedestrian Needs who crossed around the Crossing Bridge (Pt) was not related.

The relationship between Percentage (Pa / Pt) and traffic volume also turned out to be unrelated, although initially before the research was conducted it was suspected of having a strong relationship. This happens because apparently the time gap between traffic platoon caused by traffic signals around the pedestrian bridge crossing.

The relationship between Percentage (Pa / Pt) and Vehicle Speed also does not exist. Either with the speed of a motorcycle, light vehicle and heavy vehicle separately or at an average speed in two directions.

The relationship between percentage (Pa / Pt) and comparison time of crossing also apparently does not exist. As shown in Table 2., the average time for the actual crossing on the pedestrian bridge crossing is 57 seconds (at busy time), 56 seconds (at not busy) and below it ranges from 23 seconds (at busy times), and 21 (when not busy). The route under the bridge is on average faster than the route through the pedestrian bridge crossing.

The relationship between percentage (Pa / Pt) and comparison of travel time also apparently does not exist. Comparison of travel time frame shown in table 3. is from 2.5: 1 at busy time and 1.7: 1 at not busy time. The time difference on the way is not too great and may be unknown to pedestrians, so another reason must be for the low use of crossing bridges. Perhaps pedestrians think that using a pedestrian bridge crossing takes a much longer time when in fact it is not, may also be related to other influences while climbing the stairs.

Pedestrian opinions about pedestrian bridge crossing for people who do not use them ground under the bridge faster, while for people who use them because they feel safe.

Crossing bridge as a crossing beacon. It turns out from observations during the survey seen more people crossing under the bridge compared with people who crossed in further positions of the bridge.

5.1 Billboard

Many pedestrian bridge crossing are built by sponsors, especially for advertisements. This may be the reason why no criteria are accepted for the dimensions of the pedestrian bridge crossing, its materials or maintenance.

The negative impact of billboards is causing security problems for pedestrians using pedestrian bridge crossing. Often the boards are so high that they cover the bridge and consequently can invite crime.

Sponsors who build the crossing bridge seem to want to get the most profit from advertising. Consequently considerations of pedestrian bridge crossing design and maintenance (except the maintenance of the billboard it self) are of secondary importance. The difference in maintenance standards between pedestrian bridge crossing and billboards mounted on pedestrian bridge crossing is often very large.

5.2 Review of Building Structures of Pedestrian Bridge Crossing

Based on observations made on the structures and interviews with pedestrians of 3 (three) ages, children, adults and parents, as well as guidance from books on the manufacture of stairs, there is a standardization of crossing bridge designs that can provide comfort for pedestrians who use so as to increase the use of a pedestrian bridge crossing.

From the research that has been done, got the design criteria of pedestrian bridge crossing in the form of suggestion as follows:

a. For Stairs:
   - Antrede (width of the road) = 240 – 200 mm (recommended 260 mm) Plus a flat bulge = 25 mm.
   - Optrede (high footing) = 150 – 130 mm (recommended 1800 mm).
   - The slope of the stairs – $\alpha = 35° - 45°$ (recommended $35° - 38°$).
   - Wide staircase > 1100 mm (recommended 1.250 mm)

b. For Bridge Floor:
   - Net floor's width: 1700 – 1800 mm (Recommended according to need or min. 1800 mm).
   - High bridge's floor: 5.5 – 6.0 , For Banjarmasin recommended 4.5 m, because there is no plan of multilevel bus usage.
     General advice : 4.5 – 5.0 m (When there is no multilevel bus)
     4.5 – 6.6 m (When there is multilevel bus)

c. The height of the banister from the staircase = 800 – 840 mm (recommended 800 – 840 mm).

d. The location of the mouth and the ladder is placed in accordance with the available land, cultivated in an easily accessible place (avoid placement in the alley) and easy to see.

e. The number of mouths of the stairs depends on the pedestrian needs area.

f. The length of the crossing bridge depends on the existing road.

Drawings and details of the pedestrian bridge crossing and stair designs are shown in the accompanying drawings.

6 Conclusion

In general, the use of pedestrian bridge crossing in the city of Banjarmasin is very low even though this bridge is in front of the center of shops crowded with vehicular traffic and pedestrian.

Many variables are expected to show the relationship with the use of the bridge, was found no. One is the absence of a correlation between traffic (speed and traffic flow) with pedestrians using pedestrian bridge crossing. This is because the traffic platoon and time gap are large enough to allow people to cross the road.

In terms of time comparison, crossing time using a bridge is always much larger than crossing under a bridge. It seems that saving a little time is more important for pedestrians in the city of Banjarmasin than on his personal safety.

Another reluctance factor of the user is that pedestrian bridge crossing are less maintenance (dirty), often used beggars, less secure due to shielding by sponsoring boards so pickpockets and such are easy to operate.

7 Suggestion

1. One way that can be used to increase the use of pedestrian bridge crossing is by set a fence on the median, under a bridge or by set a curb side fence on both sides of the road long enough (approximately 200 m). This can prevent pedestrians from crossing the highway at any place and also to extend the pedestrian route to use the crossings.
2. Maintenance of crossing bridge condition is very important in its use. It is therefore recommended to maintain routine security and maintenance as well as cleaning. To build crossing bridges, design issues should be given greater priority.
3. Road safety education programs are required in schools (especially primary schools) with a fixed studies standard.
4. For the next pedestrian bridge crossing's work construction then there needs to be a comparative study of pedestrian bridge crossing and other crossing facilities (zebra cross, pelican), in the pedestrian efficient problem, cost, influence on traffic and so forth. There is also need for further study on the efficiency of the crossing bridge after the enactment of UULLAJR (Traffic Rules on The Highway) Num. 14 / 1992 to find out whether there is any difference with before the entry into force of UULLAJR.

References

1. Elvik, R., Mysen, B.A. & Vaa, T. *Trafikksikkerhetshåndbok (Traffic safety handbook).* Oslo, Institute of Transport Economics. Complete Norwegian version accessable at www.toi.no. (1997)
2. Elvik, R. *Which are the relevant costs and benefits of road safety measures designed for pedestrians and cyclists? Accident Analysis and Prevention.* 37-45. (2000)
3. Hidayah, Nurul K. *A Study on Effectiveness of Pedestrian Bridge Utilization.* Faculty of Civil Engineering & Earth Resources University Malaysia Pahang. Pahang. (2012)
4. Tamim, Ofyar Z. *Perencanaan dan Pemodelan Transportasi.* Second Edition, Bandung Institute of Technology. Bandung. 521p. (2000)
5. Transport And Road Research Laboratory. *Toward Safer Roads in Developing Countries: A Guide for Planners and Engineers.* Crowthorne, Berkshire. (1991)
6. Bruce, John A. *The Pedestrian.* Traffic Engineering Handbook. 3rd ed. Washington D.C. : Institute of Traffic Engineers. (1965)
7. Hartanto, Budi S. *True Bridge Bridges in Bandung Less Utilized.* Road and Transportation Engineering Magazine, no. 035, February: 44 - 46. (1986)
8. O' Flaherty, C.A. *Transportation Planning and Traffic Engineering.* Edward Arnold Ltd. London. (1997)

9. Mosabbir, MD. *Pedestrian's Behavior on Road Facilities.* UTM Press. Bangladesh. (2015)
10. Setiawan, Rudy. *Factors Affecting the Utilization of Pedestrian Bridges.* Malang. (2006)

Traffic congestion analysis using travel time ratio and degree of saturation on road sections in Palembang, Bandung, Yogyakarta, and Surakarta

Budi Hartanto Susilo[1,*], and *Ivan* Imanuel[1]

[1] Maranatha Christian University, Depart. of Civil Engineering, Bandung, Indonesia

Abstract. Traffic performance have been measured based on degree of saturation, velocity, travel time, and delays. However, they weren't enough to measure the traffic fluctuation that happen on normal days. This study aims to develop some modified analysis about traffic congestion based on the degree of saturation (DS) and travel time ratio (TTR). The travel time ratio is defined as a ratio between the travel time at peak hour and the travel time at off-peak hour in a certain road section. Traffic congestion conditions are classified into 4 categories: Peak-hour Congestion, Lengthy Congestion, Momentary Congestion, and Smooth Traffic. Based on survey results on road sections in Palembang, Bandung, Yogyakarta, and Surakarta, it can be concluded that peak-hour congestion occurs when TTR > 1,40 and DS > 0,75, lengthy congestion occurs when TTR < 1,40 and DS > 0,75, momentary congestion occurs when TTR > 1,40 and DS < 0,75, and smooth traffic when TTR < 1,40 and DS < 0,75.

1 Introduction

Traffic and road transport as mandated by Indonesian Government [1] must have criteria for a secure, safe, orderly, smooth, and integrated with other transport modes. Muhtadi [2] said that at the moment, traffic congestions often arise in large cities in Indonesia because of the increasing number of vehicles that pass by affecting the capacity of the road that had been planned earlier. One of the treatments for the congestion problems the periodical evaluation of performance of urban road traffic to ensure the optimal performance of the road.

Speed, travel time, and delays are measures commonly used as indicators of the performance of traffic [3]. In addition, traffic performance is also measured by the degree of saturation. The relationship between the flow and travel time can be expressed as a function where if the flow increases then the travel time will also increase [4]. However, the degree of saturation still can't explain more about the fluctuations of flows that occur on each day. As a result, it often occurs a smooth traffic flow on a road which has a degree of saturation above one or vice versa. This study aims to provide further understanding

* Corresponding author: budiharsus@yahoo.com

regarding the performance of traffic, especially congestion conditions that may occur on a road.

2 Desk study

The Traffic flow is affected by drivers and vehicles interacting with other vehicles, infrastructure, and surrounding environment. The volume that occurs is not always fixed but fluctuating due to several factors such as: time, composition, split portion, lane configuration, the kind of area use, road classification, road feature, the number and types of access control, shape and geometry of the streets [5]. The existence of unbalanced daily variation pattern between peak and off-peak time becomes a major concern for transportation planners experts because of congestion problems encountered in the big city usually happens at peak hours [4]. Generally the flow of traffic is divided between peak hour and off-peak hour.

Travel time is the time it takes to travel the full length of observed path. This study gives information about the running speed, the travel speed, the speed fluctuations, delays between two stations. There are two kinds of delay, namely fixed delay and operational delay. Fixed delay is a delay due to a temporary traffic signal while operational delay is a delay due to the traffic movements such as as turning vehicles, in and out activities, pedestrian crossing, parking, heavy traffic volume, insufficient capacity, and traffic accidents. [6].

According to Indonesian Highway Capacity Manual [7], Degree of saturation (DS), is defined as the ratio of the traffic flow/volume (V) and the road capacity (C) of a certain road section. The volume and capacity are both in passenger car units per hour (PCU/hour). The degree of saturation, also called V/C ratio or volume capacity ratio, is the major indicator in suggesting the traffic performance. The higher the degree of saturation value the lower the traffic performance.

$$DS = \frac{V}{C} \tag{1}$$

where,
DS = degree of saturation
V = traffic volume (PCU/hour)
C = road capacity (PCU/hour)

The congestion increases when the flow is so high resulting vehicles get very close to each other. Total congestion occurs when vehicles have to stop or move very slowly [4]. Conditions of traffic congestion are often expressed with the level of service from A through F [8].

3 Study method

This study aims to provide a modified classification on traffic congestions in order to ease us solving the congestion problems. Two parameters, travel time ratio and the degree of saturation, are used in this study as indicators of traffic congestion. Figure 1 shows basic framework of this study.

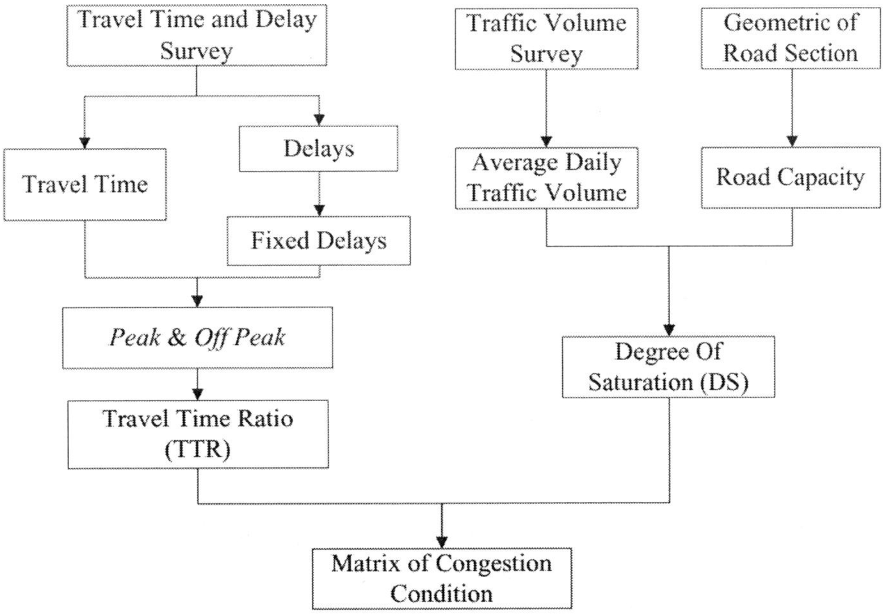

Fig. 1. Study framework

While the travel time ratio has been well known as ratio of travel time between various transportation modes [9], the proposed Travel time ratio (TTR) is a comparison between travel time on a road section during peak hours and off-peak hours. During rush hour, a large traffic volume tends to produce a lengthy travel time. It is influenced by the low speed due to the high density of traffic and the large amount of operation delays due to high level of transportation activities during rush hour. As the travel time during peak hours and off-peak hours have been recognized, TTR is proposed to be calculated as follows:

$$TTR = \frac{TT_P}{TT_0} \tag{2}$$

where,

TTR = travel time ratio

TT_P = travel time during peak hour (s)

TT_0 = travel time during off-peak hour (s)

If the number of surveys on each peak hour and off-peak hour is limited, then the travel time needs to be corrected against fixed delays such as signal controls and railway crossing. Corrections are intended to anticipate the significant influence of fixed delays on both time of the survey. For example, if there aren't any fixed delays occurred during the peak hour survey because of the green lights at all intersections while there are some during off-peak hours survey, then this may result in travel time during peak hours becoming shorter than travel time during off-peak hours.

$$TT = TS - FD \tag{3}$$

where,

TT = corrected travel time (s)

TS = real travel time (s)

FD = fixed delay time (s)

T By knowing its degree of saturation and travel time ratio, the condition of congestion that occurred on a road section can be assessed. Conditions of congestion based on the degree of saturation and the travel time ratio are classified into four types as shown in Table 1, namely:

a) Peak hour congestion: a condition of congestion where there are high traffic flow and activities at peak hours (0,5-2 hours) resulting significant delays and low speed. Both the degree of saturation and the travel time ratio on this road section are high. This condition expectedly happens on road sections with number of lane > 4 and high side frictions.
b) Lengthy congestion: a condition of congestion where high traffic flow and activities occur for several hours in a day (> 2 hours). The degree of saturation is high but the travel time ratio is low. This condition expectedly happens on road sections in downtown of major cities.
c) Momentary congestion: a condition of congestion where the traffic flow is considerably low all day but sometimes it becomes high for a short period of time. This condition expectedly happens on road sections with low capacity that a minor change in traffic flow can cause significant effects.
d) Smooth traffic: a condition where there is no congestion on a road sections all day. The traffic flows and side activities are both low and do not reduce traffic speed significantly nor cause any great operational delay.

Table 1. Matrix of congestion condition

Parameter	**High TTR**	**Low TTR**
High DS	Peak-hour congestion	Lengthy congestion
Low DS	Momentary congestion	Smooth traffic

The data were obtained from surveying on road sections in four cities in Indonesia, namely Palembang, Bandung, Yogyakarta and Surakarta. The road network includes national road network as well as urban road network. Three types of surveys had been conducted, namely travel time and delay survey, traffic volume survey, and road geometry survey. Travel time and delays survey was done to get travel time experienced by a normal moving vehicle on the road sections. This survey produced travel times and delays along with the location and cause. Traffic volume survey was done by counting the various passing vehicle on a certain spot location and generally conducted for 16 hours with the types of vehicles include motorcycles, light vehicles, heavy vehicles, and unmotorized vehicles. The geometry survey was conducted to identify the width and number of lanes, the median, the width of road shoulders/curb, and the land use/side friction along the road sections to calculate the road capacity.

4 Data and analysis

4.1 Determination of travel time ratio

Based on the results of calculations shown in Table 2 the lowest TTR on the road network in Palembang is 1.1 (Jend. Sudirman), while the highest TTR is 1.91 (Demang Lebar Daun). As the average value on all surveyed road sections is 1.57, the travel time ratios in Palembang are relatively high. Most of the delays are caused by the high density of traffic flow and oversaturated condition at the intersections.

Table 2. Travel time ratio of road sections in Palembang

Name of Road Section	Direction	Real Travel Time (s)		Fixed Delay Time (s)		Corrected Travel Time (s)		TTR
		TS_P	TS_0	FD_P	FD_0	TT_P	TT_0	
Kol. H. Burlian	South	1071	627	113	48	958	579	1.65
Kol. H. Burlian	North	865	637	84	82	781	555	1.41
Jend. Sudirman	South	1599	888	217	147	1382	741	1.87
Jend. Sudirman	North	1083	822	366	175	717	647	1.11
Residen Abdul Rozak	West	216	166	0	0	216	166	1.30
Residen Abdul Rozak	East	177	151	0	0	177	151	1.17
Basuki Rahmat	West	362	290	19	44	343	246	1.39
Basuki Rahmat	East	521	232	16	31	505	201	2.51
Demang Lebar Daun	West	649	757	64	451	585	306	1.91
Demang Lebar Daun	East	753	493	85	61	668	432	1.55

The lowest TTR on surveyed road sections in Bandung is 1.08 (Ahmad Yani and Jend. Sudirman) whereas the highest TTR is 1.25 (Soekarno-Hatta) as shown in Table 3. With the average value of 1.18, the travel time ratios in Bandung are relatively low. Most of the delays are caused by the high side frictions such as parked vehicles, pedestrian activities, and entering or exiting vehicles.

Table 3. Travel time ratio of road sections in Bandung.

Name of Road Section	Direction	Real Travel Time (s)		Fixed Delay Time (s)		Corrected Travel Time (s)		TTR
		TS_P	TS_0	FD_P	FD_0	TT_P	TT_0	
Soekarno-Hatta	East	4080	3600	1080	960	3000	2640	1.14
Soekarno-Hatta	West	5400	3720	2100	1080	3300	2640	1.25
Kebon Jati	East	314	240	16	0	298	240	1.24
Ahmad Yani	East	674	704	227	290	447	414	1.08
Ahmad Yani	West	888	652	244	128	644	524	1.23
Jakarta	West	159	129	0	0	159	129	1.23
Sudirman	West	867	795	102	89	765	706	1.08

From the results of the calculations in Table 4, the lowest TTR on surveyed road sections in Yogyakarta is 1.06 whereas the highest TTR is 1.73 (Laksda Adi Sucipto). With the average value of 1.16, the travel time ratios in Yogyakarta are relatively low. Most of the delays are caused by the high density of traffic flows.

Table 4. Travel time ratio of road sections in Yogyakarta.

Name of Road Section	Direction	Real Travel Time (s)		Fixed Delay Time (s)		Corrected Travel Time (s)		TTR
		TS_P	TS_0	FD_P	FD_0	TT_P	TT_0	
Janti	North	3180	2700	258	231	2922	2469	1.18
Janti	South	3180	2700	258	231	2922	2469	1.18
Jend. Sudirman	West	4500	4440	241	411	4259	4029	1.06
Jend. Sudirman	East	4500	4440	241	411	4259	4029	1.06
Laksda Adi Sucipto	East	2670	2340	211	269	2459	2071	1.19

Name of Road Section	Direction	Real Travel Time (s)		Fixed Delay Time (s)		Corrected Travel Time (s)		TTR
		TS_P	TS_0	FD_P	FD_0	TT_P	TT_0	
Laksda Adi Sucipto	West	2580	1680	106	253	2474	1427	1.73
Kolonel Sugiyono	East	4500	4440	241	411	4259	4029	1.06
Kolonel Sugiyono	West	4500	4440	241	411	4259	4029	1.06
Malioboro	South	4500	4440	241	411	4259	4029	1.06
Ring Road Utara	East	780	780	127	182	653	598	1.09
Ring Road Utara	West	2880	2700	198	308	2682	2392	1.12

Based on the calculation shown at Table 5, the lowest TTR on surveyed road sections in Surakarta is 1.14 (Slamet Riyadi) while the highest TTR is 2.83 (Yosodipuro). With the average value of 1.61, the travel time ratios in Surakarta are relatively high. Most of the operational delays are caused by the public transport vehicles making stops on the narrow streets.

Table 5. Travel time ratio of road sections in Surakarta.

Name of Road Section	Direction	Real Travel Time(s)		Fixed Delay Summary (s)		Corrected Travel Time (s)		TTR
		TS_P	TS_0	FD_P	FD_0	TT_P	TT_0	
Slamet Riyadi	East	485	628	27	227	458	401	1.14
Slamet Riyadi	West	613	382	167	97	446	285	1.56
Kol. Sutarto	East	148	126	46	45	102	81	1.26
Kol. Sutarto	West	374	126	64	5	310	121	2.56
Ir. Sutami	East	269	239	36	60	233	179	1.30
Ir. Sutami	West	278	238	16	72	262	166	1.58
Veteran	East	569	226	202	13	367	213	1.72
Veteran	West	428	250	140	0	288	250	1.15
Yosodipuro	East	240	207	56	74	184	133	1.38
Yosodipuro	West	427	138	62	9	365	129	2.83

4.2 Degree of saturation analysis

The degree of saturation is calculated by dividing the traffic volume by the road capacity. The used rush hour factor for general urban roads is 0.09 according to IHCM 1997 [7]. From the results of the calculations shown Table 6, the DS on road network in Palembang is relatively high with an average of 0.9. The highest degree of saturation occurs at Demang Lebar Daun, reaching 1.41 while the lowest degrees of saturation occurs at Sudirman direction, about 0.56.

Table 6. Degree of saturation of road sections in Palembang.

Name of Road Section	Direction	V (PCU/hour)	C (PCU/hour)	DS
Kol. H. Burlian	South	2506	4299	0.58
Kol. H. Burlian	North	2494	4299	0.58
Jend. Sudirman	South	3990	4153	0.96
Jend. Sudirman	North	3892	4153	0.94
Residen Abdul Rozak	West	1963	2884	0.68

Name of Road Section	Direction	V (PCU/hour)	C (PCU/hour)	DS
Residen Abdul Rozak	East	1775	2884	0.62
Basuki Rahmat	West	3322	2884	1.15
Basuki Rahmat	East	2930	2884	1.02
Demang Lebar Daun	West	6339	4481	1.41
Demang Lebar Daun	East	5771	4481	1.29

The DS of surveyed road sections in Bandung as shown in Table 7 are relatively normal with an average of 0.71. The highest DS is 1.1 (Ahmad Yani) while the lowest is 0.51 (Soekarno Hatta and Jend. Sudirman).

Table 7. Degree of saturation of road sections in Bandung.

Name of Road Section	Direction	V (PCU/hour)	C (PCU/hour)	DS
Soekarno-Hatta	East	3865	5860	0.66
Soekarno-Hatta	West	4080	5860	0.70
Kebon Jati	East	2723	4206	0.65
Ahmad Yani	East	1236	2058	0.60
Ahmad Yani	West	1128	1029	1.10
Jakarta	West	4999	5042	0.99
Jend. Sudirman	West	2162	4206	0.51

Generally shown in Table 8, the DS in Yogyakarta are relatively high with an average value of 0.89. The highest DS is 1.35 (Jend. Sudirman) while the lowest is 0.47 (Kolonel Sugiyono).

Table 8. Degree of saturation of road sections in Yogyakarta.

Name of Road Section	Direction	V (PCU/hour)	C (PCU/hour)	DS
Janti	North	3422	3935	0.87
Janti	South	2549	3935	0.65
Jend. Sudirman	West	3499	2596	1.35
Jend. Sudirman	East	2999	2596	1.16
Laksda Adi Sucipto	East	2601	2650	0.98
Laksda Adi Sucipto	West	2498	2426	1.03
Kolonel Sugiyono	East	1196	2568	0.47
Kolonel Sugiyono	West	2602	2568	1.01
Malioboro	South	2713	2400	1.13
Ring Road Utara	East	2588	4033	0.64
Ring Road Utara	West	2372	4033	0.59

As shown in Table 9, the DS in Surakarta are relatively low with an average value of 0,58. The highest DS is 0.71 (Ir. Sutami) while the lowest DS is 0.46 (Slamet Riyadi).

Table 9. Degree of saturation of road sections in Surakarta.

Name of Road Section	Direction	V (PCU/hour)	C (PCU/hour)	DS
Slamet Riyadi	East	1374	2183	0.63
Slamet Riyadi	West	998	2183	0.46
Kol. Sutarto	East	1442	2769	0.52
Kol. Sutarto	West	1789	2769	0.65
Ir. Sutami	East	1588	2411	0.66
Ir. Sutami	West	1703	2411	0.71
Veteran	East	834	1756	0.47
Veteran	West	925	1756	0.53
Yosodipuro	East	841	1529	0.55
Yosodipuro	West	1053	1529	0.69

3.3 Determination of congestion condition

The highest TTR of all surveyed road sections in 4 cities is 2.83 (Yosodipuro, Surakarta) while the lowest TTR is 1.06 (Malioboro, Yogyakarta). The average value of TTR is 1.40. Based on these results, TTR with value above 1.40 is classified "High" while TTR with value between 1.00 and 1.40 is classified "Low". The DS with value above 0.75 is classified "high" according to IHCM 1997 (Directorate General of Bina Marga, 1997) and classified low below 0.75. Therefore, the condtions of cogestion of road networks in 4 cities can be identified, according to Table 1, as given in Table 10.

Table 10. Identification of congestion condition.

Name of Road Section	Direction	DS	TTR	Congestion Condition
Kol. H. Burlian	South	Low	High	Momentary congestion
Kol. H. Burlian	North	Low	High	Momentary congestion
Sudirman	South	High	High	Peak-hour congestion
Sudirman	North	High	Low	Lengthy congestion
Residen Abdul Rozak	West	Low	Low	Smooth traffic
Residen Abdul Rozak	East	Low	Low	Smooth traffic
Basuki Rahmat	West	High	Low	Lengthy congestion
Basuki Rahmat	East	High	High	Peak-hour congestion
Demang Lebar Daun	West	High	High	Peak-hour congestion
Demang Lebar Daun	East	High	High	Peak-hour congestion
Soekarno-Hatta	East	Low	Low	Smooth traffic
Soekarno-Hatta	West	Low	Low	Smooth traffic
Kebon Jati	East	Low	Low	Smooth traffic
Ahmad Yani	East	Low	Low	Smooth traffic
Ahmad Yani	West	High	Low	Lengthy congestion
Jakarta	West	High	Low	Lengthy congestion
Jend. Sudirman	West	Low	Low	Smooth traffic
Janti	North	High	Low	Lengthy congestion

Name of Road Section	Direction	DS	TTR	Congestion Condition
Janti	South	Low	Low	Smooth traffic
Jend. Sudirman	West	High	Low	Lengthy congestion
Jend. Sudirman	East	High	Low	Lengthy congestion
Laksda Adi Sucipto	East	High	Low	Lengthy congestion
Laksda Adi Sucipto	West	High	High	Peak-hour congestion
Kolonel Sugiyono	East	Low	Low	Smooth traffic
Kolonel Sugiyono	West	High	Low	Lengthy congestion
Malioboro	South	High	Low	Lengthy congestion
Ring Road Utara	East	Low	Low	Smooth traffic
Ring Road Utara	West	Low	Low	Smooth traffic
Slamet Riyadi	East	Low	Low	Smooth traffic
Slamet Riyadi	West	Low	High	Momentary congestion
Kol. Sutarto	East	Low	Low	Smooth traffic
Kol. Sutarto	West	Low	High	Momentary congestion
Ir. Sutami	East	Low	Low	Smooth traffic
Ir. Sutami	West	Low	High	Momentary congestion
Veteran	East	Low	High	Momentary congestion
Veteran	West	Low	Low	Smooth traffic
Yosodipuro	East	Low	Low	Smooth traffic
Yosodipuro	West	Low	High	Momentary congestion

5 Conclusions

Based on the analysis, travel time ratio (TTR) along with the degree of saturation (DS) can classify the traffic congestion conditions on road sections in Palembang, Bandung, Yogyakarta, and Surakarta. Furthermore, it can be classified into 4 categories, namely:

1. Peak-hour congestion, if TTR $\geq$ 1,40 and DS $\geq$ 0,75.
2. Lengthy congestion, if TTR $<$ 1,40 and DS $\geq$ 0,75.
3. Momentary congestion, if TTR $\geq$ 1,40 and DS $<$ 0,75.
4. Smooth traffic, if TTR $<$ 1,40 and DS $<$ 0,75.

References

1. Indonesian Government, (2009) *Indonesian Laws No 22 Year 2009 about Traffics and Road Transports.*
2. Muhtadi, A., *Analisis Kapasitas, Tingkat Pelayanan, Kinerja dan Pengaruh Pembuatan Median Jalan*, Neutron, 10(1), pp. 43-54, (2010)
3. Roess, R.P., Prassas, E.S., & McShane, W.R., *Traffic Engineering*. **4th ed.**, Upper Saddle Pearson Education, Inc. (2011)
4. Tamin, O.Z., Peren*canaan dan Pemodelan Transportasi: Teori, Contoh Soal, dan Aplikasi.*, Bandung, Penerbit ITB. (2008)
5. Oglesby, C.H. & Hicks, R.G., *Highway Engineering*. **4th ed.**, New York, John Wiley & Sons, Inc. (1982)
6. Susilo, B.H., *Rekayasa Lalu Lintas*, edisi revisi, Jakarta, Penerbit Trisakti, (2012)

7. Directorate General of Bina Marga, *Indonesian Highway Capacity Manual*, Jakarta, Penerbit PU, (1997)
8. Indonesian Minister of Transportation, *Regulations of Minister of Transportation No. KM 14 Year 2006 about Management and Traffic Engineering on Roads*. (2006)
9. Susilo, Y.O. & Dijst, M., *Behavioural Decisions of Travel Time Ratios for Work, Maintenance and Leisure Activities in The Netherland*, Transport Planning and Technology, 33(1), pp.19-34. (2010)

MATEC Web of Conferences **181**, 07001 (2018) https://doi.org/10.1051/matecconf/201818107001
ISTSDC 2017

Passenger satisfaction with railway station service quality: An Ordinal Logistic Regression approach

Willy Kriswardhana[1*], *Nunung* Nuring Hayati[1] and *Septiana* Desy R[2]

[1] University of Jember, Department of Civil Engineering, Jember, East Java, Indonesia

Abstract. The objective of this paper is to examine the service quality dimensions and passenger characteristics and its effect on passenger's satisfaction on railway station. Jember Railway Station is a survey location where the questionnaires were spread. A total of 400 respondents who departure from Jember Railway Station were selected via convenience sampling method. The effects of characteristics and service quality items on the overall satisfaction were estimated using probabilistic approach. The thirteen service items extracted from Guidelines for Standardization of Railway Stations provide a representative framework of how passengers evaluate railway station service quality. This study illustrates how the probabilistic approach may be used for the direction of the effects of service quality and passengers' characteristics on the probability that a passenger presents higher level of overall satisfaction. Direction for future research is presented.

1 Introduction

Service quality is strictly linked to passenger's satisfaction [1]. Service quality is a specific judgment of services availed of while satisfaction is evaluated in broader terms [2]. Passenger's satisfaction on railway service quality has an effect on the frequency of the passenger travelling by train. Probability of passenger travelling by train will increase if the passenger feel satisfied on the railway service quality. Railway service quality is a very significant element for the travel effect, how to evaluate it is a difficult problem [3].

Period of using railway service has a positive relationship with behavior of railway staff. Frequency of travelling with train also has a positive relationship with passenger's

* Corresponding author: willy.teknik@unej.ac.id

acceptance of arrival delay of train. The need of passenger's satisfaction is important to compete with other mode of transport [4].

According to DAOP IX Jember, the number of train passengers has rapidly increased since the improvement of service quality. The evaluation of service quality leads to customer satisfaction [5]. Satisfaction is a costumer fulfillment response, which is an evaluation as well as an emotion-based response to a service [6]. The perceived service quality is not only subjective, but also based on context that will likely different according to several factors [7].

While several research studies have been conducted on airline passengers perspective, little has been published on the perspective of railway station passengers especially in Asian country, include Indonesia. Hence, this study aims to examine the effect of railway station service quality on overall passenger satisfaction. Results offer a clearer perspective for railway station administrator, enabling them to identify passenger's behavior for better service.

In this context, the objective of this paper is to examine the service quality dimensions and passenger characteristics and its effect on passenger's satisfaction on railway station

2 Methodology

This study conducts a quantitative method with self-administered questionnaire in order to make sure the respondents know the conceptual of the questionnaire. A total of 400 questionnaires were sent out among respondents at the Jember Railway Station. Satisfaction scores (scaled 1-4) were collected for the overall satisfaction and for the satisfaction with several service attributes. In addition, background characteristics of respondents are also asked for, such as trip frequency, age, genders, level of education, income, trip origin and destination. The survey included attributes related to services railway station facilities. The items on railway station facilities were aligned with Guidelines for Standardization of Railway Stations, published by PT Kereta Api Indonesia (Persero).

This study focused on service attributes that are related to passengers overall satisfaction. First step of data analysis is testing the predictor variable includes service quality dimensions. Predictor variables were tested with validity and reliability test. Significance test is done by comparing the value of r arithmetic with r table, if r-arithmetic> r-table and has positive value, then the variable is valid. While if r-arithmetic< r-table, then the variable is invalid [8]. The measurement of the reliability of the question items by spreading the questionnaire to the respondents, and the results of the score measured the correlation between the answers scores in the same question items with Cronbach Alpha (a). A construct or variable is said to be reliable if it gives a Cronbach Alpha value> 0.70 [8].

In the second step, an ordinal logistic regression model was estimated to examine service dimension and characteristics of passenger effects on satisfaction in overall. There are some advantages in using probabilistic model, especially when dealing with ordinal scales, categorical variable, and some cases where data do not satisfy linear regression assumptions [9].

Logistic regression generates an odds ratio associated with the value of each predictor. The probability of an event is defined as the probability of an arising outcome divided by the probability of an event not occurring. In general, the opportunity ratio is a set of opportunities shared by other opportunities. Because of the estimation coefficient expresses the relationship between logit and explanatory variables, their first interpretation pointing toward the effect on that probability passengers provide a higher level of satisfaction than lower level. In addition, the exponential coefficient can be used to evaluate its effect in terms of percentage change in odds [10].

Backward elimination was used as an alternative for model specification. As in many disciplines, null-hypothesis significance testing is commonly used. The test yields a statistic and probability (p-value). The null-hypothesis is rejected when the p-value is less than or equal to 0.05 or 0.01 in social science [11]. Theoretically, the p-value is a measure of continuous evidence, but in practice, it is usually trichotomized about to be very significant, slightly significant, and not statistically significant at the conventional level, with cutoffs at $p \leq 0.01$, $p \leq 0.05$ and $p > 0.10$. [12]. Most statistics handbooks present a rule of thumb of 0.1, 0.05 and 0.01 significance levels [13].

Table 1 Service quality dimensions and test result

Dimensions	r arithmetic	r table	Cronbach α
Check-in (CHK)			
Officer service (CHK1)	0.731		
Check-in process efficiency (CHK2)	0.719		
Convenience (CON)			
Availability of restaurant and minimart (CON1)	0.659		
Availability of disabled special toilet and elderly (CON2)	0.402		
Availability of breastfeeding room (CON3)	0.380		
Courtesy and helpfulness of railway station staff (CON4)	0.709	0.098	0.906
Feeling of being safe and secure (CON5)	0.585		
Basic Facilities (BAS)			
Departure lounge facilities (BAS1)	0.696		
Cleanliness of toilets (BAS2)	0.656		
Cleanliness of mosque (BAS3)	0.659		
Parking area facilities (BAS4)	0.420		
Ambience (AMB)			
Cleanliness of railway station facilities (AMB1)	0.501		
Neatness of spatial arrangement (AMB2)	0.591		

3 Result and discussion

3.1 Visitors distribution

Table 2 displays the descriptive analysis of the demographic profile of respondents, of whom 54.8% were men. A large number of respondents are from the age 18-24 (62.3%). Most participants received a monthly income of below Rp 500 (42%).

3.2 Validity and reliability analysis

The reading of Cronbach's alpha and composite reliability for each service dimensions variable is greater than 0.70, in relation to the expected factors, indicating higher reliability among the indicators. The r-arithmetic, as presented in Table 1, is greater than the r-table. It indicates that each variable is valid.

Table 2 Demographic profile of respondents

Variables		Frequency	Percentage
Gender	Male	181	45.3
	Female	219	54.8
Age	<18	62	15.5
(years old)	18-24	249	62.3
	25-34	38	9.50
	35-44	14	3.50
	45-54	10	2.5
	>54	6	1.5
Education level	Elementary	3	0.8
	Junior High School	41	10.3
	Senior High School	103	25.8
	Diploma	35	8.8
	Undergraduate	201	50.3
	Master/PhD	15	3.8
Monthly income	< Rp 500	168	42.0
(thousands)	Rp 500 - Rp 1.000	100	25.0
	Rp 1.000 - Rp 1.500	30	7.5
	Rp 1.500 - Rp 2.500	45	11.3
	Rp 2.500 - Rp 5.000	50	12.5
	Rp 5.000 - Rp 7.500	7	1.8
	> Rp 7.500	0	0

3.3 Regression analysis

Regression analysis result presents five services items and one passenger characteristic variables with significance statistic at 0.1 level. (Table 2). According to the test, the model was fitted with logistic function at a 0.05 significance level.

Table 2. Parameter estimation

		Estimate	Std. error	Wald	Sig.
Threshold	overall = 2.00	2.964	1.112	7.103	0.008
	overall = 3.00	7.505	1.193	39.552	0.000
Location	CON1	0.372	0.197	3.593	0.058
	CON4	0.522	0.278	3.526	0.060
	CON5	0.389	0.197	3.907	0.048
	BAS1	0.554	0.204	7.368	0.007
	BAS4	0.282	0.146	3.744	0.053
	Gender	-0.625	0.241	6.722	0.010

p-value<0.01; R^2 Cox and Snell 0.185; R^2 Negelkerke 0.224; R^2 McFadden 0.117

The following general estimated equation was obtained: Logit = a_j – [+0.372(CON1) + 0.522(CON4) + 0.389(CON5) + 0.554(BAS1) + 0.282(BAS4) – 0.625(gender)]. Where a_2 = 2.964, a_3 = 7.505 are the values for the intercepts a_j.

3.4 Direction of the Effects

According to the results, passengers are more likely to assign higher ranking for overall satisfaction with the railway station the getting higher his level of satisfaction with service items. These items are "Availability of restaurant and minimart (CON1)", "Courtesy and helpfulness of railway station staff (CON4)", "Feeling of being safe and secure (CON5)", "Departure lounge facilities (BAS1)", "Parking area facilities (BAS4)". These service items have positive effect, it indicates that the higher his/her satisfaction with these service items, the more likely the passenger assign a higher rating for overall satisfaction.

The non-significant effect items such as "Check-in dimensions" and "Ambience" indicate that the attributes may be generally a minor concern for passengers at railway station. Cleanliness of toilets and mosque should be the significant factor that effects the overall satisfaction. It is particularly interesting, considering the fact that cleanliness of toilet is one of the main criteria for passenger. The cleanliness of toilet is the highest value of weight in the cleanliness primary indicator [3].

Regarding the passenger characteristics, gender presented the significant effect. Gender has a negative effect on overall satisfaction. It indicates that the coding 1 (male), more likely to present a higher overall satisfaction.

4 Conclusions

The focus of research in railway station service quality has been published little on the perspective of railway station passengers especially in Asian country, include Indonesia. This study is a small piece of contribution for the development of a broader understanding of railway station service quality form perspective of passenger. The thirteen service items extracted from Guidelines for Standardization of Railway Stations provide a representative framework of how passengers evaluate railway station service quality. This study illustrates how the probabilistic approach may be used for the direction of the effects of service quality and passengers' characteristics on the probability that a passenger presents higher level of overall satisfaction.

Considering future works, earliness of arrival and trip purpose should be analyzed, regarding the earliness of arrival has a positive effect on overall satisfaction of airport services quality [10]. Furthermore, although several variables has proved to be valid, railway service quality dimension should be built from a deeper passenger behavior and expectations. Opportunities exist to further advance this research by expanding the number of variables, investigate at different location, and multiplying number of respondents since the questionnaire only distribute among 400 respondents.

Acknowledgements

We thank PT KAI DAOP IX Jember for granting access to the dataset and giving permission to distribute the questionnaires at Jember Railway Station.

References

1. Brady, Michael K., J.Joseph Jr. Cronin, Tim Brady, Michael K. Brady, and J.Joseph Jr. Cronin. "*Some New Thoughts on Conceptualizing Perceived Service Quality: A Hierarchical Approach.*" The American Avation Experience 65(3):34–49. Retrieved (http://journals.ama.org/doi/abs/10.1509/jmkg.65.3.34.18334). (2001).
2. Oliver, Richard L. "*Satisfaction: A Behavioral Perspective on the Consumer.*" Journal of Consumer Marketing 14:448. Retrieved (http://www.amazon.com/Satisfaction-Behavioral-Perspective-McGraw-Hill-marketing/dp/0071154124). (1997).
3. Eboli, Laura, Yanbing Fu, and Gabriella Mazzulla. "*Multilevel Comprehensive Evaluation of the Railway Service Quality.*" Procedia Engineering 137:21–30.(2016).
4. Kumar, P.Vima. and P. Jitin. "*A Study on Passengers' Satisfaction towards Railway Service With Reference To Coimbatore Junction.*" International Journal of Management and Commerce Innovations 2(2):578–82. (2014).
5. Ekinci, Yuksel. "*From Destination Image to Destination Branding: An Emerging Area of Research.*" E-Review of Tourism Research 1(2):21–24. Retrieved (http://ertr.tamu.edu). (2003).
6. Rust, Roland T. and Richard L. Oliver. "*Service Quality: Insights and Managerial Implications from the Frontier.*" Pp. 1–19 in Service Quality: New Directions in Theory and Practice. (1994).
7. Wilson, A., V. A. Zeithaml, M. J. Bitner, and D. D. Gremler. "*Services Marketing: Integrating Customer Focus across the Firm.*" Pp. 338–65 in Services marketing: integrating customer focus across the firm. (2012).

8. Ghozali, Imam. "*Aplikasi Analisis Multivariate Dengan Program IBM Dan SPSS.*" in aplikasi analisis multivariate dengan program ibm spss 19. (2011).
9. Akinci, Serkan, Erdener Kaynak, Eda Atilgan, and Safak Aksoy. *Where Does the Logistic Regression Analysis Stand in Marketing Literature?*: A Comparison of the Market Positioning of Prominent Marketing Journals. Retrieved (http://www.emeraldinsight.com/10.1108/03090560710737598). (2007).
10. Bezerra, George C. L. and Carlos F. Gomes. "*The Effects of Service Quality Dimensions and Passenger Characteristics on Passenger's Overall Satisfaction with an Airport.*" Journal of Air Transport Management 44–45:77–81. Retrieved (http://dx.doi.org/10.1016/j.jairtraman.2015.03.001). (2015).
11. Nickerson, Raymond S. "*Null Hypothesis Significance Testing: A Review of an Old and Continuing Controversy.*" Psychological Methods 5(2):241–301. (2000).
12. Gelman, Andrew and Hal Stern. "*The Difference Between 'Significant' and 'Not Significant' Is Not Itself Statistically Significant.*" The American Statistician 60(4):328–31. (2006).
13. Figueiredo Filho, Dalson Britto et al. "*When Is Statistical Significance Not Significant*?" Brazilian Political Science Review 7(1):31–55. Retrieved (http://www.bpsr.org.br/index.php/bpsr/article/view/154). (2013).
14. Hair, Joseph F., William C. Black, Barry J. Babin, and Rolph E. Anderson. "*Multivariate Data Analysis.*" Vectors 816. (2010).

Simulation of queue length on Amplas Entrance Toll before and after the ramp off development with Vissim software

Medis Surbakti[1,*], *Tadeus* Satria[1]

[1]University of North Sumatera, Department of Civil Engineering, Medan, Indonesia

Abstract. In 2016 Medan City administration built on off ramp at Amplas fly over, where this ramp off directly to toll booth Amplas. The faster the time needs to get to the toll booth enables the rate of arrival vehicles that impact with increasing queue length. The purpose of this study is to know the length of the queue at Amplas toll booth before and after ramp off construction with queue theory method and vissim softwaremodelings. Data acquired by recording the activities at the research site in the form of vehicles volume. The queue length obtained from the queue method with average service time 7 seconds is 6,212 meter while the result using VISSIM software is 11,988 meter. From the VISSIM software models also obtained that the length of queue after the construction of the ramp off increases compared before the construction of ramp off, from 6,67 meter to 11,988 metre.

1 Preface

1.1 Background

The increasing demand for vehicle traffic services accompanies the development of the region in Medan City. This influence has been anticipated by efforts to strengthen the road network.One of the anticipations is the construction of highway toll roads. In this case PT. Jasa Marga should provide higher reliability and ensure that vehicle operations are better than alternative roads [1, 2]. Given the toll road function must provide services in the form of smooth flow of vehicles without any obstacles, then the problems causing congestion need to be investigated. One of the factors that often cause congestion is the process of queuing at toll gates [3-5]. At the time of processing at toll booths the toll road users are required to collect entrance tickets at the entrance gate and pay the toll at the exit gate. Thus the duration of processing at the toll gate affects the smooth flow of the vehicle. In the real fact, long queues of vehicles at toll booths are often found mainly during rush hours.Some previous researches on the length of queues at toll booths indicate that the queue at toll booths occurs due to the arrival rate (flow rate) of vehicles going to toll booths is not balanced with the service level in the service substations. One of them occurs at the toll

* Corresponding author: medissurbakti@yahoo.com

booth where the queues are frequent. In addition to connecting the city of Medan other cities such as Tj.Morawa and Deli Serdang, Amplas toll booth is the fastest path to Kualanamo Airport located in Deli Serdang. This makes people more interested in using toll roads than using alternative roads. In 2016, the government made the on off ramp development on the fly over Amplas, where the ramp off will go directly to the toll booth Amplas so that people who want to go to the toll of Amplas from the direction of SM. Raja road does not need to cross intersection [6-7]. The quicker the time it takes the driver to go to the toll booth allows for an increase in arrival rate which ultimately impacts with increasing queue length. For that I as a researcher wanted to review whether there are real effects that occur after the construction of the ramp off against the length of the queue at the toll booth Amplas.

1.2 Formulating of the problem

The How the impact of ramp-off construction on the Amplas fly over the queue length at the entrance toll of Amplas?.

1.3 Research purpose

1. To determine the length of the queue at the toll booth before and after the construction of the ramp off on the amplas fly over
2. To determine the optimum service time at the toll booth Amplas
3. To analyze the capacity and service level of the toll gate entering the Amplas by using the existing system. The results of the analysis are used to determine whether at the toll gate Amplas can accommodate vehicles that come according to Minimum Service Standards (MSS)
4. Provide advice in the form of policy to reduce the length of the queue at the toll booth Amplas, where the suggestion is given in the form of simulation of queue length at the tollgate Amplas by using VISSIM software.

2 Literature review

2.1 Queuing theory

Queue occurs because the traffic flow process is disrupted by a service activity traversed. This theory provides important information needed to solve the above problems by forecasting various characteristics with queuing systems.

2.2 Calling Population

The pattern of arrival has unlimited and unlimited nature, of which the infinite nature is generally also random if the arrival is assumed to occur at a constant mean velocity and free of each other is called the Poisson probability distribution. The probability and arrival in time T is determined by the formula:

$$P(r\,,T) = \frac{e^{-\lambda T}(\lambda T)^r}{r\,!} \tag{1}$$

Where:

λ = average unified arrival time

T = time periode
e = natural logarithm number(e = 2.7182818)
r = the number of arrivals in timeT ; (n = 0,1,2....)
P(r,T) = probabilty n arrival in time T

If the arrival follows the Poisson Distribution it can be shown mathematically that the interception will be distributed according to the exponential distribution

$$P(s \leq t) = e^{-\lambda t}, 0 \leq t \leq \infty \quad (2)$$

Where:
P (s≤t) = probability where time is between arrival time unity
T = average time in system (seconds)

2.3 Level of service

The service level (μ) is the number of vehicles / persons that can be served by one service place in a given time unit. In addition to the level of service, it is also known as Time of Service (WP) which is defined as the time required by a single shelter to be able to serve one vehicle / person. Usually expressed in units of minutes / vehicles or minutes / people, so it can be concluded that [3, 4]:

$$WP = \frac{1}{\mu} \quad (3)$$

Also known as ρ notation is defined as follows

$$\rho = \frac{\lambda}{\mu} < 1 \quad (4)$$

If the value is ρ> 1, this means that the arrival rate is greater than the rate of service. If this happens, then certainly there will be a queue that will always increase in length (infinity)

2.4 Discipline queue FIFO (Fisrt In First Out)

For an example FIFO discipline is the queue of vehicles that formed in front of the toll gate.

2.5 Parameter FIFO queue discipline

The following equations (5) - (8) can be used to compute n, q, d, and w for FIFO queuing disciplines.

$$n = \frac{\lambda}{(\mu - \lambda)} = \frac{\rho}{(1 - \rho)} \quad (5)$$

$$q = \frac{\lambda^2}{\mu(\mu - \lambda)} = \frac{\rho^2}{(1 - \rho)} \quad (6)$$

$$d = \frac{1}{(\mu - \lambda)} \quad (7)$$

$$w = \frac{\lambda}{\mu(\mu-\lambda)} = d - \frac{1}{\mu} \tag{8}$$

Where:

n	= number of vehicles or people in the system (vehicle or person per unit of time)
q	= number of vehicles or people in the queue (vehicle or person per unit time)
d	= time of vehicle or person in the system (unit time)
w	= time of vehicle or person in queue (unit time)
λ	= average arrival rate
μ	= average service levels
ρ	= traffic intensity or usage factor= $\frac{\lambda}{\mu}$

2.6 VISSIM Program (visual simulation)

VISSIM is a microscopic software that serves to simulate urban traffic models and public transport operations.

This program can analyze traffic and displacements with modeling constraints such as path geometric, vehicle composition, traffic signals, stop line, driver behavior and others, Making it a useful tool for evaluating alternatives based on transport engineering as more effective and efficient decision-making steps in planning activities including simulations in the development of modes [5].

3 Research methodology

3.1 Primary data collection

Primary data collection is done at the entrance of the Amplas toll road. The survey was conducted by recording video on traffic activity at the research location. The survey was conducted by placing a video camera placed on a spot that could include the toll entrance area. The survey was conducted on Wednesday, June 7, 2017 at 07.00 - 12.00 WIB. The data obtained while conducting the survey are:

1. The volume of the vehicle leading to the toll booth where in this case is done by recording using a handycam. Then manually calculated the number of vehicles coming every 1 minute.
2. The speed of the vehicle leading to the toll booth. The speed of the vehicle is calculated by using Spot Speed method is the recording time travel of vehicles passing through the road segment of observation. Recording time travel is done by turning on the stopwatch just as the front wheel of the vehicle passes the first step of the trail following the vehicle's lane and the stopwatch is switched off to the wheel of the vehicle over the second stampede line.
3. Service time at each toll booth. Conducted when the vehicle stops in front of the substation (counter) to signify the transaction (when toll payment is in progress) until the vehicle moves to leave the substation. Service time is calculated using stopwatch

This research is done through several stages of the process. The above stages can be seen in the following figure of flow chart.

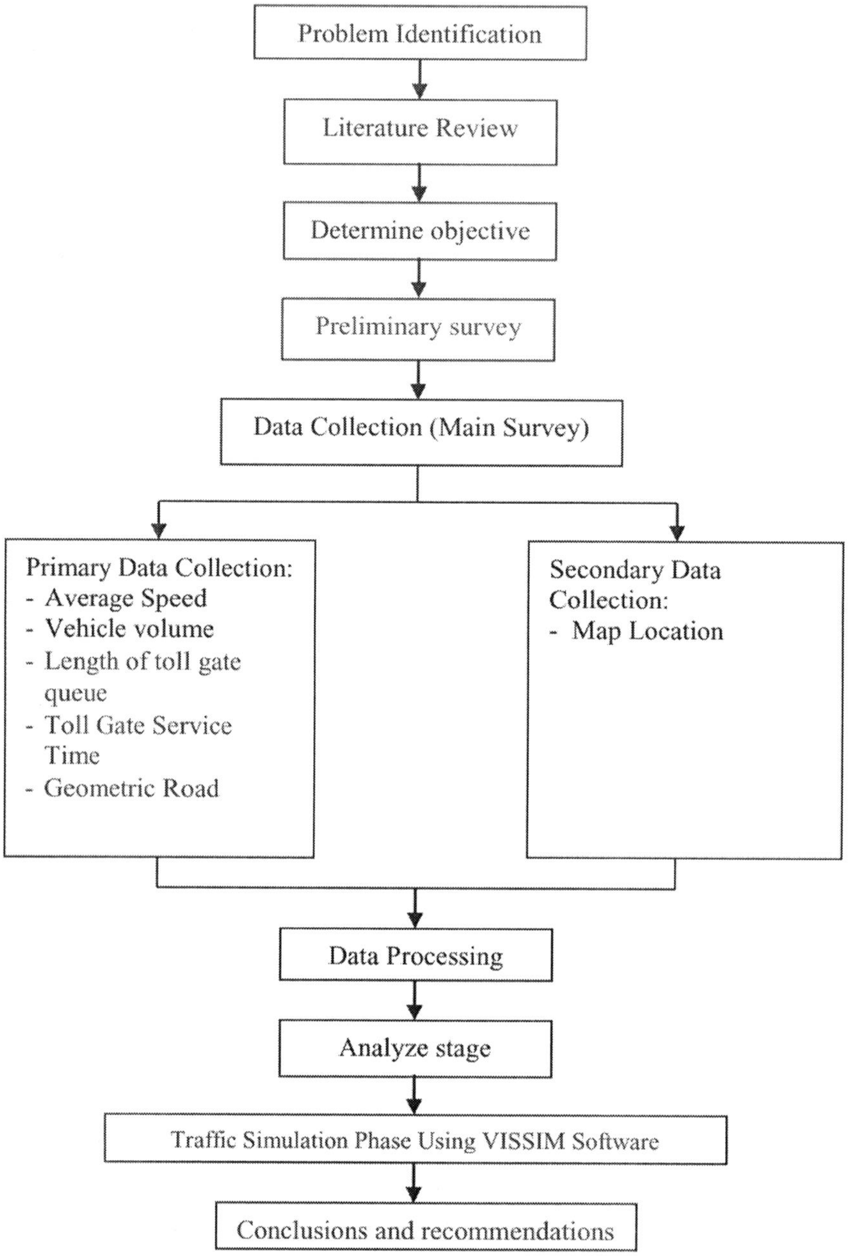

Fig. 1. Flow chart

4 Analysis and discussion

4.1 Calculation of arrival Rate

Average vehicle arrival (λ)

$$\lambda = \frac{\Sigma x}{N} = \frac{4698}{5} = 939{,}6 \approx 940 \text{ vehicle / hour}$$

Table 1. Survey data of arrival level of Amplas Toll Gate

No	Hour (WIB)	Number of vechiles
1	07.00 - 08.00	973
2	08.00 - 09.00	1004
3	09.00 - 10.00	965
4	10.00 - 11.00	919
5	11.00 - 12.00	837
Total		4698

4.2 Calculation of service level

Table 2. Data on service time average of Amplas Toll Gate

Substasion	Average Service Time (seconds)
1	6,73
2	6,82
3	7,34
Mean	6,96

The average service level of substations (μ)

$$\mu = \frac{3600}{7} = 517{,}2 \approx 517 \text{ vehicle/ hour}$$

4.3 Capacity Calculation of Amplas Toll Gate

$$\rho = \frac{940/3}{517} = 0{,}60 > 1$$

With a value of ρ > 1, the Amplas Tollgate can still accommodate the number of vehicles coming.

4.4 Calculation of queue length with queue theory

Using with ST = 7 second/vehicle
μ = 515 ; λ = 940 ; N = 3

$$n = \frac{940/3}{515-(940/3)} = 1{,}553 = 2 \; vehicles$$

$$q = \frac{(940/3)^2}{515(515-(940/3))} = 0{,}945 = 1 \; vehicles$$

$$d = \frac{1}{515-(940/3)} x \; 3600 = 17{,}851 = 18 \; seconds$$

$$w = \frac{(940/3)}{515(515-(940/3))} x \; 3600 = 10{,}860 = 11 \; seconds$$

Table 3. Calculation of queue length with combination of service time

ST	Service Rate (vehicle/hour)	Number of vehicles queuing up (vehicle)	Length of queue (m)	Number of vehicles queuing up (vehicle)	Length of queue	Waiting time (second)	Waiting time (second)
	µ	n	n	q	q	d	w
7	515	1.553	6.212	0.945	3.78	17.851	10.86
8	450	2.292	9.168	1.596	6.384	26.341	18.341
9	400	3.615	14.46	2.832	11.328	41.538	32.538
10	360	6.714	26.856	5.843	23.372	77.142	67.142
11	328	21.363	85.452	20.408	81.632	245.454	234.478
12	300	-23.5	-94	-24.54	-98.16	-270	-282

where,

n – q = 1 ; d – w = service time
1 vehicle = 4 meters
Minus sign (-) = there is a queue that will continue to grow long (infinity)

4.4 Calculation of queue length with VISSIM Software

Fig. 2. Simulation of Amplas Toll Door after ramp off development

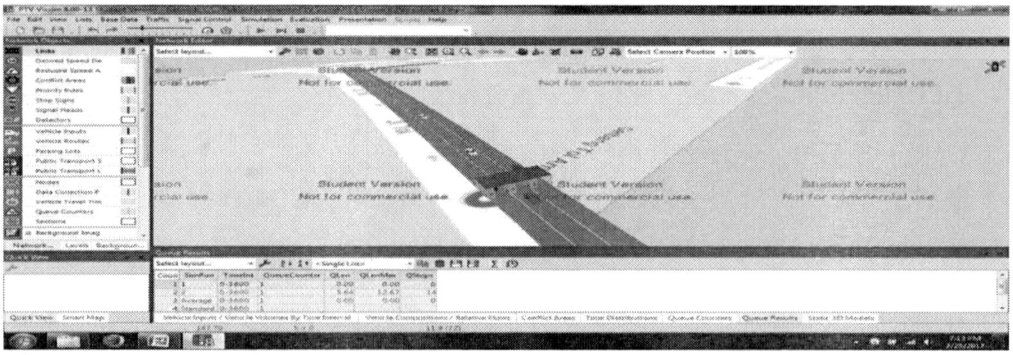

Fig. 3. Simulation of Amplas Toll Gate before ramp off development

After running the simulation, a new queue length is generated based on the output. The output of the queue length can be seen in Table 4 and Table 5.

Table 4. Queue Length of Amplas Toll Door after ramp off development

NO	Service Time	Time Interval	Minimum Queue length (meter)	Maximum Queue Length (meter)	Vehicle stop (Vehicle)
1	7 seconds	0 - 3600	11.98803	37.990954	85
2	8 seconds	0 - 3600	12.5477	37.869336	86
3	9 seconds	0 - 3600	14.1493	37.858121	85
4	10 seconds	0 - 3600	14.42131	41.486207	85
5	11 seconds	0 - 3600	15.93987	44.875732	85

Table. 5. Queue Length of Amplas Toll Door Before Ramp Off Development

NO	Service Time	Time Interval	Minimum Queue length (meter)	Maximum Queue Length (meter)	Vehicle stop (Vehicle)
1	7 seconds	0 - 3600	6.674095	14.817103	78
2	8 seconds	0 - 3600	7.37283	20.981991	78
3	9 seconds	0 - 3600	8.193706	25.884951	81
4	10 seconds	0 - 3600	8.87812	20.341288	80
5	11 seconds	0 - 3600	10.15851	23.510463	81

Based on VISSIM software modeling results obtained that the length of the queue tollgate Amplas after the construction of ramp-off increases compared to before the ramp-off. The queue length increased from 6.67 meters to 11.98 meters.

4.5 Analysis of the effectivenes of the queue length reduction policy

Polict taken:

- Increase the number of toll booths (from 3 substations to 4 substations)
- Apply "Gerbang Tol Manual" (GTO)

4.6 Addition of toll gate

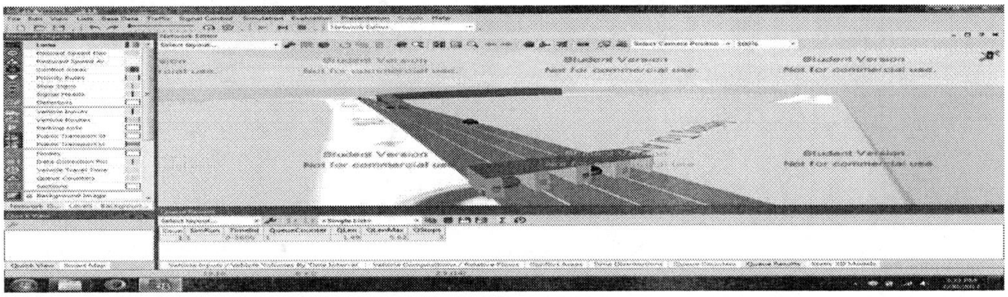

Fig. 4. Simulation of Amplas Toll Gate with 4 toll gate

Simulated with VISSIM software by inputting the same number of vehicles and service time and adding the number of toll booths to 4 pieces. After running the simulation, a new queue length is generated.Comparison of queue length results can be seen in table 6.

Table 6. Comparative Length of Toll Gate Door Amplas 3 Gardu and 4 Gardu

No	Time Service	Queue Length (meters)			
		3 substaion		3 Substation	
		Queue Theory	VISSIM Software	Queue Theory	VISSIM Software
1	7 seconds	6.212	11.988028	3.356	7.559299
2	8 seconds	9.168	12.547704	4.372	8.085288
3	9 seconds	14.46	14.149299	5.696	9.274755
4	10 seconds	26.856	14.421311	7.52	10.656138
5	11 seconds	85.452	15.939865	10.104	12.122563

4.7 Application of GTO System

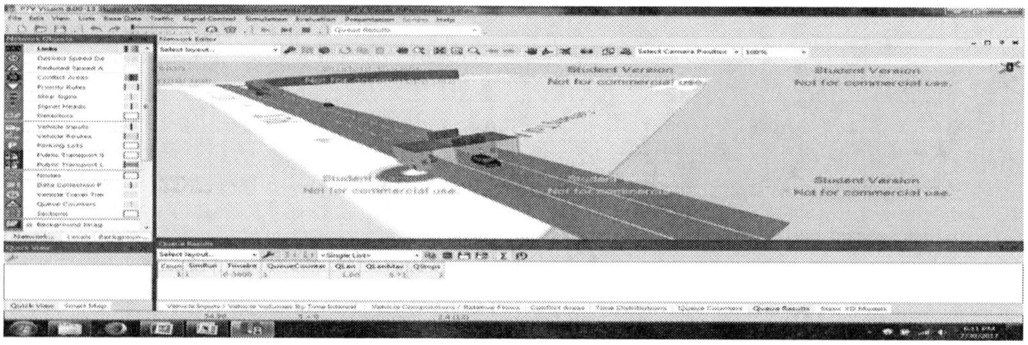

Fig. 5. Simulation of Amplas Toll Gate with GTO System

Conducted simulation with VISSIM software by applying 1 tollgate transformed into a GTO system with the assumption of time used by the driver of 3 seconds.After running the simulation, a new queue length is generated. Comparison of queue length results can be seen in table 7.

Table 7. Comparative length of Toll Gate Amplas Door with GTO System

No	Time Serive	PanjangAntrian (meter)		
		Queue Theory	VISSIM Software without GTO	VISSIM software with GTO system
1	7 seconds	6.212	11.988028	9.437607
2	8 seconds	9.168	12.547704	9.626659
3	9 seconds	14.46	14.149299	9.667078
4	10 seconds	26.856	14.421311	10.056292
5	11 seconds	85.452	15.939865	10.428499

5 Conclusions and recommendation

5.1 Conclusions

1. Based on the results of the survey conducted obtained the volume of vehicles at the substation entrance of the toll gate Amplas of 940 vehicles / hour with an average WP 6.96 seconds

2. Based on the calculation, for the current condition of service time 7 seconds obtained value service level at the toll gate of Amplas of 515 vehicles / hour / substation while the arrival of each substation of 313 vehicles / hour / substation.
3. Based on the results of queue length analysis with the average service time of 7 seconds, obtained the length of queue with queue theory analysis of 6.12 meters while the results of VISSIM modeling software queue length of 11.98 meters.
4. Based on the results of VISSIM software modeling it is found that the length of the queue tollgate quota after the build up of ramp-off increases compared to before the ramp-off. Before the ramp-off was built the queue length of 6.67 meters while after the ramp-off built the queue length to 11.98 meters.
5. In an effort to minimize the values of n, q, d, and w, it can be concluded that the priority order of policy-making is the first policy implementation of GTO system and the second priority is the policy of adding toll gate.

Recommendation

To anticipate the increase of vehicle volume, the improvement of quality that can be done is the addition of toll substations to 4 pieces of toll booths or the implementation of the GTO system on one of the toll booths.

References

1. Departemen PU DirJen Bina Marga. *Geometri Jalan Bebas Hambatan untuk Jalan Tol*, Departemen PU DirJen Bina Marga. (2009)
2. Departemen PU DirJen Bina Marga. *Manual Kapasitas Jalan Indonesia (MKJI) 1997*. Departemen PU DirJen Bina Marga. (1997).
3. Currin Thomas R. *Traffic Engineering A Manual for Data Collection and Analysis*. USA:Brooks/Cole. (1950)
4. Cahyani, Rezky D N dkk. *Analisis Kapasitas Dan Tingkat Layanan Pada Gerbang Tol Cikarang Utama*, Universitas Indonesia. Jakarta. (2013)
5. Hilmy, Ahmad and Hamid, Abdul. *Simulation of Traffic Operation and Management at Malaysian Toll Plazas using VISSIM*, Universitas Sains Malaysia. Malaysia.
6. Hobbs, F D. *Traffic Planning and Engineering*, University of Birmingham. Inggris. (1979)
7. Hutahaean, Marthyn. *Evaluasi Kapasitas dan Pelayanan Gerbang Tol Tanjung Morawa*, Universitas Sumatera Utara. Medan. (2007)

Operation ballast water of commercial vessels in Port of Tanjung Emas Semarang

Agus Tjahjono[1,*], *Wisnu* Handoko[2], *Sri* Purwantini[3]

[1]Politeknik Ilmu Pelayaran Semarang, Indonesia

Abstract. The commercial vessel uses sea water to stabilty a vessel when the vessel is not loading a cargo. The water stabilized for the vessel is known as ballast water. The activity of loading the ballast water for the ship from origin port and dissposal to the distination port has caused impact on spreading the unidenfied organism the local port. Aim of the study to determine disposal value of water ballast either from a foreign merchant vessel or domestic merchant vessel so that Port of Tanjung Emas Semarang (PTES) is able to determine a policy on ballast administration from merchant vessel include a data of Arrival and Departure Report of the Vessels (ADRV) documents for the last five years (2009-2014). Disposal ballast water of domestic vessel to PTES is average about 37,036 m^3 and increased by 76.68% in a year. Yearly ballast water disposed from the foreign commercial vessels has reached 576.045 m3 for the last 5 years. The increasing of ballast water is about 122.19%. Level of vunerability on the PTES waters is due to ballast water dissposal caused mainly from foreign commercial vessels. The PTES administration should provide water reservoir and and water ballast treatment of commercial vessels which has 51,090 m^3/per month or 81,744 kl/month.

1 Introduction

A commercial vessel, its operation uses sea water which is kept in ballast tank to maintain vessel's stability. When the cargo is empty, a commercial vessel will suct sea water from port area and it will dischrage sea water in ballast tank after reaching the next port.

Ballast water system in commercial vessel uses Ballast pump to unload or load sea water in ballast tank. In addition to increase ship's stability, sea water in ballast tank is used to get the desirable draft of a vessel, increase speed, change trim, decrease bending moment, control list during loading and unloading the cargoes and increase the vessel's manouver [1].

Water ballast which is discharged by the commercial vessels to destination port has an effect on aquatic environment, the effect at the east cost of Rhode Island has expanded invasion sea weed, migration of jellyfish, Turritopsis dohrnii at seashores in Atlantik and Passific in Panama, Florida. Origin of far away land, Italia, Mallorca, Spain, Okinawa,

* Corresponding author: a_agus_tjahjono70@yahoo.co.id

Panama Coast, have spreaded VHSV virus in 55 fish species in fish at Great Lake waters [2-4].

The impact of ballast water dicharge is also found Asian shrimp Palaemon macrodactylus at east cost of US and New York estuaries in 2001. Moreover a female shrimps which is carry the eggs is found (2001-2002, and 2008) showing the shrimps her doubled [5].

International Maritime Organization (IMO) has legalised a rule of ballast water of commercial vessel via Ballast Water Management Convention in 2004. One of the preliminary requirement is D 1 standard or first standard. It stated that vessel must exchange ballast water in the mid sea ocean range 200 mil before entering distination port or at a depth of at least 200 m in order to minimize spread of the hazard organisms in waters [6, 7].

There are three methods of ballast water exchange as follows the sequential method (the process with the emptying of the ballast water and recharged to obtain at least 95% volumetric exchange), the flow-through method (the process of replacing the ballast water so that water out through overflow or other arrangement), the dilution method (the process of replacing the ballast water filled through the top of the ballast tank with the simultaneous discharge from the bottom of the tank) [7].

The purpose of this study is to analyze the volume of the ballast water of commercial vessels disposal coming to PTES waters for a period of five years and to provide data to PTES administrator to formulate the volume of supply and storage of ballast water from commercial vessels at PTES as the cheapest option in the management of ballast water from commercial vessels.

2 Material and methods

Ballast water capacities are calculated for each commercial vessel witth Dead Weight Ton (DWT) more than 400 Gross Tonnage (GT) arrived at PTES waters according to the result from IMO for general cargo vessel 36,5%, bulk cargo 35%, liquid cargo 35%, container 30%, mixed cargo 33%, and Roll on Roll off (Ro-Ro) 33% from DWT [8-10]. Source of incoming and out going commercial vessels from PTES waters used source of Arrival and Departure Report of the Vessels (ADRV) from port harbour administration and harbour otority (KSOP) Tanjung Emas Semarang for the last five years from 2009 to 2014 [11]. PTES is managed by PT. Pelindo III at central Surabaya, and has a management area covering ports of Banjarmasin,Benoa, Samarinda, Cilacap, Padang Bai, Lembar [12]

The discharge of ballast water from a commercial vessel is known from the vessel to unload or load the cargo. If the vessel unload all cargoes in a port then to arrange the draft, either the bow or stern, the vessel has to ballasting of sea water at the port to compensate for the vessel even keel event (there is no big difference between the bow and stern draft) . Whereas if the vessel loads to the entire hatch or existing cargo space, the vessel must deballast the seawater (from the port of origin to the loading port) to compensate for the cargoes.

3 Results and discussion

The domestic commercial vessel berth at PTES in 2009, DWT is about 693,470 MT, maximum DWT accumulation was on October whereas the minimum was on November. It's appropriate to its capacities of ballast tank there at the same month has experienced the maximum and minimum condition. Discharge ballast for each month reached 4,089.75 m^3 with the minimum dispossal on February (1,286 m^3), maximum on October (8,121 m^3).

There have been any increased disposal ballast water as 60.64% in 2009. The arrival commercial vessels at PTES on November are at minimum (70 vessels) which is the smallest vessel is about (533,990 MT) and the smallest capacities are (186,080 MT), because there were more vessel loading the cargoes, there was more ballast water dispossal (6,088 m³) (Figure 1 a.)

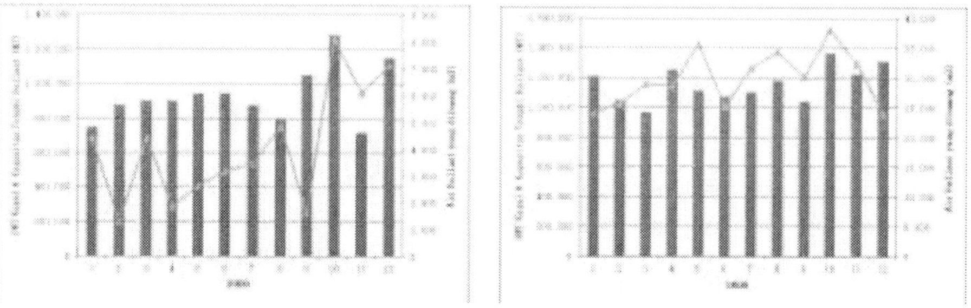

Fig. 1. DWT, ballast tank capacities and ballast water discharge of commercial vessels from a) domestic b) foreign in 2009 (DWT> 400 MT, exclude barge and tug boat)

The foreign commercial vessels coming to PTES have fewer numbers than domestic vessels, where the number of foreign commercial vessels are a maximum of 69 compared to 112 vessels. However, the average foreign vessels's DWT is 884,293.67 MT compared to domestic vessels of 693,470 MT. It also has implications for larger vessel ballast capacities at an average of 274,870 MT. The average monthly ballast water discharged per month is 29,9915 m3, a minimum of 23,978 m3 in January, and a maximum of 38,070 m3 in October which shows the vessels the most loading cargoes compared to unload.During the year there was no increase in ballast water discharges to PTES (Figure1 b.).

The monthly average of vessels entering PTES weigth 852,900 MT with the minimum month of February (599,398 MT) and the maximum in September (1,072,708 MT), it is proportional to the number of ships coming to PTES where in February there were only 74 and September 114 vessels. The implications of the number of ship DWT to PTES are the vessels's ballast capacities, minimum 206,779 MT (February) and maximum 370,448 MT (September), with monthly average ballast capacities of 295,184 MT. The discharge of ballast water to PTES is different from DWT condition and vessels's ballast tank capacities, which in May experienced maximum discharge (8,862 m3) and January minimum (802 m3) with monthly discharge of 3,291 m3. For domestic vessels in 2010, an increase in DWT and vessels's ballast tank capacities came in at 21.80% while for the discharge of ballast water to PTES there was an increase of 139.53% starting from the beginning of the month until December (Figure 2 a) .

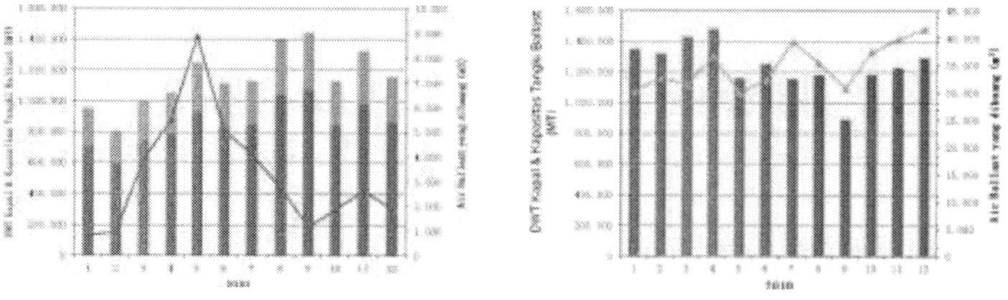

Fig.2. DWT, ballast tank capacities and ballast water discharge of commercial vessels from a) domestic b) foreign in 2010 (DWT> 400 MT, exclude barge and tug boat)

In April 2010, foreign's commercaial vessel DWT and ballast tank capacities coming to PTES experienced a maximum condition of 1,130,451 MT and 353,430 MT respectively, minimum in September with DWT vessels of 677,596 MT and ballast tank capacities of 212,693 MT. The vessel that discharges the ballast water experienced a maximum condition in December of 41,590 m3, the minimum month of May of 29,917 m3 and the monthly average of ballast water discharge of 34,776 m3. Foreign vessels coming to PTES in 2010 are decreasing, it can be seen from DWT and its ballast tank capacities are 3.71% and 6.08% respectively. However, for the discharge of ballast water from foreign vessels increased by 37.36%, it was due to an increase in the loading of cargoes for export (Figure 2 b.).

Domestic vessels entering the PTES in 2011 amounted to 1227 vessels, a maximum of 121 in December and a minimum of 84 vessels in February. The average domestic's vessels DWT for the entire year are 982,561 MT, with maximum conditions in August (1,114,851 MT), the minimum month of February being 865,342 MT. DWT and domestic's vessel ballast tank capacities are increased by 10%. It is proportional to the ballast tank capacities of 385,152 MT and 299,554 MT. Ballast water discharged to PTES on average is 2,367 m3/month, with maximum conditions in October and March of 4,605 m3 and 1,518 m3. Ballast water discharges by domestic commercial vessels to PTES increased by 44.62% (Figure 3 a.)

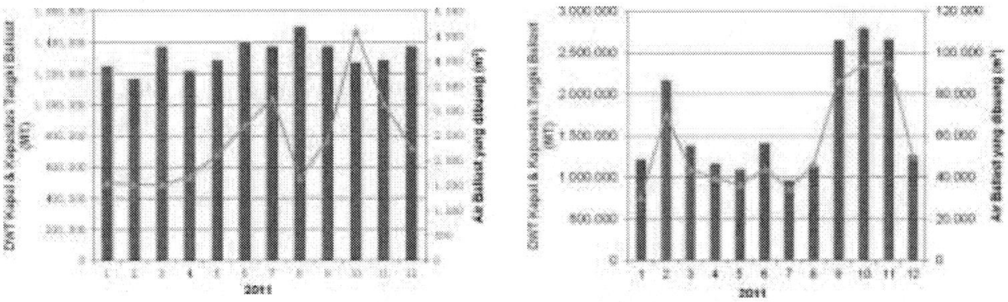

Fig. 3. DWT, ballast tank capacities and ballast water discharge of commercial vessels from a) domestic b) foreign in 2011(DWT> 400 MT, exclude barge and tug boat)

Foreign commercial vessels coming to PTES with DWT average of 1,265,368 MT in 2011, with maximum and minimum conditions in October (2,126,581 MT) and July (734,683 MT), DWT of the arrival vessels increased by 3, 94%. The average capacities of the ballast water during the entire year amounted to 395,082 MT with maximum and minimum conditions of 667,499 MT and 225,446 MT respectively. Ballast water discharge to PTES waters increased by 64.26%, with maximum and minimum conditions of 95,619 m3 and 30,561 m3 occurring in November and January. Average monthly ballast water disposal during the year by foreign vessels are 55,914 m3 (Figure 3 b.).

Domestic commercial vessels with an average DWT of 872,448 MT while maximum and minimum DWT conditions in May and June of 966,248 MT and 784,227 MT. This is in line with the ballast tank capacity of 333,989 MT and 270,489 MT. Ballast water discharges from domestic vessels in the average of 4,113 m3/month with an increase of 7.84% (Figure 4 a.).

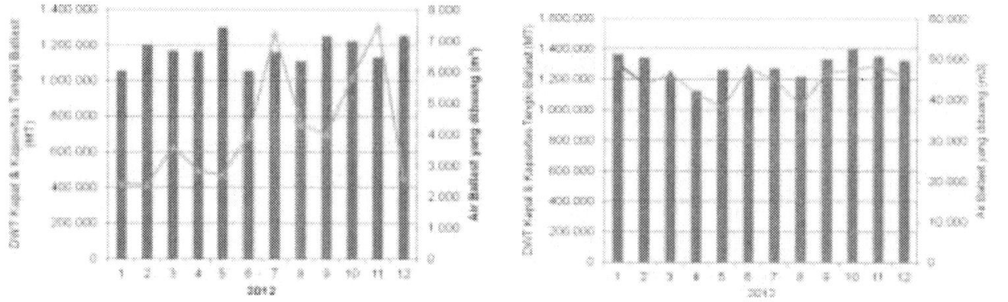

Fig. 4. DWT, ballast tank capacities and ballast water discharge of commercial vessels from a) domestic b) foreign in 2012(DWT> 400 MT, exclude barge and tug boat)

The average DWT and ballast tank capacities of foreign vessels coming to PTES amounted to 978,757 MT with maximum and minimum DWT in October and April. There was decrease of foreign vessels arrival these year is indicated by the decrease of average vessels's DWT by 3.9%. In line with this, the ballast water discharged by foreign vessels also decreased 6.68%. Average ballast water disposed of 44,783 m3, with maximum discharge conditions in January and minimum in August (Figure 4 b.)

Domestic commercial vessels berthing on PTES with average DWT of 934,149 MT/month with an increase of DWT from the beginning of the month of 20.32%. The arrival of the vessels reached a maximum number of 143 vessels in September with DWT 1,119,560 MT with the ballast tank capacities of 385,250 MT. While the minimum conditions occurred in February, where there were vessels visits of 103 vessels with DWT 738,975 MT and ballast tank capacities of 251,028 MT. Increased ballast water discharged by domestic vessels is small at 1.33%, with maximum disposal in February and a minimum in June (Figure 5 a.).

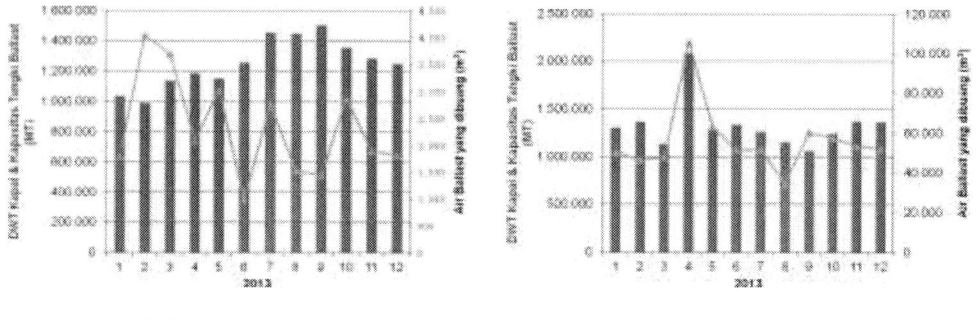

Fig. 5. DWT, ballast tank capacities and ballast water discharge of commercial vessels from a) domestic b) foreign in 2013(DWT> 400 MT, exclude barge and tug boat)

Foreign commercial vessels coming to PTES totaling 676 vessels with a mean dead weight of 1,011,526 MT, with April experiencing maximum conditions and the minimum September of 1,589,247 MT and 813,035 MT respectively. In line with the dead weight of the vessel, the tank capacities of the ballast water also follows where it has experienced maximum and minimum conditions in the same month. Ballast discharges from foreign vessels average 56,165 m3, maximum in April (105,338 m3) and August (34,479 m3). Increased discharge of ballast water from the beginning of the month was small at value 3.34% (Figure 5 b.).

The dead weight of domestic commercialvessels coming to PTES with an average of 934,149 MT, an increase in ship arrival by 20% from the beginning to the end of the month. The domestic's vessels arrival of totaling 143 vessels, the deadweight and ballast water capacities in September became the maximum (1,119,560 MT and 385,250 MT). However, this condition does not apply to ballast water discharges, due to the most loaded vessels, experiencing maximum conditions in February with a value of 4,044 m3, and a minimum in May (990 m3). The discharge of ballast water into PTES waters was a relatively small increase of only 1.33% (Figure 6 a.).

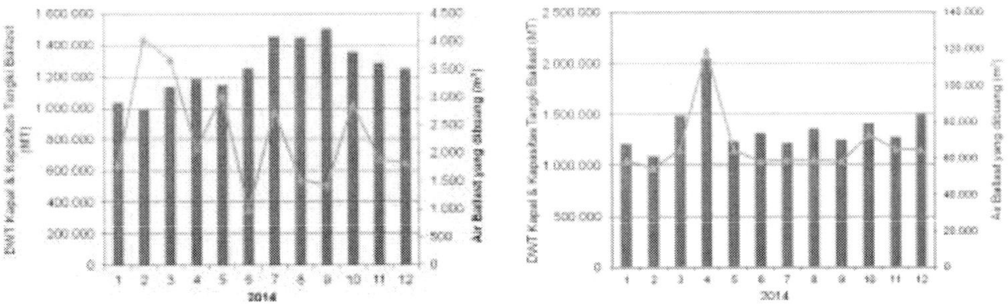

Fig. 6. DWT, ballast tank capacities and ballast water discharge of commercial vessels from a) domestic b) foreign in 2014(DWT> 400 MT, exclude barge and tug boat)

DWT of foreign vessels ranging from 830,170-1,564,907 MT with maximum and minimum conditions occurring in April and February, the vessels's dead weight coming to PTES increased 23.7% from the beginning of the month. While the vessels's ballast capacities are also proportional to the dead weight of foreign vessels ranging from 261,111-482,097 MT. The ballast water discharge experienced an increase of 9.63% with the maximum and minimum conditions occurring in April and February, where the average of ballast water discharged into PTES was 66,467 m3/month (Figure 6 b.).

The number of domestic and foreign commercial vessels arriving to PTES from 2009-2014 where for domestic merchant vessels ranges from 1043 to 1408 vessels, with the average of 1236 vessels/year. The foreign vessels coming to PTES are smaller, ranging from 676-1045 vessels with an annual average of 768 vessels. The arrival of domestic and foreign vessels to PTES during 5 years increased where for domestic ships 35% while overseas ships were 54.6% (Figure 7.).

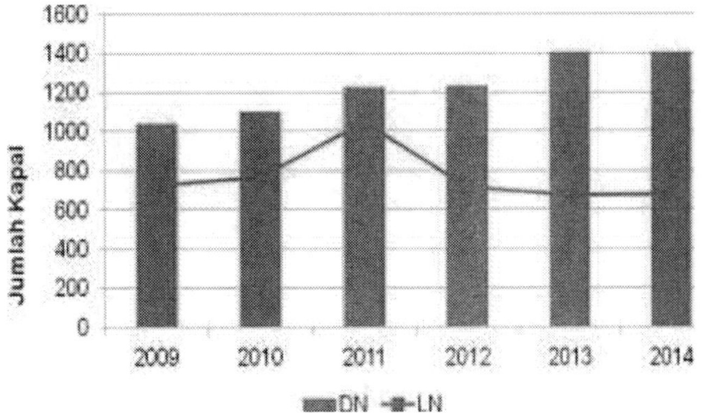

Fig. 7. Recapitulation the numbers of commercial vessels from domestic and foreign at PTES

DWT of domestic vessels coming to PTES for 5 years ranged from 8,321,640-11,790,736 MT with an increase of 41.69%. This is proportional to the number of domestic vessels arrivals which increased by 35%. The average ballast water discharge of domestic vessels in PTES is 37,036 m3/year, ranging from 27,937 m3 to 49,360 m3 during the same time period which has increased by 76.68% (Figure 8.)

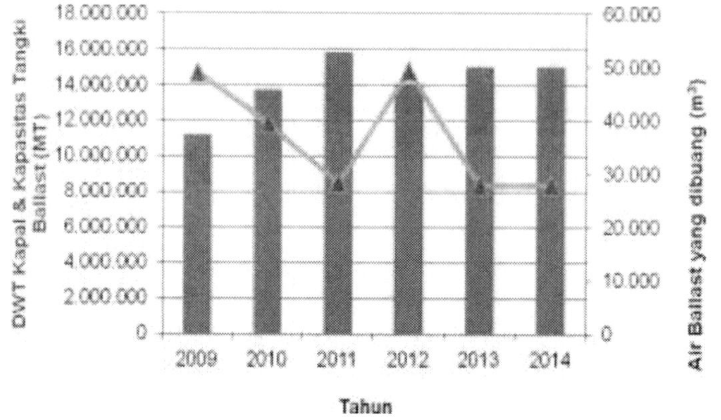

Fig. 8. Recapitulation of DWT, ballast tank capacities and ballast water discharge of domestic vessels at PTES (2009-2014)

The dead weight of foreign vessels coming to PTES ranges from 10,611,524 MT-15,184,422 MT with an average DWT of 12,257,847 MT/year. There was an increase during 5 years period of 5 years in foreign vessels arrival and dead weight in PTES amounting to 54.59% and 43.09%, respectively. In line with this, the capacities of the ballast water also increased with the same value. The average annual ballast water discharge of the foreign vessels within that period reached 576,045 m3, with the maximum and minimum discharges in 2014 and 2009. There was a significant increase of ballast water discharge by foreign vessels by 122.19%. This shows that foreign vessels coming to PTES are more cargo loading for export (Figure 9.).

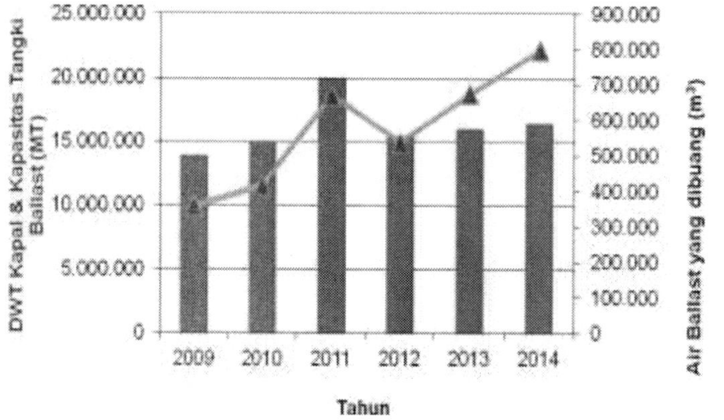

Fig 9. Recapitulation of DWT, ballast tank capacities and ballast water discharged of foreign vessels at PTES (2009-2014)

Effectiveness level of mid ocean ballast water exchange rates shows a decrease in Skeletonema costatum abundance on container vessels entering to Hong Kong port from California waters, a decrease in abundance of diatoms and dinoflagellates on commercial vessels originating from Mexico [13, 14].

The strategy undertaken by the European shipping community in response to the Ballast Water Management (BWM) Convention is to enforce this Convention as a temporary solution despite its limited effectiveness, in collaboration with agencies developing BWM Convention in Europe to help harmonize BWM requirements across Europe [15]. The study of the effectiveness of mid ocean ballast water exchange shows the greatest exchange occurs when the source of sea water comes from the port with a low salinity and the exchange will be effective when it occurs in deeper waters than the land, so the ballast water exchange does not discharge all taxa from the port of origin and unsuitable in a regional ballast water setting [16].

4 Conclusion

DWT of domestic's commercial vessels coming to PTES for 5 years ranges from 8,321,640 to 11,790,736 MT with an increase of 41.69%. The average of ballast water discharge to PTES is 37,036 m3/year, an increase of 76.68%. Arrival of domestic and foreign vessels to PTES during 5 years has increased where for domestic ships 35% while overseas ship 54.6% The weight of foreign commercial vessels coming to PTES are about 10,611,524 MT-15,184,422 MT. The average ballast water discharge of foreign vessels within 5 years reaches 576,045 m3/year. The significant increase of ballast water disposal by foreign vessels are 122.19%.

The vulnerability level of PTES waters due to ballast water disposal is mostly caused by foreign vessels compared to domestic vessels due to the foreign vessels more load cargoes compared to domestic vessels.

The administrator of PTES should provide a reception and ballast water treatment facilities from a commercial vessel with a capacity of 51,090 m3/month or 81,744 kl/month.

Acknowledgment

The authors would like to thank the authority of Port of Tanjung Emas Semarang (PTES) who has provided data especially Mrs. Erry. The authors also thank Mr. Slamet Riyadi who has helped in the formulation of the method of calculating the capacities of ballast water output. Thanks to the vessel's officer to the author who has authorized the ballast vessel ballast tank inspection.

References

1. Van Dokkum, K., *Ship knowledge covering ship design, construction and operation* : **E 2**, Dokmar, Enkhuizen. (2005).
2. Villalard-Bohnsack, M., and M. Harlin, *Grateloupia doryphora (halymeniaceae, rhodophyta) in Rhode Island waters (USA), geographical expansion, morphological variations and associated algae*, Phycologia ProQuest, 40, 4, pg. 372. (2001).
3. Miglietta, M.P., and H.A Lessios, *A Silent invasion, Biology Invasion*, 11:825-834. DOI 10.1007/s10530-008-9296-0. (2008).
4. La Rue, E.C., C.R., Ruetz III, M.B. Stacey, and R.A. *Thum, Population genetic structure of the round goby in Lake Michigan : implications for dispersal of invasive species*, Hydrobiologia, 663, 71-82, DOI 10.1007/s10750-010-0555-6. (2011).
5. Warkentine, B.E., and J.W. Rachlin, *The first record of Palaemon macrodactylus (oriental shrimp) from the eastern coast of North America*, Northeastern Naturalist, 17, 1, pg.91., ProQuest. (2010).

6. IMO (International Maritime Organization). *International Convention for The Control and Management of Ship's Ballast Water and Sediments*, 2004. International Maritime Organization, London, p 36. (2005).
7. IMO (International Maritime Organization). *Guidelines for Ballast Water Exchange* (G6). Annex 2, Resolution MEPC.124 (53). Adopted on 22 July. (2005).
8. IMO (International Maritime Organization*), Guidelines for ballast water management and development of ballast water management plans* (G 4), International Maritime Organization, London, adopted on 22 July 2005, Resolution MEPC.127.53. (2005).
9. Butron, A., Orive, E., and I. Madariaga, *Potential risk of harmful algae transport by ballast waters : the case of Bilbao harbor*, Marine Pollution Bulletin 62, page 747-757. DOI : 10.1016/j.marpolbul.2011.01.008. (2011).
10. David, M., Perkovic, M., V. Suban, and Gollasch, S, *A generic ballast water discharge assessment model as a decision supporting tool in ballast water management. Decision Support Systems 53*, page 175-185, DOI: 10.1016/j.dss.2012.01.002. (2012).
11. KAPTES (Kantor Administrasi Pelabuhan Tanjung Emas Semarang), *Data kegiatan angkutan laut dalam dan luar negeri di pelabuhan Tanjung Emas Semarang tahun 2009-2014,* Bidang Lala dan Kepelabuhanan, Kantor Administrasi Pelabuhan Tanjung Emas Semarang. (2011).
12. PT. Pelabuhan Indonesia III (Persero) Cabang Tanjung Emas, *Tanjung Emas port directory* , Surabaya, Dwitama Wukirindo. (2011).
13. Zhang, F. and Dickman, M., *Mid-ocean exchange of container vessel ballast water 1 : seasonal factors affecting the transport of harmful diatoms and dinoflagellates,* Marine Ecology Progress Series, **Vol. 176**, page 243-251. (1999).
14. Dickman, M. and Zhang, F., *Mid-ocean exchange of container vessel ballast water 2 : effects of vessel type in the transport of diatoms and dinoflagellates from Manzanillo, Mexico, to Hongkong, China*, Marine Ecology Progress Series **Vol. 176,** page 253-262. (1999).
15. David, M. and Gollasch, S., *EU shipping in the dawn of managing the ballast water issue, Marine Pollution Bulletin* **Vol. 56**, page 1966-1972, DOI:10.1016. (2008).
16. McCollin, T., Shanks, A.M. and Dunn, J., *Changes on zooplankton and diversity after ballast water exchange in regional seas, Marine Pollution Bulletin* **Vol. 56**, page 834-844, DOI:10.1016. (2008).

Identification of feeder service in every station on serving Jenggala Train

Hera Widyastuti[1,*], *Dio* Hananda Ziantono[2] , *Frans* de Jesus Babo[2] , *Jimi* Amijaya[2] , and *Rizal* Zafikhal Biad Sektyoko[2]

[1]Civil Engineering, Faculty of Civil Environmental and Geo Engineering, 6011 Surabaya, Indonesia
[2]Civil Engineering, Faculty of Civil Environmental and Geo Engineering, 6011 Surabaya, Indonesia

Abstract. Optimum performance of station affects on the level of service to the consumer that is train service users. The consumers of station or train always want to have appropriate services, from purchasing ticket in ticket counter until waiting for the train to board or arrive. But, based on earlier research, there were still much candidate of passengers looking for information about transportation upon arrival on the station. The purpose of the writing is to evaluate performance and get hints to make some improvements. While, for intermodal analysis of whole railway stations serving KRD Jenggala has been integrated with nearby terminal through the service of feeder that is Lyn.

1 Introduction

PT. Kereta Api Indonesia is now already operating pioneer trains in several regions in Indonesia [1]. The development of pioneer train is firstly operated on Java Island in East Java regions which are Sidoarjo and Mojokerto. Being named as *Kereta Api Perintis Jenggala*, this train is operating through Sidoarjo railway station, Tulangan railway station, Tarik railway station, and Mojokerto railway station.

Railway station, also as a transportation node has important role in organizing trains, excellent service is also part of PT. Kereta Api Indonesia (Persero)'s target to maximize the effort in service innovation [2]. With the existence of train that is serving through four railway stations as a node for the writters to do evaluation among stations serving Jenggala train investigating the existing facilities and services as evaluation materials about how far minimum standards in station have been applied and also capable of giving recommendation for better service performance in those railway stations. Surely consumers want to get better services, starting from ordering the tickets until near boarding and also availability of important facilities for the costumers [3].

* Corresponding author: hera.widyastuti@yahoo.co.uk

2 Methodology

Method that had been done on this preliminary identification was gathering primary data related to observing existing condition on field around the station that became stopping place for KRD Jenggala. That matter is important thing to do for identifying feeder services and connectivity supporting facilities around station that served KRD Jenggala.

3 Discussions

3.1 Intermodal Analysis

Reviewed from the aspect of population movement, the increase of urban area's residents population and urbanization making the increase of movement to or from the city. This matter gives logic consequence that it is needed for balancing between vehicle and infrastructure especially in transportation. This is meant to support the mobility of the population to do their activities. One of the method to fulfill the needs of transportation service is by providing public transportation [4,5].

KRD Jenggala is one of the rail based mass transportation that is offered by PT. KAI to support the movement of people from Sidoarjo to Mojokerto or vice versa. This train is inaugurated in November 12th 2014 to serve pioneer route in line between Sidoarjo Railway Station until Tarik Railway Station and continues until Mojokerto Railway Station. KRD Jenggala consists of two chained new AC built train made by PT. Industri Kereta Api (INKA) Madiun, which each has four diesel train units of economic class with AC built.

Availability of feeder service and supportive facility in nearby stations become the place for KRD Jenggala to stop is important component that must be noticed to ease the event of transportation movement for KRD Jenggala service users from one place to another. This matter is important to do to attract people's interest to use JRD Jenggala as primary transportation option in their daily. The availability of vehicle and infrastructure connectivity services on every station that become a place for KRD Jenggala to stop, the performance needs to be improved. Cross of train services need internal alignment and between modal transportation, distance of the hub, and central logistic toward the terminal/station and time availability for internal movement and between modal transportation. (PP 72/09) [6]. Therefore a first identification related to the connectivity on every station that becomes the place for KRD Jenggala to stop is needed. The purpose for first identification to be done is that to identify if there are feeder services and supportive facilities around the stations that become the place KRD Jenggala to stop.

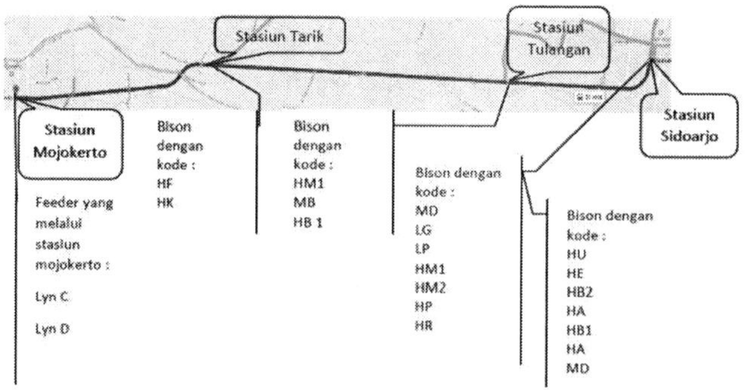

Fig. 1. Station's Feeder Network Map

Table 1. Amount of Feeder Services on Every Station.

Station Name	Station Class	Terminal Availability	Feeder Services	Feeder Total
Sidoarjo Station	Big	Yes	HU, HE, HB2, HB1, HA, MD, LG, LP, HM1, HM2, HP, HR	12
Tulangan Station	Small	Yes	HM1, MB	2
Tarik Station	Small	Yes	HF, HK	2
Mojokerto Station	Big	Yes	Lyn C and Lyn D	2

Table 2. Feeder Services Crossing on Sidoarjo Station.

No	Type of Feeder Services	Feeder Services Routes
Sidoarjo Station		
1	Lyn HA	Terminal Larangan – Jl. Diponegoro – Raya Jati – Cemengkalan – Embong Malang – Sumput – Sarirogo – Raya Wulung – Raya Saimbang – Kebonagung – Patimura – Panjuran – Sandang pangan – Jemundo – Sawunggaling – Raya Kltek – Raya Geluran – Raya Kalijaten – Ps Taman – PP
2	Lyn HB 1	Terminal Larangan – Jl.Diponegoro – Jl.Pahlawan – Raya Jati – Raya Cemengkalan – Ray Suko – Raya Lebo – Raya Pangliam Sudirman – Raya Pilang – Raya Banar – Raya Modong – Raya Kemantren – Raya Tulangan – Raya Kepadangan – Raya Kebaron – Raya Wonomlati – Raya Krembung – PP.
3	Lyn HB 2	Terminal Larangan – Jl.Diponegoro – Jl.Pahlawan – Raya Jati – Raya Cemengkalan – Raya Suko- Raya Lebo – Raya Panglima Sudirman – Raya Pilang – Raya Ketimang Raya Jimbaran – Raya Wonoayu – Raya Semambung – Raya Simo Angin-angin – Raya Tanggul – Ds. Tropodo Raya Prambon – Raya Terminal Krian- PP.
4	Lyn HE	Pasar Porong – Ray Bhayangkari – Jl.Diponegoro – Ds. Lajuk – Jololeksono- Jenggot – Ploso – Rejeni – Raya Kandangan – Raya Krembung – Lemujut – Cangkring – Bulang – Simpang – Pejangkungan – Kates – Wirobitung – Jl. Brawijaya – Gedangrowo – Jl.Mojopahit – Prambon-PP.
5	Lyn HM 1	Terminal Larangan – Jl. Diponegoro – Jl. Tthamrin – Jl. Gajahmada – Jl. Mojopahit – Jl. Bligo – Raya Candi – Raya Gelam – Sumorame – Karangtanjung – Kedondong – Sidomoro – Medalem – Pasar Tulangan – PP.
6	Lyn HM 2	Terminal Larangan – Jl Diponegoro – Jl. Thamrin – Jl. Gajah Mada – Jl. Mojopahit – Jl. Bligo – Raya candi Raya Gelam – Sumorame – Karangtanjung – Gagangpanjang – Randegan – Kategan – Kendensari – Kalisampurno – Ketapang – Pasar Porong – PP.
7	Lyn HP	Terminal Larangan – Jl. Diponegoro – Jl. Thamrin – Jl. Kombers Duriat – Jl. Kartini – Jl. Yos Sudarso – Bluru Kidul – Kemiri – Siwalan Panji – Sidomulyo – Prasung – Dukuhtengah – Damarsih – Sawoan – Kwangsan –Betro – Sedatiagung – PP.
8	Lyn HR	Terminal Larangan – Jl. Diponegoro – Jl. Pahlawan – Raya Jati – Cemengkalang – Cemengbakalan – Urangagung – Mojorangagung – Wonokasihan – Sawocangkring – Lambangan – Kemulan – Becirongengor – PP.
9	Lyn HU	Terminal Larangan – Jl. Diponegoro – Jl. Pahlawan – Jl. Teuku Umar – Jl. Sultan Agung – Jl. Pagerwojo – Entalsewu – Sidokepung – Jumputrejo- Ganting – Masangan Wetan – Bohar – Wage – Pepelegi – Medaeng – Raya Geluran – Raya Kalijaten – Pasar Taman – PP.
10	Lyn LP	Terminal Larangan – Jl. Diponegoro – Jl. Thamrin – Jl. A.Yani – Jl. Jenggolo – Raya Buduran – Tebel – Kranggan – Gemurung –Kwangsan – Pepe – Pulungan – PP
11	Lyn LG	Terminal Larangan – Jl. Diponegoro – Jl. Thamrin – Jl. A.Yani – Jl. Jenggolo – Jl. Yos Sudarso – Jl. R.A Kartini – Jl. Trunojoyo – Bluru Kidul – rangkah Kidul – Perum .Intan Permai – Gebang – PP.
12	Lyn MD	Terminal Larangan – Jl. Diponegoro – Jl. Thamrin – Jl. Gajah Mada – Jl. Mojopahit – Raya Bligo – Raya Candi – Ds. Gelam – Sugihwaras – Kedung Kendo – Durung Banjar – Durung Bedug – Modong – Kepuh Kemiri – Grabagan – Popoh – Simo Ketawang – Simo Girang – Watutulis – PP.

Table 3. Feeder Services Crossing on Tulangan, Tarik and Sidoarjo Station.

No	Type of Feeder Services	Feeder Services Routes
Tulangan Station		
1	Lyn HM 1	Terminal Larangan – Jl. Diponegoro – Jl. Tthamrin –Jl. Gajahmada – Jl. Mojopahit – Jl. Bligo – Raya Candi Kedondong – Sidomoro – Medalem – Pasar Tulangan – PP – Raya Gelam – Sumorame – Karangtanjung – PP.
2	Lyn MB	Terminal Krian – Sidomulyo – Jeruklegi – Penambangan – Bogempinggir – Pasar Tulangan – PP.
Tarik Station		
1	Lyn HF	Terminal Krian – Raya Kemangsen – Balongbendo – Bakalan – Waruberon – Jati Kemuning – Balongmace'an – Tarik- PP.
2	Lyn HK	Terminal Krian – Tropodo – Klagen – Kedungwonokerto – Klantingsari – Mergosari – Tarik – Singogalih – Kedungbocok – Miripprowo – PP.
Sidoarjo Station		
1	Lyn C	Terminal Kertajaya – Jl. Jaya Negara – Mangelo – Jl. Jaya Wardana (Jl. Wijaya Kusuma) – Jl. RA. Basuni – Jl. Sooko – Brangkal – Jl. Sooko – Jl. RA. Basuni – Jl. Mojopahit – Jl. Bhayangkara – Jl. KH. Nawawi – Jl. Cokroaminoto – Jl. Gajah Mada – Jl. Pahlawan – Jl. Jaya Negara - Terminal Kertajaya.
2	Lyn D	Terminal Kertajaya – Jl. Meri – Jl. Tropodo – Jl. Pahlawan – Jl. Gajah Mada – Jl. Cokroaminoto – Jl. KH. Nawawi – Jl. Bhayangkara – Jl. Mojopahit – Jl. Tribuana Tungga Dewi – Jl. Surodinawan – Jl. Sambiroto – Brangkal – Jl. Sambiroto – Jl. KH. Usman – Jl. Jaya Wardhana – Jl. Teratai – Jl. R. Wijaya – Jl. Jaya Negara - Terminal Kertajaya.

On the table 1, based on identification it can be inferred that Sidoarjo Station is the most station that the feeder services are going through which is 12 Lyns. The feeder services in their operation have different routes. Route tracks of each Lyn shows in the table 2 and 3.

3 Conclusions

Result of first identification that is obtained from the survey that all stations serving KRD Jenggala have operational feeders from station to nearby terminal. But the availability of this feeder is not supported with supportive facility like schedule of feeder's arrival or information board about feeder service routes that are operational in station to connect to the terminals. And to improve the performance of future services, the writers is giving advice so that there will be regulation to organize connectivity between main mode and feeder to be one company that provides transportation service. Because as for now between the owner of main mode and owner of feeder are from different companies so it is difficult to synchronize the importance between both companies without having regulation that specially organizes the connectivity between both of them.

References

1. Honing, J. *Ilmu Bangunan Jalan Kereta Api*. Jakarta: PT.Pradya Pramita. (1975).
2. Martila A. John and James C. John. *The Analysis of the Importance and Satisfaction level of the Customers*. New Jersey:Prentice Hall Inc. (1997).
3. Kotler, Philip. *Marketing Management*. New Jersey: Prentice Hall Inc. (2000).
4. Umar, H. *Metode Riset Perilaku Konsumen Jasa*. Jakarta: Ghalia Indonesia. (2003).
5. Tjiptono, J. and Chandra, G. *Service, Quality Satisfaction*. Yogyakarta: Andi. (2005).
6. Department of Transportation of Indonesia, *Regulation of the Minister No. 9 of 2011 for Standard Minimum Service of Public Transport using Train*. Jakarta (2011).

MATEC Web of Conferences 181, 08001 (2018) https://doi.org/10.1051/matecconf/201818108001
ISTSDC 2017

Determining the advertisement of tax priority on urban road based on road performance

Nunung Nuring Haryati[1,*], *Wahyu* Agus Winarno[2], and *Sonya* Sulistyono[2]

[1] Jember University, Department of Civil Engineering, Jember, Indonesia
[2] Jember University, Department of Accounting, Jember, Indonesia

Abstract. Each region has the potential to be explored as a source of income without necessarily harming people and the environment. One of them is the utilization of public space as a media campaign or advertisement. The advertisement placement should be planned to have a function, adding the aesthetic aspect of urban space and not disturbing the comfort and safety of the community. Promotional media can take advantage of active and passive outdoor space. The active outer space is a space that is deliberately created as a public outdoor space such as city park, pedestrian and road. While the passive outer space is a outer space that is not accidentally created or already exist by it, such as river banks, railroads and natural green spaces. Promotional media nowadays is not only in physical, but also in electronic form. Placement of advertisement associated with the tax burden. The amount of tax can be determined from road performance. Where the more crowded the urban space, the more people see the advertisement. This can certainly be used as one of the parameters in determining which points are having the biggest impact on tax value.

1 Introduction

City space has great potential for the media community, namely advertising. Advertisement serves as the dissemination of information for marketing purposes. Advertising is temporary (short term, less than one year) because it is usually associated with a certain moment and there are fixed or long-term installation (above one year). Strategic placement of billboards can provide an opportunity for people to gain good visual access to information resources. The existence of variety of shapes, sizes, and colors that are striking, will be interesting when it viewed at a certain distance. If the media is designed attractively with the arrangement of beautiful lights, it will further brighten the face of the city at night.

The result from the observation in the field indicated, that the condition of the billboard is not organized and ruined the city landscape. The proportion of billboards with urban space is often unbalanced. Some billboards that are temporary placement irregular, interfere with road corridors, the level of security is also less attention. Therefore, it is necessary to

* Corresponding author: nunung.nuring@unej.ac.id

arrange the billboard in order to add the beauty of the city, also the function of information to the community, with good and optimal arrangement can increase Jember Revenue.

City location mapping is divided into several zones. These zones are all potential, easily visible and located in strategic areas. However, the placement of billboards also can be classified according to the zones where this determines the rental price. The zone-determining parameters are based on the strategic location determined by the number of people passing through the site, the accessibility, and how much attention it can attract. So it is necessary to map which locations are potential laying billboards and the distribution of the zones. The use of advertisement is important in order to give information to people. However, it needs to be handling carefully to create an aesthetic and neat arrangement. For instance, the advertisement can be creating as a combination between road signs and street name plate in the same place. Most of all, the suitability of the advertisement has been done in Jember, from Hayam Wuruk road - Gajah Mada road. Some of the billboard has a different shape and size. Generally, the condition of billboard needs to get more attention in terms of aesthetic, architectural, and environment aspect, not only on business or marketing goals. It can be seen from a various kind of advertisement that do not really consider about other aspect, that is the reason why there should be a regulation that oversee the balance of structuring that might have some effects to other sector.

Based on a review of the problems that have been submitted in the previous few paragraphs, the purpose of this study is to create an optimum model placement of advertisement tax based on road performance.

2 Research methods

What is done in the placement optimization of advertisement locations are:

1. Understanding the definition of advertisement

Advertising is an information or instruction tool or communication media. Based on the content of the message, the material, the nature of the information and the technical installation, the billboard are divided into two, namely: commercial media, concerning advertisement media that give information in goods or services for the benefit of trade (private sign),non-commercial media, and regarding signs containing information services to the public (public sign).

2. Advertisement planning evaluation

In the advertisement evaluation, there are some things that can be alluded to in its implications. Here is the theory of planning billboard in terms of open space, building and massing, strategic location and the aesthetics of the city.

The definition of evaluation in Indonesian dictionary means an assessment and results. Evaluation is an attempt to document and assess what happened and why it happened, the simplest evaluation is to gather the information about the circumstances before and after the implementation of a program [1].

According to PP Number 39 Year 2006 on Procedures for Control and Evaluation of Implementation of Development Plans, in the implementation, evaluation activities can be done at different stages of evaluation at: the planning stage (ex-ante), evaluation at the on-going stage, and evaluation at the post-implementation stage (ex-post) [2].

3. Open Spaces

Open space is a room within a city or a larger area, either in the form of an area, or in the form of longitudinal areas or pathways, the use is more open nature that is basically without building, consisting of green open space and non-green open space. Meanwhile, classifications of open space into two types are elongated and cluster [3].

4. Building Use and Massing

The city masses including buildings, soil surfaces, and objects that form city spaces and patterns to define mass and shape of buildings with principles and thoughts behind the physical form of the city. Based on Long Beach Design Guidelines, the appearance and configuration of the buildings include elevation, lighting and front-of-store design. While made a synthesis shape and mass of buildings that include the scale, associated with views, circulation, the size of adjacent buildings [4]. City space is a major element of urban design, scale and sense of surroundings (sense of enclosure), also a type and space of the building.

5. Strategic location of Advertisement

Advertising always choose a strategic location. It is intended to attract as many people as possible and also to increase the clarity of details of the messages delivered. It is also intended to reduce costs as low as possible in the installation of billboards. In a variety of literatures that discusses both billboards and signage placement, at a point location, several factors of concern issues was found.

In the process of optimizing the placement of billboards to increase the PAD (locally-generated revenue) of Jember District, two models were developed using the least squares least squares (OLS) approach. The tool used to create this modeling is Minitab 17 Software. The first model is used to predict and define the priority of layout/placement of advertisements in accordance with the strategic mapping of the road segment (Strategic Road Model Model). The second model is used to predict and define priority of layout/placement of advertisements in accordance with road corridor (Strategic Model of Road Corridor).

The methods used in these two models, based on the results of the study data in the field of transportation in Jember District, the results of the studies are grouped based on roads and road corridors. The information contained in the assessment, such as the degree of saturation, travel time, velocity, and V/C ratio, is used to create an OLS model based on roads and road corridors. So, the model can be used to predict and determine the priority of layout and placement of advertisements.

Table 1. Issue of concern in strategic value factors

Source	Issue of Concern
Otto Klepper's Advertising Procedure, Russel, Thomas and Verril, Glenn, 1986	Traffic Volume, dimensions, land use, in addition to specifying the value (value) is based on; Visible billboard distance; Speed of travel; The viewpoint of the billboard; The distance between billboards.
The Appraisal of Outdoor Advertising Signs, Suute. Donald, 1994 [5]	Location (future land use tendency), billboard structure, land use lease, point of view, and type of traffic.
Signs Regulations for Small and Midsize Commu-nities, Kelly, Eric Damain and Rasso, Garry J, 1989	Technical installation, land use, and altitude.

3 Result and discussions

The procedure of calculating the rent value, and the selling value of the tax object and the strategic value of advertisement in Jember Regency as regulated in Jember District Regulation Number 28 Year 2011. The strategic value component of advertisement calculation is based on: land use, size/dimension of advertisement, viewpoint, class of the road, and price point/location of installation of billboard.

The price of the installation point of the billboard is based only on the national and regency road classes or classes 1 and 2. However, in addition to the two road classes, there are additional parameters: road function, strategic location value, delays, volume of passing vehicles, and speed of vehicle.

Overview of road conditions is divided into roads and intersections. The favorite position of an advertisement is located at the intersection. Compared to the road segment, the position of this intersection is potential in billboard placement. On the road side the rider's sight is straight and continuing to run, to be able to pay attention and read billboards smaller. At the intersection position, the average rider reduces their speed. This behavior depends on the type of intersection; the intersection in Jember is divided into 3: Signal intersection, intangible intersection and intersection with roundabout. At the signal intersection, the rider will stay for a while or stop at the red light, so the opportunity to pay attention around the intersection is bigger.

Population density effects on how big the billboard will be read by the public. The denser the existence of this, the billboard will be more and more potentially seen by many people around. Vehicle volume is also influential. The higher the volume of vehicles, the more potential to be seen by many people. In addition to volume, vehicle speed is also influential. The slower, the longer the person is at the intersection / the road, the more people who potentially see the billboard. Survey of traffic volume at intersection is done by classification. Vehicles crossing the survey sites are classified into several types, namely light vehicles (LV), heavy vehicles (HV), motorcycles (MC), and non-motorized vehicles (UM). The collection of volume data on each intersection is divided into four periods. On the effective day of the morning period is done at 06.30 - 08.30 WIB, the period of noon at 10:30 to 12:30 pm, the afternoon period at 14:30 to 16:30 pm, and the night period from 18:30 to 20:30 pm. While on the holiday period of the morning carried out at 09.30 - 11.00 pm, the period of noon at 12:45 to 14:00 pm, the period of noon at 16:15 - 17:15 pm, and the period of the night at 18:15 - 19:45 pm. Each type of vehicle has its own equivalence value (EMP).

3.1 Modeling based on road segments

The first model developed is an optimization model based on road segment. The data obtained from the transportation survey in the city of Jember, there are 11 road segments that allow optimization done with modeling using OLS. Table 2 presents the results of road performance analysis based on transportation survey results in Jember. Road performance is described in, DS, travel time, and speed.

Table 2. Road performance and strategic location

No.	Road Segment	DS	Travel Time (minute)	Speed (km/hour)	Strategic Location
1	Jalan Hayam Wuruk	0.32	0.58	60	3
2	Jalan Gajah Mada	0.34	0.57	60	3
3	Jalan Sultan Agung	0.52	0.35	60	4
4	Jalan Ahmad Yani	0.91	0.34	18	5
5	Jalan Trunojoyo	0.87	0.22	18	5
6	Jalan Kalimantan	0.58	0.33	24	2
7	Jalan Sumatra	0.44	0.15	36	2
8	Jalan Mastrip	0.48	0.36	30	2
9	Jalan Riau	0.62	0.46	30	2
10	Jalan Jawa	0.63	0.41	24	2
11	Jalan Panjaitan	0.48	0.28	48	3

In making model-based segment optimization model using OLS, then there is assumption BLUE (Best, Linear, Unbiased and Estimator) to be fulfilled so that modeling become a good model. One of the tests to be done is to determine whether the data obtained is in accordance with the assumption of normality or not. The assumption requires the distribution of data can be process properly when the distribution of the data follows a diagonal curve line, or the data spreads along the diagonal line. Based on the test results in Figure 1, the Probability Plot of each variable, either individually plot or mixed plot between variables, shows the distribution of data following the diagonal curve line, or the data spreads along the diagonal line. This means that the variables used in the modeling meet the assumption of normality, and deserve to be used for modeling.

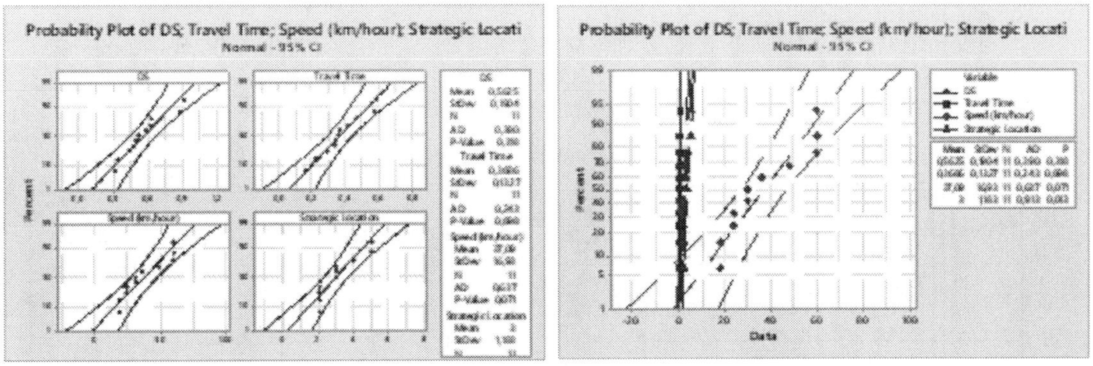

(a) (b)

Figure 1. Normal probability plot– model road segment (a) Normal probability plot individual, (b) Normal probability plot mixed

The next stage is to create a strategic model with the dependent variable / response in this modeling is the strategic location score of each road segment. While the independent variable are the degree of saturation, travel time, and speed. The assumption underlying this modeling is that the strategic location scores on the road segment are determined on all three variables. The more saturated, the longer the travel time, the slower the vehicle through the road segment, then the location is a strategic location to place the billboard. It wills makes the riders have a longer stop/slow time on the road segment. Although it has the potential for a higher billboard location, the priority is to make the priority scale to do the arrangement of the advertisement placement on the road segment. Based on the model developed below, it can be a priority order on the score of the expected strategic location so that it can be used as the guideline for the advertisement arrangement.

Based on the output of modeling using Minitab 17, the model is obtained as follows:

Regression Equation is:

Strategic Location = -5,02 + 9,34 DS - 0,86 Travel Time + 0,0832 Speed (km/hour)

Predictor	Coef	SE Coef	T	P
Constant	-5,02	1,55	-3,23	0,014
DS	9,34	1,59	5,87	0,001
Travel Time	-0,86	1,58	-0,54	0,603
Speed (km/hour)	0,0832	0,0186	4,47	0,003

S = 0,574294 R-sq = 83,51% R-sq(adj) = 76,44%

Based on these outputs, the important variables that become significant predictors of determining the strategic location is the variable DS is the variable saturation degree of road segment. This can be seen from the regression coefficient of 9.34 with a significance value of 0.004 <0.05 which means that this statistically significant variable becomes a predictor of the dependent variable (strategic location). The second variable is a significant predictor of velocity variable with regression coefficient value of 0.0832 and significance value 0.004 <0.05 which means that statistically this variable is significant to be the predictor of the dependent variable (strategic location). While the travel time variables are not statistically significant influence in determining the location of strategic placement of advertisement. This model can also obtain the ability of a model in explaining a dependent variable by looking at the value of R2 over the model. This model has a R2 value of 83.51% and R-adjusted of 76.44%. This means the model can explain the change of the dependent variable of 76.44%.

From the model that has been developed, it can be seen in the following table that presented the value of expectations (expected value) on the predictor value that can be used to create the priority of ordering/placement of advertisements as follows:

Table 3. Priority rank base on road segment model

No.	Road Segment	Expected Value	Priority Order	No.	Road Segment	Expected Value	Priority Order
1	Jalan Hayam Wuruk	2.46	8	7	Jalan Sumatra	1.92	10
2	Jalan Gajah Mada	2.66	6	8	Jalan Mastrip	1.65	11
3	Jalan Sultan Agung	4.53	2	9	Jalan Riau	2.85	5
4	Jalan A Yani	4.68	1	10	Jalan Jawa	2.51	7
5	Jalan Trunojoyo	4.41	3	11	Jalan Panjahitan	3.24	4
6	Jalan Kalimantan	2.11	9				

3.2 Modelling based on road corridors

From the survey data of transport of outer corridor group (26 outer corridors), and corridor inside (4 corridors inside), there are 16 lines of intersection data that can be processed and used to create road-based corridor modeling. Table 4 is a tabulation of road performance survey results data on each road corridor. In the street-based intersection of the road corridor, there are 3 predictors considered to affect the strategic location of DS, Delay, and LOS. Based on the results of one of the BLUE assumption tests, it shows that the variable "Delay/delay time" of the graphic shows a data pattern that is too critical, meaning there is potential for data not following the normal pattern. The standard deviation of this variable is 11.069. The abnormality of this data pattern is corroborated by the value of KS of 0.219 and the value of P-value of 0.032 is lower than 0.05 which means the data pattern is not normal. For an interval estimate of 95% we believe that the mean degree of saturation is in the range 0.396 to 0.711 where,

Table 4. Corridor road performance and strategic location

No.	Intersection Name	DS	D	LOS	Strategic Location
1	KFC [1]	0.75	18.70	2	3
2	Kenanga[1]	0.99	8.33	1	2
3	GatotSubroto[1]	0.88	22.52	3	3
4	Tembakan[1]	1.00	20.94	3	3
5	Dr. Wahidin[1]	0.11	9.26	1	2
6	PasarTanjung[4]	0.28	13.30	2	5
7	RS Subandi[3]	0.32	14.04	2	4
8	Jarwo[3]	0.47	30.53	3	4
9	PKK/ Mastrip[5]	0.77	13.97	2	2
10	RRI [4]	0.36	34.32	3	4
11	DPR [5]	0.74	42.64	4	2
12	Gladakkembar[3]	0.62	32.47	5	2
13	Polres[2]	0.34	10.55	1	2
14	Istana [1]	0.78	9.29	3	2
15	Matahari[1]	0.15	7.63	1	2
16	UntungSuropati[1]	0.31	7.54	1	2

Note: [1]Un-signalized intersection with three legs
[2]Un-signalized intersections with four legs
[3]Signalized intersections with three legs
[4] Signalized intersections with four legs
[5] Roundabout intersections with four legs

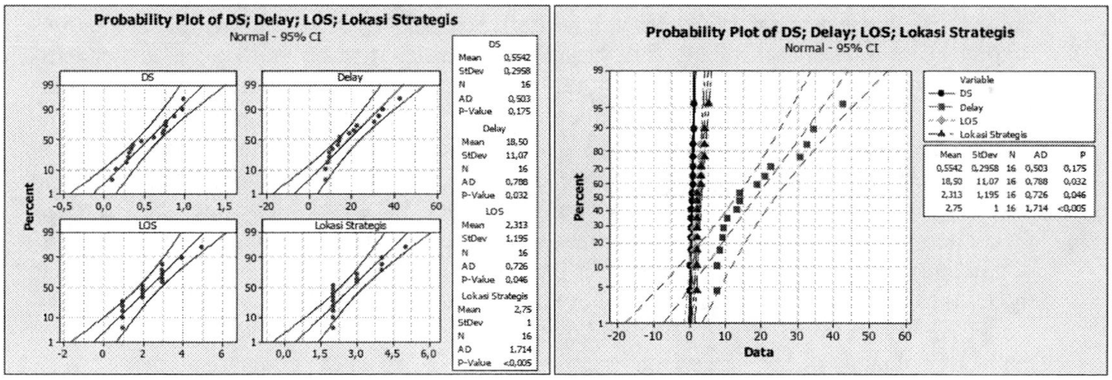

(a) (b)

Figure 1. Normal probability plot– model road segment (a) Normal probability plot individual, (b) Normal probability plot mixed

The strategic location variable shows the data pattern is too oblique to the right side, meaning there is potential data not follow the normal pattern. The standard deviation of this variable is 1. The abnormality of this data pattern is corroborated by a KS value of 0.336 and a P-value value of 0.0005 which is lower than 0.05 which means the data pattern is abnormal. For an estimated 95% interval we are convinced that the average degree of saturation is in the range of 2.21 to 3.28.

The next step is to develop a strategic location optimization model based on the road corridor. The dependent variable / response in this modeling is the strategic location score of each road corridor segment. While the independent variable is the degree of saturation, delay time (delay), and LOS. The more saturated, the longer the delay, and the slower the vehicle through the road corridor, the location is strategic to place the billboard, because it makes the riders have a longer stop / slow time on the road segment. Although it has the potential for a higher billboard location, the priority is to make a priority scale to do the arrangement of the advertisement placement on the road segment. To develop the model of the variables in this second model, it is worth noting there are two variables whose data pattern does not follow normal data pattern rules i.e.; delay and strategic location variables. To overcome these problems so that it can be done further analysis and modeling, then both variables are done with logarithmic calculation (log) on the original value. After this stage, a priority ranking can be made on the score of the expected strategic location that can be used as a guideline for the advertisement arrangement. The results of the regression model are shown in Table 5.

Table 5. The results of the regression model

Predictor	Coef.	SE Coef	T	P
Constant	0.0156	0.6189	0.03	0.980
DS	-0.3166	0.3299	-0.96	0.362
LnDS	0.1273	0.3591	0.35	0.731
Delay	-0.0156	0.0132	-1.18	0.267
LnDelay	0.9318	0.6181	1.51	0.166
LOS	-0.2612	0.1104	-2.37	0.042
LnLOS	1.2884	0.5965	2.16	0.059

Indicators of the model: S = 0.0977; R-Sq = 72.6%; R-Sq(adj) = 54.4%

Based on these outputs, consistently these results indicate that the important variable becomes a significant predictor of determining the strategic location is the LOS variable only. It can be seen from LOS regression coefficient of -1.7839 with significance 0.004 <0.054 which means statistically this variable is significant to be predictor of dependent variable (strategic location). Furthermore, comparative with variables on logarithmic calculation LnLOS obtained regression coefficient of 9.029 with a significance value of 0.068 if done on two tailed test then this logarithmic variable also statistically significant to predictor strategic location. The second test model is to use variables that all are logarithmic first. The results also show consistency that LOS variables affect the strategic location. While the other two variables, statistically have no significant effect in determining the strategic location on advertisement placement. In this model the ability of a model is to explain a dependent variable by looking at the value of R^2 over the model. This model has R2 value of 69.6% and R-adjusted of 49.3%. This means that the model can explain the change of the dependent variable by 49.3%. While the comparison model with logarithmic value has a R2 value of 72.6% and R-adjusted by 54.4%. This means that the model can explain the change of the dependent variable by 54.4%.

From the model that has been developed, then in the following table can be presented the value of expectation (expected value - EV) on the predictor value that can be used to create the priority of ordering/placement of advertisements as follows:

Table 4. Corridor road performance and strategic location

No.	Intersection Name	EV-1	EV-2	Priority Order	No.	Intersection Name	EV-1	EV-2	Priority Order
1	KFC [1]	3,40	0,52	5	9	PKK/ Mastrip[5]	3,10	0,47	6
2	Kenanga[1]	1,00	0,17	16	10	RRI [4]	3,85	0,57	1
3	GatotSubroto[1]	3,04	0,47	7	11	DPR [5]	2,30	0,35	12
4	Tembakan[1]	2,76	0,44	8	12	Gladak kembar[3]	1,92	0,29	15
5	Dr. Wahidin[1]	2,36	0,35	10	13	Polres[2]	2,53	0,38	9
6	PasarTanjung[4]	3,82	0,56	3	14	Istana [1]	2,35	0,34	11
7	RS Subandi[3]	3,84	0,57	2	15	Matahari[1]	2,10	0,31	13
8	Jarwo[3]	3,77	0,57	4	16	Untung Suropati[1]	2,03	0,29	14

Note: [1] Un-signalized intersection with three legs
[2] Un-signalized intersections with four legs
[3] Signalized intersections with three legs
[4] Signalized intersections with four legs
[5] Roundabout intersections with four legs

The data were obtained from surveying on road sections in four cities in Indonesia, namely Palembang, Bandung, Yogyakarta and Surakarta. The road network includes national road network as well as urban road network. Three types of surveys had been conducted, namely travel time and delay survey, traffic volume survey, and road geometry survey. Travel time and delays survey was done to get travel time experienced by a normal moving vehicle on the road sections. This survey produced travel times and delays along with the location and cause. Traffic volume survey was done by counting the various passing vehicle on a certain spot location and generally conducted for 16 hours with the types of vehicles include motorcycles, light vehicles, heavy vehicles, and unmotorized vehicles. The geometry survey was conducted to identify the width and number of lanes, the median, the width of road shoulders/curb, and the land use/side friction along the road sections to calculate the road capacity.

4 Conclusions

Both models are developed based on road segment and road corridor, it can be concluded that (1) in the road segment-based model, which becomes the main predictor in optimizing the determination of strategic location, the variable DS is the variable of saturation degree of road segment, followed by Variable speed. While the other two variables are the ratio and travel time is not statistically significant as a predictor in optimizing the placement of the billboard. Furthermore, from the statistical model obtained, the road segment Ahmad Yani ranks as the first priority of a strategic location that needs to be done. It optimizes the arrangement and placement of billboards; and (2) in a road-based corridor model, only LOS variables are the main predators and the only statistically significant variables that need to be considered in optimizing the layout and placement of billboards. Furthermore, from the statistical model obtained, RRI intersection is the main corridor which becomes the priority of strategic location for optimization of layout and placement of advertisement.

References

1. Robson,W.A., *Great Cities of the World: Their Government, Politics and Planning.* New York: Macmillan Publishing Co. (1995).
2. Anonim. *Peraturan Pemerintah Nomor 39 Tahun 2006 tentang Tata Cara Pengendalian dan Evaluasi Pelaksanaan Rencana Pembangunan.* Jakarta: Sekretariat Negara RI. (2006).
3. Watson, D and Alan P; Robert S. *Time Saver Standards for Urban Design.* London: Mc Graw- Hill Book Company. (2003).
4. Spereigen, Paul D. *Urban Design : The Architectur of Toen and Cities.* London: Mc.Graw- Hill Book Company. (1965).
5. Sutte, Donald T. *The Appraisal of Out Advertising Signs.* US: Appraisal Institute. (1993).
6. Anonim. *UU RI Nomor 32 Tahun 2004 tentang Pemerintah Daerah.* Jakarta: Sekretariat Negara RI. (1992).
7. Anonim. *UU RI Nomor 24 Tahun 1992 tentang Penataan Ruang.* Jakarta: Sekretariat Negara RI. (1992).
8. Dharmawan, E. *Teori dan Implementasi Perancangan Kota.* (2003).
9. Gallion and Einser. *Urban Pattern.* New York: Van Nostrand Company Inc. (1963).
10. Josep and koppleman, Lee E. 1990. *Guide For Cities.* Jakarta: Penerbit Erlangga.
11. Anonim. *Kamus Penataan Ruang.* Jakarta: Departemen Pekerjaan Umum. (2009).
12. Mandika, M. Pengaruh Signage Pada Bangunan-bangunan Komersil Terhadap Estetika Visual Koridor Jalan Pandanaran Semarang. *Tesis.* Semarang: Pasca Sarjana Jurusan Arsitektur Universitas Diponegoro. (2004).
13. Murtomo, Aji. Penataan Papan Reklame pada Penggal Jalan Hayam Wuruk Semarang. *Jurnal Ilmiah Perancangan Kota dan Permukiman.* (2007).
14. Setiawan, Altim. Kajian Faktor Nilai Strategis Lokasi Penempatan Reklame di Kota Palu. *Jurnal Ruang.* Vol 1 Nomor 1 September 2009. (2009).
15. Watson, D and Alan P; Robert S. *Time Saver Standards for Urban Design.* London: Mc Graw- Hill Book Company. (2003).
16. Wijaya, Monang and Suwando, Djoko. Evaluasi Tata Reklame Pada Koridor Utama Kota Magelang. *Jurnal Ruang.* Vol. 1 Nomor 1 Tahun 2013. (2013).
17. Zahnd, Markus. *Perancangan Kota secara Terpadu, Teori Perancangan Kota dan Penerapannya.* Yogyakarta. (1999).

Potential cargo demand of Kuala Tanjung Port as international hub port in Western Indonesia

Rosita Sinaga[1,*], *Windra* Priatna Humang[2] , and *Abdy* Kurniawan[1]

[1]Ministry of Transport, Center of Research and Development, Jakarta, Indonesia
[2]Andi Jemma University, Department of Civil Engineering, Palopo, Indonesia

Abstract. The Port of Kuala Tanjung will provide a logistical diversification opportunity in which the need for freight services for both export, import, and inter-island distribution through port will continue to increase. Seeing the magnitude of this potential, it is hoped that Kuala Tanjung Port can become an alternative International Hub Port in Western Indonesia that can provide added value for economic growth significantly. As a potential port to become an international Hub Port, Kuala Tanjung must have a large cargo demand, coming from its own hinterland as well as shifting from other nearby ports. The largest potential hinterland of Kuala Tanjung Port is in the form of oil palm and rubber plantations with derivative goods. Kuala Tanjung Port is predicted to accommodate 65% of total goods production in North Sumatra, both containerized and bulk liquid.

1 Introduction

Kuala Tanjung Port has the potential to become a hub of logistics for the Asia Pacific region (Coordinating Ministry for Economic Affairs, Republic of Indonesia). The development of the Port of Kuala Tanjung as the International Hub Port has been proclaimed by the government in the Blueprint of the National Logistics System (Sislognas) which is part of the Masterplan for the Acceleration of Economic Development and Expansion of Indonesia (MP3EI) and the Concept of Sea Tollway.

Kuala Tanjung Port will be developed as an Industrial Gateway Port that integrates a harbor with an industrial area of 3,000 Hectare and Sei Mangkei Special Economic Zone (SEZ). It has a natural depth location of -14 m LWS at a distance of 2.7 km from the coast with the waters around the port open directly to the Strait of Malacca. Kuala Tanjung Port is located on the east coast of North Sumatra Province. The location is strategic because it is on the busy path of world trade, the Strait of Malacca. In addition, the natural condition of the surrounding area is also considered the criteria as the International Hub Port, in addition to the growing industrial clusters around it [1-6].

The potential of hinterland of the Kuala Tanjung Port is enormous. Many large scale companies such as PT Indonesia Asahan Aluminum and PT Multimas Nabati Asahan and other industrial companies. The most dominant hinterland potential is the product of Palm

* Corresponding author: rosita_dephub@yahoo.com

Oil / CPO and its derivative goods, so that the Port of Kuala Tanjung can be an alternative port other than Belawan Port for the delivery of industrial products located in Batubara Regency and surrounding areas throughout Indonesia and export.

In addition, the existence of the Special Economic Zone (SEZ) Sei Mangkei in Simalungun Regency which has an area of 140 Ha and is 50 km from the Port of Kuala Tanjung will has opportunities for diversification in the field of logistics where the need of freight services for the distribution of goods through ports will continue to increase. Seeing the large potential of such demand, Kuala Tanjung Port is expected to be one of the International Hub Port in the Western Indonesia that can provide added value for economic growth significantly. This paper aims to analyze the potential of cargo (container and liquid bulk) at the Kuala Tanjung Port, so it can be predicted the capability of Kuala Tanjung Port as an alternative hub port in Indonesia.

2 Methodology

The analytical method used is descriptive analysis. Identification of hinterland area characteristics include geographical, socio-economic, PDRB, and commodity. The potential demand for sea transport generated by the hinterland area of Kuala Tanjung Port is analyzed using multiple regression. This analysis is based on the identification of potential sea transport demand in the hinterland area. To get the clarity of the potential used model time series of the last few years [7]. The general theoretical model is $P = f(X_1, X_2 \ldots \ldots)$, with X_1, X_2 is the potential variable of demand for hinterland. The regression analysis model can made modeling the relationship between two or more variables. The general relationship for multiple regression is as follows:

$$Y = A + B_1X_1 + B_2X_2 + \ldots\ldots\ldots + B_mX_m \quad (1)$$

where:

Y = the dependent variable, $X_1 \ldots X_m$ = independent variable
A = regression constant, $B_1 \ldots B_m$ = regression coefficient

3 Result and discussion

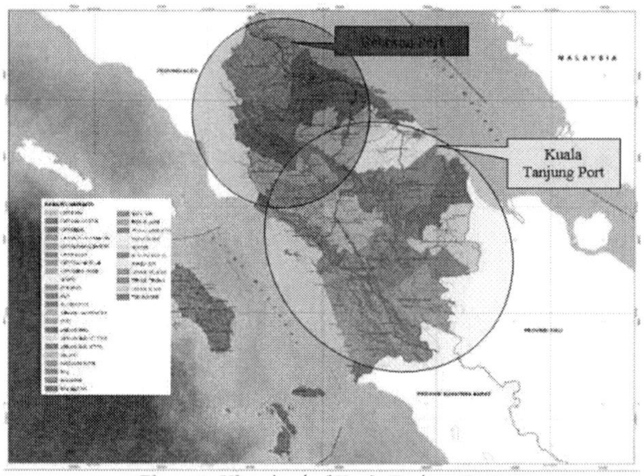

Fig 1. Hinterland of Kuala Tanjung Port

3.1 Hinterland area of Kuala Tanjung Port

The location of Kuala Tanjung Port which is quite close to the Belawan Port, hinterland area of Kuala Tanjung Port and Belawan Port will be intersected. The determination of hinterland boundary of Kuala Tanjung port is based on the accessibility level of land transportation to access the port. In addition, hinterland area is also based on the state preference (level of desire) of business people to make the Kuala Tanjung Port as a gateway activity distribution of goods that shows on Figure 1.

3.2 Formatting the title, authors and affiliations

3.2.1 Potential liquid bulk

The Bulk Liquid terminal is planned to have a production capacity of 3.5 million tons per year. Based on the result of the study before, considering the possibility of moving loading and unloading traffic from Belawan Port to Kuala Tanjung Port and based on the potential of the hinterland area of Kuala Tanjung Port and development of Sei Mangkei SEZ, the projected demand for liquid bulk as in the following table 1.

Table 1. Liquid bulk projection demand of multipurpose terminal in Kuala Tanjung Port

No	Commodity	2020	2021	2022	2023	2024	2025	2026	2027
1	Biodiesel Industry								
	FAME	500.000	500.000	500.000	500.000	500.000	500.000	500.000	500.000
	MES	50.000	50.000	50.000	50.000	50.000	50.000	50.000	50.000
	B-Caroline								
	Glycorine	47.000	47.000	47.000	47.000	47.000	47.000	47.000	47.000
2	Oleokimia Industry								
	Fatty Alcohol	80.000	80.000	80.000	80.000	80.000	80.000	80.000	80.000
	Soap Noodle								
3	Refinery Industry								
	Cooking oil	500.000	500.000	500.000	500.000	500.000	500.000	500.000	500.000
	RBD Olain								
	RBD Slarina	50.000	50.000	50.000	50.000	50.000	50.000	50.000	50.000
4	Palm Oil Industry								
	PKO	51.600	51.600	51.600	51.600	51.600	51.600	51.600	51.600
	PKM								
5	Other Oleochemical Product Derivatives								
	Fatty Ester	100.000	100.000	100.000	100.000	100.000	100.000	100.000	100.000
	Emulsifier	100.000	100.000	100.000	100.000	100.000	100.000	100.000	100.000
	Pharmaceutical								
6	CPO/PKO	1.219.742	1.181.679	1.153.325	1.011.990	926.102	1.209.533	1.195.663	1.515.293
	Total	2.698.342	2.680.279	2.431.925	2.490.590	2.404.702	2.688.130	2.674.263	2.993.820

Source : RIP Kuala Tanjung KP. 148/ 2016

3.2.2 Potential of container

The Based on the result of potential demand analysis, total demand for container in Multirpurpose Terminal is the sum of the three components (geographic area, palm oil industry and rubber of SEZ Sei Mangkei) as presented in the following table.

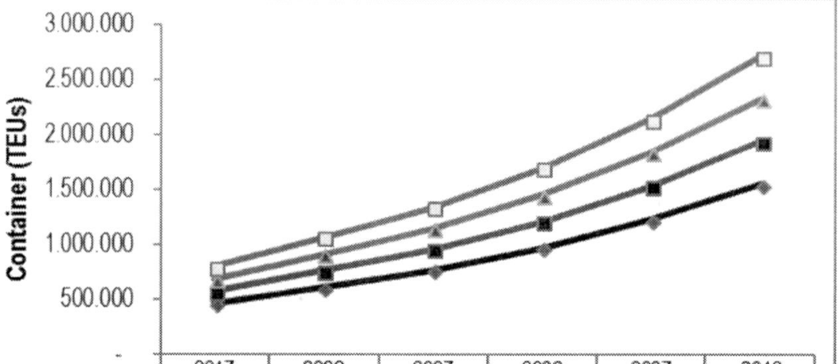

	2017	2022	2027	2032	2037	2042
pessimis,	456.650	607.486	765.909	968.851	1.225.565	1.550.301
moderate,	570.813	759.357	957.386	1.211.063	1.531.956	1.937.876
optimistic	684.976	911.228	1.148.864	1.453.276	1.838.348	2.325.451
very optimistic	799.138	1.063.100	1.340.341	1.695.488	2.144.739	2.713.026

Fig . Container demand of Kuala Tanjung Multipurpose Terminal

3.2.2 Potential cargo in terminal hub-international

Demand analysis for Container Terminal (International Hub) is done through evaluation of RIP Pelabuhan Kuala Tanjung (KP 148/2016). This review is aimed at identifying differences in assumptions about the development of the Port of Kuala Tanjung. The result of this review is the determination of development assumptions and scenarios for the Port of Kuala Tanjung

a. Probability of shifting cargoes

Stakeholder perceptions analysis is conducted with the aim of obtaining preferences from users to move from origin port to Kuala Tanjung port. This preference is expressed in units of probability of displacement. For example, if a preference score of 0.5 can be concluded that 50% of users will move from the origin port to Kuala Tanjung Port.

Belawan Port and Tanjung Priok Port are made into 2 (two) harbors of comparison with reason (1) Belawan Port is an existing port operating in North Sumatra and (2) Tanjung Priok Port is a port in Indonesia that represents an international port.

Probability of shifting used for the Port of Kuala Tanjung is carried out using the Belawan Port Curve drawn from the Tanjung Priok Port curve. Figure 2 explains that if the Kuala Tanjung Port competes with Tanjung Priok Port, then to reach the equilibrium point, Kuala Tanjung Port must be able to operate with TRT for 14 hours. While, the real condition, Kuala Tanjung Port compete directly with the Belawan Port. Therefore, the 14 hour TRT value is then inputted as an input in the transition curve at Belawan Port. By entering the TRT input for 14 hours, we have a transition probability of 65%. That is, if the Kuala Tanjung Port operates with TRT for 14 hours it will occur a transition of 65% from Belawan Port to the Kuala Tanjung Port.

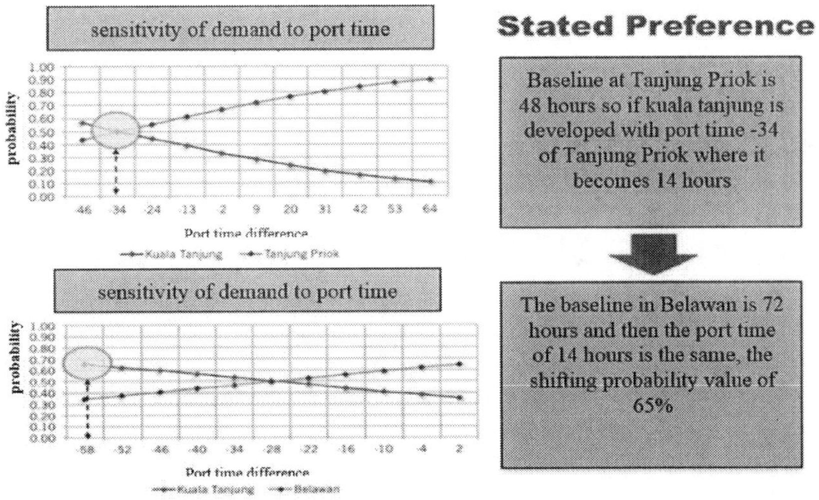

Fig 2. Probability of shifting to Kuala Tanjung Port

By assigning TRT plan of the Kuala Tanjung Port for 14 hours or in other words is 0.58 days, it is certain that Kuala Tanjung Port will be able to compete with other competitor ports such as Port of Singapore, Port of Tanjung Pelepas, Port Klang and Port of Penang for quite a long time. Because, as a comparison only, the dwelling time in Port of Singapore is 19 hours or equivalent to 0.8 days.

b. Demand based on hinterland geographic aspect

Based on geographical aspects of hinterland, Kuala Tanjung Port will be able to attract demand from several provinces, namely North Sumatra, West Sumatra, Riau, and Jambi. Demand projection is done by using multilinear regression model which variables are provincial PDRB, production of potential commodities such as palm, rubber and coffee. Here is an example of multilinear regression model development for several commodities in North Sumatera Province.

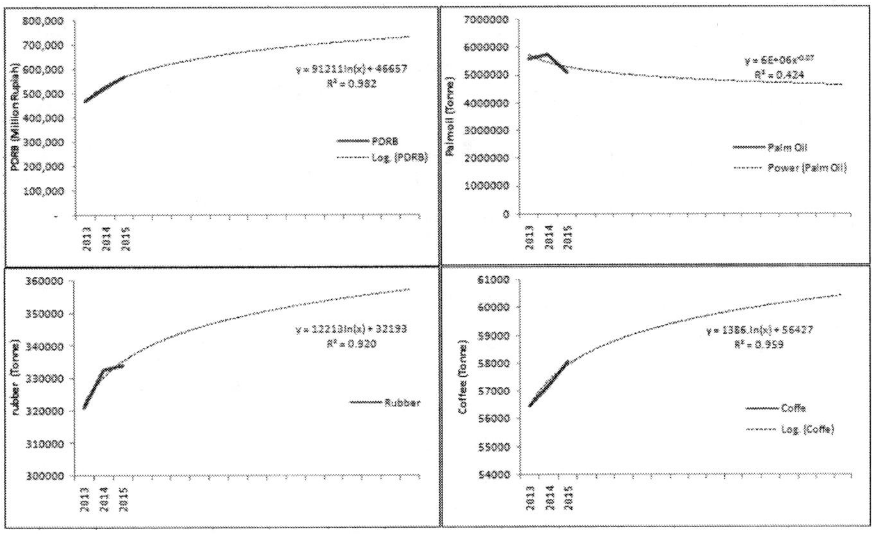

Fig. 3. Projection of demand hinterland determinant variables

Kemenhub study (2014), demand from North Sumatra is predicted to be captured by Container Terminal (International Hub) of Kuala Tanjung Port. However, taking into account the operation of Multipurpose Terminal then the prediction is modified on the basis that there will be a container demand that is shifted to the multipurpose terminal. Figure 3 shows the number of transitions that occur for each terminal.

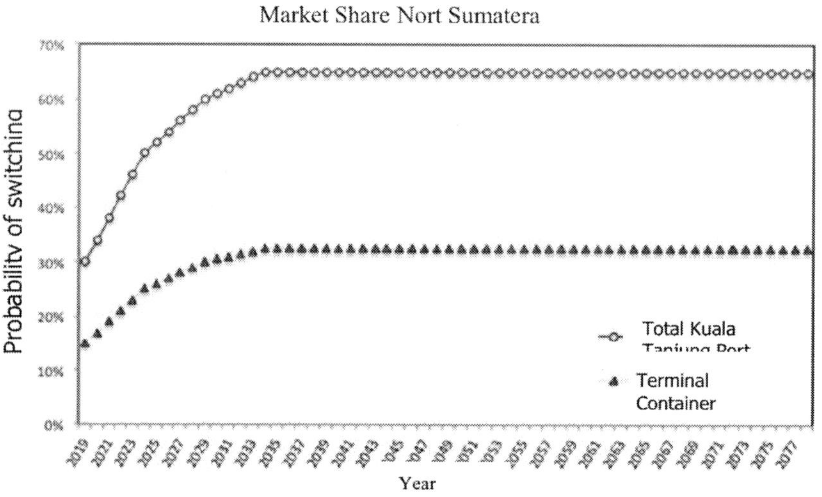

Fig. 4. Proportional shifting of North Sumatra demand to Kuala Tanjung Port

From Figure 4 can be seen that with the operation of multipurpose terminals, demand for containers will be divided between Multipurpose Terminal and Terminal Hub International. Based on preference, it is predicted that 65% of total production in North Sumatera will go to Kuala Tanjung Port and demand will be equally divided between the two terminal.

c. *Demand of SEZ Sei Mangkei*

Table 2. Projection of downstream production of oil palm industry SEZ Mangkei

No.	Palm Downstream Products	Projection of Downstream Industry Based Palm Products			Total	
		2018	2020	2025	Liquid	Solid
1	CPO	2.550.000	2.650.000	3.000.000	29.900.000	
2	CPKO	100.000	100.000	100.000	930.000	
3	Basic Oleochemistry					
	Fatty Acid	135.000	135.000	135.000		1.485.000
	Fatty Alcohol	90.000	90.000	90.000	810.000	
	Soap Noodle					440.000
	Glycerine	16500	16500	16500	148.500	
4	Oleochemical Derivatives:					
	Fatty Alcohol Amide	10.000	10.000	10.000	70.000	
	Methyl Ester Sulfonate	15.000	15.000	15.000	135.000	
	Fatty Alcohol Sulfonate				52.500	
	Fatty Ester				87.500	
5	Olefood					
	Cooking Oil	600.000	600.000	600.000	4.800.000	

	Margarine	10.000	10.000	10.000	70.000	
	Shortening	15.000	15.000	15.000	90.000	
6	Paper Industry Products	-	100.000	100.000		700.000
7	Wood Products Industry	-	-	300.000		1.500.000
8	Industrial Fertilizer Products	100.000	100.000	100.000		900.000
	Total	3.701.500	3.901.500	4.551.500	37.093.500	5.025.000

Table. 3. Projection of downstream rubber industry of Sei Mangkei

No.	Rubber Downstream Products	Unit	Projection of Rubber-Based Downstream Industry Products		
			2018	2020	2025
1	Dry Rubber Products	Ton	1.100.000	1.400.000	2.000.000
2	Motorcycle Tire (Tire R-2)	Million /Unit	34	35	37
3	Car Tires (Tire R-4)	Million /Unit	20	24	30
4	Automotive Rubber Parts	Million /Unit		10	25
5	Green Tyres	Million /Unit		15	30
6	Rubber Thread	Ton	500	1.000	2.500
7	Retread Aircraft Tires	Million /Unit	1	1	3
8	Medical Devices	Million /Unit	5	10	30
9	Natural Rubber Based Resin	Ton	1.500	1.500	3.000
	Total	Million /Unit	60	95	155
		Ton	1.102.000	1.402.500	2.005.500

Based on Sei Mangkei's development plan data, Sei Mangkei will be able to produce containers equal to 68,364 TEUs in 2019 and increase up to equivalent to 453,595 TEUs in 2078.

d. Total demand of container (multipurpose and hub-international)

The projection of demand for Container Terminal of Kuala Tanjung Port is presented in Figure 4 for 5 years demand. Alternative projection is done with a pessimistic approach, moderate, optimistic and very optimistic.

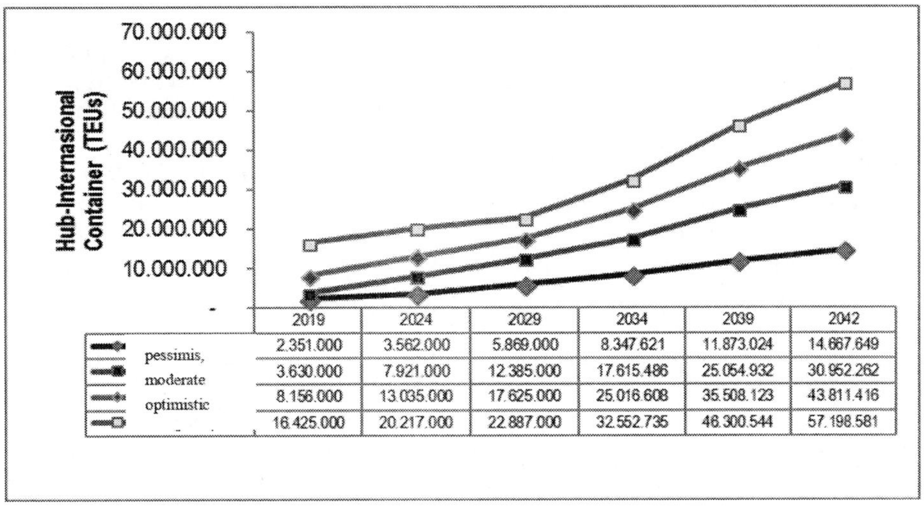

Fig. 4. Container demand projection (international hub)

Based on the results of analysis, prediction of Container Terminal (International Hub) based on a pessimistic approach will capture demand of 2.3 Million TEUs in 2019, and increase to 14.5 million TEUs in, 2042, Based on a moderate approach will be able to capture demand by 3.6 Million TEUs in 2019, and increase to 30.5 million TEUs in 2042. Based on the optimistic approach will capture demand by 8, 1 Million TEUs in 2019, and increased to 43.8 million TEUs in 2042. While based on a very optimistic approach will capture demand of 16.2 Jt TEUs early in 2019, and increased to 57.1 million TEUs in 2042. It is also known that over time the demand growth rate will decrease indicates that demand is starting to show saturation.

Table 4. Comparison of demand terminal multipurpose, demand hub-international, demand transhipment *)

Year	Demand Multipurpose	Demand Hub International	Demand Transhipment	Difference Demand Hub-International to Multipurpose Terminal	Percentage (%)		
					Demand Multipurpose to hub-internasional	Internasional Demand Transhipment to International Hub Terminal	Against Total Cargo in the Malacca Strait
2019	570,813	3,630,000	363,000	3,059,187	15.72%	10%	3%
2024	759,357	7,921,000	1,188,150	7,161,643	9.59%	15%	6%
2029	957,386	12,385,000	2,105,450	11,427,614	7.73%	17%	9%
2034	1,211,063	17,615,486	5,108,491	16,404,422	6.87%	29%	12%
2039	1,531,956	25,054,932	7,265,930	23,522,976	6.11%	29%	12%
2042	1,937,876	30,952,262	8,976,156	29,014,387	6.26%	29%	12%

Notes: *) Moderate alternatives
Source: Analysis result, 2016

4 Conclusion

The development of Kuala Tanjung Port is predicted to accommodate 65% of total goods production in North Sumatra, both containerized and bulk liquid. Predicted sea transport demand generated by Sei Mangkei SEZ and industrial area in hinterland of Kuala Tanjung Port as follows:

a. Prediction of cargo in multipurpose terminal shows that liquid demand trend to fluctuate. In 2017 is predicted to reach 2.93 million tons and then decreased by 2020 only 2.69 million tonnes and then increased again in 2027 reached 2.99 million tonnes. While the predictions of container trend to increase recorded in 2017, the prediction of container reached 570,813 TEUs, 2020 to 672,234 TEUs, then increased in 2025 to 871,476 TEUs.

b. Prediction of container in international hub port shows that SEZ Sei Mangkei region becomes the largest contributor to the production of palm oil industry in 2020 reached 3.90 million tons, increased in 2025 to 4.55 million tons. The production of rubber industry in 2020 reached 1.40 million tons and increased by 2025 to 2.05 million tons. While for 2020 container is estimated to reach 0.9 million TEUs, 2025 reaches 2.6 million TEUs, 2030 reaches 8.1 million TEUs and 2035 reaches 10.3 million TEUs.

Acknowledgment

The research was carried out with the support of data from the Centre of Research and Development of Maritime Transport, River, Lake and Ferry, the Ministry of Transportation.

References

1. Keputusan Menteri Perhubungan.KP 414 2013 tentang Rencana Induk Pelabuhan Nasional (RIPN). Jakarta. (2013)
2. Bambang Triatmojo. Perencanaan Pelabuhan, Beta Offset, Yogjakarta. (2010)
3. Bowersox, Donald J. Logistics Management 2, Second Edition, Terjemahan oleh Hasyim Ali, Bumi Aksara, Jakarta. (1978)
4. Coordinating Ministry For Economic Affairs, Republic of Indonesia. Masterplan for Acceleration and Expansion of Indonesia Economic Development. Jakarta: Coordinating Ministry For Economic Affairs. (2011)
5. Hayath, Y. Intermodality, Concept and Practice, Israel Shipping and Aviation Research Institute, LLP London. (1987)
6. Jean-Paul Rodrigue. Intermodal Terminals, Mega Ports and Mega Logistic. Department of Global Studies & Geography Hofstra University Hempstead, New York. (2009)
7. Jinca, M.Y. Antisipasi Perkembangan Teknologi Petikemas Terhadap Prasarana dan Sarana Transportasi Multimoda, Publikasi FSTPT, UGM Yogyakarta. (2008)
8. Ghozali., I. Ekonometrika, Teori Konsep dan Aplikasi dengan SPSS 17, UNDIP Semarang. (2009)

A demand and capacity analysis on bus semirapid transit network (Case: Jabodetabek public transport network)

Alvinsyah[1,*], and *Edi* Hadian[1]

[1]University of Indonesia, Civil Engineering Department, Jakarta, Indonesia

Abstract. The objective of this research is to estimate the potential demand and capacity for Jabodetabek Bus Semi rapid Transit (BST). A transport model based on four step modeling is utilized to analyze the demand at network level. The model is developed from previous works and adjusted through a calibration and validation procedure with data from the field. The base year O-D matrix is calibrated by trip length frequency distribution through a matrix balancing process. While the passenger flow is validated through transit assignment procedure with data on the existing bus routes. Based on the assumptions made and different operational characteristics and fare system scenarios, a simulation through the developed model on the proposed BST is conducted. From the analysis, the total boarding, the passenger flow on peak hour for each BST corridor are obtained. Based on the average and maximum passenger flow, the required capacity for each corridor is determined.

1 Introduction

City of Jakarta and its surrounding (bodetabek) area, consists of DKI Jakarta (Special Capital District of Jakarta) and its eight surrounding municipalities, is known as a city of chronic and severe congestion with increase of motorized vehicles. Therefore to tackle this chronic problem, a new agency called Badan Pengelola Transportasi Jabodetabek (BPTJ), responsible for all transportation issues in Jabodetabek (Greater Jakarta) area is established. One of its main responsibility is to prepare and implement Transport Master Plan for Jabodetabek Area (RITJ). Referring to the draft of RITJ [1] seems that the most ambitious and potential program that should be implemented in a very short period is its High Capacity Bus/HCB (i.e. Bus Rapid Transit/BRT) network plan as shown in Figure A-1 in the appendix. While, the whole Mass Transit Network are shown in Figure A-2 and Figure A-3 respectively in the appendix. Having reviewed the RITJ and the existing TransJakarta Busway system, and also referring to Vuchic [2], this kind of HCB system can be categorized as Bus Semirapid Transit (BST).

Assumming that the definition of BRT [3, 4] is held, and taking into account the existing physical condition for its proposed network, it seems that BPTJ needs a special treatment and very extra effort to accomplished this Plan. Since passenger demand or

* Corresponding author: alvinsyah2004@gmail.com

ridership plays a dominant role in urban mass transit system, often used as a basis reference to determine the capacity of mass transit [5-7], it is crucial to conduct a demand assessment to the proposed plan either at network level or at route level. Accordingly, the objective of this research is to conduct a demand analysis and to determine the design capacity on each proposed BST service line.

The following sections describe the research framework, assumptions and scenarios, simulation and analysis, and finally the research conclusion.

2 Research framework

This research is commenced by reviewing some previous works [1, 8-12] to prepare a transport model. In general, the preparation of public transport passenger demand forecasting model is based on network development on supply side and the projected socio-economic framework on demand side which involves network scenario development for mass transit system. By adapting the four step model from previous works [1, 8, 10, 13-16] a calibration and validation procedure, to update the model, are then conducted. The model calibration procedure is explained in a more detail in Hadian and Alvinsyah [17]. Once the updated model, commonly called as base year model, is set, then it is utilized to estimate future motorised person trips. Taking the actual mode share composition [11], this total motorised person trip is split into trips with private vehicle (i.e. car and motorcyle) and trips with public transport.

Since current public transport service is dominated by bus system and the mass transit plan in RITJ is mostly road-based system, it is necessary to conduct road network performance analysis repesented in speed on each link in the network. The estimated speed is derived from the highway equilibrium assignment simulation. A more detail explanation on this assignment procedure can be found in Alvinsyah dan Hadian [8]. This estimated speed will be used as a basis (i.e. operational speed) in estimating the potential public transport demand, especially the roadbase mass transit demand.

Having obtained the total public transport trip, then, this trip is assigned to the existing and the planned public transport network through a transit assignment procedure. In the validation process, the person trip from the O-D matrix is assigned to the transit network and adjustment to all parameter values in the generalised cost are made until the passenger flow from the assignment process is conformed with that of from field data.

In transit assignment model, the monetary cost, which is fare applied to each particular line services is converted to 'time' and is weight-combined with total travel time which consists of in-vehicle, waiting at stops, access and egress time components in the specific form called 'generalized cost' function. This function is expresed in the following [18];

$$C_{ij} = a_1 t_{ij}^{v} + a_2 t_{ij}^{w} + a_3 t_{ij}^{t} + a_4 t_{ij}^{n} + a_1 \delta^{n} + a_5 F_{ij} \qquad (1)$$

where,

t_{ij}^{v} = In vehicle time from i to j

t_{ij}^{w} = Walking time from station or to station

t_{ij}^{t} = Waiting time at the station

t_{ij}^{n} = Transfer Time at station

δ^{n} = Penalty for transfer

F_{ij} = Fare from i to j

a_1 to a_5 = Coefficients.

The macro planning software called EMME is used to execute the assignment process where the standard transit assignment model implemented in this software is based on the concept of optimal strategies which minimize transfer, waiting and in-vehicle time [18]. The most general formulation of this model is described in Spiess [19], where the cost of a strategy is the sum of link travel times, c_a, weighted by the probability of traveling on link a, and the waiting time at nodes i is weighted by the probability of traveling through node i. The assigning of the trips from all origins to destination r, according to the optimal strategy, corresponds to solving the following linear optimization problem:

$$Min \sum_{a\in A} c_a v_a + \sum_{i\in I} \omega_i \quad (2)$$

Subject to

$$\sum_{a\in A_i^+} v_a - \sum_{a\in A_i^-} v_a = g_i \quad i \in I,$$

$v_a \leq f_a \omega_i$	$a \in A_i^+$,	$i \in I$,
$v_a > 0$	$a \in A$	

Variables ω_i as in formula (2) represent the total waiting time (in person minutes) at node i. Further and detail explanation on the mathematical derivation of the above objective function can be found in Spiess [19]. Using the generalized formulation as in equation (1), any transit network T can be represented, and the mechanics that are involved in constructing the links and the nodes of the generalized network depend very much on the particularities of the transit network and the degree of aggregation considered. The fundamental concept of this strategy is based on the assumption that a passenger arrived at a station select an optimum route from a group of route randomly to reach his or her destination and board the the first vehicle arrived. This mechanism is done repetitively until reaching the final destination.

Prior to conducting a simulation, future O-D Matrix is developed by applying the gravity model and the forecasted trip generation. Then, based on several assumptions and scenarios set in this research, a simulation to ridership is conducted through the validated transit assignment procedure. Once the simulation is done, the estimated total boarding passenger and passenger profile (flow) at peak hour for each BST service line is then analyzed. Referring to the passenger profile (flow) resulted from the simulation, the maximum flow is identified and the average flow is calculated for each service line. Finally, based on these two values the design capacity for each service line is determined. Diagramatically, the research framework is summarized in Figure 1.

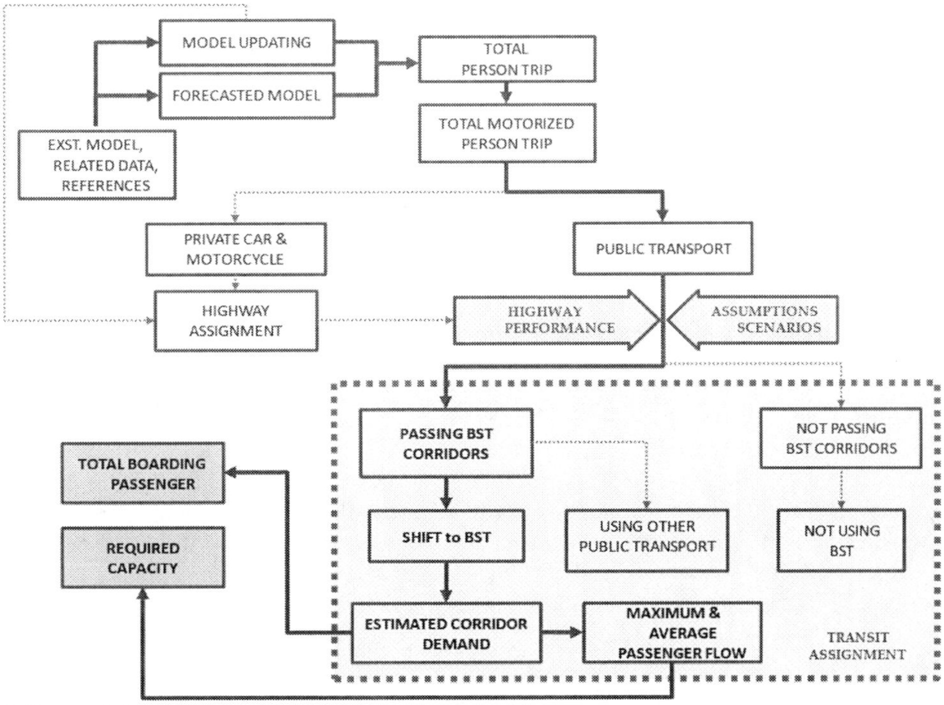

Fig. 1. Research framework

3 Assumptions and scenarios

To obtain the ridership through model simulation, some basic assumptions and scenarios related with trip generation, public transport share, and operational parameter of the system need to be established first. Referring to previous work [1], the population growth rate for Jabodetabek area from 1971 to 2005 is about of 4,7% per annum and that for DKI Jakarta is around 2,0% per annum. In 2020, the growth rate prediction is about 1,32% and 0,49% per annum for Bodetabek and Jakarta respectively. While in 2030, the prediction is about 1,03% and 0,19% per annum for Bodetabek and Jakarta respectively. Based on these population growth rate prediction, the trip growth for each district (kelurahan) in Jabodetabek area is derived and the trip growth for each district (kelurahan) in Jabodetabek area ranges from 0.77% to 1.99% for each local city in Jabodetabek with avarege of 1.38% in 2020 and 0.95% in 2030 [17].

The current public transport share adopted for this research is taken from the home interview survey conducted in the previous work [11], which is 24% from the total motorised person trip in Jabodetabek area. While, all each public transport mode parameter, the proposed BST implementation staging and tariff rate are adapted from the previous work [1]. In supply side, as in RITJ [1], the proposed BST is divided into two different services namely express and reguler BST. Since the reguler BST service route has not been defined yet, this research will only consider the express BST routes for the analysis. The supply-demand scenario developed in this research is summarized in Table 1.

Table 1. Supply and demand scenario

Scenario	Description
SD-22	Consider transport network (S) plan and potential demand (D) for the year of 2020. Fare integration applies for the **same mode technology only**, except for the existing bus system[a], while for physical integration[b] applies for all type of public transport mode, excepts for the existing bus system.
SD-23	Similar with the **SD-22**, but consider potential demand (D) for the year of 2030
SD-33	Similar with **SD-22**, but consider transport network (S) plan and potential demand (D) for the year of 2030

4 Simulation and analysis

Referring to model simulation on Jabodetabek road network, the predicted network performance is shown in Table 2. The road network performance shown in Table 2 consists of the average link speed and the average travel speed, where the former is purely based on each link speed either towards to the city center or vice versa, and the later is calculated by taking into account the traffic flow at the respected link (i.e. weighing the speed with the traffic flow). It is important to explain this a bit further because the link speed does not reflect the congestion perceived by the user in the real situation, but the average travel speed. Therefore link speed is usually higher than the travel speed as shown in Table 2.

Table 2. Highway network performance (km/hour)

	Averg. Link Speed		Averg. Travel Speed	
Region	2020	2030	2020	2030
Jakarta	23.4	19.3	13.6	10.6
Tangerang City	20.0	14.4	12.7	9.4
Tangerang District	19.4	15.7	11.2	8.8
South Tangerang City	26.1	21.2	15.4	12.6
Depok City	20.8	16.3	11.5	8.5
Bekasi City	20.8	18.7	11.4	10.5
Bekasi District	22.6	21.2	13.4	12.0
Bogor City	27.7	26.5	12.5	13.8
Bogor District	27.1	24.4	15.7	13.5
Average	23.3	19.8	13.5	10.9

By using the predicted road link speed as a basis for mix-traffic roadbase public transport design speed, a simulation to the proposed Mass Transit System in RITJ (i.e. Rail and Roadbase Mass Transit) is then conducted. From the simulation, the potential demand for the proposed Express Jabodetabek BST represented in peak hour total boarding passenger is shown in Table 3.

Besides the total boarding passenger, the simulation result also comes up with passenger profile demand (i.e. sectional passenger flow/volume) along the service line. Based on this profile demand, the maximum passenger flow (peak hour passenger per direction/phpd) on each service line are determined. Ideally the design capacity for a designated line is based on the maximum flow, hence it will guarrante that all passenger can be accommodated. Yet, by taking the maximum flow as the design capacity, there will be a situation where most of section in the line will experience quite low load factor. This is considered not good from the operational cost point of view, because financially it is not efficient. In order to avoid and compromise this undesirable situation, it is advisable to set an envelope as the design capacity which will form a lower and upper boundary. Therefore an average flow is calculated and set as the lower boundary of the design capacity, while the maximum flow is set as the upper boundary. Having calculated the average flow for each service line, the capacity need to be provided for the proposed BST is shown in Table 4.

Table 3. Jabodetabek Express BST potential demand (pax/hr)

Line_ID	Service Line O-D	Supply-Demand Scenario			Change (%)	
		SD-22	SD-23	SD-33	□SD(23-22)	□SD(33-23)
BPX1	Bekasi-Senen	32,114	33,999	14,074	6	-59
BPX2	Depok-BlokM	13,692	14,540	12,389	6	-15
BPX3	Bogor-Senen	14,512	15,184	14,493	5	-5
BPX4	Poris-BlokM	16,941	18,401	7,707	9	-58
BPX5	Cikarang-BlokM	47,811	51,420	19,533	8	-62
BPX6	Cibinong_Grogol	15,476	16,553	6,094	7	-63
BPX7	TnAbang-Bekasi	25,700	27,407	13,499	7	-51
BPX8	Depok-Senen	10,561	10,903	11,281	3	3
BPX9	Depok-TNAbang	10,771	11,260	10,730	5	-5
BPX10	Poris-Senen	4,004	4,168	1,570	4	-62
BPX11	Poris-TnAbang	3,137	3,301	1,187	5	-64
BPX12	PdCabe-Senen	11,985	12,926	8,930	8	-31
BPX13	PdCabe-TnAbang	11,385	12,327	9,049	8	-27
BPX14	PsAnyar-Senen	4,206	4,394	1,624	4	-63
BPX15	BSD City-Senen	6,633	7,077	3,110	7	-56
BPX16	BSD-TnAbang	6,138	6,575	3,592	7	-45
BPX17	Ciledug-TnAbang	7,640	8,101	3,341	6	-59
BPX18	Depok-Manggarai	6,489	6,764	7,072	4	5
BPX19	HrIndah-Cikarang	4,510	4,879	1,922	8	-61
BPX20	Kalideres-Poris Ext.	462	509	59	10	-88

Referring to Table 3, in 2020 (i.e. scenario SD-22), BST potential demand has a very wide range from 462 passenger/hour to as many as 47,000 passenger/hour. When all assumptions in scenario SD-22 hold, , there are three BST service lines, namely BPX-1, BPX-5 ans BPX-7, have a very high (more than 25,000 passenger/hour) demand and one service line, BPX-20, has very low demand.

When the 2020 network is assigned with 2030 demand, indicated with scenario SD-23, demand on all BST lines are slightly increase by 3% to 10%. Yet, in 2030, this situation is totally different when new services like Jabodetabek Regular BST network, East-West MRT corridor and LRT network [1] are operated. Simulation results yield that demand on 65% of BST service lines are significantly decreased which count for more than 40% compared with those of scenario SD-22. Having observed all BST service lines highly affected by these new services (i.e. service lines marked with bold letter and shading) in Table 3 and taking a closer look at their ittenerary in Figure A-1 as in the Appendix, seems that these affected lines mostly (11 of 13 lines) serve East-West direction trip. Therefore, to identify which new service gives most impact to the BST potential demand, a further analysis on demand shifting pattern is carried out by comparing number of passenger of each mass transit line for different supply-demand scenario as in Table 1. From the simulation result, the demand shifting pattern is depicted in Figure 5 with passenger flow map.

Table 4. Transjabodetabek Express BST estimated capacity

		Passenger/hour			
		2020		2030	
ID_KOR	Corridor	Min	Max	Min	Max
BPX1	Bekasi-Senen	9,788	12,344	2,884	3,773
BPX2	Depok-BlokM	1,917	6,402	1,780	5,574
BPX3	Bogor-Senen	2,928	3,624	3,035	3,868
BPX4	Poris-BlokM	5,108	7,560	1,932	2,684
BPX5	Cikarang-BlokM	15,678	19,276	5,488	7,673
BPX6	Cibinong_Grogol	2,236	6,573	1,151	1,854
BPX7	TnAbang-Bekasi	5,738	9,093	2,600	4,179
BPX8	Depok-Senen	1,046	3,503	1,467	3,407
BPX9	Depok-TNAbang	1,273	4,024	1,209	3,488
BPX10	Poris-Senen	533	918	114	564
BPX11	Poris-TnAbang	454	864	90	350
BPX12	PdCabe-Senen	2,336	4,606	2,187	4,290
BPX13	PdCabe-TnAbang	2,544	4,606	2,443	4,290
BPX14	PsAnyar-Senen	472	931	112	564
BPX15	BSDCity-Senen	536	2,618	359	1,076
BPX16	BSD-TnAbang	476	2,618	369	1,076
BPX17	Ciledug-TnAbang	776	1,770	1,319	1,867
BPX18	Depok-Manggarai	776	1,770	1,319	1,867
BPX19	HrIndah-Cikarang	743	2,219	158	629
BPX20	Kalideras-Poris Ext.	65	251	18	36

This shifting pattern is distinguished with two different colour, where lines experiencing reduced demand are marked with red colour, and lines having additional demand are marked with green colour. The line thickness indicates the amount of shifted passenger to or from other service line. Referring to Figure 5 and Figure A-1 to Figure A-3 in the Appendix, it is shown that all new mass transit services take some of the BST line's demand and the E-W MRT corridor consumes the biggest share of this shifted demand as indicated by thickness of the flow map. So, this situation could explain why all BST lines which serves East-West movement direction are the most suffered services in term of passenger or demand reduction. As a consequence, similar phenomenon also occurs in the estimated capacity for several lines marked with bold letter and shading indicated in Table 4. In order to make sure the consistency of BST lines having suffered from shifted demand resulted in the total boarding and the capacity estimation, a simple comparison between marked lines in Table 3 and those of in Table 4 is made. This comparison shows a conformity as shown in Table A-1 in the Appendix, where all nine BST lines that serve East-West direction movement are marked as the most suffered lines due to the operation of E-W MRT corridors.

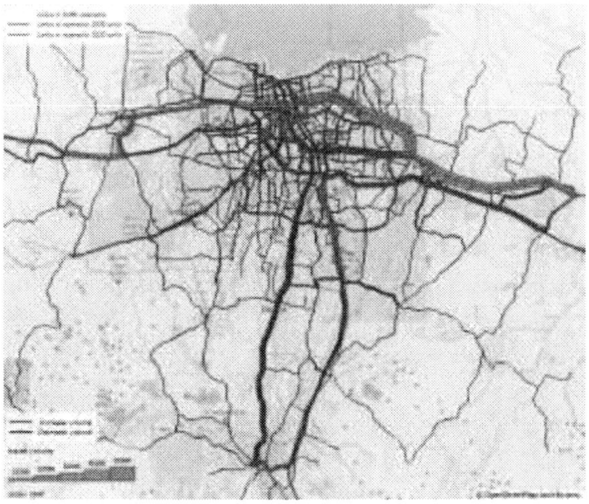

Fig. 5. Shifted demand between mass transit

5 Conclusion

A demand forecast analysis on the BST network of Jabodetabek Transport Master Plan (RITJ) is conducted and yields to the estimated ridership and required capacity for each service line proposed in the plan. Since most of mass transit proposed in RITJ is heavily on the roadbase system, as a consequent, highway average network speed is predicted and then this predicted speed is used as the roadbase mass transit design speed at the respected link. The analysis shows that in 2020, the BST lines have a very wide range of potential demand which range from 462 passenger/hour to as many as 47,000 passenger/hour. It is also indentified that three service lines have a very high demand, and one line has very low demand. Further analysis in 2030 as in scenario SD-33, comes up with totally different BST demand figures where almost all lines which serve East-West direction experience a very substantial demand drop which account for more than 40% than those of in scenario SD-23. These situation also occurs for the estimated capacity of these affected BST service lines. An additional analysis is then conducted to identify which line service causes this shifted demand, and from the simulation result, appears that E-W MRT corridor absorps the most E-W BST service lines's potential demand.

References

1. IUTRI, *Capacity And Demand Analysis Of Jabodetabek Mass Transit Network*, Final Report, JICA(2016).

2. Vuchic V. R. *Bus Semirapid Transit Mode Development and Evaluation*, Journal of Public Transportation, Volume **5**, Nomor 2 (2002).
3. TRB, *Transit capacity and Quality of Service Manual (TCQSM)*, TCRP Report 100, Transportation Research Board, Washington D.C (2003).
4. ITDP, *Bus Rapid Transit Planning Guide*, Institute for Transport Development & Policy, New York (2007).
5. Deen, T.B. and Pratt, R.H. *Evaluating Rapid Transit, in Public Transportation* edt. Gray, G.G and Hoel, L.A., Prentice Hall, Inc., Englewood Cliffs New Jersey, USA, pp. 293 – 331 (1992).
6. Vuchic V. R. *Urban Transit; Operations, Planning, and Economics*, Wiley & Sons Inc., Hoboken, New Jersey, USA (2005).
7. Wright, L. *Bus Rapid Transit, Deutsche Gesellschaft für Technische Zusammenarbeit* (GTZ) GmbH (2004).
8. Alvinsyah dan Hadian E. *Analisis Dampak Aktifitas Kawasan Reklamasi Pantura terhadap Kinerja Jaringan Jalan di DKI Jakarta*, Jurnal MTI Vol. **1** No. 1, hal. 43-58 (2016).
9. Kemenhub, *Survey Wawancara Rumah Tangga di Kota Bekasi, Depok dan Bogor*, Laporan Akhir, Badan Penelitian dan Pengembangan Kementerian Perhubungan (2014).
10. JICA, *Study on an Integrated Transport Master Plan for Jabodetabek Area*, Final Report (2004).
11. JICA, *Study On Jabodetabek Urban Transportation Implementation Plan (Review On Sitramp-2004)*, 2nd Draft Final Report (2011).
12. JICA, *A Study on Jabodetabek Public Transport Implementation Strategy (JapTraPIS)*, Final Report (2012).
13. BSTP, *Studi Penyusunan Master Plan Pola Transportasi Makro (PTM) Jabodetabek Tahap I*, Direktorat Jendral Perhubungan Darat, Final Report (2009).
14. BSTP, *Studi Pedoman Perencanaan Trayek Pengumpan (Feeder) untuk Angkutan Massal Berbasis Jalan*, Direktorat Jendral Perhubungan Darat, Final Report (2010).
15. CTS, *Pola Transportasi Makro Jakarta*, Final Report, Dishub DKI - CTS UI (2004).
16. KOICA, *Master Plan for Jabodetabek Railway, PMC Service for Master Plan and Feasibility Study for Jabodetabek Railway in Indonesia*, Final Report (2012).
17. Hadian and Alvinsyah, *Impact On Ridership Of New Railbase Transit Due To The Operation Of Extensive Bus Semi Transit Network (Case Study: Greater Jakarta (Jabodetabek) Public Transport Network)*, Proceeding of 15th International Conference on Quality in Research (2017).
18. INRO, *User Manual, EMME Suite*, Montreal (2015).
19. Spiess H., and Florian, M., , *Optimal Strategies: A New Assignment Model for Transit Networks, Transportation Research Part B: Methodological*, Volume **23B** No. 2, pp 83-102 (1989).
20. Kemenhub, *Survey Wawancara Rumah Tangga di Kota Tangerang dan Tangerang Selatan*, Laporan Akhir, Badan Penelitian dan Pengembangan Kementerian Perhubungan (2013).

Appendix

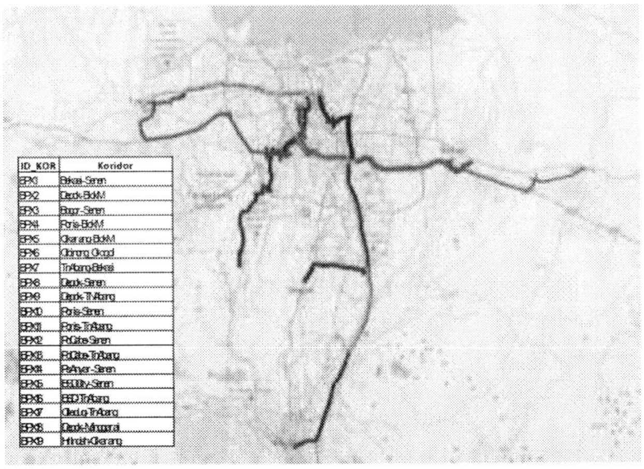

Fig. A-1. HCB network in 2020

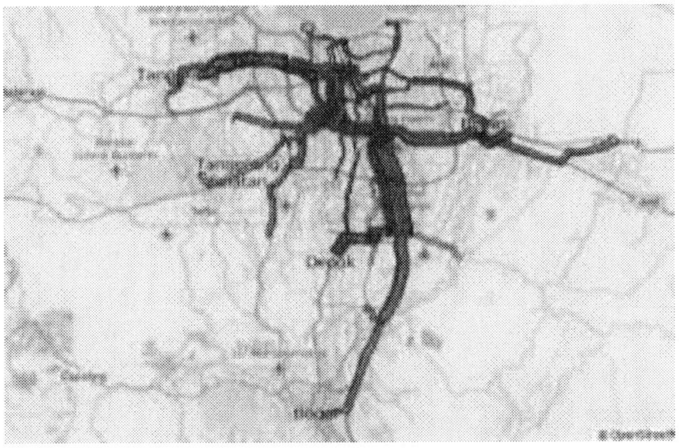

Fig. A-2. HCB network in 2030

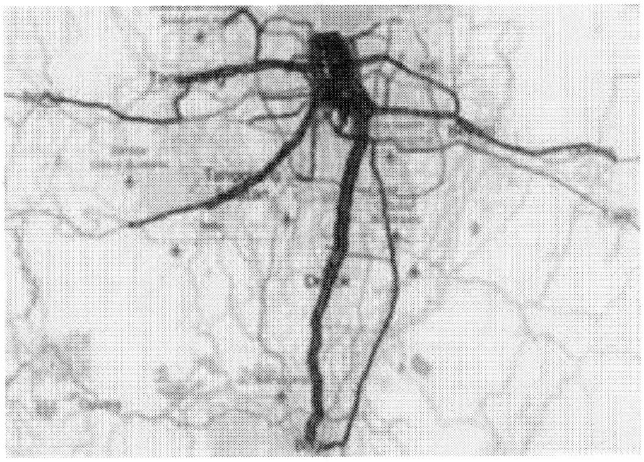

Fig. A-3. Rail network in 2030

Table A-1. List of affected Jabodetabek BST line

Line_ID	Total Boarding	Capacity Provision
BPX1	Bekasi-Senen	Bekasi-Senen
BPX4	Poris-BlokM	Poris-BlokM
BPX5	Cikarang-BlokM	Cikarang-BlokM
BPX6	Cibinong_Grogol	Cibinong_Grogol
BPX7	TnAbang-Bekasi	TnAbang-Bekasi
BPX10	Poris-Senen	Poris-Senen
BPX11	Poris-TnAbang	Poris-TnAbang
BPX14	PsAnyar-Senen	PsAnyar-Senen
BPX16	BSD-TnAbang	BSD-TnAbang
BPX17	Ciledug-TnAbang	Ciledug-TnAbang
BPX19	HrIndah-Cikarang	HrIndah-Cikarang
BPX20	Kalideres-Poris Ext.	Kalideres-Poris Ext.

Customer perspective of Trans Metro Pekanbaru service on corridor Pandau-Ramayana

Muchammad Zaenal Muttaqin[1,*], and *Abdul* Kudus Zaini[1]

[1]Islamic University of Riau, Civil Engineering Department, Pekanbaru, Indonesia

Abstract. Characteristic of public transport service in a city could be different one and another because of passenger itself. Passengers with their experience explained more about service from bus. This research has aim to know service performance from Trans Metro Pekanbaru (TMP) based from customer perspective. Factor Analysis and Gap analysis used to know dominating parameter that represent it. The result showed students and employees (77.96 %) dominated for passenger group using Trans Metro Pekanbaru. All of customer perception could divide by three group among other psychological factor, primary needs service, and transit place factor. Psychological factor became highest opportunity to represent positive value of Trans Metro performance with level of service reach until 97.23 % compared by its expectation. On the other side, transit place factor was only has 81.5 % performance, thus it became primary focus for TMP to improve their performance to better service quality in the future.

1 Introduction

Transport can be defined as means to move people or goods from one place to another place. In general, there are two groups in transport, they are private transport and public transport. Public transport was differ than private transport. Public transportation is advanced transport organization that operate accessible transport that for all citizen or apart for them based on its rules or given policy [1]. It has function to help others to reach their destination place and reduce all problem from transport sector. Nowadays, many cities in the world has been developed public transport as backbone of system transportation in their cities. One of them is Pekanbaru in Indonesia.

Pekanbaru, as a developing city in Indonesia and has its target become one of metropolitan city in Indonesia has a problem with development of public transport system. Public transit system by bus in Pekanbaru handled for all component, whether for main route, bus station and transit routes. Satisfaction in the public transit system can affect people to go by this transport modes with continuously. The problem occur that people still not want to use Trans Metro. Trans Metro user is not yet reached their target. In fact, passengers of Trans Metro just only amount for about 4000 passengers per day, still far from the target 18000-20000 passengers per day. Effect from public transport problem in

* Corresponding author: muchzaenalmuttaqin@eng.uir.ac.id

Pekanbaru could be shown with congestion in the road. Congestion in Pekanbaru always happened, especially on working hours [2]. In details, road capacity in Pekanbaru increasing by year with Level of Service 0.75 – 1.00 which shown that already saturated and make congestion especially in the primary road in Pekanbaru [3]. Related with them, objective of this study contains discover how service characteristic from Trans Metro based from customer view. Furthermore, this study could be a foundation that what aspect that Trans Metro need to improve about its quality when serve customer.

1.1 Literature review

Service has different characteristic with product. Identification of services could be shown with its intangibility. Service cannot be shown because real product of service is an effort, moral, experience, process, performance or anything that people do to their target which is customer [4]. This thing make difficult to customer to evaluate service before they make decision before they use that service. Characteristic of the service may different one and another. It was related with how service can deliver its value to the customer. So that, knowing more about service characteristic took from customer as its object.

Public transport as apart of service has same characteristic with other service. It focused to meet customer needs with perceived value for customer as primary objective of the service. Identification of the services was the first evaluation how service can meet about people needs [4]. It would describe how service is created to serve customer in their activities.

Characteristic of the service was related with service marketing. Service marketing has concept based from what service offered and what customer perceived from them. Identification is a starting point before companies offer their service to customer. This process is needed to determine what right service that would be offered as same as customer needs. Identification is most effective method to know customer needs before services would be delivered for customer [5].

Customer has experience as the result of after services from the company. Experience would identify what the services has been perceived by customer. Customer experience has relationship between company and customer. Service characteristics dominated by customer experience [6]. It would affect evaluation process for the service. Service evaluation obtained during same time about production and consumption also after purchasing process. So that, customer experience would identify what characteristic of the service based from customer satisfaction about it.

Customer experience for the service showed by their perception about the service. Perception from customer determine how performance from its service works for customer. Performance could be one important aspect to evaluate about public transit operator, not just about profitability [7]. Performance itself make effect with satisfaction from customer and identify how service delivery went to customer that create service characteristic.

Service performance can be valued with customer perspective. Customer perspective played role for both company and customer by their experiences [8]. For company side, customer perspective built value and realize them to evaluate and increase progress for their quality. On the customer side, customer perspective would make perception for the products or services also expectation of the services. Comparison between both of them would affect customer loyalty with the company's products or services, see fig.1 below.

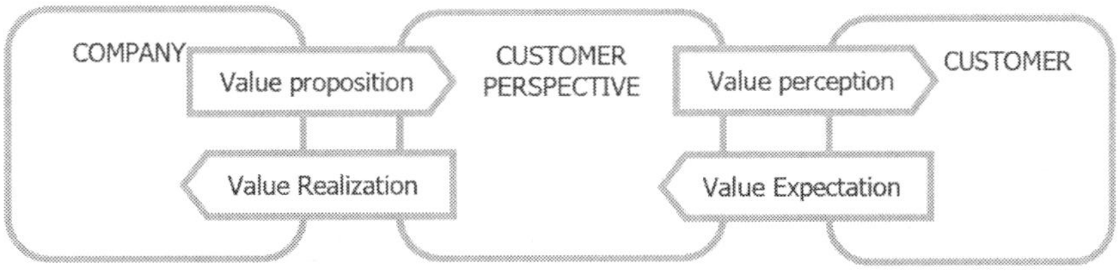

Fig. 1. Customer perspective on customer and company side

Customer perspective captured by perception from customer. Perception as one of component in experience play its role to make satisfaction and construct them to service quality. Perception was a cognitive process to know about information from its activity then give impression as unique thing from its information based on current condition [9]. Perception of service was obtained by comparing customer expectation with performance from the service. Performance of the service based on customer perspective would become a standard to know about how customer valued the service and what primary characteristic from the service that should be focused to make better quality service in the future.

2 Methodology

This study would be analyzed by quantitative approach. Quantitative approach emphasize to measure and analyze data with statistical, mathematical or numerical view and report them as generalization of the subject and explain what happened related with particular theme. Quantitative will give insight for conclusion that started from general view and end with specific as main purpose [10]. In details, this study used multivariate analysis with Exploratory Factorial Analysis and Servqual with gap analysis. Exploratory Factorial Analysis (EFA) used to collect and simplify data parameter that measured in this study. The result from this method was knowing about which parameter that dominated in TMP service especially on corridor Pandau-Ramayana. Finally, it would be know how characteristic of Trans Metro Pekanbaru (TMP) services based from customer perspective. On the other side, Servqual based from gap analysis would be used to know how fulfillment needs for customer expectation from TMP services.

Evaluate service based from passenger view. In the service process, there are two main component which consist of expected service and perceived service. In the service quality (Servqual) analysis, comparison between expected service and perceived service has known as "gap" analysis. Perceived service is what real services that given by company to the customer and expected service is what customer wants based on the service. Both of them would be evaluated and make perceived service quality which shown how service performance can fulfil customer wants [5]. Formula for gap analysis has given below,

$$SQ = P - I \tag{1}$$

where,
SQ = Service Quality,
P = Perception,
I = Importance.

This study collected data from questionnaire form that filled by passengers of TMP especially in corridor Pandau-Ramayana. Data has been collected by random sampling method on April and May 2017. Surveyor took questionnaire data from respondents who exist in the transit place when was waiting the bus and in the bus when the bus was running. Questionnaire divided by three section. First section was about data identification from passenger, second section was about passenger perception actually about TMP service and last section about what passenger expectation for TMP service.

3 Discussion

3.1 Profile of TMP corridor Pandau-Ramayana

Trans Metro Pekanbaru was a public transport in Pekanbaru which adopted from mass transportation systems in the other cities before in Indonesia, like in Jakarta with Trans Jakarta and Yogyakarta with Trans Jogja. Trans Metro Pekanbaru has operated first in 2009 based from Transportation Ministry's policy and operated just only two routes [4]. In 2012, Trans Metro Pekanbaru develop their routes become 7 routes and going to develop more based on demand and accessibility in Pekanbaru city. Corridor Pandau-Ramayana became first route that operated by Trans Metro Pekanbaru (TMP). This corridor started from Pandau Housing (border area between Pekanbaru and Kampar) and Ramayana (Business Central District of Pekanbaru), see figure 2 below,

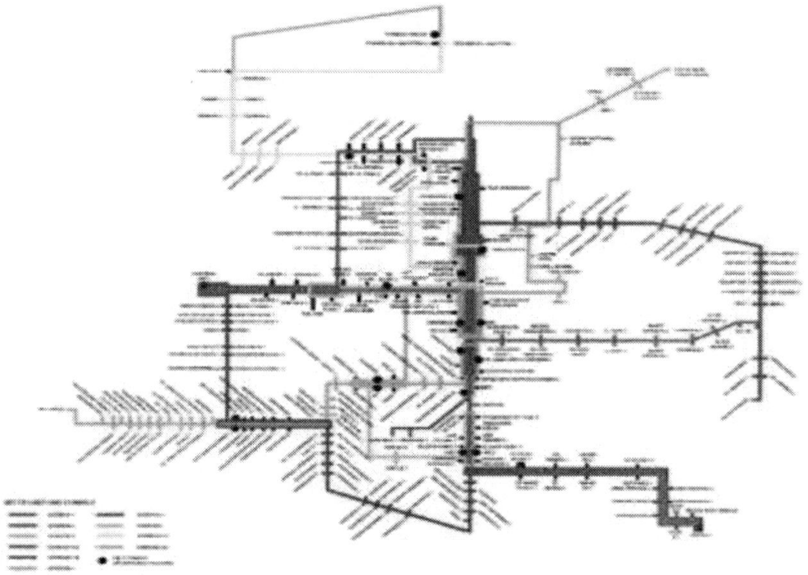

Fig. 2. Trans Metro Pekanbaru routes

This corridor is one of primary routes in TMP routes, because this passed national road in Pekanbaru and passed many office, public place, education place until government office. Thus, the corridor attracted passenger by its location and in fact, this corridor is the largest amount of passengers in TMP than other routes [11]. The bus for corridor Pandau-Ramayana has capacity 83 people and scheduled from 06.00 in the morning until 21.00 in the night (15 hours) everyday. Trans Metro Pekanbaru operated by Public transit consortium company and controlled by Transportation Ministry of Pekanbaru city [12].

4 Data and analysis

4.1 General description

This study used data from questionnaire with 249 sample from 300 sample has been distributed. Based from sample, domination of students and employee has been shown as passengers of TMP corridor Pandau-Ramayana. 40.56% passengers had profession as students and 37.4 % passengers was employees. Another passengers was entrepreneur (11.2 %) and others, like unemployment etc (14.5%). Other description showed that distance for transit place was about 100 – 500 meters. Many passengers used Trans Metro Pekanbaru where it had distance about 100 – 500 meter (51 %) from their destination. Another result, 37.3 % passengers had distance from transit place of Trans Metro less than 100 meters. This result means that passenger had tendency to use Trans Metro where accessed about 500 meters, On the other word, attraction from passenger to use Trans Metro valued high when it accessed 100 – 500 meters.

Motives from passengers analyzed to know stimulant factor from passengers before using Trans Metro Pekanbaru. Figure 3 showed that there was no domination factor that attract customer to use Trans Metro Pekanbaru. Pricing factor become highest motives why passenger want to use Trans Metro (13 %). Nowadays, passengers must pay 4000 Indonesian Rupiah for each trip using Trans Metro or 1000 Indonesian Rupiah if passenger is student below Junior High School. Another motives showed about safer, environmental friendly, reliable and safer was second highest motives about reason from passengers why using Trans Metro Pekanbaru (12%). It means about psychological factor attracted passenger to use Trans Metro as their modes. As one of modes choice factor, psychological and past behavior can highly affected for decision making process in the future [13].

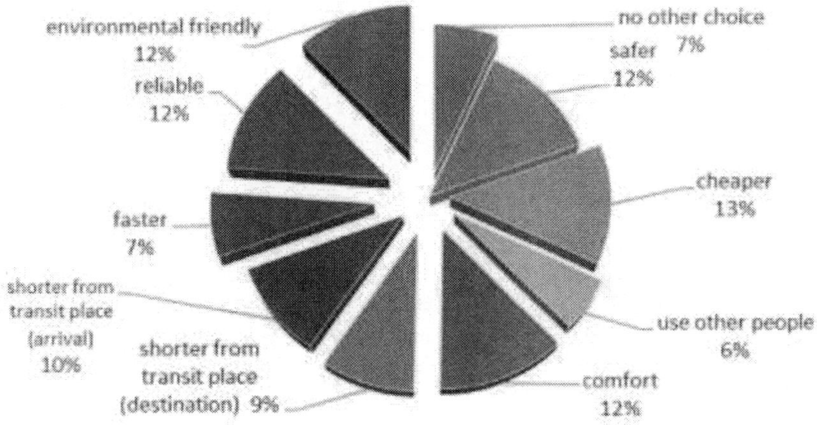

Fig 3. Motives from passenger to use Trans Metro Pekanbaru

As explained above, factorial analysis used to make advance exploration about quality service of Trans Metro Pekanbaru. In this case, parameter of perception would be grouped and defined with identical relationship between variables. Service performance captured by customer's impression when used Trans Metro Pekanbaru. Dominating parameter would be known as highest tendency how service worked for passenger. The result showed that parameter of perception could divide by three groups based on factorial analysis. There are psychological factor, primary needs services, and transit place factor (see appendix a). Psychological factor has included responsiveness, and assurance of the service.

Responsiveness and assurance were apart from dimension of service quality [14]. Responsiveness on this group contained faster services, employees are nice and friendly, employees were always ready to serve customer and good caring for passenger. Assurance on this group contained cleanliness in the bus, clear information from employee, good looking from employee. Finally, all of parameter in this group make positive value and improve psychological value for passenger when using Trans Metro Pekanbaru. Second group was primary needs service. This group contain more about what primary service that Trans Metro did to their passengers. Parameter that included in this group were about safety in the bus, sitting place facility, easy to get ticket, accessibility, punctuality etc (see appendix a). Passenger had high expectation that this parameter should be served first to them first. On the standard rules, like SK Dirjen no.687 tahun 2002, this parameter was primary parameter to evaluate public transport, especially for bus.Last group in this case was transit place factor. This group contained more about transit place condition that affect passenger to use Trans Metro Pekanbaru, like waiting time, safety, comfortability and cleanliness in the transit place.

Based from gap analysis, all parameters had negative value on gap value. It means Trans Metro Pekanbaru still unable to make positive satisfaction to passengers. Thus, it need more improvement from Trans Metro Pekanbaru about service quality. The higher service quality would make higher satisfaction from customer about the services, vice versa [5]. In this case, Trans Metro Pekanbaru need to see advance what proper improvement that can improve service quality to make positive value in gap analysis (performance is more than expectation).

Table 1. Customer perspective of Trans Metro Pekanbaru

Factors	Parameter	Component			
		Performance	Importance	Fulfillment level	Gap
Psychological factor	Faster and responsive services	7.87	7.92	99.39%	-0.05
	Good looking from employees	8.03	8.13	98.76%	-0.10
	Comfortability	8.14	8.29	98.21%	-0.15
	Employees are ready to serve customer everytime	7.98	8.14	98.03%	-0.16
	Employees are nice, friendly and polite	8.05	8.28	97.23%	-0.23
	Cleanliness In the bus	8.13	8.43	96.52%	-0.29
	Good caring for passenger	7.76	8.04	96.46%	-0.29
	Clear information from employees	8.00	8.31	96.23%	-0.31
	Driver driving attitude	7.95	8.27	96.07%	-0.33
	The bus already modern bus	7.87	8.24	95.51%	-0.37
Primary needs service	Safety in the bus	8.00	8.20	97.55%	-0.20
	Sitting place facility	7.87	8.10	97.17%	-0.23
	Easy to get ticket	7.82	8.16	95.86%	-0.34
	Accessibility	7.46	7.80	95.72%	-0.33
	Facility in the bus	7.78	8.14	95.56%	-0.36
	Operating time	7.41	7.88	93.99%	-0.47
	Clearly information in the transit place	7.27	7.73	93.98%	-0.47
	Punctuality	7.19	7.74	92.89%	-0.55
Transit place factor	Waiting time	6.47	7.57	85.37%	-1.11
	Safety in the transit place	6.21	7.41	83.75%	-1.20
	Comfortability in the transit place	5.93	7.47	79.31%	-1.55
	Cleanliness in the transit place	5.88	7.58	77.58%	-1.70

The number of group represent about how performance of Trans Metro compared with its expectation from passengers. The result showed that first group (psychological factor) has smallest gap, which means passengers valued that performance of Trans Metro Pekanbaru was in nearest value with their importance (see table 1 below). First group has level of service about 97.23 % performance compared with its importance. Trans Metro Pekanbaru should be improved more in this parameter group and promote it to other society. It would attract more people to use Trans Metro Pekanbaru because of its performance value that nearest from its importance. Another result showed that Transit place factor has smallest level of service and highest negative value for gap value. It means that passengers valued that there were more problem exists for transit place factor on Trans Metro Pekanbaru services. Thus, performance of Trans Metro Pekanbaru still lower than passenger's importance. Smallest level of service was on cleanliness in the transit place with score about 77.58 %. Passengers valued there was not clean in the transit place so that make them disappointed for the service. Trans Metro Pekanbaru had more responsibility to solve the problem for better service quality especially in the transit place.

Table 2. Performance and importance of Trans Metro Pekanbaru

Factors	**Component**			
	Performance	**Importance**	**Level of Service**	**Gap**
Psychological factor	7.98	8.20	97.23%	-0.23
Primary needs service	7.60	7.97	95.37%	-0.37
Transit place factor	6.12	7.51	81.50%	-1.39

4 Conclusions

Capturing about customer perspective on public transport services was important to see how public transport can satisfy customer and affect its service performance. Performance of the service based on customer perspective would become a standard to know about how customer valued the service and what primary characteristic from the service that should be focused to make better quality service in the future. This study used service perception from passengers as a key to captured service performance of Trans Metro Pekanbaru. In this case, students and employees dominated on passenger type of Trans Metro Pekanbaru. Parameter for perception of Trans Metro Pekanbaru divided by three groups, that is psychological factor, primary needs service and transit place factor. Trans Metro Pekanbaru should be improve their performance from negative value for all parameter on gap analysis. Based from each group, psychological factor is highly dominated on this service and became highest positive value for passenger because of its smallest negative value on gap analysis (-0.23). Improving value for this parameter could be attracted other customer to use Trans Metro and pride for the service. On the other side, transit place factor became highest negative value (-1.39). It was urgent to improve this parameter, like comfortability in the transit place (-1.5) and cleanliness in the transit place (-1.7). Customer feels that transit place factor might be a reason to will not used Trans Metro again in the future. Finally, customer perspective was a key how service performance of public transport service and could be a comparison to evaluate its performance in addition using exist rules, like company standard, or another rules that applicable.

References

1. Aarts, H., Verplanken, B., & Knippenberg, A. *Predicting behaviour from actions in the past: Repeated decision making or a matter of habit?.* Journal of Applied Social Psychology, 28(15), 1355-1374. (1998).
2. Dziekan, K. *Ease-of-use in public transportation – A user perspective on information and orientation aspects. Unpublished doctoral thesis in traffic and transport planning,* Infrastructure and Planning Royal Institute of Technology Stockholm, Sweden (2008).

3. Panjaitan, A. M. *Kajian Sistem Jaringan Jalan di Wilayah Kota Pekanbaru*. Jurnal Teknik Sipil USU, 2(1) (2013).
4. Saputra, Parada Afkiki Eko. *Evaluasi Kinerja Simpang Tiga Bersinyal Jalan Sudirman-Jalan Tuanku Tambusai Pekanbaru.S1 Thesis*. Atma Jaya University, Yogyakarta. (2011).
5. Muttaqin, M. Z *Promoting bus as alternative transport modes in the city based from customer view (case study : Trans Metro Pekanbaru for engineering students in islamic university of Riau, Indonesia)*. Master Thesis. Universitas Gadjah Mada. Yogyakarta . (2016).
6. Zeithaml, V. A., Parasuraman, A., & Berry, L. L. *Delivering quality service: Balancing customer perceptions and expectations. Simon and Schuster*. (1990).
7. Berry, L. L., & Parasuraman, A. Marketing service. *Competing through Quality*, New York. (1991).
8. Joewono, T. B., & Kubota, H *User satisfaction with paratransit in competition with motorization in indonesia: anticipation of future implications.Transportation*, 34(3), 337-354. . (2007).
9. Gentile, C., Spiller, N., & Noci, G. *How to sustain the customer experience:: An overview of experience components that co-create value with the customer*. European Management Journal, 25(5), 395-410. (2007).
10. Gentile, C., Spiller, N., & Noci, G. *How to sustain the customer experience:: An overview of experience components that co-create value with the customer*. European Management Journal, 25(5), 395-410. (2007).
11. Sultana, S., & Rana, S. *Service Quality:(Service Gap Analysis) A case study-" Komvux"*.Master Thesis in Business Administration.Master International Management of Gotland University. (2010).
12. Department of Transportation. *Data Jumlah penumpang Trans Metro Pekanbaru Tahun 2015. Bidang Angkutan Dinas Perhubungan Kota Pekanbaru*. (2015).
13. Hafis, A., Imam, R., Hakim, A., & Santoso Haryono, B. *Aktor Pelaksana Pengelolaan Transportasi Publik Perkotaan (Studi Kasus Bus Trans Metro Di Kota Pekanbaru). WACANA,* Jurnal Sosial dan Humaniora, 16(4), 171-178. (2014).
14. Aarts, H., Verplanken, B., & Knippenberg, A. *Predicting behaviour from actions in the past: Repeated decision making or a matter of habit?*. Journal of Applied Social Psychology, 28(15), 1355-1374. (1998).
15. Parasuraman, A, Zeithaml, V. A. and Berry, L. L. SERVQUAL: *A multiple-item scale for measuring customer perceptions of service quality. Journal of Retailing*. Vol. 64, No. 1. (1998).

Measurement modelling of the perceived service quality of intercity bus terminal

Ashari Abdullah[1,*], *Ananto* Yudono[1], *Sakti* Adji Adisasmita[2], and *Arifuddin* Akil[1]

[1]Hasanuddin University, Architecture Engineering Department, Gowa, Indonesia
[2]Hasanuddin University, Civil Engineering Department, Gowa, Indonesia

Abstract. There is a tendency on decreasing effectiveness on the use of intercity bus as result of relocation, this is indicated from passengers' reluctance for transit in the terminal. This research aims to develop monitoring and service evaluation indicators in intercity bus terminal. This research is to test quality of service factors out of 19 indicators. This research is conducted in Makassar Metro Terminal. The data is collected through survey to 350 passengers. Results of Exploratory factor analysis (EFA) classify the indicators into 5 factors, namely: Terminal Performance, Connectivity, Ease of Transfer, Operator Performance, Reliability service. Confirmatory Factor Analysis (CFA) is used to confirm the formed factors. Results of the CFA show that the Connectivity and Ease of Transfer have quite high factor loading each of which is by 0,93 and 0,9 compared to 3 other factors so that two factors are the priority in improving terminal service performance.

1 Introduction

Bus Terminal is one of the components of intercity transportation system. In its development, it is necessary for policies on development of terminal, both new terminal and relocation terminal so that it can respond on the city growth. The relocation is conducted to solve any traffic congestion in former location as well as to increase terminal loading capacity [1]. However, the built terminal has not many activities, this is caused by passengers' reluctance for transit activity. This quiet terminal affects on the creation of illegal terminal having operation within the city either in arterial roads or private bus pool. It leads to some impacts, one of which is decreasing terminal performance related to the less of connectivity, so that the passengers do not use the terminal / halte [2]. There is a tendency on deregulation of transportation rules on intercity bus in some countries such as German, UK, USA, and Sweden not using terminal as transit facility. This bus transports or drops off passengers in predetermined curbside and can access city road [3].

This research aims to identify any factors affecting on quality of intercity bus terminal services. There is a great limitation on researches concerning intercity terminal so that in preparing these research indicators, there are literature studies having no cover on bus terminal [4], it is not only caused by very limited sources, but also by transfer facilities such

* Corresponding author: ashari.arsitek@gmail.com

as interchange [5-11]. Although in the literature study, there are broad reviews on its scope not only on terminal, but in each prepared category indicators, it has included terminal/ transfer facility service quality indicator. From the literature studies, there is only studying on intercity terminal with Regression Model analysis method. From results of literature studies, there is none studying on quality of intercity bus terminal service [4].

2 Desk study

Out of some literature studies [5], studied on key factor an efficient transport interchange using 37 indicators divided into 8 categories. From the research results, the transfer information and condition are the main factor to improve transport interchange performance such as travel information, accessibility, coordination between transportation operators. [6] studied on intermodal transfer using 15 indicators divided into 7 categories. Coach Terminal using 22 indicators divided into 7 categories [4]. Interurban Interchange using 28 indicators divided into 8 categories [8]. LRT using 37 indicators of service quality divided into 8 categories [9]. Railway using 26 indicators using 7 factors [10]. Station using 16 indicators divided into 5 categories [12]. Service quality for railways passenger using 20 indicators divided into 5 categories [11]. Interchange using 40 indicators divided into 6 categories [7]. Intercity bus using 22 indicators divided into 4 categories [13]. Intercity of bus used 9 indicators [14].

Table 1. literature review in transportation

Author(s)/Year/Type Transfer Facility/Country, Analysis Method	Category Of Service Quality
Hernandez & Monzon (2016), Interchange, Finland-Spain- UK, PCA	Information, Transfer Condition, Safety & Security, Emegency Situation, Desain & Image, Enviroment Quality, Service & Facilities, Comfort Of Waiting Time
Silva and Bazrafshan (2013), Intermodal Transfer, Portugal, EFA & SEM	Information, Convenience & Easy Of Use, Security, Facilities Quality, Global Comfort, Global Securyty, Global Satifaction
Yatskiv et al. (2009), Coach Terminal, Latvia, Regression Model	Accesibility, Information, Time, Costumer Service, Comfort, Relibility/saftey, Enviroment
Dell'Asin (2014), Interurban Interchange, Madrid, MCA-CA	Availibility, Accessibility, Information,Journey time, Costumer Care, Comfort, Safety, Enviromental Impact
De Oña, Et Al. (2016), LRT, Spain, EFA-SEM	Avalibility, Accesibiity, Information, Timelines, Attention To Client, Comfort, Safety, Enviromental Pollution
Eboli, et al. (2016), Railway, Italy, Multi Level Fuzzy	Saftey, Cleaniness, Comfort, Service, Information, Personnel, Other
Iseki and Taylor (2010), Station, US, CSI-IPA, OLR	Aminities, Information, Connection & Realibility, Acces, Security & Safety
Nathanail (2008), Railway, Hellenic, Multicriteria Analysis	Itenerary accuracy, System Safety, Cleaness, Passenger Comfort, Servicing, Pasengger Information
Hine & Scott (2000), Interchange, UK, qualitative Method	Waiting Environment, Information Provision, Service, Costumer Care, Ticketing, Queuing
Wen et.al (2005), Intercity bus, Taiwan, EFA-SEM	On-board amenity,crews' attitude, station performance, operational performance.
Bordagaray et al. (2013), Interurban buses/Spain, Ordered Probit Model	Waiting time, journey time, reliability, vehicle occupancy, driver kind- ness, comfort, price of the ticket, quality of the vehicle and available information.

From results of literature study in this research, it is prepared 19 indicators to be tested to prepare the service quality factor using EFA and CFA. Benefits of this research will provide description on evaluation on service quality of intercity bus terminal. This research is conducted in Makassar Metro Terminal, Southeast Sulawesi, Indonesia. The Makassar Metro Terminal is a relocated terminal in 2003, this terminal was formerly located in Panaikang Terminal in 5 kilometers from the city center and it was located in suburb located in 15 kilometers. In its initial operation, this terminal is quite active providing services for intercity passengers, but along the time the passengers and operators tend to conduct the transit activity outside the terminal, either in Bus Pool, in informal terminal or directly door to door service. There are various policies taken to effective the terminal, but until now, there is no significant improvement.

In this research, this questionnaire is tested face validity, content validity. Assessment point of service quality uses Likert scale of 1-7 with "1 = very unsatisfied, 4= neutral, into 7 = very satisfied". The researcher received 350 questionnaires. The sample size for maximum likelihood (ML) estimation should be at least 15 times the number of observed variables [15, 16].

Tabel 2. Characteristic of respondent

Atribut	Frequency	Percent
Gender		
Male,	153	46,2%
Female	178	53,8%
Age		
17-25 years	230	69,5%
26-40 years	84	25,4%
41-65 years	16	4,8%
65 or More	1	0,3%
Travel Purpose		
Go home	172	52%
Private	83	25,1%
Study	17	5,1%
Bussiner	35	10,6%
Others	24	7,3%
Occupation		
Wiraswasta	41	12,4%
Government employee	30	9,1%
Student	179	54,1%
Private employee	46	13,9%
Others	35	10,6%
Education		
SD-SMA	105	31,7%
College Student	108	32,6%
D1-D3	29	8,8%
S1	70	21,1%
S2	14	4,2%
S3	3	0,9%
Others	2	0,6%

The first factors consisting 6 indicators namely environment quality, safety, lightning, facility, brokers and security that are categorized as Terminal Performance, this is in line with the factors prepared [13]. The second factors consist of four indicators namely support of public transportation, access from the Street, location, availability and clarity of information that are categorized as Connectivity. The third factors consist of four indicators namely porter, terminal staff service, Ease of carrying goods, Coordination transport Operators with administrator that are categorized as Easy Of Transfer. The fourth factors consist of three indicators namely Ease of purchasing tickets, availability of passenger bus/cars, and crew & driver services that are categorized as Operator Performance. The fifth factors consist of two indicators namely waiting time and punctuality that are categorized as Reliability Service. Out of the formed factors, it can be seen that the passengers are less satisfied on the terminal performance with the average value is ranging from 3,5378-3,9698 and are quite satisfied on the operator performance with the average value is ranging from 5,0816-5,4562 meanwhile related to the connectivity, easy of transfer and reliability passenger's service, these tend to be neutral with each average value is ranging from 4,3172-4,4924, 3,6224-4,3837, 4,2205-4,2749. The reliability test uses alpha Cronbach's with the value above 0,7 as acceptable [17]. The alpha Cronbach values are ranging from 0,735-0,848, so that it is considered acceptable the criteria.

2.1 Structure equation model

Tabel 3. Exploratory factor analysis

No	Factor (a , CVE) Indicator	Mean	Factor Loading	Communalities
	Terminal Performance (α = 0,805 CVE= 15,601%)			
P12	Environment Quality (Cleanliness ,Pollution , Temperature)	3,5378	0,763	0,678
P11	Safety	3,9335	0,743	0,644
P10	Lighting	3,7734	0,707	0,635
P13	Facilities (Toilet, Seat ,Shops)	3,7372	0,699	0,586
P16	Brokers of purchasing tickets	3,9698	0,547	0,444
P9	Security	3,8610	0,505	0,552
	Connectivity (α =0,735 CVE= 13,798%)			
P3	Support of Public Transportation	4,3172	0,787	0,678
P4	Access From the Street	4,4924	0,729	0,658
P2	Location Of terminal	4,4743	0,632	0,544
P1	Avaibility & Clarity Of Travel Information	4,4804	0,430	0,397
	Ease Of Transfer (α =0,801 CVE= 12,800%)			
P8	Porter	3,6224	0,819	0,755
P7	Terminal staff Service	4,2024	0,700	0,681
P6	Ease of carrying goods	4,1692	0,593	0,610
P5	Coordination transport Operators with administrator	4,3837	0,557	0,646
	Operator Performance (α = 0,802 CVE=12,286%)			
P15	Ease of purchasing tickets	5,4562	0,862	0,772
P14	Availability of passenger bus / cars	5,1299	0,794	0,744
P17	Crew and driver service	5,0816	0,651	0,614
	Realibility Service (α = 0,848 CVE=10,188%)			
P18	Waiting Time	4,2749	0,885	0,850
P19	Punctuality	4,2205	0,860	0,799

Based on results of EFA, it is formed 5 factors consisting of Terminal performance, Connectivity, Easy of transfer, Operator performance, and Reliability Service. Then, it is conducted the model measurement forming covariance between factors. The first order model results in standardized factor loadings ranging from 0,51-0,88 with the acceptable value is above 0,50 [18], any values below this value is recommended to be eliminated (Hair et al., 2010, Janssens et al., 2008). The model fit indices for the first order (X2= 475,882, X2/df=3,328, GFI=0,868; AGFI= 0,824; TLI= 0,845 NFI=0,826; CFI=0,870; PNFI=0,728: PGFI=0,653 and RMSEA = 0,084; SRMR=0,066) does not acceptable. Therefore, it is necessary to conduct the modification indices to improve the model fit indices. After applying the modification indices covariance then it is examined on the Standardized Residual covariance values exceeding 2,58 and its results of P9, P10, P16,P7 are eliminated from the model.

The model fit indices for results of first order modification are (X2= 174,824, X2/df=2,241, GFI=0,934; AGFI=0,899; TLI= 0,932; NFI=0,913; CFI=0,949; PNFI=0,679: PGFI=0,607; and RMSEA = 0,061; SRMR=0,047). The AGFI value is lower then recommended however, it is still considered as the marginal model [19].

The calculated validity constructs are convergent validity and discriminant validity. Source of reference value to calculate Convergent validity is standardized loading factor, the lowest value to be recommended is 0,5 or more, the ideal value is 0,7 or more, the lowest AVE value is 0,5 or more [20, 21]. Meanwhile to calculate the discriminant validity through AVE estimates for two factors also should be greater than the square of the correlation between the two factors to provide evidence of discriminant validity [22, 23]. Another method is used to test the discriminant validity if the average variance extracted (AVE) is greater than the maximum shared squared variance (MSV) or average shared squared variance (ASV). Discriminant validity was present when the correlation between two constructs was lower than the recommended value $r \leq 0.85$ [24]. Composite reliability should be 0.7 or higher to indicate adequate convergence or internal consistency CRE> 0.7 [25]. Composite reliability and AVE are calculated based on the formulation [23].

Table 4. Results of discriminant validity test for service quality factor

Factor	AVE	CRE	MSV	ASV	1	2	3	4
Terminal Performance	0,55	0,78	0,35	0,2				
Connectivity	0,45	0,76	0,72	0,37	0,52			
Easy Of Transfer	0,51	0,80	0,72	0,38	0,59	0,85		
Operator Performance	0,70	0,82	0,27	0,19	0,24	0,53	0,52	
Realibility Service	0,73	0,84	0,19	0,15	0,34	0,44	0,40	0,38

Based on results of construct validity as seen in table 4, it is obtained that there is a factor having low AVE value, namely connectivity by 0,45 however if referring that the AVE value is taken from the factor loading value, based on this that the factor loading that can be in the terminal is greater by 0,5 then the factors are still inputted in the interpretation. Meanwhile for the Discriminant validity, it can be seen that there are two factors with smaller AVE values than MSV namely Connectivity and Easy of transfer, if using the argument [24] stating that the highest factor correlation value is 0,85 then it is considered to acceptable the criteria. The lowest Composite reliability value is 0,78 and the highest one is 0,94. The values are greater than 0,7 then these factors acceptable the composite reliability. As conclusion, modification of model fit indices acceptable the criteria, also by the convergent validity and discriminant validity considered acceptable by recommended values. This modification model is then used for the second order model as seen in figure 1.

3 Study method

This research is to test the factors forming from 19 indicators of service quality of intercity bus terminal. Its initial indicators are tested using exploratory factor analysis, the results obtain 19 indicators that are classified into 5 categories namely terminal performance, connectivity, easy of transfer, operator terminal, reliability service. Results of EFA are then analyzed using SEM to confirm the forming factors. There are some researches almost similar to the forming factors in this research, namely [13] studying on intercity bus service, there are two similar factors, namely terminal performance and operator performance. The factors from a research [9] are accessibility that is similar to connectivity and timeliness that is similar to reliability service. The factor formed from a research [5] is transfer condition that is similar to easy of transfer. This word difference is because the transfer condition refers more to condition while the easy of transfer refers to activity purpose.

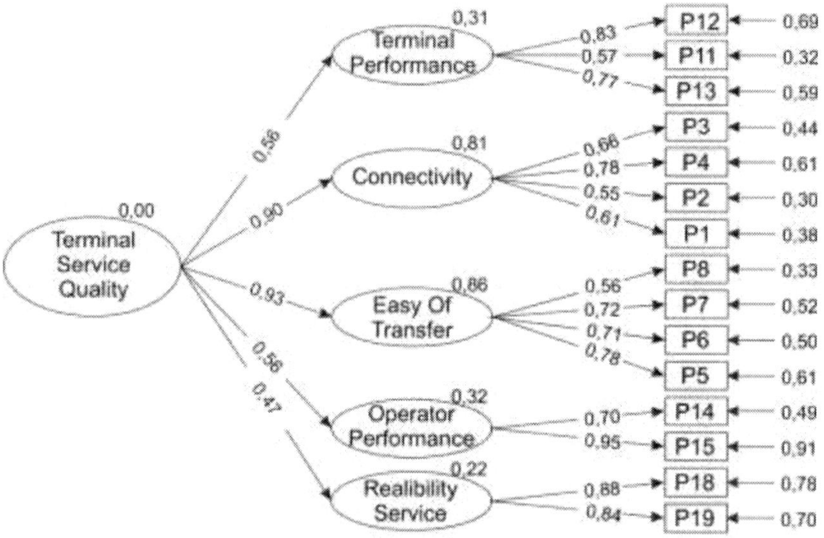

(X2= 189,367, X2/df = 2,309, GFI=0,929; AGFI=0,896; TLI= 0,928; NFI=0,906; CFI=0,944; PNFI=0,708: PGFI=0,635 ;dan RMSEA = 0,063; SRMR=0,0521)

Fig. 1. Model result of second-ordered confirmatory factor analysis

Based on results of CFA, out of 19 indicators, there are 4 indicators to be eliminated to obtain model fit indices. The indicators are because the standardized covariance error value is more than 2,58. Based on the factor loading value in CFA in this research, it is stated that there will be an improvement on terminal service by prioritizing respectively on easy of transfer, connectivity, operator performance, terminal performance and reliability service. From the ease of transfer, the high factor loading value is P4 related to terminal Access from the Street, this is because currently the terminal is located 500 meters from arterial roads. The connectivity indicator having high factor loading is P5 Coordination transport Operators with administrator. The operator performance indicator having high factor loading is P15 namely ease of purchasing tickets, although average passengers are quite satisfied with this indicator. The terminal performance indicator having high factor loading is P12 namely environmental quality, average passengers are quite satisfied with this indicator, this is in line with the condition that there are some points in the terminal providing less attention at cleanliness, although in passenger circulation, it can be seen an effort to improve the environmental cleanliness. At last, there ability service indicator having high factor loading is P18 namely waiting time, the passengers while waiting for the departure can spending their time with available facilities, so that the passengers are not bored at waiting [5].

4 Conclusions

Terminal development in Indonesia based on regulations directs the terminal location on suburb and enters arterial road, so that there is similar problem faced in the terminal namely the passengers are reluctant to transit in the terminal because of poor location and lack of public transport services in the city. Therefore, it is necessary to study in further the policy of relocating the terminal in the suburbs, the location selection does not only consider in the suburban area but also on its accessibility. If the terminal has been relocated then there are two policies that can be taken to improve the terminal services performance, namely:

1. Ease of transfers: this service priority is to ensure that passengers who will transit can conduct any activity smoothly and without any constraints or interference, this can be done first by creating effective coordination between operators and administrator so that it can work together in a synergistic manner. Second, it is through proactive and friendly attitude from terminal management staff assisting passenger transit activities. Third is to minimize passengers' difficulty in carrying their goods such as provide of trolley and porter.
2. Connectivity: this service is done firstly by increasing connectivity of selected location which is easily accessed one from supporting roads or locations. Second, there is provided information about passenger travel need related to arrival, departure schedules, bus availability and others. Third, there is adequate provision of alternative transportation means in the city so that it can support terminal services.

For further researches, it is necessary to conduct in some relocated terminal so that it can draw proper conclusion concerning any factors giving impacts on quality of terminal services as well as determine on a variety of problems faced as a result of relocation.

References

1. Dimitriou, H.T. A *Developmental Approach to Urban Transport Planning: An Indonesian Illustration*, Oxford University Press, Kualalurnpur, Malaysia. (1993),
2. Ceder, Avishai. 2007. Public Transit Planning and Operation: Theory, Modeling and Practice. Transport. Elsevier: Butterworth- Heinemann.
3. Augustin,K.,Gerike,R.,Sanchez,M.J.M.,Ayala,C.*Analysis of intercity bus markets on long distances in an established and a young market: The example of the U.S. and Germany*. Res. Transp. Econ.48,245–254, Elsevier ,(2014).
4. Yatskiv, Irina, Vaira Gromule, Nadezda Kolmakova, and Irina Pticina.. "*Development Of The Indicator Of Service Quality." Proceedings of the 9th International Conference "Reliability and Statistics in Transportation and Communication" (RelStat'09), 21–24 October 2009, Riga, Latvia, P. 124-133. ISBN 978-9984-818-21-4 Transport and Telecommunication Institute, Lomonosova* 1, LV, no. October: 124–33. (2009).
5. Hernandez, S., Monzon, A., *Key factors for defining an efficient urban transport interchange: users' perceptions*. Accepted in J.Cities. (2015).
6. Abreu e Silva, J., Bazrafshan, H., *User satisfaction of intermodal transfer facilities in Lisbon, Portugal: analysis with structural equations modeling*. Transp. Res. Rec. J. Transp. Res. Board 2350, 102–110.(2013)

7. Hine, J., Scott, J.,. *Seamless, accessible travel: users' views of the public transport journey and interchange.* Transp. Policy 7 (3), 217–226. (2000).
8. Dell'Asin, Giulia. *"Key Quality Factors And Travellers Profiles At Interurban Interchanges."* Thesis Doctoral, 53722. (2014).
9. De Oña, J., de Oña, R., Eboli, L., Forciniti, C., Mazzulla, G., *Transit passengers' behavioural intentions: the influence of service quality and customer satisfaction.* Transp. A Transp. Sci. 12 (5), 385–412. (2016).
10. Eboli, Laura, Yanbing Fu, and Gabriella Mazzulla.. "Multilevel Comprehensive Evaluation of the Railway Service Quality." Procedia Engineering 137. Elsevier B.V.: 21–30. (2016).
11. Nathanail, E., *Measuring the quality of service for messengers on the Hellenic railways.* Transp. Res. Part A 42, 48–66. (2008).
12. Iseki, H., & Taylor, B.D. Style versus service? *An analysis of user perceptions of transit stops and stations.* Journal of Public Transportation, 13(3), 23–48. (2010).
13. Wen, C.-H., Lan, L., Cheng, H.-L., *Structural equation modeling to determine passenger loyalty toward intercity bus services.* Transp. Res. Rec.: J. Transp. Res. Board 1927, 249–255. (2005).
14. Bordagaray, M., dell'Olio, L., Ibeas, A., Cecín, P. *Modelling user perception of bus transit quality considering user and service heterogeneity.* Transp. A: Transp. Sci., 1–17. (2013).
15. Stevens, J., *Applied Multivariate Statistics for the Social Sciences.* Lawrence Erlbaum Associates, NJ. (1966).
16. Golob, T.F., *Structural equation modeling for travel behavior research.* Transp. Res. Part B: Methodol. 37 (1), 1–25. (2003).
17. Wittwer, R. *Cluster-analytical-creation of a typology of young adults' travel behavior in Germany. España: Presented at the XI Congreso de Ingeniería del Transporte*, CIT 2014, Santander. (2014).
18. Bagozzi, R. P., & Yi, Y. *On the evaluation of structural equation models. Journal of the Academy of Marketing Science*, 16(1), 74-94. (1988).
19. Hair, J. F. J., Black, W. C., Babin, B. J., Anderson, R. E., & Tatham, R. L. *Multivariate data analysis (6th ed.). Upper Saddle River, NJ*: Pearson Prentice Hall. (2006).
20. Anderson, J. C., & Gerbing, D. W, *Structural equation modeling in practice: A review and recommended two-step approach. Psychological Bulletin*, 103(3), 411- 423. . (1988a).
21. John, G., & Reve, T. *The reliability and validity of key informant data from Dyadic relationships in marketing channels.* Journal of Marketing Research, 19(4), 517- 524. (1982).
22. Bagozzi, R. P., & Phillips, L. W. *Representing and testing organizational theories: A holistic construal. Administrative Science Quarterly*, 27(3), 459-489. (1982).
23. Fornell, C., & Larcker, D. F. *Evaluating structural equation models with unobservable variables and measurement error.* Journal of Marketing Research, 18(1), 39-50. (1981).
24. Kline, R. B. *Principles and practice of structural equation modeling (2nd ed.). New York, NY: Guilford Press.* (2005).
25. Hair, J. F. J., Black, W. C., Babin, B. J., & Anderson, R. E. *Multivariate data analysis: A global perspective (7th ed.). Upper Saddle River, NJ*: Pearson Education. (2010).

26.Janssens, W., Wijnen, K., Pelsmacker, P. D., & Kenhove, P. V. *Marketing research with SPSS. Harlow, England; New York: Prentice Hall/Financial Times*. (2008).

27.Kaiser, H. F., & Rice, J. *Educational and psychological measurement. Little Jiffy, Mark* IV, 34, 11. (1974).

28.Kim, J. A *multidimensional and hierarchical model of integrated service quality for the Internet golf stores. Retrieved from* http://dl.acm.org.ezproxy.lincoln.ac.nz/citation.cfm?id=979102. (2003).

29.Land, L., and C. Foreman. Neighborhood Intermodal Transfer Facilities. Report no. NCTR-392-16. *Research and Special Programs Administration, U.S. Department of Transportation, Florida Department of Transportation and University of South Florida, Center for Urban Transportation Research, and National Center for Transport Research, Tampa,* Fla., (2001).

30.Tabachnick, B. G., & Fidell, L. S. *Using multivariate statistics (5th ed.). Boston*: Pearson/Allyn & Bacon. (2007).

School bus operational performance in Malang City

Septiana Hariyani[1,*], and *Dadang* Meru Utomo[1]

[1]Brawijaya University, Department of Urban and Regional Planning, Malang, Indonesia

Abstract. Based on the preliminary observation not all students are interested in using the school bus, despite the school bus's service since 2014 in Malang city. So, this study evaluates school bus operation performance. This study used on board survey to calculate the load factor and travel time of school bus. While static survey is used to collect waiting time for passenger. The results showed that load factor departure and load factor arrival over 70%. Average school bus's travel time departure for six routes was 44-58 minutes, and average travel time arrival for 37-65 minutes for six routes. The last is waiting time for passenger 5-14 minutes.

1 Introduction

There are five categories of home-based trip destination, namely: 1. To workplace, 2. To school or university (the movement of the purposes of education), 3. To shops, 4. To social interests, and 5. To recreation areas [1]. The movements require a mode of transport to move called transportation, whether it is a private mode or a mode of public transport. One form of public services needed by the community is a transportation service [2]. Transport is an activity of the movement of people and goods from one place (origin) to another (destination) by means of a vehicle. Movement to the school also requires transportation either public transport or private transportation, but there is a special transport that serves students is a school bus.

The results of previous research indicate that the school bus load factor is still low i.e. 0.5 [3]. Speed of school bus trip is reported around 40 km/h with travel time of 40-60 minutes. School bus service time is in the morning and afternoon corresponding to the time to depart and go home from school and has a waiting time of passengers between 5-10 minutes.

Another research indicates that the school bus's level of service in Malang City needs to get more attention to improve the attributes: adequate space, and seating capacity in order to increase its performance [4].

Building upon the previous research, this study aims to evaluate the load factor, travel time of school bus, and waiting time for passenger of Malang's school bus service, since its first operation in 2014.

* Corresponding author: septianahariyani@ub.ac.id

2 Methods

This study was conducted on six school bus routes in Malang City, namely route A, B, C, D, E, and route F. Route A : Tlogomas – Tugu, route, route B : SPBU Mergan – Tugu, route C : T. Mardyopuro-Tugu, route D : Sawojajar (Madyopuro Terminal) – Tugu, route E : National electricity company (PLN) Gadang – Tugu, and route F : Samsat Kacuk – Tugu. The six school bus routes and school distribution can be seen in Figure 1.

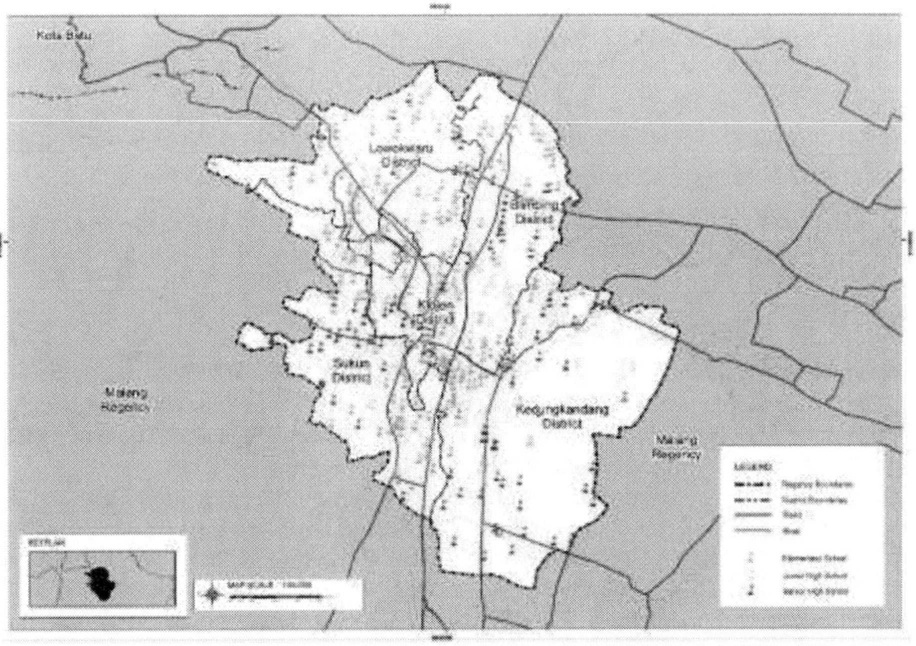

Fig. 1 The six school bus routes and school distribution in Malang City

On board survey was used to calculate the load factor and travel time of the school bus service. While static survey was used to collect waiting time for passenger. Then the data was put into calculation bellow:

2.1 Load factor

The standards set by the Directorate General of Land Transport, 1996 (Dirjen Perhubungan Darat) for the load factor value is 70% (0.7) and there is a 30% (0.3%) reserve to accommodate the possibility of passenger spikes, and at this level the inroads of passengers inside the vehicle are still acceptable [5].

$$\mathbf{LF = \frac{Vp}{Cb} x100\%} \quad (1)$$

Note:
LF : *Load factor*
Vp : Average Passenger Volume in the bus
Cb : bus capacity

2.2 Travel time

Travel time can be defined as the time taken to travel a certain distance and will have a relationship associated with the average speed. Travel time is an indicator that determines the service level of a bus operation. It is clear from the obligations of the bus operator to supply the existing demand, as an indicator of the level of service [6].

If it assumes that the arrival time of the passengers on the bus departure site, the same or different, assuming a constant headway, it can be concluded that the average waiting time for passengers is:

$$w = \frac{h}{2} = \frac{60}{2f} \tag{2}$$

Note :
w = average passenger waiting time
h = headway (minutes)
f = one way bus frequency (bus/hour)

3 Result

Geographically Malang is located at coordinates 112° 06 '- 112° 07' East Longitude and 7° 06 '- 8° 02' South Latitude. The city of Malang is located in East Java Province with a population of 900,468 inhabitants [7]. The total area of Malang is 110.06 km^2. Malang is the second largest city in East Java after Surabaya.

The school bus is a transportation facility provided by the Education Office of Malang City, where the school bus consists of 5 school buses from the government, 1 school bus from Corporate Social Responsibility (CSR) and 1 reserve school bus also from CSR. Procurement of this school bus by the government provides funds worth 4 billion rupiah for 5 units of buses with the brand IZUSU with type IZUSU / NQN 71 EC. Ez-1 with each bus worth 800 million rupiah per unit. The procurement of school buses is prioritized for less fortunate students living in the suburbs. The buses can be seen in Figure 2.

School bus procurement began on December 29, 2014 and commenced operations on January 12, 2015, although there was a polemic with the public transport driver who refused and disagreed with the existence of the school bus, the procurement of the school bus is expected to reduce the use of motor vehicles by students and the number of congestion due to transportation. The procurement of the school bus was distributed to six existing routes in Malang City, where the six school bus units were spread with a capacity of 35 passengers seated with 10 passengers standing for the school bus from the municipal government while the school bus obtained from CSR there were only 25 passengers seated and 20 passengers to stand (Figure 3).

Fig. 2 School bus in Malang City

Fig. 3 Inside of school bus

Evaluation of operational performance of school bus in Malang is done by knowing travel time, and load factor. The results of the operational performance evaluation will be used as a basis for improving the operational performance of school buses in Malang.

Load factor is used to measure the capacity of passengers in each journey so as to know the maximum capacity of a mode, besides the load factor serves as a guide to know the availability of the fleet is sufficient, still lacking, or exceeding the needs of a track and an indicator of the efficiency of a route. The seating capacity on the School Bus is 35 seats with 10 stands. The standard for the load factor values established by the Directorate General of Land Transportation is 70% and there is a 30% reserve to accommodate in case of a passenger surge. Load factor work is differentiated every school bus that operates. The data used to calculate the load factor is the number of passengers and the capacity of the school bus passengers.

Based on the load factor calculation on each route the excess of passengers has been set by the Directorate General of Land Transport where the standard for load factor value is 70% and there is 30% reserve to accommodate passenger spike. This excess of passengers is influenced by some behavioral factors of travelers for poor students in the selection of modes, especially the school bus. Factors that affect, among others, 42% of students who use school bus have a pocket money ranging from 5000-10000, where the students prefer school buses because it is not necessary to pay to board the school bus so they no longer need to spend additional costs, As many as 88% of students using school buses travel single trips where they directly choose to go home, as well as a school rule that prohibits students from bringing the vehicle to school by as much as 70% thus requiring students with private vehicles not to use it to school. Load factor of school bus showed in Table 1.

Travel time can be defined as the time taken to travel a certain distance and will have a relationship associated with the average speed used to travel a certain distance. Factors that affect the travel time of a mode of travel speed, route length, travel time, and waiting time. Based on the result of the travel time calculation on each route, both the travel time and the return time of the average travel time of the school bus takes ± 50 minutes to reach the end point of the route where as many as 54% of students who use school buses need travel time ranged Between 10-15 minutes to get to school, 28% takes 15 minutes and the remaining 18% takes time to go to school ranging from 5-10 minutes.

Table 1 School bus's load factors (LF)

	LF departure	**LF arrival**
School bus route A	140%	94%
School bus route B	123%	120%
School bus route C	82%	68%
School bus route D	109%	97%
School bus route E	106%	89%
School bus route F	89%	103%

However, after the interview about the school bus time desired by the students is as much as 48% of students who use school bus wanted travel time 10-15 minutes, 40% of students want travel time 5-10 minutes and the rest of> 15 minutes as much as 12%. The travel time of the school buses on each route is not in accordance with the demand of the students who use it therefore needs consideration to maximize the travel time of the school bus in accordance with the demand desired by the students.

Table 2 School bus's travel time (TT)

	TT inbound (minutes)	**TT outbound (minutes)**
School bus route A	45	65
School bus route B	57	45
School bus route C	58	54
School bus route D	44	39
School bus route E	45	38
School bus route F	46	37

Table 3 Waiting time for passenger

	Waiting time for passenger (minutes)
School bus route A	12
School bus route B	10
School bus route C	11
School bus route D	5
School bus route E	10
School bus route F	14

4 Conclusion

Since the first commencement in 2015, the utilization of Malang City's school bus service has increased as seen in the relatively high load factor (more than 70% as set by the Directorate General of Land Transport). With most riders coming from middle income families, despite its rising popularity, there are several points that require improvement to increase the riders' satisfaction. The first point is the necessity of evaluating the route length to provide shorter travel time. Currently, riders expect shorter travel time of 5-10 minutes, while the actual travel time reaches almost an hour. The second is the urgency of providing more fleets to allow shorter waiting time. Based on the findings, this research suggests the local government to consider more effort to provide a better school bus service, not only to improve riders' experience, but also to reduce transportation problem in general.

Reference

1. Tamin, OZ. *Perencanaan dan Pemodelan Transportasi* – Edisi Kedua. Bandung: Penerbit ITB. (2000).
2. Warpani, Suwardjoko P. *Pengelolaan lalulintas dan angkutan jalan.* Bandung: Penerbit ITB (2002).
3. Prasetyo, T, Djakfar, L, and Abusini, S, *Evaluasi dan Potensi Pengoperasian Bus Sekolah (Studi Kasus: Bus Halokes Kota Malang),* Media Teknik Sipil Volume 13, Nomor 2, p.185-192 (2015).

4. Hariyani, S. *School bus's level of service in Malang City*. IOP Conf. Series: Earth and Environmental Science 70 (2017) 012023.
5. Direktorat Jenderal Perhubungan Darat. (1996).
6. Morlok, E. K. *Introduction to Transportation Engineering and Planning*. Mc. Graw-Hill Kogakuha. (1978).
7. Office of Demographic Affairs Malang City, Juli (2017)

The application of random regret minimization on commuter's mode choice behaviour: Model-Fit comparisons with Rum-Modelling (case: comparison between Matsuyama and Yogyakarta)

Audinda Virsa Leinia[1,*], *Imam* Muthohar[1] , and *Ibnu* Fauzi[1]

[1]Universitas Gadjah Mada, Civil and Environmental Engineering, Yogyakarta, Indonesia

Abstract. Modelling the mode choice behaviours of travellers is a key to design effective transport management policies, particularly in shifting travellers to public transport. Abundant studies have analysed the impact of level of services on mode choice preferences through its Random Utility Maximization (RUM), but the possibility of minimalize the regret have been overlooked. This paper will discusses the possibility of using generalised Random Regret Minimization (G-RRM) model on choosing transportation modes. The study is performed in two cities for comparison: Jogjakarta in Indonesia and Matsuyama in Japan. A stated preference (SP) survey isconducted, in which respondents choose Bike or Bus under hypothetical situations. As the result of RUM revealed that travellers prefer the transportation mode with more ensuring level of service. While an empirical proof of concept, the G-RRM model is estimated on a stated mode choice dataset, and its outcomes are compared with RUM and RRM counterparts.

1 Introduction

Many modern cities are facing traffic problems due to rapid urbanisation. Increasing population induces increasing number of mobility demand, which results in problems in transportation. Particularly in developing Asian cities, the traffic problems have existed for many years. Growing economies have enabled the residents to own private vehicles, such as motorbikes and cars. This change makes the individuals more dependent on the private mode of travel and causes serious traffic jam on a daily basis. The problem of traffic congestion stems to disrupted traffic conditions, such as traffic accidents; and delay in mobility of people and goods, which cause huge economic loss every year.

A public transportation could give an outstanding solution to the above problems by attracting the travellers of private vehicles to mass transportation modes. However, the benefit of the public transportation systems is not easily attained. In general, private

* Corresponding author: audinda.virsa@gmail.com

vehicles are more attractive mode of transportation because of their comfort and speed. To devise an effective public transportation policy, it is essential to understand the relationship between travellers' mode choice decisions and the factors affecting it.

Abundant studies have analysed the impact of level of services, on mode choice preferences using Random Utility Maximization. However, the possibility of minimalize the regret have been overlooked. To the best of authors' knowledge, Chorus, et al. [6] has been empirically comparedbetween the application of Random Utility Maximization and Random Regret Minimization on consumer choice modelling. They identified that the RUM and RRM perform equally well on some aspects, but there are possibility of compromising effect on using RRM Although the study presents promising findings, it does not quantitatively evaluate the impact of using RRM while modelling the consumer choices.

2 Overview of Study Site

2.1 Yogyakarta, Indonesia

Yogyakarta is one of the most growing cities in Indonesia, and the mobility demand is dramatically increasing. In order to meet the demand, the government provided a public transportation system by operating Transjogja Bus in 2008. However, since its start, the number of passengers has been decreasing [2]. According to Department of Transportation of Yogyakarta, the designed time headway of Transjogja Bus is 16 minutes, and the targeted punctuality tolerance level is 5 minutes with average speed is 25 km/hour. In reality, a lot of schedules are missed from its tolerance level because of many obstacles in a road lane, which shuts away the passengers.

2.2 Matsuyama, Japan

Matsuyama has a well-developed transport network as the prefecture capital. It has a network of suburban railways run by the Iyo Railway. Iyo Railway also operates a system of trams and bus, which serve as the main modes of public transportation in the city. The public transportation network in Matsuyama is almost the same with Yogyakarta. Nevertheless, the majority of travellers tend to use private vehicles, such as bicycles, motorbikes or private cars.

3 Methodology

3.1 Questionnaire individual attribute

The first section is devoted for asking individual attributes of the respondents. The questions are designed to comprehend their socio-demographic characteristic, daily travel behaviours and past experiences in using public transportation. The list of questions includes the following:

1. Age
2. Gender
3. Occupation
4. Disposable income per month

The questions related to their travel behaviours seek for the daily commuting characteristics and the past experiences as listed below:

1. Daily mode of transportation for their commuting
2. Travel distance
3. Travel time
4. Travel cost
5. Private vehicle ownership (number of their own private vehicles, including car, motorbike and bicycle)
6. Motorbike availability
7. Experience of encountering crime such as thieves or gender harassment in using public transportation (never, once, twice, or more)
8. Experience of encountering or self-involvement in traffic accident in using private vehicles (never, once, twice, or more)

3.2 SP survey

The Stated Preference (SP) survey presents hypothetical travel scenarios where the respondents are required to choose either Bus (as a public transportation) or Motorbike (as a private vehicle) under various conditions. The factors considered in this study are listed below:

1. Travel Cost (TC)
2. Travel Time (TT)
3. Maximum Delay Time (Max. DT)
4. Average Waiting Time at the bus stop (Ave. WT)
5. Access Time to the bus stop (AT)
6. Frequency of traffic accidents while using the motorbike (Acc.)

3.3 Experimental design

Scenarios are created by assigning two levels of values to each of the six attributes. According to Rushing,H., Karl, A,et.al [8]Each of the six attributes has two cases, which creates 64 scenarios ($= 2^6$) in total. In order to ease the answering load for respondents, the scenarios are decreased based on 2^{k-2} fractional factorial design [3], where k represents the number of attributes. The 2^{k-2} fractional factorial design is constructed by listing down a basic design consisting of the scenarios consisting of a full factorial in $k-2$ attributes. Then the remaining two attributes are added. Thus there will be 16 scenarios.

Table 1. Levels of attributes for Yogyakarta

Attribute	Bus	Motorbike
TC (Rp)	1000	2000
		3000
Acc. (accident/month)	0	1
		2
TT (minutes)	20	15
	25	
Max Delay (minutes)	4	5
	6	
Ave WT (minutes)	3	0
	5	
AT (minutes)	5	0
	8	

Table 2. Levels of attributes for Matsuyama

Attribute	Bus	Motorbike
TC (Yen)	50	100
		150
Acc. (accident/month)	0	1
		2
TT (minutes)	20	15
	25	
Max Delay (minutes)	4	5
	6	
Ave WT (minutes)	2	0
	4	
AT (minutes)	4	0
	6	

3.4 The-RUM model

According to Hensher,D.A [7], logistic regression is part of regression analysis that is used when dependent variable is dichotomous which represents whether an event appears or not. This regression forms a dependent variable as a linear combination of independent variables. The linear function, a utility function, is then transformed to probability using logit function. To predict if an alternative is chosen based on model, the value of utility function should be contrasted by alternative choice, transformed into a probability with score between 0 and 1.

Based on Issac K,et.al [3] citated from Ben-Akiva,et.al [4], the assumption of the model is that respondents would choose the travel mode for the greatest utility under certain condition, and the utility function of it is composed by fixed term and random term. The formulas are as follows:

$$U_{in} = V_{in} + \varepsilon_{in} \quad (1)$$

$$V_{in} = \beta_0 + \sum_{k=1}^{K} \beta_k X_{ink} \quad (2)$$

where $U_{in}$is utility function of the $i-th$ travel mode chosen by the $n-th$ respondent ($Bus\ (i=1), Motorbike\ (i=2)$); $V_{in}$is the fixed term of utility function (U_{in}); $\varepsilon_{in}$is the random term of utility function (U_{in}); $K$is the number of attributes (which is also called characteristic variables) affecting on the mode choice of respondents; $\beta_k$is the undetermined coefficients; X_{ink} is the $k-th$ factor of the $i-th$ travel mode chosen by the $n-th$ respondent.If the random term ε_{in} follows the Gumbel distribution and all variables are independent each other, the probability for the $n-th$ respondent choosing $i-th$ travel mode, P_{in}, is defined by the following equations:

$$P_{in} = \frac{\exp(V_{in})}{\sum_{i=1}^{2} \exp(V_{in})} \quad (3)$$

3.5 The-RRM model

The RRM model has been designed to incorporate the notion of regret-based decision making in non-risky choice models. The RRM model hypothesizes that, when confronted with a choice set, the decision-maker chooses the alternative from the set that has minimum regret. The regret of alternative *i* is described by the sum of binary regrets where alternative *i* is compared to every other alternative in the choice task on each attribute. Overall regret is increasing with the number of attributes on which alternative *i* is outperformed as well as with the number of alternatives by which alternative *i* is outperformed (as denoted by the summation over j≠*)*, and the importance of the attribute (as denoted by *δm*).

$$RRi = Ri + \varepsilon i = \sum_{j \neq i} \sum_{M=1}^{m} ln(1 + exp[\beta m\ .(Xjm - Xim)]) + \varepsilon i \quad (4)$$

RRi: denotes the random (or: total) regret associated with a considered alternative i
Ri : denotes the 'observed' regret associated with *i*
εi : denotes the 'unobserved' regret associated with *i*
βm : denotes the estimable parameter associated with attribute *Xm*
Xim, Xjm : denote the values associated with attribute *Xm* for, respectively, the consideredalternative i and another alternative *j*

3 Result

3.1 Attributes of respondents

The Socio-demographic attributes of the respondents are summarised in Table 3. In both cities, the majority of respondents are male aged 18-24 years old with an occupation as student. The disposable monthly income varies, but ranges in an average student level in both cities.

Table 3 Socio-demographic attributes

Attributes		Yogyakarta	Matsuyama
Gender			
	Female	36.00%	18.00%
	Male	64.00%	82.00%
Age			
	18-24	94.00%	88.00%
	25-49	6.00%	12.00%
Occupation			
	Student	90.00%	100%
	Full-time worker	4.00%	
	Self-employed	4.00%	
	Others	2.00%	
Disposable income			
	(Indonesian Rp)		
	<500,000	20.00%	-
	500,000 - 1,000,000	16.00%	-
	1,000,000 - 1,500,000	28.00%	-
	1,500,000 - 2,000,000	16.00%	-
	>2,000,000	20.00%	-
	(Japanese Yen)		
	<50,000	-	52.00%
	50,000-100,000	-	36.00%
	100,000-150,000	-	10.00%
	>150,000	-	2.00%
	N	400	400

Based on travel behaviour factors, the significance factor that could distinguished between Yogyakarta and Matsuyama respondents as Table 4.

3.2 RUM-Model

In order to create the valid input data, test of significance is conducted to figure out the attributes that affect the preference to the choices. The affecting factors are identified from the statistical test such as Wald-test for each variable. In the Wald-test, the null-hypothesis is $\beta_i = 0$ (Not significantly affecting), while the alternative hypothesis is $\beta_i \neq 0$ (significantly affecting). According to Ben-Akiva, et.al [5], the Wald-test calculation with α

= 0.05 and DF = 20 will has critical t-value as 1.96.The table 5 and 6 will show the significant affecting factors in both cities.

Table 4 Mode choice behaviours

Attributes		Yogyakarta	Matsuyama
Daily mode of travel			
	Car	6.00%	2.00%
	Bicycle	10.00%	88.00%
	Motorbike	84.00%	10.00%
Daily travel distance			
	<3km	46.00%	90.00%
	3-6km	24.00%	10.00%
	6-9km	14.00%	0%
	>9km	16.00%	0%
Daily travel time			
	<5 minutes	14.00%	44.00%
	5-10 minutes	42.00%	52.00%
	10-15 minutes	20.00%	4.00%
	>15 minutes	24.00%	0%
Daily travel cost			
	(Indonesian Rp)		
	0 - 1,000	4.00%	-
	1,000 – 3,000	32.00%	-
	3,000 – 5,000	28.00%	-
	5,000 – 7,000	22.00%	-
	>7,000	14.00%	-
	(Japanese Yen)		
	0 - 50	-	98.00%
	50 – 100	-	2.00%
	>100	-	0%
Private Vehicle Owned			
	0	6.00%	32.00%
	1	86.00%	58.00%
	2	4.00%	8.00%
	3 or more	4.00%	2.00%
Motorbike Availability			
	Yes	88.00%	24.00%
	No	12.00%	76.00%
Crime in public transport			
	Once	10.00%	2.00%
	Twice	2.00%	0%
	Three times or more	2.00%	0%
	Never	86.00%	98.00%
Traffic accident in Private			
	Once	34.00%	34.00%
	Twice	10.00%	0%
	Three times or more	36.00%	0%
	Never	20.00%	66.00%
	N	400	400

Table 5. RUM-Model estimation result (Yogyakarta)

	Coefficient	Wald	p-value
TC	-0.004	1.961	0.041
TT	-0.272	19.964	0.000
Mode_1	-3.914	30.643	0.000
Dist_2	0.947	7.191	0.007
Dist_3	0.609	2.928	0.008
BK	-2.197	8.695	0.003
ExAcc_2	0.693	5.636	0.018
ExCrime	-0.260	0.559	0.015
Constant	5.228	24.093	0.000

Table 6. RUM-Model estimation result (Matsuyama)

	Coefficient	Wald	p-value
TC	-0.011	9.328	0.002
TT	-0.061	2.732	0.038
Acc	-0.539	9.328	0.002
Dur_1	0.505	5.03	0.025
OwnVeh_1	-0.471	3.992	0.046
BK	-2.095	55.874	0.000
Constant	2.765	9.257	0.002

Based on tables above, the utility function equation in Jogjakarta using binary logistic regression model is

$$\ln\left(\frac{P_{bus}}{1-P_{bus}}\right) = -0.004TC - 0.272TT - 3.914Mode\_1 + 0.947Dist\_2 + 0.609Dist\_3 - 2.197BK + 0.693ExAcc\_2 - 0.260Crime + 4.484$$

Where :
Pbus : Probability that Bus is chosen
TC : Travel Cost of Bus in a Scenario
TT : Travel Time of Bus in Scenario
Mode_1:Motorbike as respondent daily mode of transportation
Dist_2 : Daily travel distance about 3 – 6 km
Dist_3 : Travel distance more than >6 km
BK : Bikeavailable(yes), The respondent has a motorbike and/or could use it
ExAcc_2: The experience of traffic accidents
ExCrime: The experience of encountering a crime on public transportation

Based on tables above, then the utility function equation in binary logistic regression model that suits in Matsuyama is:

$$\ln\left(\frac{P_{bus}}{1-P_{bus}}\right) = -0.011TC - 0.061TT - 0.539Acc + 0.505Dur\_1 - 0.471OwnVeh\_1 - 2.095BK + 2.765$$

Where :
P_{bus} : Probability that Bus is chosen
TC :Travel Cost of Bus in a Scenario
TT : Travel Time of Bus in Scenario
Acc : Number of accident on private vehicle in scenario
Dur_1 : Travel duration that takes more than 5 minutes
OwnVeh_1: One vehicle ownership
BK : The respondent hasa motorbike and/or could use it

The goodness fit between Yogyakarta and Matsuyama also has different result due to the limitation number of respondent, but both model in Yogyakarta and Matsuyama has been fulfilled the criteria by the Omnibus test that state the significance of the model which is lower than 0.05 as α-value and also HosmerLemeshow value that is bigger than 0.05 which means that the model chosen is fit. Yet the variance in Indonesia can represent 54.7 % to the respondent variance, while in Matsuyama only represent 47.0%. In overall percentage, both are more than the null hypothesis which is 50%, the overall percentage in Matsuyama is 64.8% and in Yogyakarta is 83.9%. The -2 Log Likelihood shown in Yogyakarta is 686.335 which if compared to Chi-critical ($\alpha = 0.05$ and DF = 20) which is 422.695 (Chi-square of Omnibus Test),then the value of -2LL is bigger and stated on rejection area, which means Ho is rejected and proved that the unrestricted model is better than restricted model. While in Matsuyama the -2LL value which is 1000.201 is also higher than Chi-critical (108.835). Both cities stated the unrestricted model is better than restricted model. Restricted model is a model with only estimating the constant parameter, while the parameter inside scenario is not being used, while unrestricted model in involving socio-demograph factors, travel behaviour factors and all attributes in scenario.

3.3 RRM-Model

The respondents' characteristics and the SP survey results are analysed for estimating the mode choice model. The RRM-Model will be differs in result with RUM since in RRM estimated attributes will not involving the socio-demograph and travel behaviour attributes, the proceeded attributes are the SP scenario only. Since the RRM model is not fit with the dummy variables in socio-demograph and travel behaviour attributes. The result in RRM-Model using Biogeme will consist of coefficient parameter, statistical t-test and statistical p-value as mentioned below:

Table 7. RRM-Model estimation result (Yogyakarta)

Name	Value	Robust t-test	p-value
TC (X1)	-0.005	0.119	0.039
TT (X2)	-0.372	29.778	0
Mode_1 (X3)	-3.414	30.654	0
Dist_2 (X4)	0.931	7.281	0.008
Dist_3 (X5)	0.591	2.998	0.009
BK (X6)	-2.375	8.594	0.003
ExAcc_2 (X7)	0.658	5.364	0.019
ExCrime (X8)	-0.315	0.578	0.014
gamma	4.918	24.135	0
-Init log likelihood:		-277.259	
-Final log likelihood:		-678.457	
Rho-square		0.357	

Table 8. RRM-Model estimation result (Matsuyama)

Name	Value	t-test	p-value
TC (X1)	-0.015	9.192	0.003
TT (X2)	-0.049	2.894	0.038
Acc (X3)	-0.521	9.772	0.002
Dur_1 (X4)	0.499	5.031	0.023
OwnVeh_1 (X5)	-0.456	3.679	0.041
BK (X6)	-2.101	55.473	0
gamma	2.923	9.599	0.001
-Init log likelihood:		-997.108	
-Final log likelihood:		-1001.121	
Rho-square		0.441	

Based on tables above, the utility function equation in Jogjakarta using binary logistic regression model is:

Ln (1+exp(Pbus))= -0.005TC -0.372 TT
-3.414Model+0.931 Dist2 +0.591Dist3 -2.375BK +0.658ExCrime -0.315 ExAcc2+4.918

Where :
Pbus : Probability that Bus is chosen
TC : Travel Cost of Bus in a Scenario
TT : Travel Time of Bus in Scenario
Mode_1:Motorbike as respondent daily mode of transportation
Dist_2 : Daily travel distance about 3 – 6 km
Dist_3 : Travel distance more than >6 km
BK : Bikeavailable(yes), The respondent has a motorbike and/or could use it
ExAcc_2: The experience of traffic accidents
ExCrime: The experience of encountering a crime on public transportation

Based on tables above, then the utility function equation in binary logistic regression model that suits in Matsuyama is:

Ln (1+exp(Pbus))= -0.015TC -0.049TT
-0.521Acc+0.499Dur1
-0.456OwnVeh1-2.101BK +2.923

Where :
Pbus : Probability that Bus is chosen
TC :Travel Cost of Bus in a Scenario
TT : Travel Time of Bus in Scenario
Acc : Number of accident on private vehicle in scenario
Dur_1 : Travel duration that takes more than 5 minutes
OwnVeh_1: One vehicle ownership
BK: The respondent hasa motorbike and/or could use it

3.4 Comparison RUM-Model and RRM model

There are slight differences between 0-LL in Yogyakarta and Matsuyama, but the value of Rho-square are quite different value in both cities. The significant factors both in Yogyakarta and Matsuyama are the same. In terms of model fit, RUM-Model generally fit the data significantly. In context of SP survey, the "opt out" or "none of these seems to performs better in RUM. The model also has an ability to provide correct choice probability predictions for no-choice option.

While the RRM-Model, Generally fit the data significantly. In context of SP survey, the "i am indifferent" seems to performs better in RRM models. The model has no ability to provide correct choice probability predictions for no-choice option, but it could capture compromising-effect, it recognizes alternatives with mediocre performance.

Table 9. Differences goodness of fit (Yogyakarta)

Parameters	RUM	RRM
Final LL	686.335[a]	678.457
Rho-square	0.547	0.357

Table 10. Differences goodness of fit (Matsuyama)

Parameters	RUM	RRM
Final LL	1000.201[a]	1001.121
Rho-square	0.470	0.441

Based on the model-fit test as the comparison of choices in actual site and choices in the observation, the proportion could be seen in Table 11 for Yogyakarta model and Table 12 for Matsuyama model.

Table 11. Site comparison in Yogyakarta

		ACTUAL	
		Mode 1	Mode 2
OBSERVATION	Mode 1	82.63%	17.37%
	Mode 2	27.12%	72.88%

Table 12. Site comparison in Matsuyama

		ACTUAL	
		Mode 1	Mode 2
OBSERVATION	Mode 1	65.87%	34.13%
	Mode 2	42.06%	57.94%

The difference in goodness of fit as in Table 11 and 12, as expected, is in favour of the RUM-model, which is a sign that RRM indeed has some difficulty with handling imputed choice sets. As in the model fit of site comparison in Yogyakarta, the approaching value for bus is 82.63% while in Matsuyama is 65.87%.

4 Conclusion

In transportation system, every aspects should be considered to affect the preference. In this model there are 12 variables, with 4 socio-demograph factors (age, gender, occupation, income per month) and 8 travel behaviour factors (daily mode of transportation, travel distance, travel duration, travel cost, vehicle ownership, availability of motorbike, self-experience on criminality and accident).

In terms of RUM-Model and RRM-Model that involves dummy variables using SPSS, factors that is significantly affecting the preference in Yogyakarta are are Travel Cost, Travel Time, Daily Mode of Transportation, Travel Distance, Availability of using Motorbike, Self-experience of Crimininality on public transportation and Self-experience of Accident on Private Vehicle. While in Matsuya the significant factors are Travel Cost, Travel Time, Safety while using the mode of transportation, Travel duration to campus, Private Vehicle Ownership, and Availability of using Motorbike.

The comparison analysis between RUM and RRM using Biogeme have slight difference, in terms of model-fit parameter and statistics test, and also the value of goodness of fit due to RUM ability to predict the probability of "no choice" option. The differences might getting large if the SP survey attributes level are vary and also the number of respondent is larger. The further statistical check like, t-test comparison and p-value is needed whether RRM and RUM model are fit to be used in terms of Mode Transportation Choices.

References

1. Adamowicz, W., Louviere, J., & Williams, M. Combining revealed and stated preference methods for valuing environmental amenities. *Journal of Environmental Economics and Management, 26*, 271-292. (1994)

2. Annual Report Jasa Marga. Access October 2016, *http://www.jasamarga.com/download/AR_jm2016.pdf* .(2011)
3. Baidoo,Issac K, & Nyarko, Eric. *Stated Preference Modeling for a Prefereed Transportation Mode*. Ghana: iiste.org. (2015)
4. Ben-Akiva, M., and S. Lerman. *Discrete Choice Analysis*. London: MIT Press, (1985).
5. Ben-Akiva, M. &Morikawa, T. Estimation of switching models from revealed preferences and stated intentions. Transportation Research Part A: General, 24(6), 485-495. (1990)
6. Cranenburgh, S., Guevara, C., & Chorus, C. New insights on random regret minimization models. *Transportation Research Part A: Policy and Practice,* 74, 91-109. doi: 10.1016/j.tra.2015.01.008. (2015)
7. Hensher,D.A., Rose,J.M., & Greene, W.H. *Applied Choice Analysis: A Primer* (Vol.1). Cambridge: Cambridge University Press. (2005)
8. Rushing,H., Karl, A., &Wisnowski,J. *Design and Analysis of Experiments by Doughlas Montmogery: A Supplement for Using JMP*. North Carolina, USA: SAS Institute Inc., (2013)
9. Susilo, Y.O., Joewono, T.B. &Santosa, W. *An Exploration of Public Transport Uses' Attitudes and Preferences towards Various Policies in Indonesia: Some Preliminary Results*. Journal of the Eastern Asia Society for Transportation Studies, 8. (2010)
10. Train, K. *Discrete choice methods with simulation*. United States: Cambridge university press. (2009)

Urban spatial structure re-identification as an approach to solving urban transport problem (A case study in Malang)

Dadang Meru Utomo[1,*], and *Septiana* Hariyani[1]

[1]Brawijaya University, Urban and Regional Planning Department, Malang, Indonesia

Abstract. In an effort to solve the congestion problem, the local government of Malang City has taken various steps. However, unfortunately, the result has not been effective. This shows that the efforts that have been taken so far are sectoral, therefore a more comprehensive approach is needed in order to reach a more effective result. This research aims at re-evaluating the determination of Malang City's urban spatial structure through the delineation of narrower units of area using the centrality index method. The research suggests that urban spatial structures have a very clear pattern of association with congestion locations. It can be concluded that the local government of Malang City needs to put more consideration on evaluating the urban spatial structure to solve the problem of transportation which is ever-growing the in second largest urban area in East Java Province.

1 Introduction

Traffic problems that haunt almost all big cities in Indonesia, including Malang in East Jawa Province, are mostly due to the imbalance between the growth of private vehicle use and the inadequate provision of transport infrastructure. As the second biggest urban area in the province, Malang has attracted high traffic for its strategic location adjacent to the country's most touristic city of Batu, its wide stretch of natural beaches in its southern part, and its vibrant city life. The ever-growing traffic has admittedly brought transport problems in need of serious attention.

The local government of Malang, reportedly, has attempted several different approaches to solving its transport problem. The approaches include traffic management strategy, provision of new road infrastructure, as well as public transport provision [1]. However, it seemed that the aforementioned transport solutions have not been fully effective, as seen in many prevalent traffic hot-spots during the city's busy hours and especially during the weekends. Also, the ownership of private vehicles has been reported to grow steadily in the last couple of years [2]. Another problem related to transport in Malang is the increase of road accidents, according to official reports. It is therefore an urgent situation for the local

* Corresponding author: dadang.utomo@ub.ac.id

government and transport planners to think of a more comprehensive and sustainable approach to stopping the problem to cause further externalities.

Several scholars have come to an idea of looking at the problem from a more holistic perspective, which is by evaluating the urban spatial structure. An effective spatial structure of a city will lead to a less-decentralized pattern of movement towards a single city center [3]. A spatial urban structure approach is an alternative to improving environmental qualities through the provision of different urban facilities and infrastructure, one of which is the efficiently planned transportation to support sustainability [4]. In addition, the correlation between urban spatial structure and transport has been justified through empirical research; traffic generation models help explain the spatial distribution of space in an urbanized area [5]. Furthermore, the miles driven and travel choices depend on the spatial structure in which a person lives [6]. These explanations justify the fact that a positive correlation between urban spatial structure and travel patterns does exist. In an Indonesian context, supports that there has been a scientific correlation between travel patterns and the spatial distribution of land-use which forms the urban structure [7]. And then comes the question, which urban structure is the most effective for alleviating transport problems in cities?

A multinucleated or polycentric type of urban spatial structure leads to an efficiency of energy used in transport, compared to a more single-centered type of spatial structure [8]. In the same light, the research find that the evenly distribution of facilities, argued as an effective urban structure, allows shorter trips, cuts down the consumption of fuel energy, and minimizes the potential of accidents [3,4]. In many cases, it also increases the number of mode shifts to public or active transport. Building upon these findings, it can be said that urban spatial structure could be a more appropriate approach to a better transport treatment, thus alleviate the ever-worsening transport problem in Malang.

In terms of development, Malang has undergone a significantly rapid urbanization in the last couple of decades. Sub-urbans, known as kecamatan, have been provided with a wide range of facilities with supposedly better road infrastructure. This has arguably been a strong reason for the increase of movements within the city. With so many points of traffic congestion in Malang, it seems that there has not been an adequate prioritization of improvement. This is probably due to the lack of consideration of urban structure in formulating the strategy to improve the transport condition.

A clear picture of the city's urban structure is explained in the Spatial Plan document (RTRW). The classification of the structure hierarchy is based on the population density, road infrastructure availability, and the availability of facilities. Kecamatan Klojen has been designated as the city centre, whilst the other five kecamatan being sub-centres. This classification of urban structures, theoretically, should be able to evenly distribute movement, thus avoid overburdening any particular roads. However, looking at the transport problems which are apparent in many parts of the city, it is hypothetically stated that the existing urban structure should be carefully reclassified. This paper, therefore looks at assessing the urban spatial structure to suggest a better solution for the city's transport problem.

2 Methods

In assessing the accuracy in the classification of urban structure in Malang, this research adopts the similar approach to what the RTRW is using, known as the centrality index. This approach calculates the number of facilities and population density in a given unit of area. The unit of area, noted as BWP, with higher calculation of centrality becoming a unit of area with a higher hierarchy of structure. Tabel 1 explains the delineation of each BWP in

Malang. This research assumes that the current designation of BWPs is too vague, thus prioritization of transport solution seemed to be inaccurate.

2.1 Location

In the hope to formulate a better transport prioritization, this research re-classifies the afore-mentioned BWP classification into smaller-sized BWPs. The reclassification is done based on the sub-urban administrative adjacency. Table 2 explains the reclassification of unit of area in Malang.

Table 1. Malang's delineation of sub-urbans

Number of BWP	Sub-urban area coverage
1	Merjosari, Dinoyo, Sumbersari, Ketawanggede, Jatimulyo, Lowokwaru, Tulusrejo, Mojolangu, Tunjungsekar, Tasikmadu, Tunggulwulung, Tlogomas, Penanggungan
2	Bunulrejo, Purwantoro, Pandanwangi, Blimbing, Purwodadi, Polowijen, Arjosari, Balearjosari
3	Bandungrejosari, Bakalan Kerajan, Mulyorejo, Bandulan, Tanjungrejo, Pisang Candi, Karangbesuki
4	Kasin, Sukoharjo, Kidul Dalem, Kauman, Bareng, Gading Kasri, Oro-Oro Dowo, Klojen, Rampal Celakat, Samaan
5	Kedungkandang, Sawojajar, Madyopuro, Lesanpuro, Cemorokandang, Jodipan, Polehan, Kesatrian
6	Arjowinangun, Tlogowaru, Wonokoyo, Bumiayu, Buring, Mergosono, Kotalama, Kebonsari, Gadang, Ciptomulyo, Sukun, Bandungrejosari

Table 2. Proposed Malang's delineation of sub-urbans

Number of BWP	Sub-urban area coverage
1	Dinoyo, Ketawang Gede, Merjosari, Tlogomas, Penanggungan, Sumbersari
2	Tulusrejo, Tunggul Wulung, Tunjung Sekar, Jatimulyo, Lowokwaru, Mojolangu, Tasikmadu
3	Polowijen, Purwodadi, Arjosari, Balearjosari, Blimbing, Bunulrejo, Purwantoro, Pandanwangi
4	Polehan, Jodipan, Ksatrian, Sawojajar
5	Lesanpuro, Madyopuro, Cemoro Kandang, Kedungkandang
6	Kauman, Klojen, Sukoharjo, Kasin, Rampal Celaket, Kidul Dalem
7	Bareng, Gading Kasri, Oro-Oro Dowo, Sama'an
8	Bandulan, Karang Besuki, Mulyorejo, Pisang Candi, Tanjungrejo, Bakalankrajan
9	Kebonsari, Sukun, Ciptomulyo, Mergosono, Bandungrejosari, Gadang, Kota Lama
10	Arjowinangun, Tlogowaru, Wonokoyo, Bumiayu, Buring

3 Result and discussion

The Spatial Plan of Malang (RTRW Kota Malang 2010-2030) has divided its area into six units of area based on the spatial distribution of public facilities, population density, and the availability of transport infrastructure. Table 3 explains in more detail the calculation of the BWP hierarchy based on the classification in Malang's RTRW.

Table 3. The actual structure centrality index calculation

BWP	Population density		Facility density		Road infrastructure availability		Indeks Sentralitas	
	Score	**Class**	**Score**	**Class**	**Score**	**Class**	**Score**	**Rank**
1	89	Low	16	Low	161,36	High	0,88	3
2	90	Low	18	Low	177,18	High	1,01	2
3	82	Low	17	Low	143,88	High	0,73	4
4	143	Medium	27	Medium	155,22	High	2,25	1
5	65	Low	10	Low	132,90	High	0,72	5
6	56	Low	8	Very Low	91,67	High	0,55	6

Based on the calculation, it can be seen that BWP 4 is statistically positioned at the highest hierarchy of centres of the city's spatial structure, while the remaining five BWPs are designated as sub-centres. For the purpose of delivering transport strategies, this current classification seems to be too vogue as it fails to fully identify specific locations of transport hot-spots. Logically, a greater generation of traffic which means a greater possibility of traffic congestion will be attracted towards the higher hierarchies. However, the current hierarchy calculation does not accurately show transport improvement priority. For example, the heavy traffic in the area of Ketawanggede and Dinoyo of BWP 1 is admittedly much more in need of transport solution than in Blimbing of BWP 2. As being hypothesized at the beginning of this research, this relatively inaccurate method of delineation contributes to the inaccurate strategy of transport. Therefore, this research hopes to assess the urban spatial structure using smaller units of area. Table 4 explains the detailed calculation of proposed urban structure.

Table 4. The actual structure centrality index calculation

Zone	Population density		Facility density		Road infrastructure availability		Indeks Sentralitas	
	Score	**Class**	**Score**	**Class**	**Score**	**Class**	**Score**	**Rank**
1	104	Medium	18	Low	155.09	High	13.10	5
2	77	Low	15	Low	165.49	High	11.68	6
3	93	Low	18	Low	177.18	High	19.51	2
4	138	Medium	26	Medium	219.33	High	8.24	9
5	41	Low	5	Very Low	106.03	High	9.30	8
6	111	Medium	24	Medium	150.56	High	34.91	1
7	119	Medium	31	Medium	160.20	High	11.02	7
8	82	Low	17	Low	143.98	High	13.90	3
9	147	Medium	17	Low	159.05	High	13.28	4
10	23	Low	4	Very Low	65.84	Medium	6.96	10

Based on Table 4, it can be projected that traffic problems may concentrate the most in zone 6, zone 2, and zone 8 as they rank 1, 2, and 3, respectively. Figure 1 and 2 depict the current urban spatial structure and the proposed one, respectively.

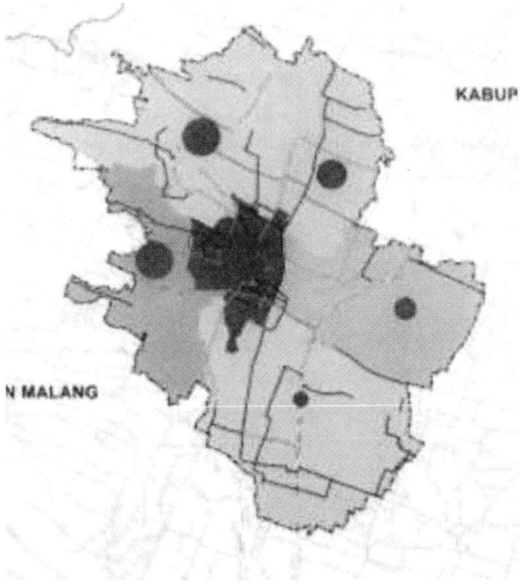

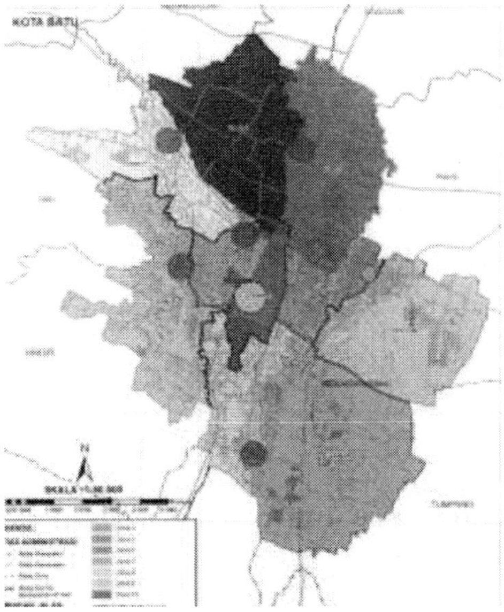

Fig. 1. Malang's urban spatial structure according to RTRW

Fig. 2. Malang's proposed urban spatial structure

Unlike the former hierarchy calculation, the proposed calculation provides a more accurate positioning of transport problem projection (Figure 2). Following the city's CBD (Zone 6) as the most congested area is the area surrounding Universitas Brawijaya (Zone 2) and the area around Dieng Plaza (Zone 8). These zones, based on a recent survey of traffic congestion points, are amongst the worst. During the city's peak hours and weekends, the roads' level of service reach the level D or E. Figure 3 explains the spatial distribution of traffic congestion in Malang.

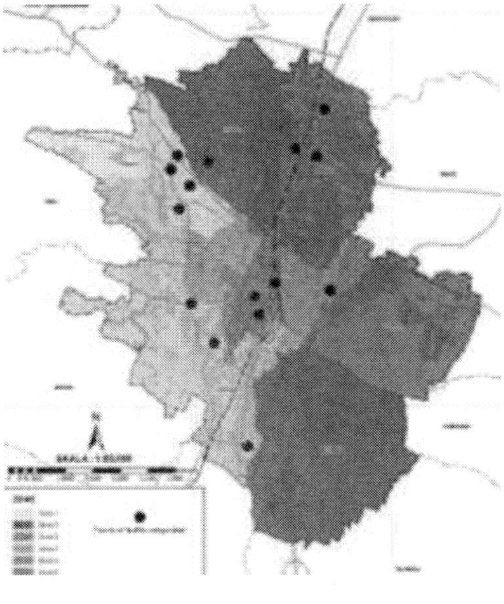

Fig 3. Points of traffic congestion in Malang

Figure 3 shows that there is a link between urban spatial structure and transport movement. It appears that the identified congested locations are ones that function as activity centers in the urban structure hierarchy (Figure 2). By comparing the two figures, it can be said that the theories that postulates the relationship between the urban spatial structure and travel movement patterns are valid enough to be applied in Malang. Therefore, this could be the basis for determining the appropriate transport strategy to handle the problems that occur. Strategies could be prioritized to aim the designated centres according to the centrality index (Table 4).

4 Conclusion

Using a similar approach in determining the hierarchy of urban spatial centre to the RTRW document, this paper re-assessed the hierarchy based on a smaller size of units of area. With more centres determined, this research allowed a more detailed trajectory of travel movement within the city. For instance, the area around Universitas Brawijaya, which was originally over seen, was now obvious in need of a more serious transport solution due to its relatively higher rank in the centrality index. Looking at the points of traffic congestion in Malang, and building upon the findings of this research, it can be suggested to the local government of Malang that there must be a prioritization of transport improvement accordingly to the hierarchy of urban spatial structure.

Reference

1. Ekawati, N., Soeaidy, M., & Robawanto, H *Kajian Dampak Pengembangan Pembangunan Kota Malang Terhadap Kemcetan lalu Lintas (Studi pada Dinas Perhubungan Kota Malang)*. Jurnal Administrasi Publik (JAP), 129-133. (2014).
2. Dinas Perhubungan Kota Malang. *Tataran Transportasi Lokal. Malang*: Dinas Perhubungan Kota Malang. (2015).
3. Setiawan, B. *Pengaruh Struktur Ruang Kota Terhadap Pola Pergerakan di Kota Semarang dan Kota Surakarta. Skripsi*. Tidak dipublikasikan. Semarang: Universitas Diponegoro. (2004).
4. Tondabola, L. *Pengembangan Struktur Ruang: Mereduksi Mobilitas Perkotaan*. Jurnal Media Matrasain, 73-79.Ridwan, *M. Measurement Scale of Research Variables* (pp. 24-31). Bandung: Alfabeta. (2015).
5. Curtis, C. *The Windscreen World of Land Use Transpot Integration*. Jurnal Town Planning Review, 423-453. (2005).
6. Bento, A., Cropper, M., Mobarok, A., & Vinha, K. *The Effects of Urban Spatial Structure on Travel Demand in The United States*. Jurnal Review of Economics and Statistic, 466-478. (2005).
7. Tamin, Ofyar Z. *Transport Planning and Modeling*. Bandung: ITB Bandung. (2000).
8. Greene, D. *Urban Subcenter: Recant Trends in Urban Spatial Structure*. Jurnal Economic Literature, 1-13. (2001).

What's the cause of the decrease in the number of angkot's passengers for LDG route in the city of Malang

Budi Sugiarto Waloeya[1,*], *Imma* Widyawati Agustin[1], and *Dadang* Meru Utomo[1]

[1]Brawijaya University, Department of Urban and Regional Planning, Malang, Indonesia.

Abstract. Increasing the number of population every year in the city of Malang makes the increase also the number of private vehicle users who trigger the occurrence of congestion and lack of public interest to ride public transport, especially in terms of operation and service of public transportation. The main purpose of the research is to explore factors causing the decrease of public transport passengers for LDG route. The research used Importance Performance Analysis. The results showed that the main factors causing the decrease of the number of passengers on the LDG route in Malang City based on the public perception such as driver behavior, the condition of the transport (inside and outside), transportation hygiene, the availability of place to put luggage, the availability of seating, the availability of route / route information LDG transport, duration of stop transportation time at the terminal, and length of travel time.

1 Introduction

Transportation is one of activities that has an important role in daily life. That is because transportation is closely related to the location and extent of human activities, goods and services. Transport also has an important role for the Indonesian people caused by several factors such as the geographical situation of Indonesia which consists of thousands of small islands and large, the waters of which comprise the bulk of the sea, rivers and lakes are in desperate need of their freight carried by land, water and air to reach all areas of Indonesia. However, behind the important role that transportation there are also some very crucial problems caused by lack of transportation planning precisely. One of the problems is congestion. One of the causes of congestion in urban areas is the increasing tendency of transportation service users to use private vehicles compared to public transport [1]. In addition to the improving economic conditions leading to higher levels of private vehicle ownership, the declining role of public transport is also due to the low level of public transport service itself. The low level of service usually involves inadequate facilities and infrastructure, long travel time, number of passengers exceeding transport capacity, low level of comfort, and difficult accessibility to certain areas.

* Corresponding author: budieswe@yahoo.co.id

Mass transit commonly used by people in Indonesia is urban transportation ("Angkot"). City transport is a transport from one place to another within a city by public bus or public passenger car bound in the route (Ministerial Decree No. 35 in 2003). The public transportation that people want is orderly, smooth, safe, convenient and fast transportation. This is one of the options in developing urban transport system, especially mass transportation to remain in demand by the community. Development of urban transport also has a mission that must be able to reduce congestion, able to maintain the quality of the environment and affordable by all levels of society [2]. Malang City is one of the cities that has many centers of activity. It has a high rate of economic growth and it is a city that has a high number and density of the population. This is because Malang City is a city of Education filled with many students who come from various regions. These things certainly have a major influence on the transportation system in the City of Malang. The increasing number of population growth based on statistical data of Malang City in 2014 is 0.8 per cent per year. In addition, the growth of private vehicles in Malang last five years is an average of 8.5 per cent per year [3].

According to the Head of Department of Transportation of Malang City, the number of transportation that operates in Malang City has decreased. The total number of public transportation, which initially amounted to 2,500 units, now still operates only 1,500 units. Even from 1,500 units is only 70 per cent were frequent and feasible to operate. This is due to the lack of passengers. Based on the previous research, it is known that there is a decrease in the number of passengers in LDG route in Malang City by 73 per cent, from 612 passengers to 167 passengers [4].

Fig. 1 Condition of angkot (LDG route) in the city of Malang
Sources: Field Survey, 2017

Some urban transport carrying or transporting passengers exceeds a predetermined capacity in the standard of a maximum of 12 persons. The condition of the urban transport itself (LDG route) that is still less well maintained affects the comfort of the passengers while it is in the transport and also can reduce the passengers' interest to choose LDG route transport (Figure 1).

Figure 2 shows passengers rarely cause drivers have to wait long for passengers are transported not a bit. It is precisely to influence the waiting time of passengers who indirectly passengers who had boarded the transport had to wait for other passengers, even the waiting time can be up to 20 minutes. Thus, the waiting time of LDG passengers is quite long compared to the standard according to the World Bank that is more than 20 minutes.

Fig. 2 Angkot were parked and waiting for passengers at Pasar Besar Road
Sources: Field Survey, 2017

2 Methods

This type of research is reviewed based methods [5], the study was classified into quantitative and qualitative research. It is classified as quantitative because the data is in the form of numbers which is intended to interpret directly and analyzed through the number, level and scale as in the calculation of operational performance, financial performance and calculation of IPA. This research is also qualitative because in this research also uses data related to public perception, as well as graphs and diagrams to describe the characteristics of passengers.

2.1 Location

The location of the study is the road or corridor passed by the LDG route of public transport ("Angkot") round-trip (Table 1 and Figure 3).

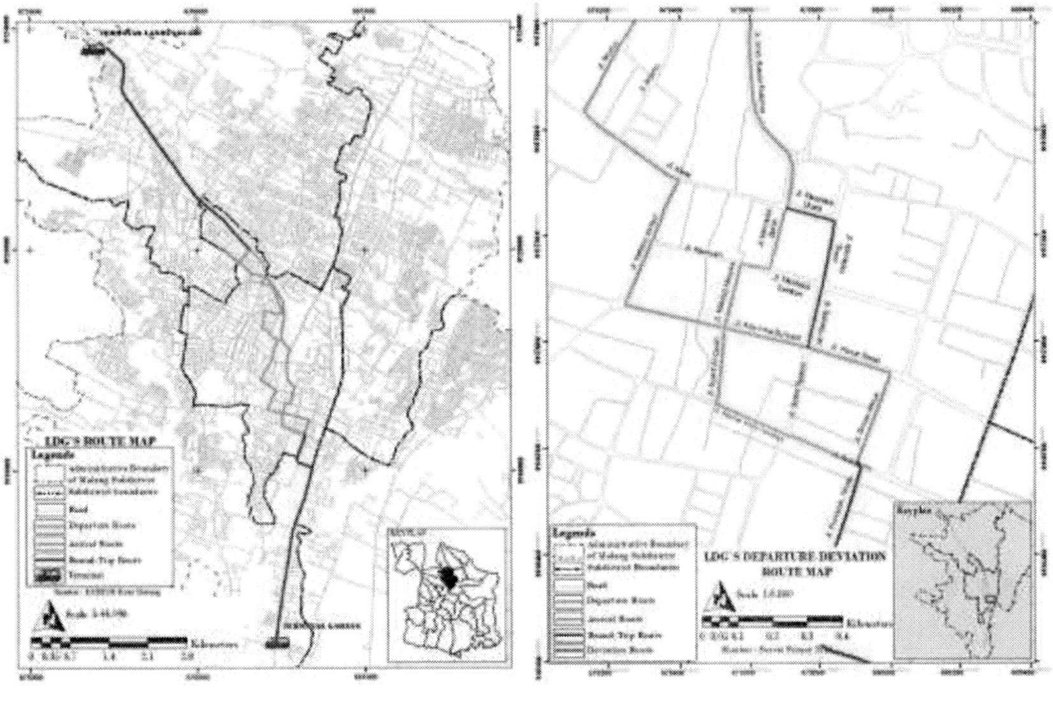

Fig. 3. LDG route of angkot round-trip in the city of Malang and deviation route

Table. 1 LDG route of angkot round-trip in the city of Malang

Route	Departure	Arrival
LDG	Landung Sari Terminal – Raya Tlogomas - Mayjen MT. Haryono – Mayjen Panjahitan – Brigjen S. Riyadi – Jend Basuki Rahmat – Merdeka Barat – Kauman – KH. Wachid Hasyim – Ade Irma Suryani – Pasar Besar – Sersan Harun – Kyai Tamin - Prof. Moh. Yamin – Sartono SH – Kol. Sugiono –Gadang Terminal	Gadang Terminal – Kol. Sugiono – Sartono SH – Prof. Moh. Yamin – Kyai Tamin – Kapten Piere Tendean - Syarif Al Qodri – Ade Irma Suryani – KH. Hasyim Ashari – Kawi – Bromo – Buring – Cerme – Guntur – Anjasmoro – Raung - Ijen – Bandung – Veteran – Terusan Bogor – Mayjen Panjahitan – Mayjen Haryono – Raya Tlogomas - Landung Sari Terminal

2.2 Sample

The population is the whole of the research subject. The sample is partially or representative of the population studied. Sampling of urban transport (Angkot) passengers can be done by interviewing urban transport (Angkot) passengers using accidental sampling method, ie taking samples with certain considerations that are not designed meeting first and easiness to get the necessary data, for example, easy to find, easy to reach or coincidence [5]. In this study the population taken is the entire population in Malang City. The calculation of the sample of city transport is calculated using the Bernoulli formula as follows:

$$n = \frac{\left[Z_{\frac{\alpha}{2}}\right]^2 \cdot p\,(1-q)}{e^2} \quad (1)$$

Information:
n = total of minimum sample
Z = the level of significance ($\alpha = 0.05$) so obtained (Z = 1.96)
e = level of error (10%)
p = the proportion of the number of questionnaires considered correct (50%)
q = the proportion of the number of questionnaires considered to be false (50%)

Determination of samples of urban transport drivers can be conducted by interviewing directly to the driver of urban transportation for LDG route in Malang City. Determination of sample from the owner of urban transportation can be conducted by interviewing directly to the owner of urban transport LDG route in Malang.
Thus obtained results of the calculation as follows:

$$n = \frac{\left[Z_{\frac{\alpha}{2}}\right]^2 \cdot p\,(1-q)}{e^2}$$

$$n = \frac{[1{,}96]^2 \cdot (0{,}5)\,(0{,}5)}{(0{,}1)^2}$$

$$n = \frac{3{,}8416 \times (0{,}25)}{0{,}01} = 96.04 \sim 96$$

Based on the sample calculations using the Bernoulli formula, the minimum sample obtained for this study is as much as 96 respondents.

2.3 Data colection

2.3.1 Primary survey

The collection of primary data is the data collected by direct observation to locations that have been previously. The techniques used in this survey include [6]:

1. Observation or Field Survey: Observation is a direct observation to the object of research and the systematic recording of matters relating to research activities carried out. The observations in the form of identification of potential problems that exist in the study area. This observation technique is used to find out the performance of LDG route service including passengers up and down point, number of operating fleet, travel time, load factor, and headway.
2. Interview: Interview is a means of collecting data used to obtain information directly from the source.
3. Questionnaire: The questionnaire is a list of questions given by researchers to others who are willing to provide a response (respondent) in accordance with the request of researchers. The purpose of distributing this questionnaire is to find complete information about a problem from the respondent. Questionnaires are divided into two types: open questionnaires and closed questionnaires.
4. Documentation: The documentation is intended to obtain data directly from the research include relevant books, legislation, activity reports, photos and data relevant to the study. In this study, documentation will be collected in the form of photographs during field observation (photos of LDG transport, terminal, photograph of field survey activities, etc.).

2.3.2 Secondary survey

Table 2 Institutions and data required

<table>
<tr><th>No.</th><th>Institutions</th><th>Data</th><th>Function</th></tr>
<tr><td>1</td><td>Regional Development Planning Board (BAPPEDA) Malang</td><td>• SHP Map of Malang City
• Spatial Plan of Malang City</td><td rowspan="3">• References
• Profil
• Used in making alternative programs in the form of policies and related issues that will be submitted to the experts / stakeholders.</td></tr>
<tr><td>2</td><td>Central Bureau of Statistics (BPS) Malang</td><td>• Malang City in Figures
• Distrcit in Figures</td></tr>
<tr><td>3</td><td>Department of Transportation (DISHUB) or the Department of Traffic and Road Transport (DLLAJ) Malang</td><td>• Number of city transportation fleet in Malang city 2017
• Route of urban transport route in Malang
• Number of fleet per route of transportation of city of Malang
• Mileage distance every urban transport trajectory in Malang City</td></tr>
</table>

Secondary data collection is conducted by obtaining data from related institutions, studying scientific papers, literatures, reports, as well as other literature related to the issues discussed in order to obtain the theoretical foundation in the discussion. This study will be obtained from the initial hypothesis is based on conditions on the ground and do a comparative study between the field and the theories of the existing literature.

2.4 Importance performance analysis

Importance Performance Analysis (IPA) in this study is used to determine the level of satisfaction and the interests of public transport passengers based on the criteria that have been prioritized by the stakeholders that will be used in determining the variables related strategies any transport service is lacking and needs to be enhanced or prioritized handling.

Table 3 Variable of public transport service assessed by public transport passengers for LDG route

Criteria	No	Variable
Safety	1	Avoidance of accidents
	2	Free from crime or free from crime in Public Transport (Angkot)
Comfort and Load Factor	3	Duration of passengers waiting for LDG transport
	4	Driver behavior (reckless / kebut-kebutan)
	5	Driver service to passengers
	6	Ease of passengers while up and down from transport
	7	Protection from bad weather
	8	Transport conditions (inside or outside)
	9	Cleanliness in transport
	10	Availability of place to put luggage
	11	Seating availability (not crowded)
	12	Condition and quality of seating
	13	Ease of reaching the stop of transportation (bus stop)
	14	Interesting stops
	15	Raise and lower passengers in designated places
Reliability	16	Operational schedule / transport schedule has a fixed schedule
	17	The availability of public transport that serves passengers at all times
	18	There are stops / halte
	19	Availability of route information / LDG transport routes
	20	Transportation operates through route / route that has been determined
Travel Time	21	The number of LDG transport that operates / serves passengers at any time
	22	The length of time stops transport in the terminal
	23	The length of the driver up / down passenger
	24	Transportation time duration (short)
Tariff	25	Tariffs from LDG transport (the amount of expenses that passengers spend on travel destinations)
	26	Costs are relatively cheap / affordable
	27	Tariff suitability of transportation with travel time
	28	Provision of special tariff for children and students

This is the phase of the IPA method in detail from the weighting stage to the stage of determining the variables based on each IPA quadrant.

1. Weighting: The scale used is the Likert scale (5 levels) which is generally used in research is the measurement of attitudes, opinions and perceptions of a person or group of people about social events or symptoms [6].
2. Conformity Level: User satisfaction is illustrated by the degree of conformity between the perceptual assessment of the quality and the importance level assessment of aspects of service performance.
3. Cartesian diagram: X axis filled the score of service / implementation quality level, and Y axis filled by score of importance level.

Cartesian diagram is a structure that is divided into 4 sections bounded by two lines that intersect perpendicularly at the points (X, Y), where X is the average of the average score of the level of perception / satisfaction of users of all the items in service. Y is the average of the average score of importance of all factors affecting service user satisfaction (Figure 4).

The value of each factor is:

$$x_i = \frac{X_i}{n} \quad (2)$$

$$y_i = \frac{Y_i}{n} \quad (3)$$

The determinant as the midpoint on the X and Y axes:

$$\bar{x}_{max} = \frac{x_1 + x_2 + x_3 + \ldots + x_n}{n} \quad \ldots\ldots (4)$$

$$\bar{y}_{max} = \frac{y_1 + y_2 + y_3 + \ldots + y_n}{n} \quad \ldots\ldots (5)$$

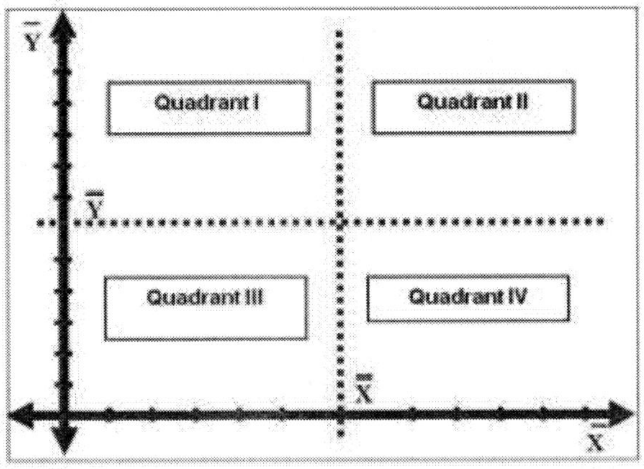

Fig. 4 Cartesian diagram

The following is an explanation of the four quadrants contained in the IPA method [7]:

1. Quadrant I: Concentrate Here

The factors in this quadrant are considered to be a very important factor by consumers but the current conditions are not satisfactory so the management is obliged to allocate adequate resources to improve the performance of these various factors. The factors in this quadrant are a priority to be improved.

2. Quadrant II: Keep Up The good Work

Factors in this quadrant are considered as supporting factors for customer satisfaction so that the management is obliged to ensure that the performance of the institutions they manage can continue to maintain the achievements that have been achieved.

3. Quadrant III: Low Priority

Factors in this quadrant have a low level of satisfaction and at the same time is considered not very important for consumers, so the management does not need to prioritize or pay too much attention to these factors.

4. Quadrant IV: Possible Overkill

The factors in this quadrant are considered less important and management needs to allocate resources related to those factors to other factors that have higher handling priorities that still require improvement.

3 Result and discussion

The result of IPA is a quadrant divided into 4 (four) sections bounded by two intersecting lines (Figure 5 and Table 2). The X axis is obtained from the average of the satisfaction rate score ($\bar{\bar{X}}$) of 3.58 whereas the Y axis is derived from the average of the average interest rate score ($\bar{\bar{Y}}$) of 4.09.

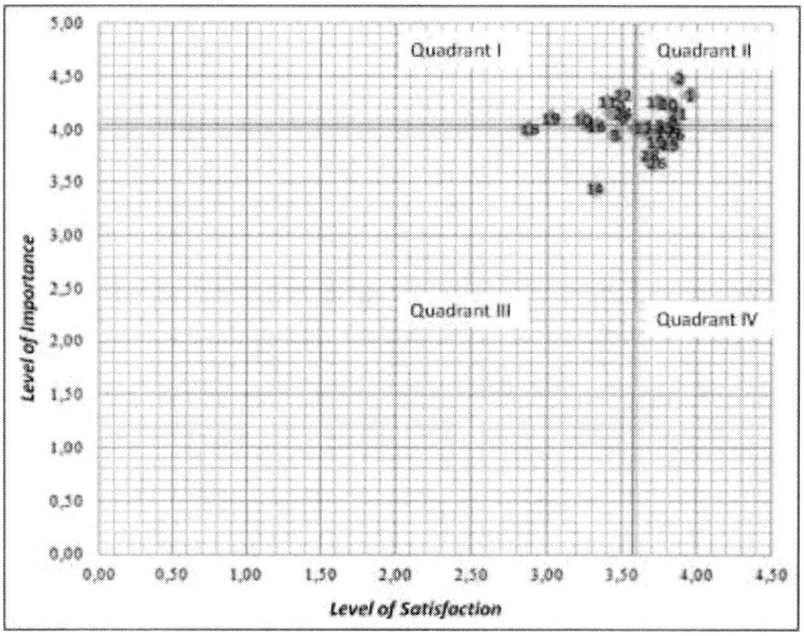

Fig. 5 IPA diagram

Variables or attributes in quadrant I by society (passengers of LDG transport) is a variable with the main priority that must be improved and improved performance. This is because the variables contained in quadrant I is a component of LDG transport service which is considered as a factor with a very high level of importance but in its implementation has not been satisfactory or not in accordance with the wishes of the community.

The criteria of improvement of public transportation service (LDG route) in Malang City according to the community needs to be prioritized handling that consists of:

1. Comfort with criteria that need to be improved include driver's behavior in driving, transport conditions (inside or outside), cleanliness of transport and availability of place to put luggage.
2. Load Factor with components such as seat availability in the sense that passengers do not sit squashed.
3. Reliability includes the availability of route information / LDG transport routes.
4. Travel time with components that need to be repaired is the length of time to stop transport in the terminal and the length of time travel or transport time in the vehicle (transport) is short.

Table 4 Variable coordinate points existing in quadrant I

No.	Sub variable	X	Y
1	Avoidance of accidents	3.96	4.33
2	Free of crime or criminality in transport	3.89	4.49
3	Protection from bad weather	3.84	4.05
4	The availability of public transport that serves passengers at all times	3.73	4.27
5	Transportation operates by route / route that has been determined	3.81	4.25
6	The number of LDG transport that operates / serves passengers at all times.	3.88	4.17

4 Conclusion

The main factors causing the decrease of the number of passengers on the LDG route in Malang City based on the public perception such as driver behavior, the condition of the transport (inside and outside), transportation hygiene, the availability of place to put luggage, the availability of seating, the availability of route/route information LDG transport, duration of stop transportation time at the terminal, and length of travel time.

Reference

1. Tamin, Ofyar Z. *Transport Planning and Modeling*. Bandung: ITB Bandung. (2000).
2. Rahma, *NThe role of the Department of Transportation in Improving Public Service in the Field of Transport City (Study at the Transport Department of Malang City).* Journal of Public Administration (JAP), Vol.1, No. 7, pp 1296-1304. (2013).
3. *Central Bureau of Statistics.* Statistics of Malang City. Malang: Central Bureau of Statistics. (2014).
4. Irjaya, D. *Impact Relocation of Terminal Gadang Malang City to Operational Cost of Vehicle and User of City Transportation*. Malang: Brawijaya University. (2013).
5. Arikunto, P. D. *Research Procedures A Practice Approach.* In P. D. Arikunto, Research Procedures A Practice Approach (pp. 130-142). Jakarta: PT Asdi Mahasatya. (2006).
6. Ridwan, *M. Measurement Scale of Research Variables* (pp. 24-31). Bandung: Alfabeta. (2009).

7. Ong, J. O. *Analysis of Customer Satisfaction with Importance Performance Analysis at SBU Laboratory* Cibitung PT Sucofindo (PERSERO). J @ TI Undip Vol IX, 5-9. (2014).

Bearing capacity characteristic of subgrade layer quicklime treated laterite soil

Zubair Saing[1,*], *Lawalenna Samang*[1], *Tri Harianto*[1], and *Johannes Palanduk*[1]

[1]Hasanuddin University, Department of Civil Engineering, Gowa, Indonesia

Abstract. This study aimed to analyze bearing capacity characteristics of laterite soil with quicklime stabilization as road subgrade layer. The laterite soil obtained from three different locations with three different ferrous (FeO) content. Soil samples were conduct by mixing laterite soil with 3%, 5%, 7%, and 10% lime content on Proctor initial condition, then cured for 3, 14, and 28 days before tested for soil bearing capacity using CBR laboratory test. Result showed that the lime treated laterite soil increases the bearing capacity for all types of laterite soil significantly in line with incrasing of lime content on 28 days curing time (11,67% to 51,67%, 14% to 53,33%, and 14,17% to 71,67%, respectively for LH1, LH2 and LH3). The increasing of soil bearing capacity due to pozzolanic strength gain. Based on these results, the laterite soil has potentially to be used as road subgrade layer.
Keywords: Road foundation, laterite soil, quicklime stabilization, soil performance

1 Introduction

Indonesia's current development has generated efforts to manage economic resources in the community. Local natural resources are very potential to require innovation for its utilization. This is mainly due to the high competition and cost to obtain these materials. In addition, various studies are underway to find a better way to achieve this goal, a way that will cost less and be more economical when compared to using materials that naturally meet the requirements standards. In line with this, the material required for infrastructure development intended to increase, especially for construction and road foundation material. Certain areas have limited resources to qualified material, especially material (grade) A and B. One such area is particularly the Sorowako in South Sulawesi, which has been buying material from other areas to be used as road foundation material. On the other hand, the local content materials has potential to be developed as a replacement material for road subgrade layer, one of them is laterite soil.

Laterite soil containing the clay minerals are relatively high mainly illite and montmorillonite, so greater damage potentially if carried out construction work on the soil like this [1]. Clay minerals and metals are high, it can be used for various needs of both the construction, industry, and others. Laterite soil is soil groups result from highly weathering

* Corresponding author: zubairsaing.ummu@gmail.com

zone, formed from the concentration of hydrated iron oxide and aluminum, soil characteristics of this type have a hard, impenetrable, and very difficult to change if the conditions dry [2]. Laterite have a wide variety of red, brown to yellow, fine grain sized residual soil with a granular texture have a mild form nodular and well cemented. The correct use of the term laterite is a compact rock formations vesicular iron (a massive vesicular or concretionary ironstone formation). Laterite based on hardening as "freeic" for hard soil cemented iron rich, "alcrete" or bauxite for hard soil cemented aluminum rich, "calcrete" for hard soil calcium carbonate rich, and "silcrete" for soil in silica rich. Other definitions are based on the ratio of silica (SiO2) to oxide (Fe2O3 + Al2O3), for laterit such comparisons between 1.33 and 2.0, while above 2.0 is not laterit.

Research on lateritic soil has been widely applied, especially in countries that there are many types of soil such as in Asia and Africa. The higher of clay minerals content in laterite soil causing the strength decreases [3]. Laterite soil stabilization with lime will increase the strength becomes 2-3 times [2]. Addition of a little lime and cement highly efficient, improve laterite soil by adding 3% only 2% of cement and lime [1]. Laterite soil stabilization using polymer solution (GKS) has increasing of soil strength with increasing curing time after 7 days [4]. Laterite soil stabilization using ash cane fiber is very effective in stabilizing and strengthening the geotechnical properties of soil [2]. Laterite soil stabilized with ash cane and cement is very effective as a stabilizing agent in the content of 6% and 5% of the cement to reinforce the geotechnical properties of soil, and can be used for construction road as a subbase [5]. Stabilization using corn cob ash (CCA) has caused the maximum dry density decreases in ash content of 1.5% CCA, optimum water content increased in the binder content of 0 to 7.5%. CBR value increased at the CCA content of 1.5%, and then decreased with the addition of CCA. Unconfined compression strength also increased on 1.5% CCA content, and decreased with the addition of CCA [6]. Laterite soil stabilization using sodium silicate liquid indicate that the addition of sodium silicate over 9% decrease the compressive strength of the soil [7]. Laterite soil characteristics as a coating material that is fed gasoline, hydraulic conductivity increases in line with the increase in hydraulic gradient [8]. Lateritic soil with high liquid-limits is commonly used for roadbed construction. However, lateritic soil has properties that are sensitive to moisture, and therefore a common method of treating the soil is by adding lime. However, it is difficult to mix lime with lateritic soil homogeneously in the field as lateritic soil is prone to agglomeration. Lime (Ca(OH)2) is added to one of the groups of soil samples and the other group is left untreated. The maximum CBR value is not consistent with maximum dry density. Moreover, the CBR values of soil has been treated with lime are higher than the untreated soil for an agglomerate size that ranges from 0.5 mm to 2 cm. Beyond this range, the addition of lime does not improve the lateritic soil. Compaction status and water intrusion are two important influential factors on CBR values [9].The potentially of laterite soil with lime and cement stabilization as a road foundation [10-15].

The growing research on the utilization of laterite soil, indicating that the soil is very potential to be used in various constructions. However, still not much research on laterite soil with higher iron metal content (ferro laterite soil). This study was conducted to analyze the bearing capacity characteristics of laterite soil with quicklime stabilization as road subgrade layer.

1.2 Material and method

The material used in this research is laterite soil with higher ferrous content from Sorowako East Luwu Regency South Sulawesi on coordinate S 2°56'21,16" and E 121°36'26,54" as shown in Figure 1. Tests of physical and mechanical properties of the soil were conducted in laboratory according to American Standard for Testing and Materials (ASTM), as shown in Table 1. Soil stabilization using quicklime with CaO = 97,8% and silica oxide (SiO_2) = 2,2%. The addition of lime with compositions of 3, 5, 7, and 10% at maximum dry density from Proctor standard test. Sample was cured for 3, 14, and 28 days before CBR laboratory testing, with the total number of samples tested was 108. Microstructural testing were conducted to knows the micro condition of treated and untreated soil. This test using X-ray diffraction (XRD) for minerals content and Scanning Electron Microscope (SEM) for photomicrograph. While chemical content using Energy Dispersive Spectorscopy (EDS) test.

Table 1. The standard test used

Type of Testing	**ASTM Standard Number**
Grain Size Analisys	C-136-06
Liquid Limit (LL)	D-423-66
Plastic Limit (PL)	D-424-74
Plasticity Index (IP)	D-4318-10
Spesific Gravity (Gs)	D-162
Water Content (Wc)	D-2216-98
Uniconfined Compression Test (qu)	D-633-1994
Compaction Test	D-698
CBR Laboratory Test	D-1833

Figure 1. Sampling location of laterite soil

2 Discussion

Testing of the soil basic properties including; specific gravity, soil moisture content, soil density, Atterberg limits, and soil grain size analysis. Tests carried out on the laterite soil for three types with three different locations using ASTM standard test such as Table 1. Results of laboratory tests for physical and mechanical properties as well as chemical and mineral properties as shown in Table 2, 3, and 4.

Table 2. Physical properties of laterite soil

No	**Soil Characteristics**	**Unit**	LH1	LH2	LH3
1	Specific gravity (Gs)	-	2,73	2,62	2,66
2	Water content (w)	%	20,26	22,25	18,86
3	Sieve analysis				
	a. gravel	%	-	-	-
	b. sand	%	7,69	5,11	8,25
	c. Silt/clay	%	92,31	94,89	91,75
4	Consitency limits				
	a. Liquid limt (LL)	%	65,98	68,73	67,77
	b. Plastic limit (PL)	%	35,92	37,96	36,86
	c. Plasticity Index (PI)	%	30,06	30,77	30,91
5	Standard Proctor test				
	a. Maximum dry density (□$_{d\ maks}$)	kN/m^3	16,91	16,92	16,89
	b. Optimum water content (w_{opt})	%	17,23	16,72	16,53
6	Unconfined compression strength (q_u)	kN/m^2	71,44	128,88	75,61
7	California Bearing Ratio (CBR)				
	a. CBR unsoaked	%	11,77	22,99	22,80
	b. CBR soaked	%	7,67	10,33	8,50
8	Direct shear test				
	a. Cohesion (C)	kN/m^2	15,2	16,3	14,7
	b. Internal friction angle (□)	(°)	18	20	19

Table 3. Chemical properties of laterite soil obtained from EDS test

Element (%)	**Laterite Soil**		
	LH1	**LH2**	**LH3**
MgO	2,33	0,83	1,28
Al_2O_3	4,41	5,73	8,45
SiO_2	12,58	2,28	3,71
K_2O	0,1		
TiO_2	0,08		
FeO	80,5	86,55	84,88
SO_3		1,05	
CaO		0,25	
MnO		0,24	
NiO		2,78	1,38

Table 4. Mineral content of laterite soil obtaned from XRD test

Mineral Content (%)	**Laterite Soil**		
	LH1	LH2	LH3
Hematite HP, iron(III) oxide	13	7	1
Kaolinite	20	8	67
Illite-montmorillonite (NR)	60	80	18
rutile HP	1	2	11
Magnesium Silicate	6	3	3

Based on the results of Tables 2, it is known that laterite soil grains are dominated by silt/clay material of 91,75-94.89%, with plasticity index of 30,66%-30,91%. These results indicate that laterite soils are included in clay classification with high plasticity (A-7-6 according to AASTHO and CH according to USCS). While based on chemical composition on Table 3, showed that laterite soil was dominated by iron oxide content of 80,50%-86.55%, and based on showed on Table 4, laterite soils are dominated by illite-montmorillonite minerals of 18%-80%.

Subsequently, bearing capacity of lime treated laterite soil obtained from laboratory CBR test. The change of CBR value lime treated laterite soil of 3, 5, 7, and 10% with curing time of 3, 14, and 28 days as shown in Figure 2.

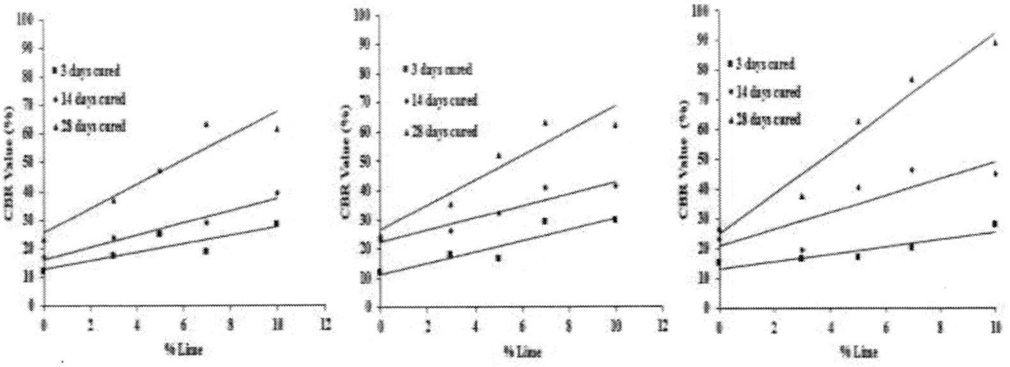

Figure 2. Correlation of CBR value with lime addition and curing time of laterite soil

Based on Figure 2, it can be seen the change of CBR value of laterite soil with addition lime content and curing time. Increasing of lime content and curing time caused the increasing of soil CBR value. Pozzolanic strength gain due to pozzolanic reaction in soil cause the increasing of soil strength and soil bearing capacity. Increased of CBR value lime treated laterite soil are 11.67% to 51.67% for LH1, 14% to 53.33% for LH2, and 14.17% to 71.67% for LH3. The increase in strength and bearing capacity of soil occurs due to the addition of lime to the soil undergoing three processes, namely the hydration of lime hydration, cation exchange reactions, and the reaction of pozzolanic carbonation [16] The process of lime hydration is a chemical reaction between lime with water in which calcium hydroxide or lime hydration ($Ca(OH)_2$) is formed. These reactions include the exchange of calcium ions (Ca^{2+}), adsorption of $Ca(OH)_2$ by particles and cementation on inter-particle contact calcium silicate. The cation exchange reaction includes the substitution of Na^+ and H^+ ions in the soil with Ca^{2+} ions from lime. The clay particles continuously absorb $Ca(OH)_2$ until saturated. This exchange reduces plasticity, improves performance, and soil strength. This reaction occurs immediately when mixing the stabilization material into the soil. Pozzolanic reactions and carbonization include the reaction between clay particles with $Ca(OH)_2$ produced by lime. This contributes to the long-term strength of pozzolanic materials. The strength of the degraded soil increases with the timing of this reaction. Calcium hydroxide in the soil-water reaction with silicate and aluminate in the soil form a sterilized material or a binder comprising calcium silicate hydrate (CSH). In line with the occurrence of a pozzolanic reaction, the CASH is slowly transformed into a crystalline phase in the form of calcium silicate hydrate (CSH) and calcium aluminum hydrate (CAH) which suffers from hardening over time to form a permanent mixture that binds soil particles. This condition causes an increase in soil strength and soil bearing capacity. This condition also seen from the results of microstructural (SEM Photomicrograph) tests as in Figure 3.

Figure 3. SEM Photomicrograph of laterite soil

Based on Figure 3, it can be seen that untreated laterite soil has large micropore structure, after treated the flocculation and agglomeration structures form calcium silicate hydrate (CSH) ions that cover the pores, resulting in smaller pores and soil fabric become more solid [7].

3 Conclusions

In this paper, three different ferro laterite soils treated with lime for road subgrade layer was studied. The main aim was to analyze the bearing capacity characteristics of laterite soil with quicklime stabilization as road subgrade layer. Based on physical properties, the laterite soil was A-7-6 for AASHTO soil classification system and CH for USCS soil classification system. An increasing percentage of the lime addition (CaO) and increased curing time lead to increased soil bearing capacity, on 10% lime mixture and 28 days curing time increase three time than untreated soil. Based on these results, the treated laterite soil has potentially to be used as road subgrade layer, but it should be tested in detail of physical model before field implementation.

References

1. F.H.M. Portelinha, D.C. Lima, M.P.F. Fontes, C.A.B. Carvalho, Modification of a Lateritic Soil with Lime and Cement: An Economical Alternative for Flexible Pavement Layers, Soils and Rocks, São Paulo, 35(1): 51-63, January-April, 2012, pp 51-63. (2012)
2. Amu, O.O., Ogunniyi, S.A. and Oladeji, O.O., Geotechnical properties of lateritic soil stabilized with sugarcane straw Ash, American Journal of Scientific and Industrial Research, Science Huβ, 2(2) pp. 323-331 (2011)

3. Sree Danya, Ajitha, A.R, Evangeline, Y. Sheela., The Suitability and Lime Stabilization Requirement of Some Lateritic Soil Samples as Pavemen, Int. J. Pure Appl. Sci. Technol., 2(1), pp. 29-46. (2010)
4. Aminaton Marto, Nima Lativi, Houman Souhaei, Stabilization of Laterite Soil using GKS Soil Stabilizer, EJGE Journal, Vol **18**, Bund. C, pp 521-532 (2013)
5. Kiran, S.P., A.N Ramakrishna, Shrinivas. H.R, Stabilization of Lateritic Soil by using Sugarcane Straw Ash and Cement, Journal of Civil Engineering Technology and Research, 2 (1) pp.615-620. (2014)
6. Yinusa A. Jimoh, O. Ahmed Apampa, An Evaluation of the Influence of Corn Cob Ash on the Strength Parameters of Lateritic Soils, Civil and Environmental Research, 6 (5) No.5. (2014)
7. Nima Latifi, Amin Eisazadeh, Aminaton Marto, Strength behavior and microstructural characteristics of tropical laterite soil treated with sodium silicate-based liquid stabilizer, Environ Earth Sci 72:91–98 DOI 10.1007/s12665-013-2939-1, Springer-Verlag Berlin Heidelberg. (2014)
8. Wisley Moreira Farias, Geraldo Resende Boaventura, Éder de Souza Martins, Fabrício Bueno da Fonseca Cardoso, José Camapum de Carvalho and Edi Mendes Guimarães., Chemical and Hydraulic Behavior of a Tropical Soil Compacted Submitted to the Flow of Gasoline Hydrocarbons, Environmental Risk Assessment of Soil Contamination, Intech, pp 638-655. (2014)
9. Yunzhi Tan, Mozhen Hu, Dianqing Li, Effects of agglomerate size on California bearing ratio of lime treated lateritic soils, International Journal of Sustainable Built Environment, 5, pp. 168–175, Elsevier B.V. All rights reserved. (2016)
10. Saing, Z., Samang, L., Haryanto, T. and Patanduk, J., "Microstructural and Mechanical Characteristic of Potential Ferro Laterite Soil as Subbase Material", *International Journal of Innovative Research in Advance Engineering*, 3 (2), pp 42-48. (2014)
11. Saing, Z., Samang, L., Harianto, T. and Patanduk, J., "Quick Lime Stabilization of Ferro Laterite Soil to Subgrade Material Implementation", *International Journal of Science and Research*, 5 (11), pp 582-586. (2016)
12. Zubair Saing, Lawalenna Samang, Tri Harianto and Johannes Patanduk, "Mechanical Characteristic of Ferro Laterite Soil with Cement Stabilizationas a Subgrade Material", *International Journal of Civil Engineering and Technology*, 8(3), pp. 609–616. (2017)
13. Zubair Saing, Vertical Deformation of Lime Treated Base (LTB) Model of Laterite Soil using Numerical Analysis. International Journal of Civil Engineering and Technology, 8(5), pp. 758–764. (2017)
14. Zubair Saing, Vertical Deformation of Lime Treated Base (LTB) Model of Laterite Soil using Numerical Analysis. International Journal of Civil Engineering and Technology, 8(5), pp. 758–764. (2017)
15. Saing Z, Samang L, Harianto T, Patanduk J, "Study on Characteristic of Laterite Soil with Lime Stabilization as a Road Foundation, *International Journal of Applied Engineering Research*, 12(14), pp. 4687-4693. (2017)
16. Jaritngam S, O. Somchainuek, and P. Taneerananon, "Feasibility of Laterite-Cement Mixture as Pavement Base Course Aggregat", *International Journal of Science and Technology Transaction of Civil Engineering* Vol. **38**. No. C1+, pp. 275-284. (2014)
17. Olugbenga O Amu, Oluwole F.B., dan Iyiola A.K., The Suitability and Lime Stabilization Requirement of Some Lateritic Soil Samples as Pavemen, Int. J. Pure Appl. Sci. Technol., 2(1), pp. 29-46. (2011)

Impact of compaction test on mineral texture breakage in Tanjung Bunga Beach, Makassar

Hamid Umar[1,*], *Dadang A.* Suriamihardja[2], *Lawalenna* Samang[3], and *Ulva Ria* Irfan[1]

[1]Hasanuddin University, Department of Geology Engineering, 90245 Makassar, Indonesia
[2]Hasanuddin University, Department of Geophysics, 90245 Makassar, Indonesia
[3]Hasanuddin University, Department of Civil Engineering Department, 90245 Makassar, Indonesia

Abstract. Rate of mineral deterioration in Tanjung Bunga Beach, Makassar city sediment, as a coastal city area that temporarily evolved with infrastructure development, through a Compaction test has found damage mineral from cracks, split and shatter. The aim of this research is to classify the compaction test from mineral as result texture breakage from compression test. Tanjung Bunga beach sediments based on grain size was classified from fine sand, medium sand and coarse sand. Petrographic analysis of 15 samples of in three drilling sites with 5 meters depth, found quartz mineral composition from 20 – 25%, hornblende bertween 5 – 20%, pyroxene 5 – 15%, pagioclast 5 – 15%, orthoclast 5 – 15%, biotite 10 – 20%, and opaque minerals 10 – 25%. Value of Compaction Test laboratory Fine Sand has 4 - 14.4 Div, Medium Sand has 6.7 - 20 Div, and Coarse Sand has 3.2 - 24 Div. Petrographic analysis after Mineral Compaction Test on fine sand was cracked up to 2 – 20%, split 2 – 12%, and shattered 2 – 10%, medium sand was cracked mineral up to 2 – 15%, split up to 1 – 13%, and shattered up to 1 – 5%, and coarse sand up to 1 – 10% cracked minerals, split up to 2 – 15%, and shattered up to 2 – 7%. High percentage of crack minerals were found in fine sand with low Compaction Test values and coarse sand with low crack mineral and high Compression Strength values. More cracked mineral was found in Quartz and no opaque minerals were shatter. Mineral pyroxene is founded as most shatter shape and orthoclase mineral with the most split shape.

1 Introduction

Tanjung Bunga beach area, Makassar area in development concept of "Water Front City", is part of "Waterfront Development" as development concept of coastal area with the construction of settlements infrastructure and coastal tourism. The burden in coastal sedimentary area may damage the minerals, change size and shape of minerals that will reduce the ability, formation strength of land, and coastal area functions.

Tanjung Bunga beach, characterized as spit beach were changing in gradual, vertical sediment composed by coastal sediment and a thickly fluvial-deltaic layer, a resistivity of 0.1 - 402 Ohm m, with value of 1.3 Φ - 2.4 Φ (fine sand - medium sand), Coastal topography is relatively sloping and affected by the tides. Lands of coastal areas, generally

* Corresponding author: hamidumar@gmail.com

high levels of rotting minerals than weathered mineral, and generally this kind is rich in kaolinites mineral types and quartz [2].

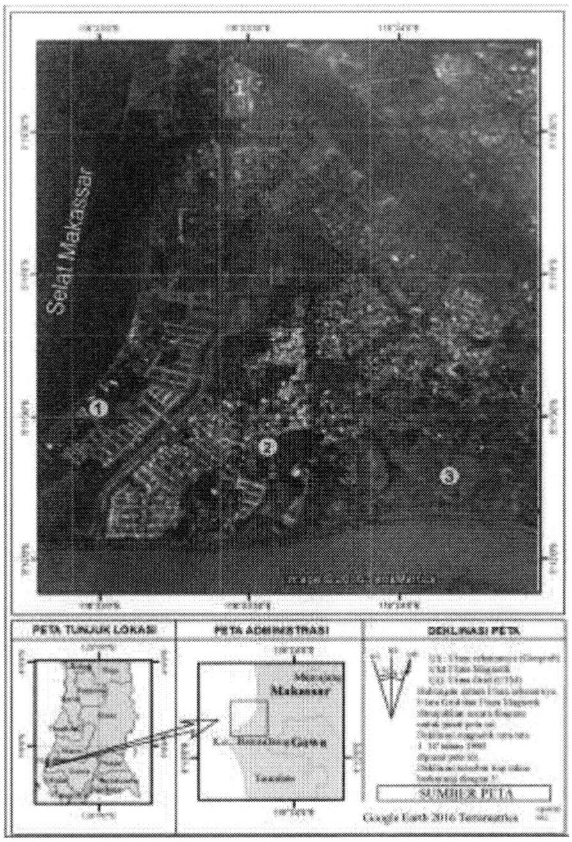

Figure 1. Map of research area and drilling point

Coastal plain of Tanjung Bunga is part of the Jeneberang River, containing minerals which are Quartz, Hornblende, Pyroxene, Orthoclase, Plagioclase, Biotite, and Opaque minerals in larger quantities than other minerals, with sizes from fine to coarse sand and rounded to sub rounded shape [3].

Quartz mineral (SiO2) is one of the main mineral constituents of sedimentary rocks, and as the most stable mineral found in clastic sedimentary rocks. Quartz minerals also found in coastal sediments and constitute up to 65% in sandstones and up to 30% in claystone [4]. The most composition of sandstones is a stable mineral, such as quartz, [5] which is part of the stable mineral group.

The aim of this research is to classify the compaction test from mineral as result texture breakage from compression test. Laboratory Compaction Test as one of the method of testing the ability and strength of the soil with loading, followed by analysis of the level of damage of minerals before and after loading with a magnification polarization microscope up to 100x conducted on thin section samples. In this study used 15 sediment samples at 3 points, which generally show the damage as well as changes size and shape surfaces of minerals from rupture, split to shattering, which is found on sediment samples of fine sand, medium sand and coarse sand.

2.1 Field research

Determination of location at 3 drilling points, based on survey and analysis of coastal sedimentation profiles which is conducted until a depth of 5 meters by using Hand Auger. Sampling, observation and measurement of grain size on each sample are taken with the addition of depth every 50 cm.

2.2 Analysis of field data

The samples were taken from 3 drilling points representing the top and bottom of each hole were 6 samples of fine-grained sediment, 4 samples of medium-sized sediment and 5 samples of coarse sand-size sediment, samples taken through sieve analysis using mesh size 60, then dried and grouped according to the size of the mineral grains. Furthermore, it is cleaned from dirt and clay minerals, then do the preparation becomes "smear slide" and "Thin Section" with a thickness of 0.003 mm, for petrographic analysis before loading and continued by testing loading with laboratory Compaction Test. Sediment samples from the loading test with Compaction Test are then repeat "Thin Section" preparation.

2.3 Analysis of laboratory data

Petrographic analysis of the sample before and after loading with laboratory Compaction Test, using a 50x-100x magnification polarization microscope, to determine the composition and amount of minerals. Which is broken from the crack level, split to shreds. The Smear Slide analysis of the sample was also observed using a 50x polarization magnification microscope to determine the grain size and grain shape of the mineral. Testing of loading with Compaction Test of the SNI standard laboratory was conducted to determine the value of change until the sample state had cracks occurring per unit time and level loading of samples.

3 Result and analysis

Sand is a non-clastic grain material came from a rock source, the sand is a naturally formed grain material and is composed of small sized materials comprising mineral and rock particles [6]. To observe the characteristics of soil, whether it is residual soil or sediment deposition, depends on two factors, which is; the natural characteristics of the soil particles themselves (such as the size, shape and composition of minerals) and the second one is particle content on the soil. In other words, these factors refer to the composition and structure [7].

The results of the research on sediment samples in the Tanjung Bunga area on Makassar showed that the minerals grain size was fine sand (0.125-0.25 mm), medium sand (0.25-0.50 mm) and coarse sand (0.50 - 00 mm)., (Wentworth scale, 1922), in Folk (1974). Shape of the minerals is sub rounded - rounded, and the composition is Quartz 20 - 25% Biotite 10-20%, Orthoclast 5-15%, Pyroxene 5-15%, Plagioclas 5-15%, Hornblende 5-20% and opaque minerals 10 -25%. Value of Compaction Teston fine sand (PH) 4 - 4.14 Div, medium sand (PS) 6.7 - 20 Div and coarse sand (PK) 3.2 - 24 Div.

The results of the loading test with CBR laboratory tests found the percentage of mineral damage from the cracking, diking, and crushing was not the same on the sediment of fine, medium, and coarse sands. Damage rate of cracked mineral is almost same in fine, lower than coarse sand. Change of mineral area before and after loading with CBR laboratory test is more on coarse sand than fine sand, there was significant change between value increases of CBR test with change level of mineral area.

Mineral composition are Quartz, Pyroxene, Hornblende, Plagioclase, Orthoclase, Biotite and Opaque mineral with different percentages on fine, medium and coarse sand sediment materials. Quartz minerals has the largest cracks percentage and Opaque mineral was the smallest. Orthoclase, Biotite and Plagioclases was the highest shatters minerals and the other minerals are generally dikes [3].

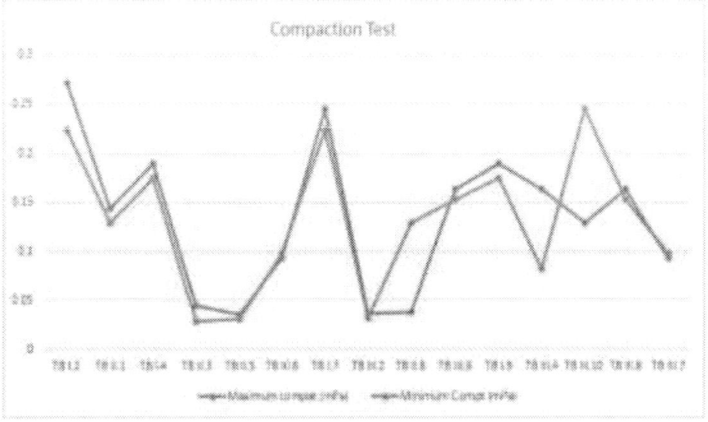

Figure 2. Value of loading of fine sand, medium sand, and coarse sand

The graphic chart show that Compaction laboratory test value indicate increase from low to high on fine sand, medium sand, until coarse sand, thereby Compaction laboratory test on coarse sand is higher than fine sand and medium sand. On the graphic show that relationship between load value Compaction laboratory test is significance with mineral grain size from fine sand, medium sand until coarse sand.

According to mineral breakage stage analysis from crack, torn apart until broken after load test with Compaction laboratory Test for sample group from fine sand, medium sand until coarse sand, there are mineral breakage percentage from crack, torn apart until dike, different on fine sandm medium sand and coarse sand, there are relationship between mineral breakage percentage with Compaction laboratory Test (figure 3, 4 & 5).

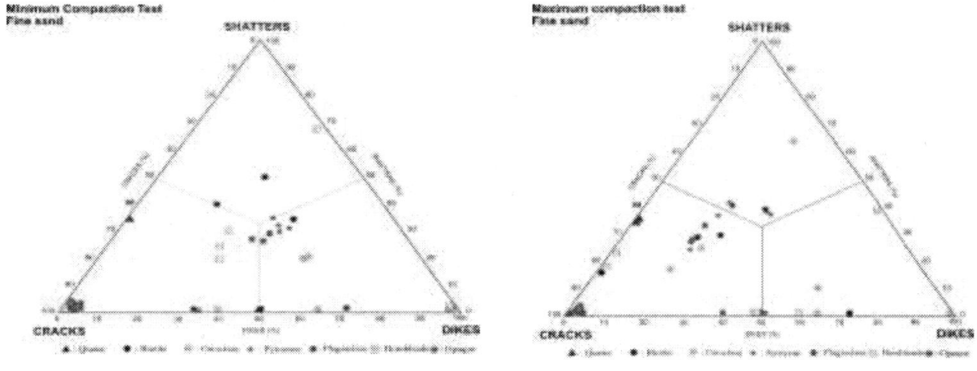

Figure 3. Damage percentage of fine sand

partly biotite, hornblende and pyroxene mineral with torn apart achieve between [illegible] on opaq mineral, crack is less than 45%. Mineral with broken between 45 – 85% torn apart and crack achieve 20 – 50% is orthoclase. Shatters mineral between 60 – 80% is partly on Biotite, Pyroxene mineral, quartz and orthoclase mineral achieve less than 60%.

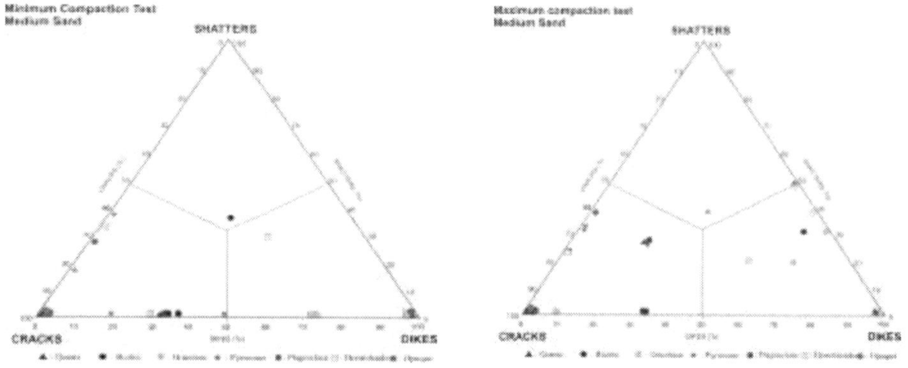

Figure 4. Damage percentage of medium sand

The mineral Quartz mineral damaged achieve between 32-52%, while mineral Plagioclase and hornblende damaged achieve between 50-100%. Cracked minerals reach 65 - 100% consisting of quartz minerals, Plagioclase and, hornblende and Opaque Minerals cracked up to 35%, as well as on Orthoclase minerals that dikes 100%. Biotite and Orthoclase are crushed up to 45%. Mineral Biotite, Pyroxene and Plagioclase. Which dikes in between 30-70%, from the amount of defective biotite minerals reaches 67 - 100%, whereas in the mineral pyroxene up to 53 - 100%.

In the encountered mineral cracking reaches 65 - 100% consisting of Quartz mineral, Plagioclase, Hornblende and Biotite. Quartz damaged achieve between 40-80%, whereas the Plagioclase minerals and mineral Hornblende total damage at between 50-100%.

In mineral Biotite, Pyroxene and Plagioclase are dikes between 40-60% and the crack reaches 62 - 85%. Damaged biotite minerals reach between 67 - 100%, whereas in mineral pyroxene reach 100%. In Biotite and Plagioclase minerals, which have been crushed to 50%, and Orthoclase minerals reach 100%. Another thing encountered in the mineral Orthoclase also dikes between 60- 100%, even up to 60% over 60%. In opaque minerals, gain rupture to 35%.

There are torn apart mineral achieve 50 – 100% consist of quartz, plagioclase, pyroxene and biotite mineral. Along with 20 – 50% torn apart are biotite and pyroxene mineral. Plagioclase 50 – 80% torn apart and crack achieve 50 – 75%, and this mineral broken until 50%. On quartz mineral damage total achieve 40 – 80%, on plagioclase and hornblende mineral damage total achieve 50 – 100%. On biotite mineral damage total is 67 – 100%, on pyroxene and plagioclase mineral damage total achieve 100%. On opaque mineral that crack until 100% from mineral damage total 45%.

This data show that broken mineral amount is more happen on fine sand and torn apart mineral on coarse sand. There are high crack mineral amount on fine sand with Compaction Test value lower and coarse sand with low crack mineral amount with high Compaction Test value. Quartz mineral much more that crack and no opaque mineral that broken, pyroxene mineral much more broken along with orthoclase mineral much more torn apart.

Triangle analysis above show that between Compaction laboratory test with damage mineral level that show high CS test value increase on coarse sand, inversely proportional with crack and broken mineral much more on fine sand with low CS test value. That thing show that Compaction Test value on coarse sand higher than fine sand and medium sand, increasing Compaction Test is not significance and negative relationship.

This thing support by geotechnic index characteristic different from soil that almost multiple diverse and mineral proportion different [2] also give different result, along with change degree that change because of load depend on gradation, soil component, mineral reduction characteristic, amount and load average and tenacity degree from particel individual load set to constant, porocity change average, pore number and weight unit depend on soil permeability.

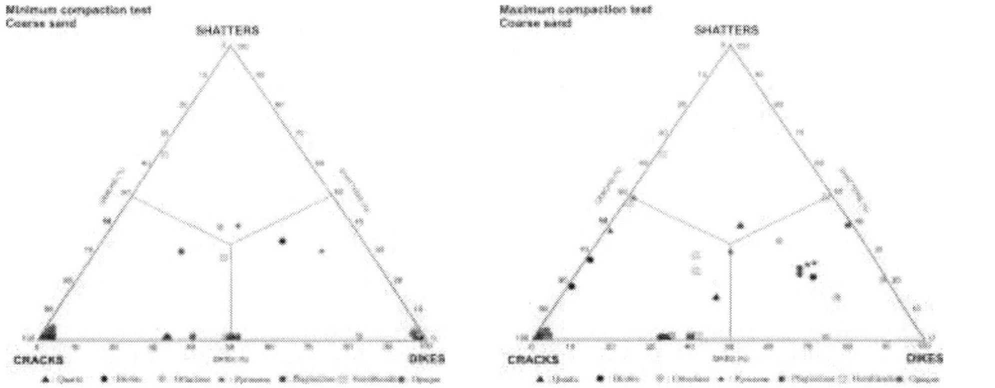

Figure 5. Damage percentage of coarse sand

Particel individual here is minerals that show microstructural characteristic, like mineral composition, grain size, and damage on mineral, considerable by some writer to become micostructure parameter to determine sedimen and rocks mechanic habit characteristic. Microstructure point to characteristic on observed surface in thin section with using microscope on milimeter scale and some writer using texture and microstructure as synonim [8].

Accroding to petrography analysys mineral damage level find on this research from crack, torn apart and broken on micro µm scale with 50x zoom until 100x (Figure 6), and mineral physical change on this matter as mineral area wide (Figure 7).

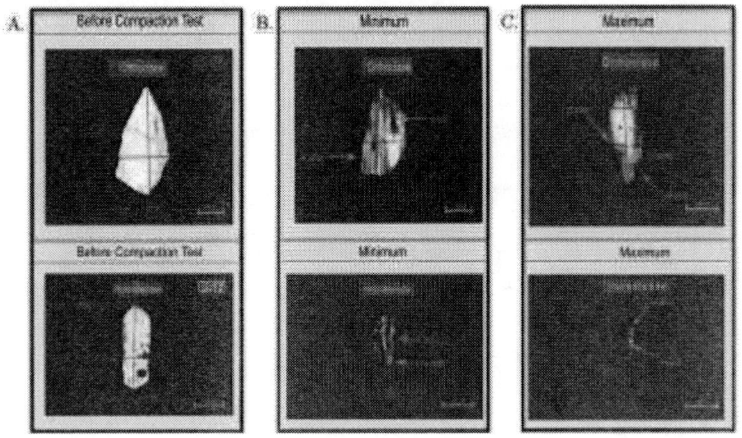

Figure 6. Condition from mineral Orthoclase and Hornblende before compaction test (A), mineral Orthoclase and Hornblende in minimum test (B), mineral Orthoclase and Hornblende in maximum value test (C).

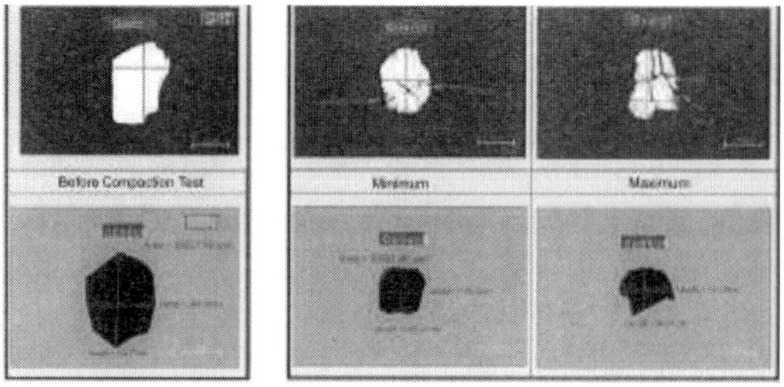

Figure 7. Photomicrograph of monomineral Quartz and Opaque mineral before barrier (A) and after barrier (B) in 100x magnification

According to mineral size change on thin section fine sand sample, medium and coarse, before and after Compaction Test load, show mineral are wide size change on fine sand lower than medium and coarse sand (Fig 6 – 7).

On the graphic difference mineral area wide size change from before Compaction Test load and after Compaction Test load, show that difference area wide size quartz and opaque mineral lower than other, this thing show that quartz and opaque mineral have hardness level higher than other, so that load value on Compaction Test related positive on the mineral. This research result also show that effect of load with Compaction Test give influence on mineral damage lower than its percentage on coarse sand with high Compaction Test value and mineral damage with high percentage on fine sand with low Compaction Test value.

Table 1. Difference change of mineral size area before and after barrier with CS laboratory test

Mineral	Fine Sand	Medium Sand	Coarse Sand
Quartz	3009.97	2300.2	1006
Biotite	2189.3	2020.93	3001.51
Orthoclase	4160.3	2221.44	999.99
pyroxene	1463.01	3363.57	4948
Plagioclase	2017	4395	1482.29
Hornblende	1406.4	2399.52	11015
Opaque	2712.97	1132.36	2964

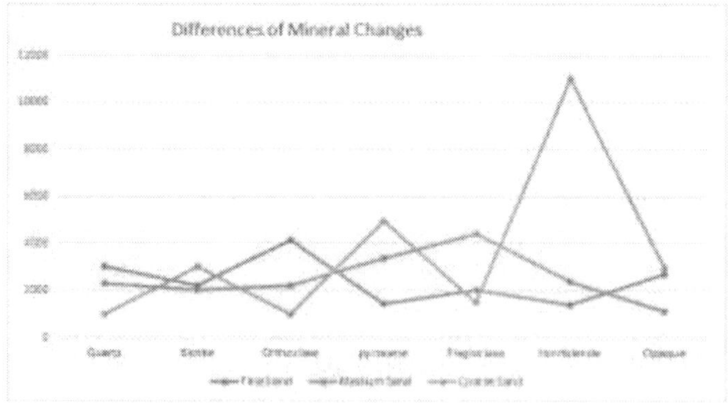

Figure 8. Difference graph of mineral size

especially importantlygiven were stressing is a very important factor to determining the soil or sediment characteristics and refers to normal soil and excessively of consolidated soil material [9].

4 Conclusions

The result of barrier test with Compaction from laboratory test showed the damage percentage of minerals from cracked, split, and shattered are not same on the fine sand, medium sand, and coarse sand sediment. Damage level of cracked and shattered minerals in fine sand is much higher than in coarse sand were Compaction Test value on fine sand is lower than the coarse sand. mineral changes area before and after barrier with Compaction Test laboratory, more occurred in coarse sand than fine sand, the increase of Compaction Test value with change level of mineral area.

Mineral composition has aree Quarts, Pyroxene, Hornblende, Plagioclase, Orthoclase, Biotite and Opaque mineral with different percentages on fine, medium, and coarse sand sediments. In Quartz minerals with has high crack and opaque mineral was lower percentage. Biotite, Plagioclase, and Orthoclase are minerals whose has highest shattered percentage were the other minerals are dikes.

Acknowledgements

Thank you for Werdi, Bayu and Irfan who has contribution during this research.

References

1. Langkoke, R., Apriani, S. Sebaran Mineral Berat Endapan Pasir Pantai Tanjung Bunga Kecamatan Mariso. Prosiding Hasil Penelitian Fakultas Teknik Unhas. ISBN: 978-979-127255-0-6. (2010).
2. Langkoke, R., Rochmanto. B. Pengaruh Karakteristik Pantai Estuari Makassar Terhadap Perkembangan Kota Pantai. Prosiding Seminar Nasional: Teori dan Aplikasi Teknologi Kelautan 2010.ITS – Surabaya. ISSN: 1412 – 2332, 9 – 10 Des. 2010. (2010).
3. Umar, H. and Al Fuad. Study on Characteristic and Distribution of Mineral Deposits of Sediment At Jeneberang Estuary of Gowa, South Sulawesi Province, Proceeding of the 1st, International Conference Civil and Infrastructure. (2015).
4. Boggs, S. Principle of Sedimentology and Stratigraphy.3 ed. Prentice Hall, Inc.Uper Saddle River: New Jersey 07548. (2001).
5. Pettijohn, F. J. Sedimentary Rocks.3rd edition. Harper & Row Publishers Inc, New York. (1975).
6. Roy, T. K. Influence of sand on strength characteristics of cohesive soil for using as subgrade of road, Proceedings – Social and Behavioral Sciences 104 (2013) 218-224 (2013)
7. Wesley, L. Behaviour And Geotechnical Properties Of Residual Soils And Allophane Clays. Obras y Proyectos 6. 5-10. (2009).
8. Duarte, M.T., H. Y. Liu, S. Q. Kou, P. -A. Lindqvist, and K. Miskocsky., 2005. Microstructural Modeling Approach Applied To Rock Material. JMEPEG 14: 104-111. (2005)

interrelationships of some sensitive clays from Eastern Canada. Can. Geotech.J. 21, (1984) 530-540. (1984).
10. Suriamihardja, D.A, Samang, L, Ria, Ulva., Umar, Hamid. Effect of Barrier On Mineral Change In Tanjung Bunga Beach Makassar. International journal of civil engineering and technology (IJCIET) vol. 8, issue 7, july 2017,pp, 911-919. (2017).

Level vulnerability damage of pavement using Pavement Condition Index method

Siti Nurjanah Ahmad[1,*], *Tri* Harianto[1], *Lawalenna* Samang [1], and *Muralia* Hustim[1]

[1]Hasanuddin University, Civil Engineering Department, Makassar, Indonesia

Abstract. This study aims to assess the vulnerability of road pavement damage using the Pavement Condition Index (PCI) method, including the direction of handling pavement road damage. Assessment of the level of road surface damage is based on the type of damage, the degree of damage, and amount of damage. Field observations and surveys were conducted on several Collector roads in Kendari. The mean PCI score as an indicator of damage to the Mokodompit road segment shows moderate damage conditions with a mean the value PCIaverage = 53 and surface damage area = 34.419% and the countermeasures according to the Bina Marga Standard recommend patching and adding pavement layers. Assessment of vulnerabilities Manunggal road damage indicates moderate damage to the value PCIaverage = 49 With the extent of surface damage = 34,419% and the response according to Standard by filling a crack with a mixture of liquid asphalt and sand.

1 Introduction

Road transport infrastructure plays an important role in supporting all human activities, the availability of adequate road network and able to connect the flow of transportation of goods and services more efficient and cheaper than other areas to enter into the city or vice versa. And the increase of traffic volume in each road segment is not accompanied by the addition of road length. Until now, Indonesia has only 296,968 km of road length (Minister of PUPR Basuki Hadimuljono, 2016), on the other hand the total road damage in Indonesia along 1,300 Km (2016) is a condition of vulnerable road damage. This is due to several physical factors / structures, environmental factors, social, and economic factors that badly affect the efforts to prevent and overcome the damage paved roads.

Similarly, the state of highways in Kendari city in post-flood 2013 and 2017 many broken collector roads, among others, on the road along the road Mokodompit 2.3 Km and, among others, on the road along the road Mokodompit 2.3 Km and Manunggal road along the 1.6 km suffered damages, including mild or severe damageThe road damage is quite disturbing the smooth flow of traffic there, head towards the city of Kendari and vice versa. The cause of pavement damage on this road segment is due to poor subgrade, excessive vehicle loads, unfavorable climatic influences (the amount of rainfall), Road materials that do not meet established standards, poor implementation of work and unavailability of drainage on the left and right side of 2 (two) road segments.

* Corresponding author: nurjanaharifuddin@gmail.com

Observing the strategic issues of road pavement damage, this study aims to assess the vulnerability of pavement damage by applying the method Pavement Condition Index and seek guidance for handling the damage of pavement layer and provide scientific input to local government especially related technical institution, so that in the implementation of repair of road damage will get more optimal result. In this study, we specifically surveyed and assessed the damage of pavement on the type of crack damage, wavy, decreased road shoulder, transverse and lengthwise damage, damage caused by water puddle and pothole damage.

2 Literature review

Road infrastructure in all areas will suffer damage, disturbance or deterioration of the condition, its quality when used to serve the activities of transporting people (passengers) or goods and services [1]. It is very necessary for pavement construction handling well, especially bending Pavement construction, including the building of new pavement or road improvements should be handling maintenance / rehabilitation periodically. In this condition the ability of the subgrade to support the load can be said to be very small. While the vehicle will still pass, due to the burden of vehicles pressing the surface of the road then there is release of the bond between the grains on the ground, and will cause the surface of the road to be broken and disappear. Well this is the initial process of road damage. Therefore almost every rainy season will appear many roads are damaged, ranging from a small hole to a very large hole. This results from an imperfect drainage condition.

According Sukirman [2] generally damage pavement effect to the road construction is not only caused by one factor, but a combination of many factor that are interconnected. The damage may be caused by:

1. Traffic, that can be an increase in load and load repetition and fatigue cracks load, resulted in layers of asphalt as well as deformation at all levels
2. Water can come from rain water and road drainage system is not good.
3. Pavement construction materials caused by the nature of the material it self or can be caused by a processing system that is not good.
4. Climate because Indonesia has tropical climate, where air temperature and rainfall are generally high, which can be one cause of road damage.
5. Unstable ground ground conditions caused by poorly implemented systems, or caused by subgrade poorly ground nature.
6. The compaction process over the underlying soil layers.
7. Damage to the pavement can occur due to vehicle load, on each layer of pavement voltage and strain occur.

2.1 Concept pavement condition index (PCI)

To determine the value of the Pavement Condition Index is the level of pavement surface conditions and Measure the usefuly of function referring to the conditions and damage to the pavement surface. PCI was developed to provide an index of pavement structure integrity and its surface layer conditions. The damage data obtained can provide information on the causes of damage and whether the damage is related to the load or climate. By conducting periodic condition surveys, the pavement condition information can be useful for predicting future road pavement performance, but can also be used as a more detailed and focused measurement input. By looking at various types of damage to the pavement construction, the Pavement Condition Index method classifies the types of pavement damage as follows:

a) Deformation Damage is wavy damage, groove, collapse, sungkur, fluffy, bumps and descend.
b) Cracked Damage is the longitudinal, transverse, diagonal, block and crocodile damage.
c) Damage to surface texture is damage to loose granules, bledding, slippery aggregates, and exfoliation.
d) Damage to the pothole is damage to patches and railroad crosses.
e) Damage at the edge of the pavement is damage; Edge crack/break and decline shoulder.

This PCI method is a numerical index whose value ranges from 0 to 100. A value of 0 indicates that the pavement is very damaged and the value 100 indicates perfect pavement. This PCI method is based on visual pavement condition survey results by looking at the type of damage, level of damage, and size identified during the survey of the condition. The degree of damage used in the calculation of the Pavement Condition Index method is *low severity level (L), medium severity level (M),* dan *high severity level (H).*

Table 1 Relationship between PCI value and road conditions

Value PCI	Damage Conditions
0 - 10	*failer*
11 - 25	*Very Poor*
26 - 40	*Poor*
41 - 55	*Fair*
56 - 70	*Good*
71 - 85	*Very Good*
86 - 100	*Excellent*

2.2 Assesment pavement condition index (PCI)

In analyzing road damage by PCI method is done by sequential way as follows [3-9]:

2.2.1 Density

Density is the percentage of area or total length of one type of damage to the total width or length of the road section measured in meters. Thus, the damage density can be expressed by the Equation Value equation (Deduct Value, DV).

The formula looks for density values:

$$Density = Ad/As \times 100\ \% \text{ Atau} \tag{1}$$

$$Density = Ld/As \times 100\ \% \tag{2}$$

Where :
Ad = Total damage area for each level of damage (m^2)
Ld = Total length of damage type for each level of damage (m)
As = The total area of the segment unit (m^2)

 https://doi.org/10.1051/matecconf/201818111003

2.2.2 Reduced value (DV)

The Reduction Value is a deduction value for each type of damage obtained from the density relation curve and the severity of the damage, is Low (L) is the degree of minor damage, medium (M) Moderate damage rate, dan Hight (H) high levels of damage.

2.2.3 Total deduct value (TDV)

Total Deduct Value (TDV) is the total value of the individual Reductions for each type of damage and the extent of damage existing in a research unit.

2.2.4 Finding Value q

The requirement to determine the value of q is determined by the amount of the individual Reducer value greater than 5 in each segment of the road segment under study.

2.2.5 Calculating value correct deduct value

CDV value can be sought after the value of q is known by summing the Reduce value Further deducting the amount of the subtraction value in the graphic image CDV corresponds to the value of q obtained.

2.2.6 Calculating the value of pavement conditions

After the CDV is obtained, the PCI for each sample unit is calculated using the equation:

$$PCIs = 100 - CDV \qquad (3)$$

The PCI value of pavement as a whole on certain road segments is:

$$PCIf = \frac{\sum PCI(s)}{N} \qquad (4)$$

Where :

$\sum PCI(s)$ = PCI for each segment unit
CDV = CDV of each sample unit.
PCIf = Value PCI average of all research areas.
PCIs = Value PCI For each sample unit
N = Number of sample units

2.2.7 Corrected deduct value (CDV)

Dikoreksi Deduct Value (CDV) Obtained from the curve of the relationship between the value of TDV with the value of CDV with the selection of curve curve in accordance with the number of individual values of deduct value that has a value greater than 2.

2.2.8 Classification of pavement quality

Value of (PCI) for each research unit can be known pavement unit quality pavement segment based on certain condition that is excellent, very good, good, fair, poor, very poor and failed [10].

2.3 Standard of handling of road damage

The handling for road damage in this study is guided by the Road Maintenance Manual [11] issued by the Directorate of Highways and the Routine Maintenance Manual for National Roads and Provincial Road Volume II Title Standard Repair Method [12] issued by the Directorate of Highways. Both manuals describe all types of damage, causes of road damage, the process of handling road damage and road maintenance processes.

3 Result

3.1 Assessment rate of road damage

Based on road damage assessment using PCI method on Mokodompit road and Manunggal road segment with the formula: PCI - 100 - CDVmax obtained average PCI value on both roads there is difference of value of road damage condition in table 2 as follows:

Table 2. Result of calculation of road damage condition by PCI method

No	Station (m)	Mokodompit Road		Manunggal Road	
		Value PCI	Road Condition	Value PCI	kondisi jalan
1.	0 + 000 - 0 + 100	38	Poor	48	Fair
2.	0 + 100 - 0 + 200	90	Perfect	68	good
3.	0 + 200 - 0 + 300	25	Poor	40	Fair
4.	0 + 300 - 0 + 400	42	Fair	32	Poor
5.	0 + 400 - 0 + 500	81	Very Good	56	Good
6.	0 + 500 - 0 + 600	38	Poor	41	Poor
7.	0 + 600 - 0 + 700	68	Good	36	Poor
8.	0 + 700 - 0 + 800	88	Perfect	35	Poor
9.	0 + 800 - 0 + 900	65	Good	78	Very Good
10.	0 + 900 - 1 + 000	20	Very Poor	65	Good
11.	1 + 000 - 1 + 100	78	Very Good	36	Poor
12.	1 + 100 - 1 + 200	66	Good	31	Poor
13.	1 + 200 - 1 + 300	35	Poor	69	Good
14.	1 + 300 - 1 + 400	51	fair	57	Good
15.	1 + 400 - 1 + 500	40	Fair	70	Good
16.	1 + 500 - 1 + 600	80	Very Good	24	Poor
17.	1 + 600 - 1 + 700	30	Poor		
18.	1 + 700 - 1 + 800	34	Poor		
19.	1 + 800 - 1 + 900	68	Good		
20.	1 + 900 - 2 + 000	34	Poor		
21.	2 + 000 - 2 + 100	47	Fair		
22.	2 + 100 - 2 + 200	51	Fair		
23.	3 + 200 - 2 + 300	40	Fair		
amount		*1209*		*786*	
Average PCI value for ***road. Mokodompit*** is = *∑PCI / n =1209 / 23 = 52,56 =* ***53*** With condition value = ***Fair***					
Average PCI value for ***road. Manunggal*** is = *∑PCI / n =786 / 16 = 49,125 =* ***49*** With condition value = ***Fair***					

3.2 Types and percentage of road damage level

Type of damage that occurred on the road Mokodompit damage pothole with damage area of 39.625m2 with the percentage of damage 34.419%, alligator cracking cracks with damage area of 14.875m2 with the percentage of damage 12.921%, patches with an area of damage of 8,688 M2 with damage percentage 7,547%, roadside crack with damage area equal to 11.000m2 with percentage of damage 9,555%, weathering and loose grane with damage area 3,75 m2 with damage percentage 3,257%, city-box crack with damage area 13 , 75m2 with damage percentage 11,943% , and damage of rutting with damage area equal to 23,438m2 with damage percentage 20,359%. But the most dominant type of damage is hole damage (pothole).

As for the road of Manunggal there is damage to the type of hole (pothole) with damage area of 39,625 m2 with damage percentage 34,419%, crack lengthwise or transverse with damage area equal to 39,625 m2 with damage percentage 34,419%, alligator cracking 14,875m2 with damage percentage 12,921%, road shoulder damage with damage area equal to 39,625m2 with percentage of damage 34,419%, patch with damage area 8,688m2 with damage percentage 7,547%, crack edge / side of road with damage area equal to 11.000m2 with percentage 9,555% damage, rutting with damage area equal to 23,438m2 with damage percentage 20,359% and damage of basin with damage area equal to 39,625m2 with damage percentage 34,419%, but the most dominant type of damage is damage crack length / transverse.

Table 3. Result of calculation of road damage condition by PCI method

No	Type of Damage	Mokodompit Road		Manunggal Road	
		Area (m2)	% kerusakan	Area (m2)	% kerusakan
1.	Hole (*pothole*)	39,625	34,419	13,87	19,403
2.	Cracks extend and transverse	----	----	15,00	20,984
3.	*Alligator cracking*	14,875	12,921	14,55	20,354
4.	Path / descent of road shoulder	----	----	5,75	8,044
5.	Patch Roadside / side fracture	8,688	7,547	7,313	10,230
6.	Roadside / side fracture	11,000	9,555	3,25	4,547
7.	Weathering and granular discharges	3,75	3,257	----	----
8.	Crack the boxes	13,75	11,943	----	----
9.	*rutting*	23,438	20,359	9,75	13,640
10.	Basin	----	----	2,00	2,798
Amount		***115,126***	***100***	***71,483***	***100***

On both sides of the road that suffered the most severe damage and need serious attention, so that damage does not increase if not quickly repaired. Damage to both roads causing inconvenience for motorists who use the road, either damaged holes light, medium or heavy hole. This occurs as a result of the development of other types of damage that are not immediately addressed, the effects of weather (especially rain/ flood) and vehicle traffic that exceeds the implied load that accelerates the formation of holes.

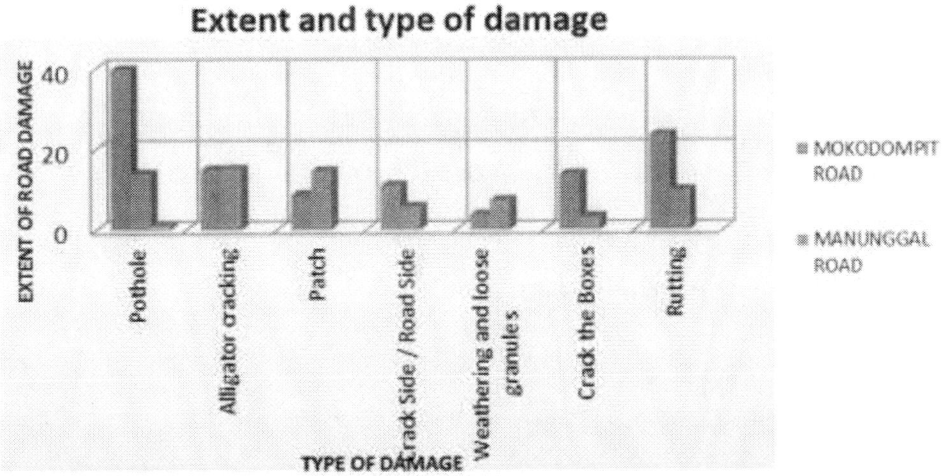

Fig.1. Area and type of road damage

3.3 Direction for improvement and handling of road damage

Based on the analysis with PCI method, the condition of the road pavement of each segment unit obtained in the field can be used to know the most vulnerable road pavement damage and to determine the priority of handling the damage by prioritizing the damage handling on the road pavement which has a pavement damage condition value which Greatest.

On Mokodompit road, the most vulnerable type of damage is damage of hole 39,625 m2 with damage percentage 34,419%, repair effort or handling according to Bina Marga regulation (1993) the corrective action is by hole patching and addition of pavement layer. While on Manunggal road type of the most vulnerable damage is the damage of crack lengthwise/transverse with damage area equal to 39,625 m2 and percentage of damage 34,419% improvement effort done by filling gap with liquid asphalt mixture and sand.

4 Conclusion

Based on the analysis results obtained conclusion as follows:
1. The average PCI score as an indicator of damage to the Mokodompit road segment showed moderate damage conditions with an average PCI value = 53 and surface damage area = 34.419% and the Assessors of susceptibility to road damage indicated moderate damage with a mean PCI value = 49 with area surface damage = 34.419%.
2. Countermeasures for Mokodompit road according to Bina Marga standard recommends to patch and add pavement layer and handling for Manunggal road by filling crack gap with liquid asphalt mixture and sand.
3. The practical implications of assessing the vulnerability level of pavement damage on collector roads become a consideration in the formulation of the improvement and maintenance of urban collector roads.

References

1. Bustan, B., Sammang, L., Nurali, Ramli, I. *Risk level assessment on road construction's contractors using cultural-professionalism based approach*. International Journal of Civil Engineering and Technology. (2016)
2. S, Sukiman. *Bending Pavement Highway*. Bandung: Nova. (1999)
3. Fadly, Ahmad. *Evauasi level of damage to the road surface of your content-paguyuman based Pavement Condition Index (PCI) method*. Position ATPW seminar. Surabaya (2013)
4. Directorate of Highways, 1983. *Routine Maintenance Manual for National and Provincial Roads Volume II Title of Standard Improvement Method No. 002 / T / Bt / 1995*. Dep. Pek. General. Jakarta.
5. Directorate of Technical Affairs. *Survey of paved road conditions in urban areas, urban governance and rural governance department of residential and infrastructure Jakarta area*. (2002)
6. Amin, Khairi. *Evaluation of type and extent of damage by using PCI method (Case study on Soekarno-Hatta Street)*. Journal of Bengkalis State Polytechnic. (2012)
7. Febry, Noval T. *Evaluate the level of road damage using the method of pavement condition index (pci) case study Jl.Purwekerto-Ajibarang Kab. Banyuwangi*. National Symposium on Applied Technology (SNTT) to-4. (2016)
8. B, Margareth E. *Comparison of Bina Marga Method and PCI Method (Pavement Condition Index) in the assessment of Pavement Conditions*. (2015)
9. A, Suswandi. *Evaluation of road damage with PCI method (Pavement Condition Index) to support decision making (Case Study: South Ring Road of Jogyakarta)*. Civil Engineering Forum Magazine No. XVIII, September 2008. (2008)
10. Hustim, Muralia. *Study of development of road network of regency in pinrang district of south sulawesi based on Multi Criteria Analysis Method*. Tri Sakti University, Jakarta.(2012)
11. Directorate of Highways. *Road Maintenance Manual No.03 / MN / B / 1983*. Jakarta. (1983)
12. Directorate General of Highways. *Technical Guidance Survey of Road Engineering Planning No. 013 / T / Bt / 1995*. Dep. Pek. General. Jakarta. (1995)

Time	Movement of Vehicles from Land Use (V_i)	Continuous Flow Movement ($V_{eksternal}$)	Total Volume Movement	Capacity	Ratio V/C	LOS
06.00 - 07.00	11739	5663	17402.0	5023.2	3.46	F
07.00 - 08.00	8815	5887	14701.5	5023.2	2.93	F
08.00 - 09.00	8199	5093	13292.1	5023.2	2.65	F
09.00 - 10.00	7977	4663	12640.2	5023.2	2.52	F
10.00 - 11.00	7970	4404	12374.1	5023.2	2.46	F
11.00 - 12.00	7640	4037	11677.4	5023.2	2.32	F
12.00 - 13.00	8047	3978	12025.2	5023.2	2.39	F
13.00 - 14.00	8313	4008	12321.2	5023.2	2.45	F
14.00 - 15.00	7887	4183	12070.0	5023.2	2.40	F
15.00 - 16.00	7804	4386	12190.7	5023.2	2.43	F
16.00 - 17.00	11553	4498	16051.1	5023.2	3.20	F
17.00 - 18.00	12253	4866	17118.8	5023.2	3.41	F
18.00 - 19.00	11777	4579	16355.9	5023.2	3.26	F
19.00 - 20.00	7578	4040	11617.5	5023.2	2.31	F

Performance of pre-stressed concrete sheet pile after long-term exposure in the marine splash environments

Rita Irmawaty [1*], *Daisuke* Yamamoto[2], *Hidenori* Hamada[2], and *Yasutaka* Sagawa[2]

[1]Hasanuddin University, Civil Engineering Department, 92171 Makassar City, , Indonesia
[2]Kyushu University, Civil and Structural Eng. Dept., 744 Fukuoka City, Japan

Abstract. Pre-stressed concrete (PC) sheet piles are commonly used for river revetment, retaining wall, etc. However, it has not been used in port and harbor structure due to corrosive environmental conditions, which is severe for steel corrosion in concrete structure. Performances of PC sheet piles were evaluated after 12 years exposure in the marine splash environment. An evaluation of mechanical properties, carbonation depth, porosity, chloride content and PC steel condition were carried out. Overall, all specimens showed satisfactory performance: only a few concrete corner and edge chips, low chloride content and low rate of corrosion were observed. Investigation results indicated that BS5 (B = GGBFS, S = steam curing, and cover thickness of 5 cm) showed the best performance (more durable), even though its strength was lower than the other.

1 Introduction

Pre-stressed concrete (PC) sheet piles are commonly used for river revetment, retaining wall, etc. However, it has not been used in port and harbor structure due to corrosive environmental conditions, which is severe for steel corrosion in concrete structure. Sheet pile rarely used in marine environment due to thin cover concrete, which is high possibility to corroded. Steel corrosion may occur directly through oxidation of metals or by penetration of aggressive substances through pores or cracks of the concrete cover [1], [2]. Such as reinforced concrete in general, deterioration due to reinforcement corrosion becomes the most important durability issue nowadays.

Chloride-induced corrosion is one of the main mechanisms of deterioration affecting the long-term performance of concrete structures. In Japan, a large majority of structures are built either near the costal or indirect contact with seawater. The durability of reinforced or pre-stressed concrete structure depends on the resistance of concrete to chloride penetration [3], [4]. Naturally concrete provides physical and chemical protection to the reinforcing steel from chloride penetrating. A chemical protection normally provided to all embedded steel due to the high-alkaline nature of Portland cement concrete. In addition, a physical protection is also provided by the concrete cover with adequate thickness. However, this

* Corresponding author: rita_irmawaty@yahoo.co.id

protection is largely depending on the permeability of the concrete and the cover thickness of steel bars [5], [6].

Concrete strength over 50 MPa seems to be one solution to achieve low permeability and high durability for long-term exposure in marine environment. In addition, partial replacement of cement by Blast Furnace Slag (BFS) in producing High Strength Concrete (HSC) has been established. The objective of this study is to obtain experimental data to discuss an adequate cover thickness of PC sheet pile under marine conditions and effectiveness of BFS to enhance chloride resistance. In this study, PC sheet pile model specimens were exposed under marine splash environment during 12 years.

2 The material and method

2.1 Materials

The specimens were made with High Early Strength of Portland Cement (HSPC), Ground Granulated Blast Furnace Slag (GGBFS), river sand, crushed stone, with additional HRWR+AE admixture to improve the performance of concrete. Material properties used in this study are described in Table I and II.

Tabel 1. Physical and Chemical Compositions of Cements [7]

Items	HSPC	GGBSF
Density, g/cm^3	3.14	2.92
Blaine fineness, cm^2/g	4550	6020
MgO, %	1.4	6.2
SO_3, %	2.9	0.1
LOI, %	1.3	0.2
Total alkali, %	0.56	-
Ion chloride, %	0.004	0.005

Tabel 2. Properties of Materials [7]

Material	Specification
Fine aggregate	River sand Density = 2.62 g/cm^3
Coarse aggregate	Crushed stone Density = 2.62 g/cm^3, MSA = 20 mm
HRWR+AE admixture	Polycarboxylate ether based
AE admixture	Alkylcarboxylic based

2.2 Mix Proportions

The mixture proportions of concrete are summarized in Table III. The water to binder ratio (w/b) was 32% with target slump and air content about 8.0 ± 2.5 cm and 4.0 ± 1.0%, respectively. Both air-entraining admixture; and air-entraining and water-reducing admixture were used based on the cement mass. Compressive strength design was 70 MPa.

Tabel 3. Mix Proportion [7]

Material	Mix-1	Mix-2
Sand-aggregate ratio s/a, %	45.3	44.9
Water, kg/m^3	150	150
Cement, kg/m^3	469	234
GGBFS, kg/m^3	-	234
Fine aggregate, kg/m^3	784	770
Coarse aggregate, kg/m^3	950	949

2.3 Specimen Design

There are four sheet pile specimens, name as "PN3", "BS5", "PN5" and "PS7". The name index is indicated by [P for HSPC, B for GGBFS, N for water curing, S for steam curing and the number (3, 5 and 7) indicate cover depth in cm]. Details of mix and curing condition of specimens are described in Table IV.

Tabel 4. Mix and Curing Condition of Sheet Pile Model

Specimen	Mix condition	Curing Condition	Thickness (cm)
PN3	Mix-1	Condition-2	7
BS5	Mix-2	Condition-1	11
PN5	Mix-1	Condition-2	11
PS7	Mix-1	Condition-1	15

Note:

- Curing condition-1: 1 day steam, followed by 2 days wet, then covering by vinyl sheet.
- Curing condition-2: 3 days wet, then covering by vinyl sheet.

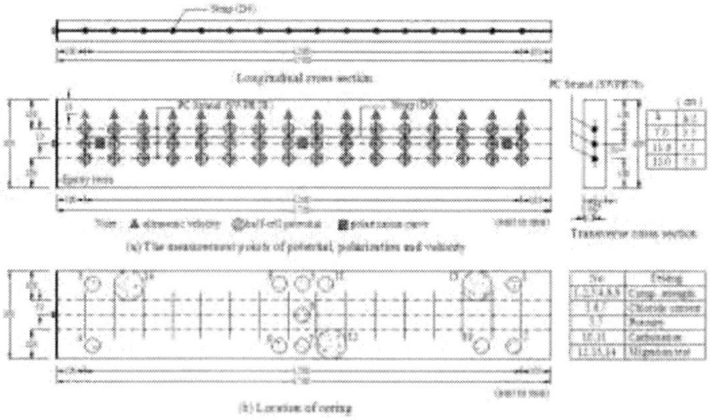

Fig. 1 Sheet pile specimen model geometry and measurement point

Figure 1 shows sheet pile specimen model geometry and measurement points of half-cell potential, polarization curve, ultrasonic pulse velocity and core sampling. As shown in this figure, PC sheet pile dimension in 1,700 mm x 300 mm, and height varies from 70, 110 and 150 mm with cover thickness is 30, 50 and 70 mm, respectively. The variation in thickness aims to evaluate the effect of cover depth as natural protection on concrete to chloride ion penetration. In center line of specimen, three PC strands (SWPR7B) with distance of 100 mm from the edges were embedded. In order to fix the position of strands, 16 straps were placed with spacing of 100 mm.

2.4 Curing Condition and Exposure Test

The main difference lies in curing, that is after casting steam curing or water curing for a day, followed by vinyl sheet covering for 2 months (see Table 4). Then accelerated deterioration test was conducted in the laboratory, as repeated wet-dry cycle. One cycle was 3.5 days, consisted of 1 day wet (sprayed warm seawater of 40°C) and 2.5 days air dry, continued until 220 cycles (770 days all). The number of 220 cycles is estimated to be equal to 3 years in the real-time exposure study [4].

Fig. 2 Sheet pile specimen

Fig. 3 Seawater splashing condition

Furthermore, specimens were placed in an open space and exposed to seawater splashing. The specimens were subjected to 2 cycles of wetting and drying a day. One cycle was seawater splashing for 4 hours and drying for 8 hours; during 12 years. Figure 2 and 3 show sheet pile specimens in the exposure site and condition when seawater is splashing.

2.5 Method of Evaluation

After 220 cycles of accelerated test, several tests were conducted. Details of testing and the results have been published [7]-[9]. Furthermore, in this paper we will present condition after 12 years of continuous exposure under marine splash environments. First, specimens were transferred from the exposure site to the laboratory and cleaned. Then specimens were investigated for corrosion of steel bar by using half-cell potential and anodic polarization curve by contact method. The half-cell potential was measured at interval 100 mm for each strand; whereas polarization curve was only checked at three points (15 cm from both edges and in the middle of specimen). The measurement was conducted with the silver/silver chloride electrode after 1 hour of pre-wetting.

In addition, ultrasonic pulse velocity (UPV) was measured by the direct method with path length of 100 mm. The UPV reading was conducted at 16 points along sheet pile. After that, core samples of 50 mm in diameter, were taken for compressive strength, chloride analysis, porosity and carbonation depth testing. Compressive strength was measured according to JIS A 1108. Testing was conducted on six core samples for each specimen. At the same time, a static elastic modulus was also conducted.

Chloride ion concentration was measured at certain depths of core samples (50 mm in diameter) based on JCI-SC4. Cores were taken from three points at midspan of the sheet pile. Then cores were cut into 5 pieces (thickness varied from 10 to 20 mm) and crushed into powder. Porosity of the mortar samples was measured at depth of 30 to 40 mm of core samples. Then pore size distribution of the samples was evaluated by MIP (Mercury Intrusion Porosimetry). Carbonation depth was evaluated on freshly cut surface of core samples after spraying a 1% phenolphthalein solution.

Effective chloride diffusion was evaluated by electric migration technique according to JSCE-G571-2003, where core samples (100 mm in diameter) were taken, and cut 30 mm from the surface of the sample. Then concrete sample divides the anode (0.5 mol/L of NaCl solution) and cathode (0.3 mol/L of NaOH solution) of a migration cell. The constant DC potential of 9 Volts was applied across the cell, and negatively charged chloride ions migrate from the cathode toward the anode through the concrete pore solution.

3 Results and discussion

3.1 Visual observation

Observations were made by washing the exterior surface of the specimens, then carried out checks on all corner, edge and surface of the specimen whether the steel corrosion has occurred. If steel corrode, there are usually signs of deterioration on the concrete surface such as rusting, cracking and spalling. From visual inspection, there were no adequate information on the corrosion risk of steel embedded in concrete, and only a few concrete corner and edge chips were observed. This was caused by the removal of surface mortar and fine aggregate as a result of seawater splashes. Degradation phenomenon such as small cracks or leaching of rebar rust is not identified.

3.2 Mechanical Properties

Compressive strength of concrete specimens after 12 years exposure in the marine splash environment is shown in Figure 4. The value is based on average of 6 core samples for each specimen. It is found that strength of concrete decreased slightly. Decrement in strength was observed in all specimens with maximum 4.7% reduction (in case of PN5). There is no significant difference in strength decreasing between steam and water curing.

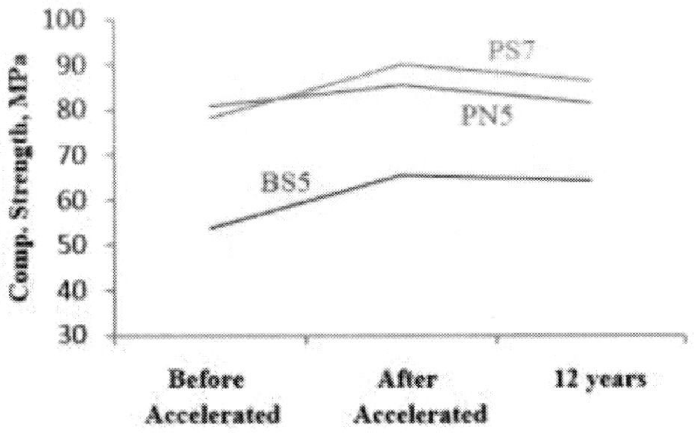

Fig. 4 Compressive strength of concrete

The pulse velocity varies between 4.50 and 4.63 km/s with coefficient of variation of 0.81 to 1.44% as shown in Figure 5. A good correlation between compressive strength and pulse velocity value is found, where the higher of strength, the higher value of pulse velocity are observed. On the contrary, from Figure 6, it is found that the elastic modulus of concrete increased (maximum 11%) after 12 years exposure, and no significant variation between compressive strength and elastic modulus data was observed. It confirms that the degradation of concrete properties in macroscale is not significantly occured.

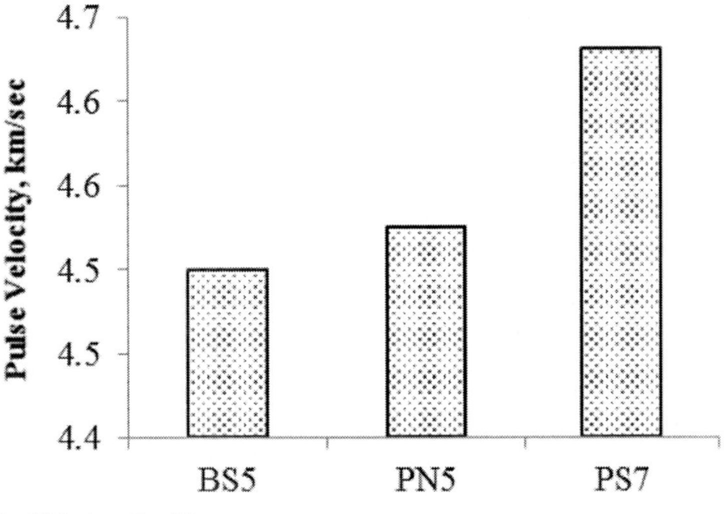

Fig. 5 Pulse Velocity after 12 years exposure

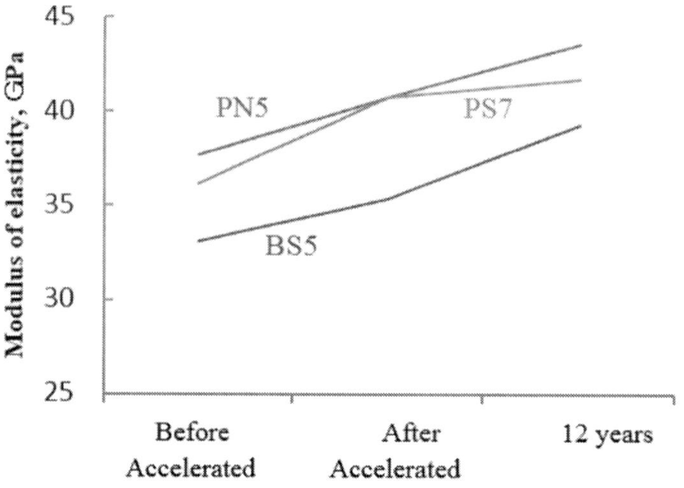

Fig. 6 Elastic modulus of concrete

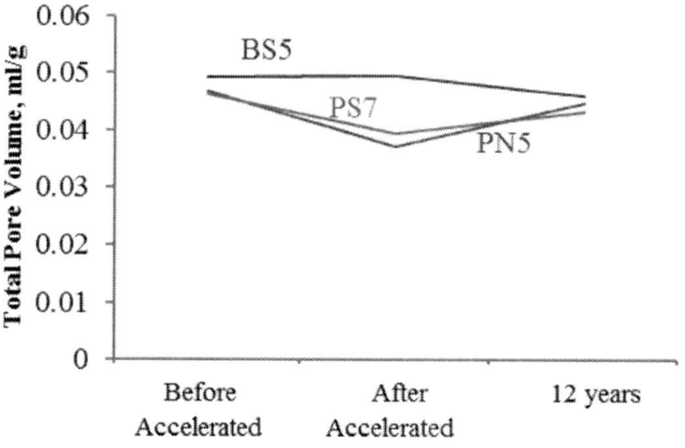

Fig. 7 Pore volume

3.3 Porosity and Depth of Carbonation

Pore volume and pore size distributions of the mortar samples collected at 30-40 mm depth of the specimens are shown in Figure 7 and Figure 8 where porosity and pore size distribution in a certain range show almost the same trend for all specimens. It is found that porosity of PN5 and PS7 increase after 12 years exposed to a corrosive environment, and vice versa in BS5 as shown in Figure 7. It seems that partial replacement of cement by GGBFS causes a reduction in the pore volume, and may also cause a reduction in the connectivity of the capillary pore. The capillary pore defined here is the porosity of pores whose diameter ranging from 0.01 to 5 μm [10]. In addition, the capillary pore distribution was slightly higher for BS5 than the other, but it did not affect significantly.

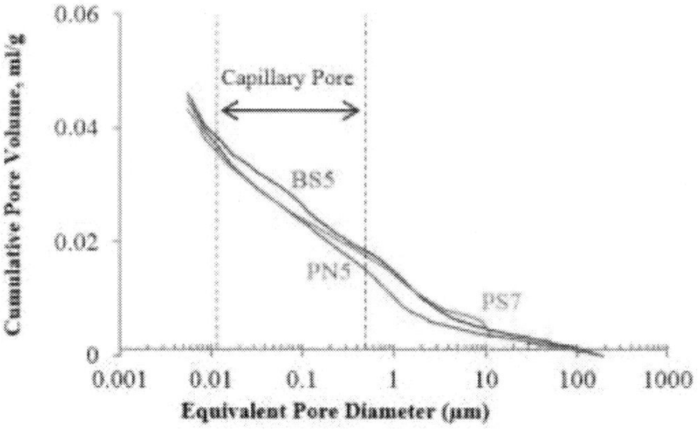

Fig. 8 Porosity of mortar samples

Carbonation depths of the core samples were not identified in this investigation. This is caused by the compactness of concrete where the total pore volume was very small (about 10%).

3.4 Chloride Content in Concrete

Total amount of chloride ion as a percentage of cement mass at 30, 50 and 70 mm depth from the surface of specimen is shown in Figure 9. The amount of chloride content varied between 0.134% and 0.365% at the depth of 30 mm, whereas at 50 mm depth showed a little amount which varies between 0.069% and 0.118%. It is found that sheet pile composed of GGBFS shows the lowest amount of chloride content compared with others at the same depth. It indicates that the microstructure of concrete using GGBFS is significantly improved and can provide a good barrier against chloride-inducement.

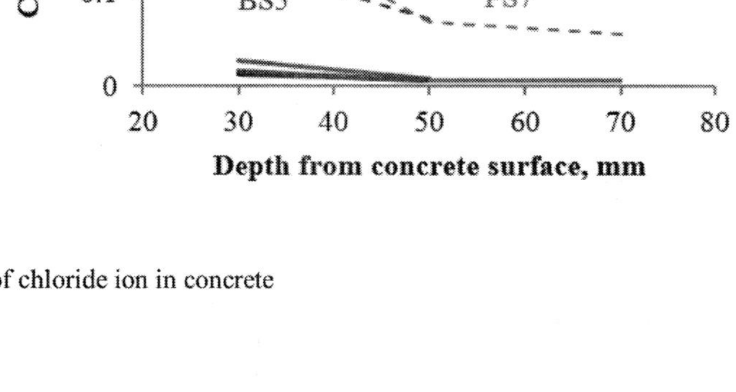

Fig. 9 Average of chloride ion in concrete

Numerous studies have shown that when present in sufficient quantities, chloride ions will promote the corrosion of steel in concrete [1], [2]. If the chloride content of 0.4% (by mass of cement) in the concrete is assumed as a threshold value for the initiate of steel corrosion, nowadays position of 0.4% chloride content in specimens of BS5, PN5, and PS7 are in the depths of 1.12cm, 2.7cm and 2.02 cm respectively from the surface.

3.5 Chloride diffusion coefficient

The diffusion coefficient (D) and concrete cover combine to provide a measure of the resistance of concrete to chloride ingress. Both the apparent (Da) and effective (De) diffusion coefficient were measured, and the results could be seen in Table V. The apparent diffusion coefficient (Da) was measured from non-steady-state of chloride test using Fick`s second law, while De was measured from steady-state of chloride test using Fick`s first law.

Tabel 5. Diffusion Coefficient (D) of Specimens

Name	Da (cm^2/year)	De (cm^2/year)
BS5	0.1435	0.120
PN5	0.2294	0.379
PS7	0.2136	0.399

From the Table V, BS5 showed the lowest diffusion coefficient. It means BS5 has good resistance to chloride ingress as compared to the other.

A. PC Strand Monitor

Natural Electrode Potential Test: Figure 10 shows the natural electrode potential after 12 years exposure in the marine splash environment. Significant change in natural electrode potential of strands occurred in PN3 and showed increasing potential for corrosion. On the other hand, irrespective of cement type, PC sheet pile height of 11 and 15 cm have potential value in the range of -56 to -101 mV:CSE/25°C. According to ASTM C876-91 (1999), potential less negative than -200 mV:CSE indicates 90% probability of no corrosion.

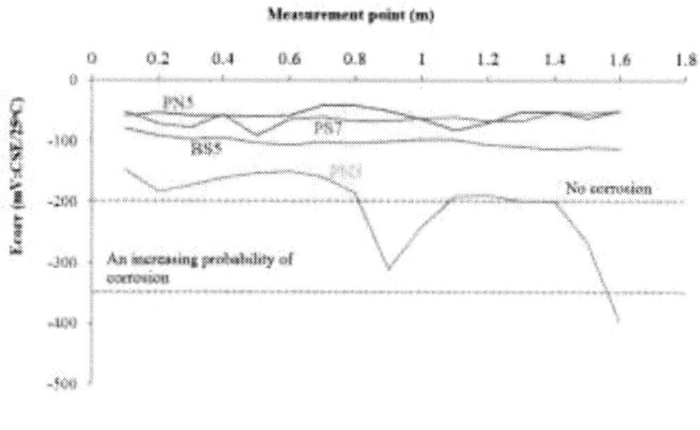

Fig. 10 Natural electrode potential

As comparison, PN3 with thickness of 7 cm is used. From the Figure 10, it can be seen that the potential value decreased sharply in some points. Then potential value in the range of -150 to -400 mV:CSE/25°C, indicates an increasing probability of corrosion. This is caused by the presence of small cracks in some parts of sheet pile as a result of the strand corrosion development on the edge whose no cover concrete. It can be concluded that PC sheet pile with thickness of 7 cm only has a 20-year service life, and it is recommended to increase the thickness to 11 cm to improve the durability.

Tabel 6. Average Polarization Resistance of Steel [2]

Speci men	Polariza tion Resistance (kΩ.cm^2)	Corrosio n Penetration (μm/year)	Rate of Corrosion
PN3 BS5	245 564	1.46 0.539	Low/mode rate Passive
PN5	228	1.34	Low/mode rate
PS7	203	1.6	Low/mode rate

From the view point of polarization resistance of steel (Rct) as described in Table VI, BS5 showed passive corrosion rate with value larger than 250 kΩ.cm^2, whereas PN3, PN5 and PS7 showed a low/moderate corrosion rate. It indicates that BS5 (50% partial replacement of cement by GGBFS) provided higher resistance to chloride ingress than the other.

Polarization Curve: Anodic polarization curve of each specimen is shown in Figure 11. PC strands were embedded in concrete exhibited various polarization curve with potential between 0.2 and 0.6 V, and current density were the range between 1 to 10 μA/cm^2, and 10 to 100 μA/cm^2. Based on the passivation film associated with anodic polarization curve, BS5<PS7<PN5 can be classified into grade 3 and PN3 is grade 1 as a degree of passivity exist [2].

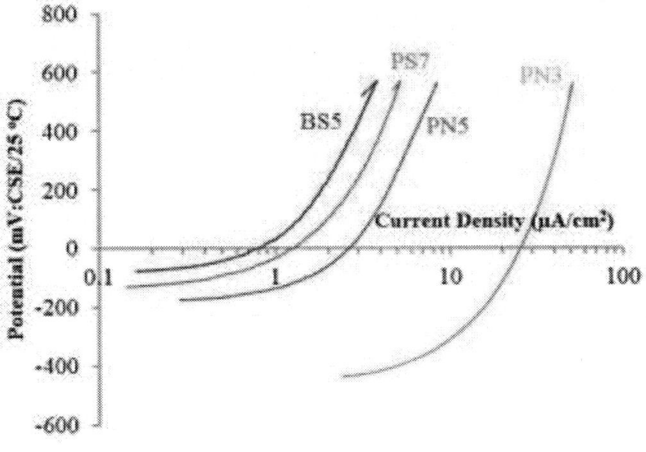

Fig. 11 Anodic polarization curve

Average of corrosion potential is sequenced as BS5<PS7<PN5<PN3 with values ranging from -125 to -370 mV:CSE. Irrespective of the cement types and curing conditions at the early life of concrete, for the same cover thickness, PN5 exhibited more potential corrosion compared to BS5. It shows that GGBFS is a good barrier against chloride-induced due to compactness of microstructure, hence it can reduce the risk of corrosion. In addition, results

from both anodic polarization curve and natural electrode potential test showed that PN3 has started to corrode after 12 years exposure.

Verification of Rebar Corrosion: From Figure 12, it can be seen that there was a little corrosion on the ends of strand caused by the absence of cover concrete and was characterized by the presence of leaching of rebar rust in the concrete. However, this little corrosion did not affect the whole strand. While in PN3 as shown in Figure 12 (c), on the edge of strand whose no cover, strand was corroded along the 5 to 7 cm and the ends of strand have been ruined. Total area of corrosion was observed after removing all concrete is approximately 1.01%.

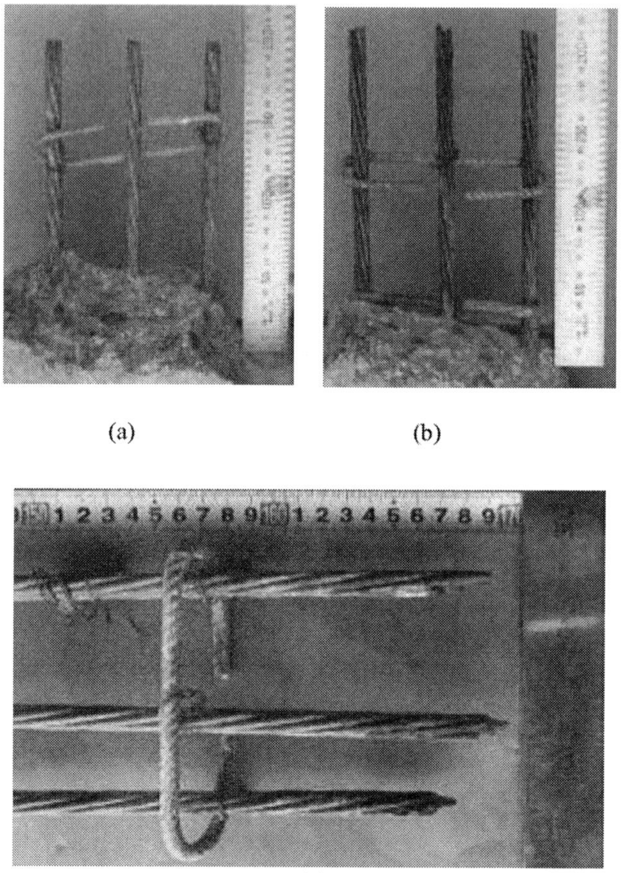

(a) (b)

(c)

Fig. 22 Verification of rebar corrosion by removing cover concrete
(a) BS5; (b) PN5; (c) PN3

4 Conclusion

In order to evaluate the performance of PC sheet pile under marine splash environment, repeated of wet-dry cycle were conducted. The results obtained may be summarized as follows:

1. Compressive strength decreased slightly after prolonged exposure in the marine splash environment while the elastic modulus kept increasing.
2. The pore volume increased both in PN5 and PS7, but decreased in BS5. Meanwhile, pore size distribution showed a similar trend in the range of capillary pore.

3. Carbonation were not identified in this investigation.
4. Chloride content in specimens is less than 0.4% by mass of cement which indicates that corrosion has not yet occurred.
5. All PC strand embedded in concrete exhibited good passivity and only a little corrosion on the ends of strand caused by the absence of cover concrete, except PN3 that showed increasing potential for corrosion. It is recommended to increase the thickness of PC sheet pile become 11 cm to improve the durability.
6. Specimen BS5 showed the best performance, which can provide a good barrier against chloride-induced due to compactness of microstructure and can reduce the risk of corrosion.

Nomenclature

B GGBFS (Ground Granulated Blast Furnace Slag)
N water curing
P HSPC (High Early Strength Portland Cement)
S Steam curing

Acknowledgment

We would like to thank The Port and Airport Research Institute Japan, who has prepared the specimen.

References

1. T. Amino, N. Otsuki, et al, "*The influence of Chloride Content at the Rebar on Corrosion Progress in Concrete of Dolphins*", in Proc. 34th Conference on Our World in Concrete & Structures, Singapore, (2009).
2. N. Otsuki, "*A study of Effectiveness of Chloride on Corrosion of Steel Bar in Concrete*", Report of Port and Harbor Research Institute:127-134, (2003).
3. Seung Lee, Subbiah Karthick, et.al, "*Long-term corrosion performance of blended cement concrete in the marine environment – A real-time study*", *Construction and Building Materials*, Vol. 154, pp. 349-360, Nov. (2017).
4. M. Robert, et al, *"Durability of Precast Prestressed Concrete Piles in Marine Environment: Reinforcement Corrosion and Mitigation-Part 1*", Final Report, pp. 2-32, June (2011).
5. M. Santhanam, M. Otieno, "5-Deterioration of concrete in the marine environment", *Marine Concrete Structures, Design, Durability and Performance*, pp 137-149, (2016).
6. P.C. Aitcin, "The durability characteristics of High Performance Concrete: a review", *Cement & Concrete Composites*, No. 25, pp. 409-420, (2003).
7. H. Hamada, H. Yokota, Y. Kikuchi, "*Durability Evaluation of Pre-stressed Concrete Sheet Pile under Marine Corrosive Environments by Acceleration Test", Technical Note of The Port and Airport Research Institute,* No. 1097, June 2005 (in Japanese). (2005).
8. H. Hamada, K. Nakahashi, R. Kuroki, "*An Acceleration Test for Chloride Ion Permeability and Corrosion of Steel Embedded of High Strength Concrete", Annual Papers of JCI*, Vol. 21, No. 2, pp. 979-984, (1999) (in Japanese).
9. N. Yuasa, H. Hamada, Y. Miyazaki, "An Acceleration Test of PC Sheet Pile for Durability Evaluation Under Marine Environment", *Annual Papers of JCI*, Vol. 21, No. 2, (1999), pp. 1027-1032 (in Japanese).

10. Lu Cui, J.H. Cahyadi, "Permeability and pore structure of OPC paste", *Cement and Concrete Research*, Vol. 31, pp. 277-282, 2001.
11. P. K.. Mehta and PJM. Monteiro, *Concrete: Microstructure, Properties, and Materials,* 3rd ed", Mc Graw Hill, pp. 442, 2006.
12. H. Hamada, Y. Sagawa, T. Ikeda, R. Morikawa, *"Several Factors Affecting the Anodic Polarization Curve of Steel Bars Embedded in Mortar", in Proc.* 6th *International Conference on Concrete Under Severe Conditions, Environment and Loading*, Mexico, (2010).

Sustainable road drainage system: Experimental model

Andung Yunianta[1,*], *Suripin*[1], and *Bagus* Hario Setiadji[1]

[1]Universitas Diponegoro, Civil Engineering Department, Semarang, Indonesia

Abstract. Roadway as part of land transportation is very important to actualize human or commodity movements. Roadway should be able to show good performance. It is known that inundation at roadway is very serious problem, and often causes disruption to road users. Conventional drainage system that has been applied with the concept of discharging or draining water to the drainage channel has not been able to evade rain water that falls on the road surface. To date, the load of drainage channel is increasing. This is contributed by the decrease of the drainage function in order to flow water through the channel. To overcome the problem, a new concept of eco-drainage, that is a concept to minimize the change of water discharge and pollutant as an impact of region development, is introduced. To do so, an experimental method with laboratory test of aggregate and soil will be conducted, followed by developing a roadway drainage model. This study aims to develop a sustainable road drainage system by increasing the ability of the drainage system in infiltrating water that related to road construction to soil in the means of providing artificial facilities. This artificial facility includes side channels, filter layers and infiltration wells. At the end of this study, the expected result is the ability to measure the efficiency and effectiveness of the performance of the model.

1 Introduction

Roadway represents an infrastructure that functions as media of land transportation movement, either motorized or non-motorized vehicles, and also pedestrian. The roadway infrastructure is generally designed to be able to cater the vehicle movements for certain design period by considering various factors, such as construction material, and also the number of vehicle load repetition that will work on it.

In service period, beside vehicle load repetition, natural factor in the form of rainfall can cause the presence of surface water run-off on roadway and the surrounding unpaved parts. The road surface will increase waterproof surfaces, therefore, high rainfall can increase the volume of surface water run-off volume and as a result, inundation on roadway surface might be occurred if road drainage does not function well. This can be contributed by waste, such as organic waste (leave, plants, and so on), and in-organic waste produced by

* Corresponding author: Andung.ay@gmail.com

human, causing to heap or fulfill road drainage channel. Therefore, it could cork the road drainage channel and result overflow water when it rain is fall. This all happen due to the understanding that surface run-off has to discharge immediately through drainage channel. If something occurs on the channel, it is the beginning of the continuous destruction of the road.

One solution offered is what people called as Sustainable Urban Drainage Systems (SUDS) i.e. a system which consists of one or more structures developed to manage surface water run-off [1]. The basic concept of this system is to improve water usefulness, to minimize the loss, and also to improve environmental conservation [2]. One of the efforts is to detain or reduce the speed of the surface water run-off when rainfall and immediately infiltrate the water into soil. As a whole, SUDS can be categories into several rainfall retention facilities, as seen in Figure 1.

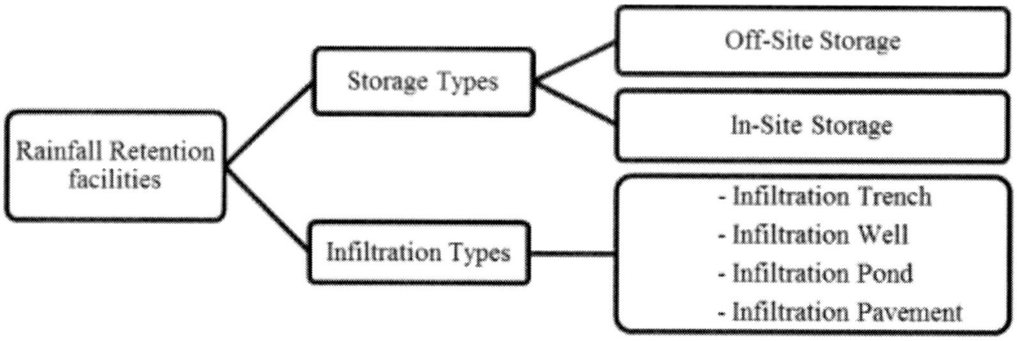

Fig. 1 Classification of Rainfall Retention Facilities in SUDS.

This aim of this research is to propose a design of sustainable road drainage system, in means of three facilities (side channel, filter layer and infiltration well), to overcome especially water inundation problem that is suitable for Indonesia condition. This paper is a documentation of the proposed of experimental model of a design of sustainable road drainage system, that is still in the development stage of the structure of the infiltration well.

2 Sustainable road drainage system concept

To date, the concept of drainage system has developed broadly into a concept of integrated water management which tends to detain rainwater in the place it falls. One concept is known as Low Impact Development (LID), that is, an environmental-friendly drainage systems concept that maintains the natural and hydrology conditions like the condition early before development. It is a concept to maintain and also imitate existing natural process to minimize the change of water discharge and pollutant as an impact of region development.

This environment-friendly drainage system is defined as an effort to manage excess water by infiltrating naturally as much as possible water into soil or flowing the water to river without exceed previous river capacities. There are several factors required to be evaluated in this concept, which cover proposed road drainage system, hydrology and hydraulics, soil permeability, infiltrate, aggregate gradation selection, and infiltration well.

2.1 Road drainage

According to new Indonesia Road Design Guide, in planning road drainage system, it should consider the existence of surface and sub-surface water, so that the proposed road drainage system can be divided into surface drainage system and sub-surface drainage system [3].

In general, to design road drainage system, it is commenced by plotting the evaluated road route onto topography map to know the service area so that it can be predicted the location required for drainage structures (drainage channel, infiltration well, and so on). In the design, it should consider surface or sub-surface water flow by following the design guide without disturbing the road construction stability.

2.2 Hydrology and hydraulics

In hydrology, one should concern about the water resources, the potency of floods, the cycles of the water process (evaporation, precipitation, infiltrate and out flow). On the other hand, people learn about the structure required to transport the water in hydraulics. It can be opened or closed channel and factors that affect the selection between them. The interaction between hydrology and hydraulics will enable to provide effective and efficient drainage system.

2.3 Soil permeability

Permeability is defined as a characteristic of pore materials that enable water to pass pore cavity. Soil pore interact each other so that water can flow from one point with high energy to another point with low energy. In the soil, the possible water stream could be laminar or turbulent. The restraint of the water flow could be contributed by soil type, granular size and shape, mass density, and also the geometry form of pore cavity. Temperature is also very influence to water restraint (surface tension and viscosity).

2.4 Infiltration

Infiltration is the flow of water into the soil through the soil surface. In the ground, water flows in lateral direction, as inter-flow to the spring, river or lake. It can also flow in the ground vertically, known by percolation. The movement of water in the soil through the pores of the soil is affected by the force of gravity and capillary force. The force of gravity causes the flow always go to the lower places, while the capillary force causes the water to move in all directions. Capillary water is always moving from the wet to dry areas. Dry soil has a larger capillary force than wet soil. This force decreases with the increase of soil moisture. In addition, the capillary force works stronger on the ground with fine grains of clay rather than coarse grained soil such as sand.

2.5 Aggregate and aggregate gradation selection

Aggregates are commonly defined as hard rock formations of the earth and shafts. ASTM 1974) defining aggregate as a material which consist of solid mineral, in the form of a large size and/or fragments. Beside aggregate properties, aggregate gradation becomes one important factor in filter layer consisted of aggregates. This is because it could affect the ability of the filter layer (in terms of velocity and volume) to flow in the water.

2.6 Infiltration well

Infiltration well is one of technical water conservation in the form of well with selected depth, filled with infiltration materials (such as sand, stone, fibre of palmtree) layer by layer up to it reaches same level with surface, in which it functions as relocation place and at the same time as a place where water infiltrate into soil [4].

3 Thinking framework and research methodology

The thinking framework of this research can be illustrated in the flow chart as shown in Figure 2.

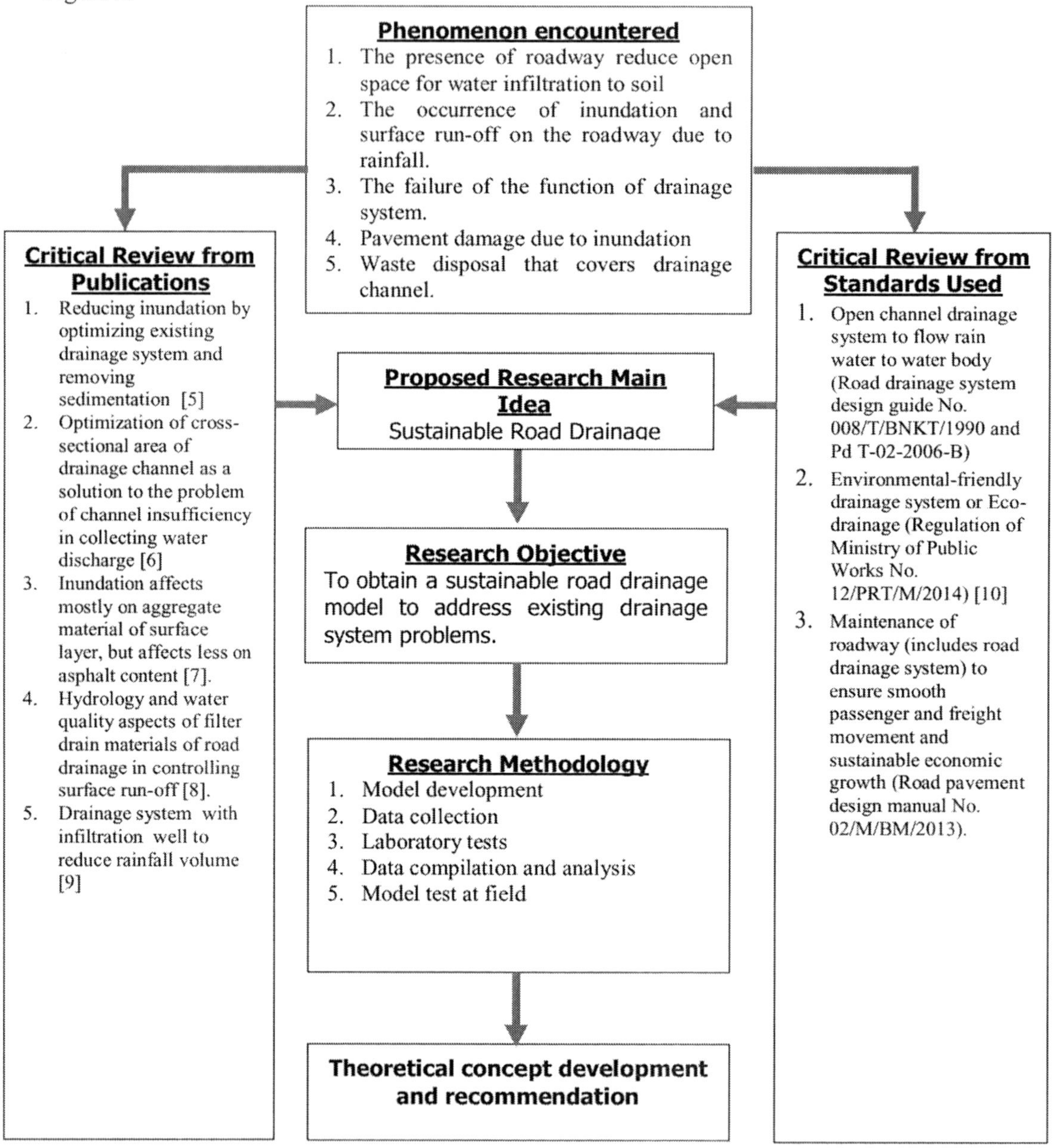

Fig. 2 Thinking framework of the research

According to Figure 2, the research methodology of this research is as follows.

3.1 Model development

In this research, the model of sustainable road drainage system is as illustrated in Figure 3. The model of sustainable road drainage system is developed using the following criteria:

3.1.1 Drainage roadway channel

a. The channel has a trapezium form and made of pairs of aggregate with mortar or concrete;
b. The floor of the channel is not hardened at certain location to enable water infiltrates to soil.
c. At the point of channel wall near infiltration well, inlet pipe is installed integrated to infiltration well, attached above and under channel wall
d. Aggregate (that pass all laboratory tests) is placed as filter in the channel.

3.1.2 Infiltration well

a. Infiltration well is made of concrete buis with 1-m diameter
b. The depth of infiltration well is influenced by the depth of ground water surface, i.e. shallow infiltration well if ground water surface is deep or a deep infiltration well if ground water surface is shallow.

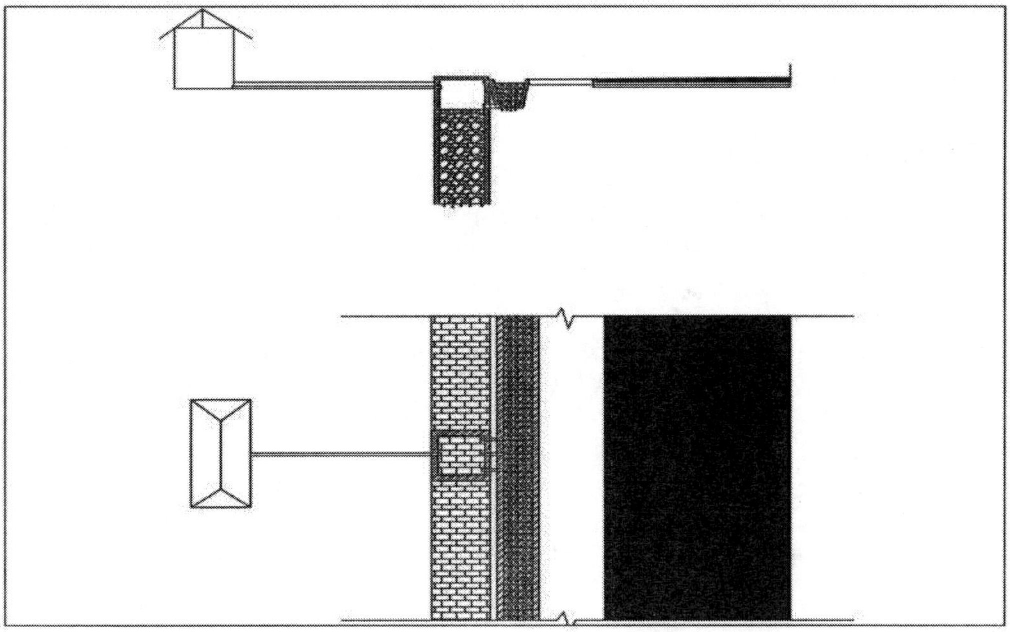

Fig. 3 Proposed Road Drainage System Model

This research is a quantitative type that used to research on a population or specific samples. Data collection is conducted by using research instruments, and analysis is conducted quantitatively/statistically with the aim to test the predefined hypothesis [5].

Table 1 Dependent and Independent Variables

Dependent Variable	**Independent Variable**
Rainfall	Hydrology analysis
Run-off debit	Run-off flow
Channel dimension	Hydraulic analysis
Permeability level	Soil permeability
Aggregate gradation	Aggregate selection
Depth of infiltration well	Level of ground water surface

There are several independent variables and their corresponding dependent variables used in this research (see Table 1). The values of both variables were observed and measured, to understand the correlation and the effect of each independent variable to its corresponding dependent variable, and vice versa.

3.2 Data collection

There are two types of data collection that are proposed to be performed. They are:
1. Primary data, which consists of:
   a. measurement of existing drainage channel dimension and velocity of the water flow,
   b. documentation of existing drainage system condition and the surroundings
   c. interview to related stakeholders (highway authority, academics, and so on).
2. Secondary data, which consists of collection of several recorded data as follows.
   a. rainfall data
   b. area of evaluated region
   c. slope of existing channel
   d. number of population of evaluated region
   e. topography of evaluated region

3.3 Laboratory Test

There are several laboratory tests will be performed in the next future. They are: (i) aggregate test, (ii) soil permeability test, and (iii) test of road drainage system model.

The aggregate tests aim to know the aggregate quality used for filter at drainage system. The aggregate tests consist of several tests as follows.
1. Sieve analyse, that is conducted to determine the gradation of the aggregate. Different aggregate gradation will be used in the test of road drainage system model to simulate the best aggregate gradation used as filter in drainage system.
2. Specific gravity and absorbtion of coarse aggregate. This test is to determine the specific gravity of aggregate (bulk, saturated surface dry/SSD and apparent), and also to know level absorption of aggregate.
3. Flakiness and elongation index, this test is used to limit the number of thin and long aggregate used in the proposed aggregate gradation.
4. Los Angeles aggregate abrasion test, to minimize the number of aggregate that tend to crushing, degradation and disintegration due to abrasion and impact force.

In soil permeability test in laboratory, there are four different tests to determine coefficient permeability of the soils [6] as follows:

1. Constant head permeabiltas test

2. Falling head permeability test
3. Determination of permeability coefficient using consolidation test.
4. Capillary horizontal test.

Test of road drainage system model aims to measure the level of water infiltration. To do so, models of road drainage system with aggregate as filter are prepared in laboratory. A simulation will be conducted by using different aggregate gradations as filler to determine the suitable aggregate gradation used with different slopes of the channel and also different amount of surface run-off. The test is carried out by measuring several parameters regarding to the performance of the model, such as time of water infiltration, the presence of inundation on top of drainage channel, the quality of water obtained from the outlet of infiltration well, and so on. It is expected that different model can be developed based on different condition encountered in the field. The illustration of typical cross-section of the model is as depicted in Figure 4.

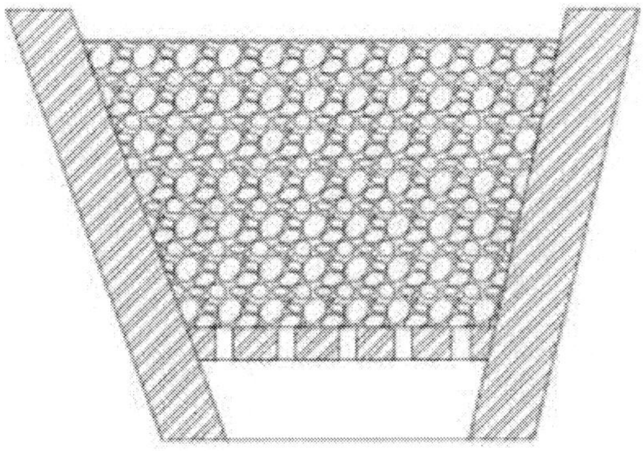

Fig. 4. Typical Cross Section of Road Drainage System Model with Aggregate as Filter

3.4 Evaluation and Validation Model

To evaluate the performance of the model, it is necessary to conduct validation process by applying and examining all components of the proposed drainage system, i.e. side channel, filter layers and infiltration well, at different conditions that commonly encountered at the field. Validation process is especially conducted to evaluate the level of water infiltration in the aggregate filter and also the velocity of water flow in the infiltration well. This step represents the period to find out the backwards of the proposed model for future development.

4 Conclusion

This paper proposed a new concept of environmental-friendly or sustainable road drainage system in which the surface water run-off is infiltrated through an integrated artificial facility consisted of side channels, filter layers and infiltration wells, instead of discharging it through side channel into water body. This new concept is also to be able to minimize the change of water discharge and pollutant as an impact of roadway operational and maintenance work. It is expected that the proposed integrated artificial facility can help the roadway free of problems due to moisture, such as inundation and flooding.

Reference

1. Estonia Latvia Programme, *Handbook on Sustainable Urban Drainage Systems, European Union*. (2013).
2. Suripin, *Sustainable Urban Drainage System, Andi Publisher*, Yogyakarta, Indonesia (in Bahasa Indonesia). (2004).
3. *Ministry of Public Works, Road Drainage System Design Guide*, Pd.T-02-2006-B (in Bahasa Indonesia). (2006).
4. Kusnaedi, *Infiltration Well for Urban and Rural Settlements*. PT. Penebar Swadaya, Jakarta, Indonesia (in Bahasa Indonesia). (2002).
5. Liany, A.H, *Optimization of Road Drainage System as an Alternative Solution to Inundation,* Journal Tekno Sipil, Vol.12 No.16, Sam Ratulangi University, Manado, Indonesia (in Bahasa Indonesia). (2014).
6. Syapawi, A. *Study of Drainage and Flood Water Inundation Problems on Jalan Kemang Manis, Journal Pilar Teknik Sipil, Polsri*, Palembang, Indonesia (in Bahasa Indonesia). (2014).
7. Chairuddin, F, *Experimental Assessment of the Impact of Rainwater Inundation on Asphalt Pavement Structures (Case Study Dr. Wahidin Sudiro Husodo Street, Makassar City),* Proceedings of KonTeks 7, Sebelas Maret University, Surakarta, Indonesia (in Bahasa Indonesia). (2013).
8. Nursetiawan, *Performance Review of Drain Filters on Highway as One Kind of Sustainable Drainage System Forms*, Institut Teknologi Nasional, Bandung, Indonesia (in Bahasa Indonesia). (2006).
9. Mahardika, A.R, Yulianto, F.A, Suripin, and Budieny, H., *Design of Environmental – friendly Drainage System on Semarang - Solo Toll Road, Undergraduate thesis,* Departement of Civil Engineering, Faculty of Engineering, Diponegoro University, Semarang, Indonesia (in Bahasa Indonesia). (2015).
10. Ministry of Public Works, *Regulation of Ministry of Public Works* No. 12/PRT/M/2014 (in Bahasa Indonesia). (2014).

Clustering ferry ports class-I based on the ferry ro-ro tonnages and main dimensions

Syamsul Asri[1,*], *Wahyuddin* Mustafa[1], *Mohammad* Rizal Firmansyah[1], and *Farianto* Fachruddin Lage[1]

[1]Hasanuddin University, Department of Naval Architecture Gowa, Indonesia

Abstract. Capacity constraint of port for Ferry Port Class I is not clearly stated in the Indonesian National Order of Port Affairs. It only said that ferry ship to be serviced is ships with more than 1000 gross tonnages (GTs). The word more than 1000 GTs is unclear as there are ships with up to 12500 GTs to be serviced in this port. This research introduces cluster of Ferry Port Class I. Ship main dimension and tonnage of sample ships is being used as a basis for the port clustering. The result shows that Ferry Port Class I should be divided into three main Ferry Port Classes I (Class I_A, Class I_B and Class I_C). This new cluster of Ferry Port Class I can be used as a reference in updating the rules regarding ferry port class and as a reference in the design of ferry port main facilities.

1 Introduction

Ferry transport service is one of the three types of transport services in Indonesian waters. As an extension of main road transport service, ferry transport service must be developed to strengthen the intra and inter connectivity of Indonesian regions especially for inter islands transport service. Transport services in a ferry transport service route is organized by supply some ferry ro ro ships to be operated regularly and continuously. Since the types of ship to carry passengers and vehicles to service the route is ferry ro ro ship, then the ferry port facilities are equipped with some facilities to support ro-ro (roll on – roll off) loading and unloading activities of passengers, vehicles and their cargos.

Ferry ports can be classified into three ferry port classes which are Ferry Port Class I, Ferry Port Class II and Ferry Port Class III [1]. The consideration for classification of ferry ports is mainly based on the port capacities which are expressed in ship gross tonnage (GT). Ferry port capacities for each class then can be divided into Class I for ship with tonnage more than 1000 GTs, Class II for ship with tonnage between 500 GTs and 1000 GTs and Class III for ship with tonnage less than 500 GTs.

There are four ferry transport services in Indonesia which can be categorized as the main routes i.e, Merak-Bahauheni, Ketapang-Gilimanuk, Padangbai-Lembar and Bajoe-Kolaka. Although all of these ferry ports can be categorized as Ferry Port Class I but the capacities of the port on each ferry port is different each other. The maximum gross

* Corresponding author: author@email.org

tonnage of ferry ro ro which service those routes is different as well. The maximum gross tonnage which service the route is represents the port capacity of each port in each route. The capacity of each port from the biggest capacity is Merak-Bakauheni ferry port with 12498 GTs of capacity, Ketapang-Gilimanuk ferry port with 876 GTs of capacity, Padangbai-Lembar ferry port with 2916 GTs capacity, and Bajoe-Kolaka ferry port with 1504 GTs as the smallest capacity [2].

2 Research Methodology

Based on the previous descriptions, the stages for further classification of the Ferry Port Class I can be divided into first, the determination of the sample ship main dimension. The ship main dimension information will be used to determine the ship sample tonnage. Finally, the classification of the Ferry Port Class I will be determined. This classification will be based on the ship main dimension and ship tonnage.

2.1 The determination of the ship sample main dimension

The ship sample main dimension will be determined with the following steps.

a. The determination of the ship sample length between perpendicular (L_{BP}) and ship breadth (B) can be done using equation (1) and (2) [3].
For L_{BP} determination:

$$L_{BP} = (\Sigma (C_{vi} L_{vi}) + D_{fa} (C_v - 1)) / L_r \quad (1)$$

Where:

C_{vi} = the number of columns for each type of vehicles
Lvi = the length of each vehicle types
Dfa = the distance between two vehicles
Cv = the number of vehicles columns
Lr = ratio between the length of vehicles area (Lv) with the length of the ship length between perpendicular (L_{BP}). Lr is determined to be 0.9.

For B determination

$$B = \Sigma(R_{vi} B_{vi}) + D_s (R_v - 1)) + 2 D_{vp} + 2 D_{ps} \quad (2)$$

Where:

Rvi = the number of rows for each type of vehicles
Bvi = the breadth of each vehicle types
Ds = the side distance between two vehicles
Rv = the number of vehicles rows
Dvp = the distance between the outer vehicles side on the outer row with the ship inner partition.
Dps = the distance between the ship inner partition with the ship breadth side.

According to the technical guidance on the minimum service requirements of ship on the river, lakes and between islands [4], the value of Dfa, Ds, and Dvp, can be determined to be 0.3 meter, 0.6 meter, and 0.6 meter respectively. Meanwhile, the value of *Dps* is determined to be 1.5 meter. The above mentioned vehicles for the determination of ship sample length and breadth is from vehicles in group IV, which is cars with the car length up to 5 meters [5] and breadth is 2.1 meter [6].
The determination of a ship length combination is based on a ratio between ship length between perpendicular (L_{BP}) and ship breadth (B) in which for ferry ro-ro in Indonesia is in a range of 3.0 and 6.5. The biggest L_{BP} value is not exceeding the biggest ship ferry

ro-ro length in Indonesia which is 131.67 m. The minimum ship sample length and breadth was determined for the ship sample tonnage which is more than 1000 GT.

b. Ship height determination can be done using the equation (3). In this equation, the ship height (H) dimension is the ratio between ship length between perpendicular (L_{BP}) and ship height (H) as a function of a ratio between ship length between perpendicular (L_{BP}) and ship breadth (B).

$$LBP/H = 5.2483\ (LBP/\ B)^{\ 0.6792}\ ;\ S\ = \pm\ 1.592 \tag{3}$$

Based on the deviation standard (S) in equation (3), there are three alternatives of ship height (H) which is determined for each ship sample length-breadth variation. They are the highest height (H_H), intermediate height (H_M) and the lowest height (H_L).

c. The determination of ship draught (T) can be done by calculating the ratio between ship breadth (B) and ship draught (T) (*B/T*) as a function of the ratio between ship breadth (B) and ship height (H) (*B/H*) as can be seen in the following equation (4).

$$B/T = 1.1765\ (B/H)^{1.1617}\ ;\ S\ = \pm\ 0.569 \tag{4}$$

Based on the deviation standard (S) in equation (4), there are three alternatives of ship draught (T) which is determined for each ship sample breadth-height variation. They are the highest draught (T_H), intermediate draught (T_M) and the lowest draught (T_L)

2.2 The determination of the ship sample tonnages

After the ship main dimension has been determined, ship gross tonnage (GT) of each ship sample can be calculated with the following steps:

a. The calculation of ship volume tonnage (VT) which consist of ship hull volume (VH) and ship superstructure volume (VS) can be done using the equation (5), (6) and (7) below [7]:

$$VH = 1.04\ LBP\ B\ T\ CB\ (1.25\ H/T - 0.25) \tag{5}$$

$$VS = LBP\ B\ (0.0036\ LBP\ B\ + 0.6687) \tag{6}$$

$$VT = VH\ +\ VS \tag{7}$$

Note: the value of ship block coefficient (CB) which being used in the calculation of the ship hull volume is 0.60.

b. The calculation of the ship gross tonnage (GT) using equation (8) as in the following [5]:

$$GT = (0.2 + 0.02\ \log10\ VT)\ VT \tag{8}$$

Ship tonnage volume (VT) in this calculation is according to the ship tonnage volume in equation (7).

2.3 Further classification of the ferry port class I

After the The last stages for the classification of the Ferry Port Class I is the determination of cluster of the Ferry Port Class I. This classification is based on the group of the ship sample main dimension and ship sample tonnages.

a. Ship sample classification which is based on the ship main dimension and ship tonnage were conducted through the determination of interval class as a result of the ship length between perpendicular (L_{BP}) with the ship breadth (B) (L_{BP} x B) using the equation 9, 10 and 11. The result of multiplication between ship breadth and ship length (L_{BP} x B) can be used as a representation of ship dimension and tonnage because ship height (H) was calculate as a function of ship length (L_{BP}) and ship breadth (B) (refer to equation 3), while ship draught (T) was calculated as a function of ship breadth (B) and ship height (H) (refer to equation 4).

$$r = db - dk \tag{9}$$

$$k = 1 + 3,3 \log n \tag{10}$$

$$p = r / k \tag{11}$$

where:
r = interval
db = the biggest data value
dk = the smallest data value.
k = number of class
n = number of data
p = length of the class interval.

b. Identification of the biggest ship sample main dimension and tonnage on each interval class.
c. Determining the cluster of the Ferry Port Class I by minimizing similarity and maximizing the capacity differences for ships service between the port classes.

3 Result and analysis

3.1 The main dimension of the sample ships

By using equation (1) and (2), the ship sample length between perpendicular (LBP) and the ship sample breadth (B) can be calculated and the result can be seen in Table 1. There are 14 groups or series of the ship lengths (with code Ln), and 5 groups or series of the ship breadth (Bn) in the table. The number on the ship length code and the number of the ship breadth code are both show the number of columns and rows of vehicles in group IV which can be loaded on the ship sample.

Tabel 1. Ship sample length and ship sample breadth

Length		Breadth (B_n) (m)				
Code (L_n)	L_{BP} (m)	B_3	B_4	B_5	B_6	B_7
L_9	52,667		14.400	17.100		
L_{10}	58,556		14.400	17.100		
L_{11}	64,444	11.700	14.400	17.100	19.800	
L_{12}	70,333	11.700	14.400	17.100	19.800	22.500
L_{13}	76,222	11.700	14.400	17.100	19.800	22.500
L_{14}	82,111		14.400	17.100	19.800	22.500
L_{15}	88,000		14.400	17.100	19.800	22.500
L_{16}	93,889		14.400	17.100	19.800	22.500
L_{17}	99,778			17.100	19.800	22.500
L_{18}	105,667			17.100	19.800	22.500
L_{19}	111,556			17.100	19.800	22.500
L_{20}	117,444				19.800	22.500
L_{21}	123,333				19.800	22.500
L_{22}	129,222				19.800	22.500

The number of the ship sample in this analysis and being the basic for the determination of ship tonnage is 405 ships sample. These ships can be categorized as follows. The number of ship in the category of ship length and ship breadth which is the ship with a variation of different ship length and breadth is 45 ships sample. On each ship length and breadth variation, there is an addition of 3 ships height variation (calculated using equation 3). Hence the total of the sample ships with this additional criterion is 135 ships sample. With the addition of criterion of the last ship draught variation to the previous criteria (calculated using equation 4), then the total ship samples will be 405.

In Figure 1, it shows the result of the calculation to determine ship sample height (H) and draught (T) as described previously. However, the figures only show the intermediate height (HM) as well as intermediate draught (TM). The highest and the lowest value in the interval are not shown.

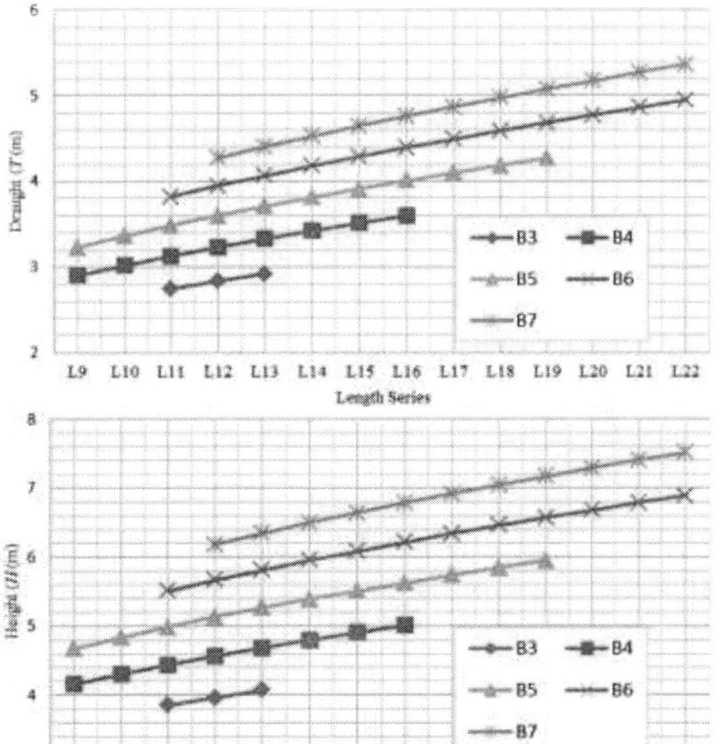

Figure 1. Ship sample height and draught (H vs L and H vs D)

4 Ship sample tonnages

As have been mention earlier, there are 405 variations of ship sample main dimension are being used in this calculation. Gross tonnage for each ship sample is calculated using equation 5, 6 and 7. Interval for ship sample gross tonnage is 1.169 GTs with LBP = 64.444 meter, B = 11.700 meters, H = 3,508 meters, T = 2,785 meters for the smallest ship sample gross tonnage up to the biggest gross tonnage of 14,351 GTs of the ship sample with its main dimension of LBP = 129.222 meters, B = 22.500 meters, H = 8.314 meters, T = 5.249 meters. Figure 2 shows a graph for the result of ship sample gross tonnage calculation.

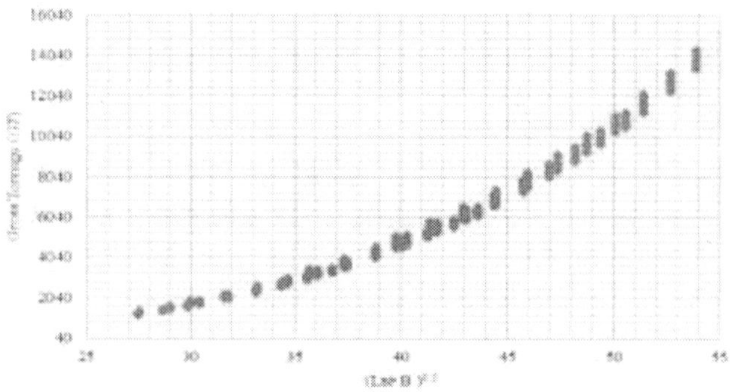

Figure 2. Ship sample gross tonnage

In figure 2, a ship with its length and breadth is more than other ship is tending to have more gross tonnage value. But some ships with the same length and breadth have different gross tonnage value as they are have different height. The same thing happens for the sample ships with the same length and breadth but different draught. They are having different tonnages as well. For the ships with the same length, breadth and height, the smaller the ship draught, the biggest gross tonnage it has.

5 Clustering the ferry port class I

For the 45 variations of ship sample length (LBP) and breadth (B) which being used in this research (Table 1), the biggest and smallest result of multiplication between ship length and breadth is 2908 m2 and 754 m2. By applying equation 9, 10 and 11, the number of ship sample main dimension can be divided into 6 classes with the length of the class interval is 359 m2. Based on these interval classes, the ship sample main dimension with the biggest tonnage on each interval classes are shown in Table 2 below.

Tabel 2. Ship sample main dimension with the biggest tonnage in each interval class

Interval class	Ship main dimension				Ship tonnage (GT)
	L_{BP} (m)	B (m)	H (m)	T (m)	
A	129,222	22,500	8,314	5,249	14351
B	111,556	22,500	8,021	5,062	11105
C	93,889	22,500	7,703	4,858	8260
D	105,667	17,100	6,436	4,064	5982
E	70,333	19,800	6,540	4,121	3996
F	64.444	17.100	5.722	3.606	2629

In Table II, it can be seen that a ship with bigger tonnage does not have bigger ship main dimension elements to compare with a ship with smaller tonnage. As in interval class C, the ship with the biggest tonnage has smaller ship length than the ship length of a ship with the biggest tonnage in interval class D. Further, ship with the biggest tonnage in interval class D has smaller ship breadth, height and draught than the ship breadth, height and draught of a ship with the biggest tonnage in interval class E.

As a consequence, the design of main facilities of port area cannot be based on the ship main dimension with the biggest tonnage. Hence, there must be clarity on the biggest ship main dimension elements of a number of ship main dimensions in each interval class. Tonnage and ship main dimension elements in each interval class are shown in Table 3.

Table 3. Tonnage and ship main dimension elements in each interval class

Interval Class	Biggest Tonnage GT)	The biggest ship main dimension elements			
		L_{BP} (m)	B (m)	H (m)	T (m)
A	14351	129.222	22.500	8.314	7.083
B	11105	129.222	22.500	8.021	7.746
C	8260	111.556	22.500	7.703	6.338
D	5982	105.667	22.500	7.357	6.007
E	3996	93.889	19.800	6.540	5.358
F	2629	82.111	17.100	5.772	4.708

Though there are 6 interval classes or 6 levels of ship tonnage, ship length and breadth level are only 5 and 3 levels (Table 3). Based on the clustering principle which is minimizing similarities and maximizing the differences between classes, cluster of ship main dimension and tonnage for the determination of cluster Ferry Port Class I is divided into three clusters as can be seen in Table 4. Tonnage and ship main dimension in interval class A, B, C and D are combined and become one cluster of port (Port Cluster IA) for servicing ships in tonnage range of 4400 GTs up to 14351 GTs while in interval classes E and F, they are become Port Cluster IB and IC respectively and servicing ships in tonnage range 3000 GTs up to 4400 GTs for port class IB and 1000 GTs up to 3000 GTs for port class IC.

Table 4. Cluster of ferry port class I

Port Cluster	Ship tonnage (GT)	The biggest ship main dimension elements			
		LBP (m)	*B* (m)	*H* (m)	*T* (m)
I_A	$4400<GT\leq14351$	129.222	22.500	8.314	7.083
I_B	$3000<GT\leq 4400$	93.889	19.800	6.540	5.358
I_C	$1000<GT\leq 3000$	88.000	17.100	5.835	4.834

The range of ship main dimension and tonnage in Table 4 become the capacity constraints of servicing ships on each cluster in Ferry Port Class I. The constraints are based on following reasons:

a. On the ships groups with the ship breadth of 17.1 meters, there are ship with the tonnage of 3030 GTs (or 3000 for simplification). Up to this tonnage, the biggest ship length to be serviced is 88.0 meters.

b. On the ships groups with the ship breadth of 19.8 meters, there are ship with the tonnage of 4389 GTs (or 4400 for simplification). Up to this tonnage, the biggest ship length to be serviced is 93.889 meters.

Figures and tables, as originals of good quality and well contrasted, are to be in their final form, ready for reproduction, pasted in the appropriate place in the text. Try to ensure that the size of the text in your figures is approximately the same size as the main text (10 point). Try to ensure that lines are no thinner than 0.25 point.

6 Discussion

Based on the basis for determination of water area need of Ferry Port [8], the water area of ferry port was determined as a function of ship main dimension. Elements of ship main dimension which will be the variables for the port water main facilities consist of:

a. Ship length (L) which being variable for port length, area for ship (length and breadth) berthing, diameter of turning basin and diameter of dock basin.

b. Ship breadth (B) which being variable for the width of water ways

c. Ship draught (D) which being variable for the depth of the port basin

As being described in the previous section, the range of the ship main dimension with tonnage between 1.169 GT and 14.351 GT is big. It covers ship length (LBP) from 64.444 m up to 129.222 m, ship breadth (B) from 11.70 m up to 8.314 m and ship draught from 2.785 m up to 5.249 m. As a consequence, Ferry Port Class I should be divided into three clusters as shown in Tabel 4.

The division of the Ferry Port Class I into three sub classes is needed to strengthen the regulation on the determination of the capacity of the ferry port main facilities for ship service. The port capacity constraint which is in ship tonnage (GT) must be clarified with the biggest ship main dimension elements consist of ship length, ship breadth, ship height and ship draught that can be serviced. The ship main dimension elements as referred is not the biggest ship main dimension with the biggest tonnage in each cluster but instead, it is the biggest ship main dimension elements from a number of ship main dimension variation in each cluster.

Further, clustering the Ferry Port Class I into three sub classes is needed in order to strengthen the regulation on the placement of ship in a ferry route which is suitable to its Ferry Port Class. On the other hand, the clarity of cluster to be serviced and ferry port capacity will increase the smooth and safety operation of ship in the ferry port area. The determination of the pier length, the area for ship berthing, diameter of turning basin and diameter of dock basin which is based on the biggest ship length overall (LOA) to be serviced is to guarantee the smooth and safety operation for berthing process and maneuverability of other ship. The width of water ways in turn must be based on the biggest ship breadth to be serviced as well in order to anticipate the sea current driving force and sea wave on ship at the time of ship entering and exiting ferry port area. In the end, the depth of the port basin must be based on the biggest ship draught to be serviced to prevent ship to run aground in the ferry port area.

7 Conclusion

This research introduces a new cluster for Ferry Port Class I. In the Decree of Minister of Transportation of Republic of Indonesia [1], the Ferry Port Class I is for ship with tonnage more than 1000 GTs. The word "more than 1000 GTs" is unclear as in fact there is a big range of ferry ro-ro with more than 1000 GTs from 1000 GT itself up to 12500 GTs. Hence there is no clear and specific classification for Ferry Port Class I.

In this research, the further classification of the Ferry Port Class I have been analyzed and introduced. This classification has clarified on what ship tonnage and what ship main dimension which can be put on service on certain ferry route which appropriate with the ferry port capacity. The analysis is based on the ship main dimension and tonnages. The result of the analysis shows that the Ferry Port Class I must be divided into three clusters which is:

a. Ferry port class IA for servicing ship with the capacity up to 14351 GTs and with the ship length (LBP) is 129.222 meters or with the ship breadth (B) is 22.5 meters.

b. Ferry port class IB for servicing ship with the capacity up to 4400 GTs with the ship length (LBP) is 93.889 meters or with the ship breadth (B) is 19.8 meters.

c. Ferry port class IA for servicing ship with the capacity up to 3000 GTs with the ship length (LBP) is 88.0 meters or with the ship breadth (B) is 17.1 meters.

References

1. Decree of Minister of Transportation of Republic of Indonesia Number KM 53 of 2002 on the National Order of Port Affairs (in Indonesia).

2. Directorate General of Land Transportation, Directorate of LLASDP, 2012, The Map of Ferry Routes, The Ministry of Transportation of the Republic of Indonesia, Jakarta (in Indonesia).
3. Asri, S, Pallu, M, S, Thaha, M, A, and Misliah, 2015, "Model Design of Inter-Island Ships Based on Transport Demand and Port Facility", International Journal of Engineering Research and Technology, Vol. 4, Issue 12, pp. 643-651.
4. Decree of Directorate General of Land Transportation, Number AP 005/3/13/DRPD/94 regarding Technical Guidance on Minimum Service Requirements of River, Lake and Ferry Service (in Indonesia).
5. Decree of Minister of Transportation of Republic of Indonesia Number PM 18 Year 2012 regarding Changes on the Decree of Minister of Transportation of Republic of Indonesia Number KM 58 Tahun 2003 regarding Mechanism for Determining and Formulating the Calculation of Ferry Service Tariff, 2012, News of the State of the Republic of Indonesia, Year 2012, Number 367 (in Indonesia).
6. Decree of the Republic of Indonesia Government Number 55 Year 2012 regarding Vehicles, 2012, Sheets of the State of Republic of Indonesia Year 2012 Number 120 (in Indonesia).
7. Asri S, Wahyuddin, Firmansyah, M, R, and Djalante, A, H, 2016 "Mathematical Model Development to Estimate Gross Tonnage of Ro-Ro Ferry", International Journal of Engineering Research and Technology, Vol. 5, Issue 12, pp. 104-107.
8. Decree of Minister of Transportation of Republic of Indonesia No. 52, year 2004 regarding Ferry Port Management (in Indonesia).
9. Decree of Minister of Transportation of Republic of Indonesia Number PM 8 Year 2013 regarding Ship Measurement (in Indonesia).
10. Hakim, A, 1997. Introduction to Statistics. First Edition. Citra Media. Surabaya (In Indonesia).

Simple macro aggregate transport demand modelling in river transport study

R.Didin Kusdian[1,*]

[1] Sangga Buana YPKP University, Department of Civil Engineering, Bandung, Indonesia

Abstract. Sustainability and development of river transport can reduce the burden of road transport, where the development of road transport requires clearing new land is increasingly expensive and development is quite expensive. To estimate the amount of potential movement of people and goods through river mode choice model can be derived based on a comparison of risk at each mode, whereby the greater the risk will be even smaller portion of the mode selected. From the results of the model calculation of the aggregate distribution modes, obtained through a portion of the potential movement of the Batang Hari river in Jambi, Sumatra Island, Indonesia is 13, 89% for the movement of people and 14.85% for goods.

1 Introduction

At least not in the last five years has conducted studies on the potential transport stream continues in Indonesia, and until now has not been around the area that have the potential transport streams surveyed and analyzed potential. Some rivers that have been the object of research include Indragiri River, River Seputih, Onion River Bones, Mesuji River, River Sekampung, Batang River (on the island of Sumatra), the Barito River, River cascade Right, Left cascade river, Amandit river, Sungai Batang Alai, Tabalong River Left, Right Tabalong River, River Tapin, Amandit River, River Balangan, Melawi River, Pawan River, River Sambas, Kendawangan River, Porcupine River, the River Kapuas, Mahakam River, River Sesayap, Kayan River, River Sembakung (in Borneo Island) [1].

If viewed from the side of the history of the movement of people across the region in line with the time line of the history of civilization and technological development of transportation, will be remembered that it used the transport stream earlier than other types of transportation in the world.

Advances in technology and the development of infrastructure construction vehicle motion media vehicles have gradually eliminate vehicle transport water through the river to reach the center of activity in the mainland, where the movement of tangible media stream to connect to the port on the coast as the gateway transportation between islands and even between continents.

* Corresponding author: kusdian@yahoo.com

The rapid growth of industrial combustion four-wheeled vehicles and two-wheeler, has been welcomed by the traveling consumer uptake rapidly growing vehicle markets as well, one of the advantages that attract consumers is its 'door to door', making it easier for individuals to move and save time. But when it continues to grow, now arrived at a situation where the users of the road (motion media) each other have slowed down movement of vehicles which can be enjoyed respectively. The volume and density of vehicles on the road continues to increase, the frequency of the number of events increased congestion on the roads as well, especially in densely populated areas.

Maintenance, improvement, and construction of new lanes on road infrastructure cost which cannot be said to be cheap, the increased vehicle traffic will accelerate the deterioration of the road surface, damage to the road surface will also hinder the flow of vehicles that run on it. The situation above be positive reasons why river transport should be prevented from extinction, river water body is a medium that has been provided by the natural motion that should be utilized as a medium water movement of vehicular traffic. With the use of the river as a media vehicle motion is expected that the distribution of the load the vehicle is moving, so that the density can be reduced, or the need for emergency opening of land for widening and the addition of new roads could be reduced [2].

The transfer of the vehicle into the river, especially for goods vehicles, that certain types of goods, such as commodities or raw materials such as energy crops rubber, palm oil and coal mining [3]. Sure to other types of commodities can be further investigated. Many aspects of research that needs to be done regarding the possibility of sustainable development of river transportation, few studies have been conducted by the author work well in a team as well as involvement in individually.

Among those who have begun to put forward the result is about how the sustainability of the system using a traditional cruise ship review the point of timber to the Port of Sunda Kalapa [3], the issues raised in this study is that the use of timber ships will be threatened to stop due to increasing scarcity and increasingly difficult, even impossible, to obtain a special wooden materials suitable for building ships, namely ironwood. But then submitted that this does not mean that the activities of traditional community-based transportation system, it should cease to operate as part of the national transportation system in Indonesia. Still need further study how the sustainability of the system and the organization of community-based shipping operations, although not using timber ship again.

To be more purposeful planning follow-up policy development and construction management on a river cruise system, aspects of the required study and apply the result is about how to estimate quantitatively the potential movements using river transportation system. In this paper aspects of modeling to estimate the potential amount of movement needs to use river transportation, particularly the movement of goods, a major topic of discussion and debate.

2 Methodology

This paper is part of research results by using a combined method of field surveys, literature surveys, and continuous analysis of the study reports based sectoral government and university research-based studies. The establishment of the model through a process of formulating a model phenomenon that formed sentences that fit the logic and common sense, then lowered into the mathematical model, then the mathematical model implemented in the EXCEL computer program package and analyzed again according to common sense and observations in the real world, or field , in this case in the cities that passed the flow of the great river, the city of Jambi in Sumatra and Kalimantan's Kutai Kartanegara City, both in Indonesia.

3 Data collection

The data in the background of the discussion contained in this paper is comprised of a team of survey data, data from personal observations of the author and secondary data sourced from Indonesia's national statistical agencies and data from previous research reports and studies, both project-based sectoral government (Ministry of Transport of the Republic of Indonesia and the Ministry of Public Works), as well as university-based research project (Directorate General of Higher Education Ministry of National Education Republic of Indonesia).

4 Modeling potential need transportation planning through movement for river

Modeling the movement needs to transport planning is an approach to phenomena involving spatial, population activities in the region, and the movement of people and / or goods vehicle over a network using motion media or vehicles that have been built over the area modeled for the area concerned. Based approach to modeling done by the system, the system boundary is a boundary where space area region. Sub space systems area is the area fractions within the area is termed the zone.

Movement is modeled with a matrix-based. So, in the process will occur interplay or interaction between the zones. Approach such a system closer to reality than the simpler approach of segment-based motion media. Terminology model into the main frame of the study was a Phase Four Transportation Planning Model.

The model consists of the generation phase of the movement-the movement of the distribution phase-phase mode selection-stage route or traffic loading [4].

4.1 Available data movement requirements

As mentioned above, when the study was conducted previously done modeling work and planning that has resulted in estimates of movement between the zones in the basin that were reviewed, the origin destination matrix between the zones. But from previous reports that exist, have not discussed the details of how the amount of movement of goods using river transport. This amount is considered in this study as a potential, where the numbers need to be estimated.

For that needs to be made between the model estimates the distribution of movement through motion media road and through motion media streams with vehicles such as boats or river boats.

4.2 Establishment division of mental models moda

The definition of mental models of the way of thinking here is the statement formed in the sentence in accordance with its meaning or logical reason. The statements formulated in this regard is that generally describes how transport users will prefer to use certain kinds of modes based on their characteristics. Model established in this study is a macro model that is aggregated.

That is the tendency to choose a group rather than individually. Requirements modeling is faced for the sake of space-based planning or region, which has been divided into small parts that zone. Data movement needs you have available is the data movement between the zones, without specifying the distribution for each mode of transportation.

Each of these modes, i.e. modes of trucks by road and boat modes through the river, will have attributes viewed as obstacles or risks trips paid users.

Simple statement that applied at the macro and aggregate are:
1. Portions election of each mode in value such that the attributes or characteristics of the offered modes risks.
2. Comparison of a portion of the selection modes with other modes will be inversely proportional to the attributes or risks at each mode of transportation.
3. Number of servings election all modes of transport is 100%.

4.3 Moda distribution mathematical model postulations

Researchers have tried several mathematical models to describe the mental models of the three items above, but in this paper will only put forward a model of the easiest and simplest way to describe the portion of the macro is the selection of aggregate, which can be expressed mathematically by equation (1)

$$Pm = (1/Rm) / [\Sigma (1/Rm)] \quad (1)$$

Where:
Pm = portion of the selected mode m
Rm = the amount of risk offered modes m

This model applies to straight mileage identical for each mode, where the risk is offered modal travel time and vehicle operating costs. If the same origin-destination of the infrastructure mode and other modes differ in the distance, according to the distance of each motion media, it needs to be taken into account proximity correction

4.4 Test of mode distribution mathematical model with artificial data

The model that has been postulated as in equation (1) in section 4.3, before being applied to the case of actual planning, need to be tested with artificial numerical data in order to know if it is in accordance with the mental model formulated in section 4.2. For this, the division will count the selected portion of the two modes are available in parallel lines in a region of space that can be used to move from one point of origin to a destination point movement of the same movement.

4.5 Use of mathematical models for estimating distribution mode selected river transport potential

From all the studies on the potential transport stream that has been done in Sumatra, Borneo, Java, and Papua, in Indonesia, it will not entirely be studied as an example of application in the real world for this paper. In this paper will only be presented in one court case, which is in the Batang Hari river which crosses the city of Jambi in Jambi Province, Sumatra,Indonesia.

4.5.1 Identification of the number of modes

Availability of premises available modes related infrastructure required for each mode of transport, as well as the pattern of freight business government and local communities. For the study area the last mode of transport are four types: motorcycles, private cars, public transport passenger cars, and the potential for passenger transport streams. As for the transport of goods only two modes are possible trucking of goods by road and transport goods by boat through the river.

4.5.2 Identification of risk at each mode

Risks that are offered by each mode may consist of more than one factor. For simplification model of the factors considered are the same factors present in each mode are just different scale numbers, and all units can be converted at the same time so it makes sense to be a factor. In the application of this mode selection model numbers used for comparison of risk is termed a combined risk

$$Rm = (Tm \ . \ Vt) + (Fm. \ Cf \qquad (2)$$

Where:
Rm = risk to offer a combined mode m
Tm = time need by mode m for a distance of 1 km on motion media
Vt = value of time (taken equal for all modes)
Fm = fuel required mode m for a distance of 1 km on motion media
Cf = the price of fuel per km for each mode

4.5.3 Mode choice model in straight river section parallel with roads

In the simplest case can be exemplified for the state of infrastructure in the field with a condition in which the land and rivers or adjoining canals straight and parallel, for example, taken along the 1 km, see Figure 2.

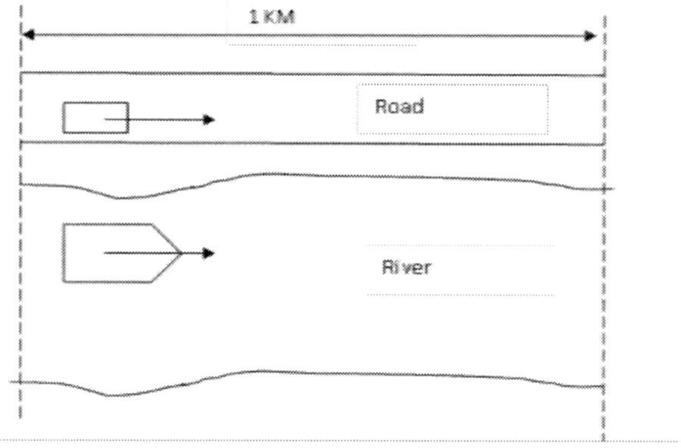

Figure 2. Case of 1 km parallel segment of road and river

Model calculations to include the risk of the price scale as indicated in section 4.6.2 above, was selected to produce a portion of each of the transport modes as shown in Table 1, and for the transport of goods are listed in Table 2.

Table 1. Mode sharing for passenger transport

Mode (m)	Rm	1/Rm	Pm
Passenger Boat	3.699,78	0,00027	26,79%
Motor Cycle	5.500,03	0,000182	18,02%
Private Car	5.500,05	0,000182	18,02%
Public Mini Bus	2.666,67	0,000375	37,17%
	$\sum(1/Rm)$	0,001009	100,00%

Table 2. Mode sharing for freight transport

Mode (m)	Rm	1/Rm	Pm
Freight Boat	3.919,11	0,000255	28,34%
Truck	1.550,00	0,000645	71,66%
	∑(1/Rm)	0,0009	100,00%

Variables declared to risk modes per 1 km as has been previously calculated, need to be corrected by calculating the actual distance to the point of origin and the point of common purpose pursued by their respective modes. The real risk is multiplied by the actual distance corresponding geometric conditions within the media path motion. Having calculated by considering the actual distance to the sample segment as in Figure 3, obtained portions modal options as listed in Table 3 for passenger transport, and in Table 4 for transport of goods.

Table 3. Mode sharing for passenger transport with real geometric distance

Mode (m)	Rm	1/Rm	Pm
Passenger Boat	8.392,80	0,000119	13,89%
Motor Cycle	5.500,03	0,000182	21,20%
Private Car	5.500,05	0,000182	21,20%
Public Mini Bus	2.666,67	0,000375	43,72%
	∑(1/Rm)	0,000858	100,00%

Table 4. Mode sharing for freight transport with real geometric distance

Mode (m)	Rm	1/Rm	Pm
Freight Boat	8.890,32	0,000112	14,85%
Truck	1.550,00	0,000645	85,15%
	∑(1/Rm)	0,000758	100,00%

4.6.5 Estimation results number of quantitative potential river transport

The approach estimates a portion of the selected mode is selected by multiplying the portion of each mode for each cell in the matrix OD matrix, this can be expressed by equation (3).

$$Tidm = Tid.\ Pm \tag{3}$$

Where:
Tid = number of moves from the zone i into zone d
Tidm = number of moves from the zone i into the zone d using mode m
Pm = portion of the selected mode m

As the final results of the quantitative estimates of potential transport by river, the River Batanghari in Jambi Province, Sumatra, in Indonesia, is as shown in Table 13 for passenger transport and Table 5 for transport of goods.

Table 5. OD matrix of passenger demand estimation for Batanghari River transportation (person/year)

	Kerinci	Merangin	Sarolangun	Batang Hari	Muaro Jambi	Tanjung Jabung Barat	Tanjung Jabung Timur	Bungo	Tebo	Jambi	Kota Sungai Penuh	Pesisir Selatan	Solok	Sawahlunto/Sijunjung
Kerinci			1.235	882	954	532	475	2.188	975	1.633			10864	5.615
Merangin			5.073	2525	2.169	1318	1.179	3.867	1597	3.947			10630	6.633
Sarolangun	1.688	5.333	-	2761	2.106	1.164	1.043	2.154	965	3.967			6060	3.944
Batang Hari	1.100	2.422	2.518	-	4.912	1849	2.720	1.743	1.365	11.780			4955	3.520
Muaro Jambi	1.367	2.392	2.208	5847	-	3857	4.511	1.957	2.283	37.864			7431	5.335
Tanjung Jabung Barat	706	1.345	1.130	1968	3.570	-	11.618	1.442	3.023	6.430			5702	4.256
Tanjung Jabung Timur	478	913	788	2197	3.168	8817	-	874	1.302	4.326			3387	2.443
Bungo	3.019	4.105	2.175	1929	1.884	1500	1.198	-	2.504	3.792			13084	8.522
Tebo	1.281	1.613	927	1438	2.092	2993	1.699	2.383	-	4.289			9593	6.424
Jambi	2.035	3.785	3.637	11780	32.934	6042	5.357	3.426	4.071	-			9966	7.046
Kota Sungai Penuh														
Pesisir Selatan														
Solok	11.466	8.793	4.768	4274	5.575	4622	3.617	10.195	7.855	8.596			-	41.143
Sawahlunto/Sijunjung	7.143	6.491	3.671	3592	4.735	4081	3.087	7.856	6.222	7.189			48674	-

*City of Sungai Penuh and Pesisir selatan are not passed by the Batanghari River.

Table 6. OD matrix of freight demand estimation for Batanghari River transportation (ton/year)

	Kerinci	Merangin	Sarolangun	Batang Hari	Muaro Jambi	Tanjung Jabung Barat	Tanjung Jabung Timur	Bungo	Tebo	Jambi	Kota Sungai Penuh	Pesisir Selatan	Solok	Sawahlunto/Sijunjung
Kerinci	-		2.565	1.308	1.697	1.184	1.057	4.218	1.446	2.421			22.141	9.993
Merangin		-	9.449	3.980	4.351	2.455	2.197	7.757	3.203	6.786			18.276	12.355
Sarolangun	2.994	6.756	-	4.897	4.001	1.475	1.585	4.098	1.344	6.061			7.678	7.495
Batang Hari	2.097	3.959	5.145	-	8.697	3.778	5.187	2.611	1.859	16.045			9.449	6.712
Muaro Jambi	2.369	3.040	3.317	9.135	-	5.349	7.297	2.713	3.694	43.751			12.880	9.246
Tanjung Jabung Barat	934	2.267	1.632	3.553	5.587	-	13.984	1.736	3.639	10.834			6.863	6.147
Tanjung Jabung Timur	871	2.208	1.653	4.727	7.343	17.514	-	1.598	2.156	8.593			8.409	4.852
Bungo	3.844	6.653	3.777	2.457	2.181	2.258	1.525	-	3.479	4.829			16.662	11.840
Tebo	2.226	2.804	1.859	2.307	3.077	5.208	3.179	4.142	-	8.027			19.258	11.864
Jambi	2.933	5.455	3.844	16.977	47.463	5.805	7.205	4.279	5.085	-			14.363	8.123
Kota Sungai Penuh														
Pesisir Selatan														
Solok	26.625	14.584	8.699	10.633	12.946	10.732	8.998	21.982	16.986	17.109			-	102.359
Sawahlunto/Sijunjung	14.206	11.189	6.815	5.239	9.418	5.953	4.098	10.417	12.376	10.486			77.448	-

4 Conclusion

The conclusion from the above discussion can be summarized as follows:

1. The existence of large rivers are navigable on the larger islands, namely for Indonesia is Sumatra, Java, Borneo and Papua, should be used to preserve and develop water transportation system of the sea to the land.
2. Sustainability and development of river transport can reduce the burden of road transport, where the development of road transport requires clearing new land is increasingly expensive and development is quite expensive.
3. Reducing the burden of vehicle traffic on the road network caused by the division of the transport stream, especially for truck traffic, can reduce the cost of road maintenance.

4. To estimate the amount of potential movement of people and goods through river mode choice model can be derived based on a comparison of risk at each mode, whereby the greater the risk will be even smaller portion of the mode selected.
5. If you have data available OD matrix, the OD matrix for each mode can be calculated by multiplying the mode selected portion of each scale matrix cells that expressed the amount of movement between zones pair of origin and destination zones. The number of potential movement using motion media streams, can be estimated by this method.

Acknowledgements

The author would like to thank the Directorate of Research and Community Services, the Directorate General of Higher Education, Ministry of National Education, Republic of Indonesia, roomates has funded this research through the National Strategic Research funding scheme, No. DIPA. 0541/023-04.1.01/00/2011, dated December 20, 2011, No. SP2H 104/SP2H/PL/Dit.Litabmas/IV/2011. dated 14 April 2012.

Also tank to the Directorate of Transport and Crossing the River, the Directorate of Land Transportation, Ministry of Communications REPUBLIC INDONESIA The author would like to thank the Directorate of Transport and Crossing the River, the Directorate of Land Transportation, Ministry of Communications of the Republic of Indonesia.

References

1. Kusdian, R. Didin, *Borneo River Transportation Sustainability,* in International Seminar on Regional Economic Development through Science, Technology, and Art Redstar, Medan Institute of Technology, pp. 1-10, (2012).
2. Kusdian, R. Didin, *Potential Revitalization River Transport in Lampung Province,* Journal of Transportation Studies Forum Transportation between Higher Education (FSTPT), Volume **(8)** Number 2, pp. 77-52, (2011) (in Indonesian).
3. Kusdian, R. Didin, Abubakar, Bakhtiar (2012), *The Sustainability of Wooden Ships in Indonesia*, Proceedings of "The Second International Conference on Port, Coastal, and Offshore Engineering (2nd ICPCO)", Bandung , 12-13 November 2012).
4. Tamin, Ofyar Z. Planning, *Modeling, and Transportation Engineering, Theory, Sample Questions, and Applications*, Publisher ITB, Bandung pp 74-80, pp 388-433 (in Indonesian), (2008)
5. Department of Transportation of the Republic of Indonesia, *Study of River and Lake Transport Mode Potential Survey in West Sumatra Province and Jambi Province*, Final Report (in Indonesian), (2012).

AUTHOR INDEX

AUTHOR INDEX